Circulating Tumor Cells

Circulating Tumor Cells

Finding Rare Events for A Huge Knowledge of Cancer Dissemination

Special Issue Editor

Catherine Alix-Panabieres

MDPI • Basel • Beijing • Wuhan • Barcelona • Belgrade

Special Issue Editor
Catherine Alix-Panabieres
University Medical Center of
Montpellier, IURC, Laboratory
of Rare Human Circulating
Cells (LCCRH)
France

Editorial Office
MDPI
St. Alban-Anlage 66
4052 Basel, Switzerland

This is a reprint of articles from the Special Issue published online in the open access journal *Cells* (ISSN 2073-4409) from 2019 to 2020 (available at: https://www.mdpi.com/journal/cells/special_issues/CTCs_cancer).

For citation purposes, cite each article independently as indicated on the article page online and as indicated below:

LastName, A.A.; LastName, B.B.; LastName, C.C. Article Title. *Journal Name* **Year**, *Article Number*, Page Range.

ISBN 978-3-03928-698-0 (Pbk)
ISBN 978-3-03928-699-7 (PDF)

Cover image courtesy of Catherine Alix-Panabières.

© 2020 by the authors. Articles in this book are Open Access and distributed under the Creative Commons Attribution (CC BY) license, which allows users to download, copy and build upon published articles, as long as the author and publisher are properly credited, which ensures maximum dissemination and a wider impact of our publications.
The book as a whole is distributed by MDPI under the terms and conditions of the Creative Commons license CC BY-NC-ND.

Contents

About the Special Issue Editor ix

Catherine Alix-Panabières
"Circulating Tumor Cells: Finding Rare Events for a Huge Knowledge of Cancer Dissemination"
Reprinted from: *Cells* **2020**, *9*, 661, doi:10.3390/cells9030661 1

Kumuditha M. Weerakoon-Ratnayake, Swarnagowri Vaidyanathan, Nicholas Larkey, Kavya Dathathreya, Mengjia Hu, Jilsha Jose, Shalee Mog, Keith August, Andrew K. Godwin, Mateusz L. Hupert, Malgorzata A. Witek and Steven A. Soper
Microfluidic Device for On-Chip Immunophenotyping and Cytogenetic Analysis of Rare Biological Cells
Reprinted from: *Cells* **2020**, *9*, 519, doi:10.3390/cells9020519 6

James W. T. Toh, Stephanie H. Lim, Scott MacKenzie, Paul de Souza, Les Bokey, Pierre Chapuis and Kevin J. Spring
Association between Microsatellite Instability Status and Peri-Operative Release of Circulating Tumour Cells in Colorectal Cancer
Reprinted from: *Cells* **2020**, *9*, 425, doi:10.3390/cells9020425 31

Claudia Hille, Tobias M. Gorges, Sabine Riethdorf, Martine Mazel, Thomas Steuber, Gunhild Von Amsberg, Frank König, Sven Peine, Catherine Alix-Panabières and Klaus Pantel
Detection of Androgen Receptor Variant 7 (*ARV7*) mRNA Levels in EpCAM-Enriched CTC Fractions for Monitoring Response to Androgen Targeting Therapies in Prostate Cancer
Reprinted from: *Cells* **2019**, *8*, 1067, doi:10.3390/cells8091067 44

Sara R. Bang-Christensen, Rasmus S. Pedersen, Marina A. Pereira, Thomas M. Clausen, Caroline Løppke, Nicolai T. Sand, Theresa D. Ahrens, Amalie M. Jørgensen, Yi Chieh Lim, Louise Goksøyr, Swati Choudhary, Tobias Gustavsson, Robert Dagil, Mads Daugaard, Adam F. Sander, Mathias H. Torp, Max Søgaard, Thor G. Theander, Olga Østrup, Ulrik Lassen, Petra Hamerlik, Ali Salanti and Mette Ø. Agerbæk
Capture and Detection of Circulating Glioma Cells Using the Recombinant VAR2CSA Malaria Protein
Reprinted from: *Cells* **2019**, *8*, 998, doi:10.3390/cells8090998 67

Afroditi Nanou, Leonie L. Zeune and Leon W.M.M. Terstappen
Leukocyte-Derived Extracellular Vesicles in Blood with and without EpCAM Enrichment
Reprinted from: *Cells* **2019**, *8*, 937, doi:10.3390/cells8080937 88

Eva Obermayr, Christiane Agreiter, Eva Schuster, Hannah Fabikan, Christoph Weinlinger, Katarina Baluchova, Gerhard Hamilton, Maximilian Hochmair and Robert Zeillinger
Molecular Characterization of Circulating Tumor Cells Enriched by A Microfluidic Platform in Patients with Small-Cell Lung Cancer
Reprinted from: *Cells* **2019**, *8*, 880, doi:10.3390/cells8080880 104

Laure Cayrefourcq, Aurélie De Roeck, Caroline Garcia, Pierre-Emmanuel Stoebner, Fanny Fichel, Françoise Garima, Françoise Perriard, Jean-Pierre Daures, Laurent Meunier and Catherine Alix-Panabières
S100-EPISPOT: A New Tool to Detect Viable Circulating Melanoma Cells
Reprinted from: *Cells* **2019**, *8*, 755, doi:10.3390/cells8070755 116

Loredana Cleris, Maria Grazia Daidone, Emanuela Fina and Vera Cappelletti
The Detection and Morphological Analysis of Circulating Tumor and Host Cells in Breast Cancer Xenograft Models
Reprinted from: *Cells* **2019**, *8*, 683, doi:10.3390/cells8070683 129

Mohammed Nimir, Yafeng Ma, Sarah A. Jeffreys, Thomas Opperman, Francis Young, Tanzila Khan, Pei Ding, Wei Chua, Bavanthi Balakrishnar, Adam Cooper, Paul De Souza and Therese M. Becker
Detection of AR-V7 in Liquid Biopsies of Castrate Resistant Prostate Cancer Patients: A Comparison of AR-V7 Analysis in Circulating Tumor Cells, Circulating Tumor RNA and Exosomes
Reprinted from: *Cells* **2019**, *8*, 688, doi:10.3390/cells8070688 148

Areti Strati, Michail Nikolaou, Vassilis Georgoulias and Evi S. Lianidou
Prognostic Significance of *TWIST1*, *CD24*, *CD44*, and *ALDH1* Transcript Quantification in EpCAM-Positive Circulating Tumor Cells from Early Stage Breast Cancer Patients
Reprinted from: *Cells* **2019**, *8*, 652, doi:10.3390/cells8070652 159

Bianca C. Troncarelli Flores, Virgilio Souza e Silva, Emne Ali Abdallah, Celso A.L. Mello, Maria Letícia Gobo Silva, Gustavo Gomes Mendes, Alexcia Camila Braun, Samuel Aguiar Junior and Ludmilla Thomé Domingos Chinen
Molecular and Kinetic Analyses of Circulating Tumor Cells as Predictive Markers of Treatment Response in Locally Advanced Rectal Cancer Patients
Reprinted from: *Cells* **2019**, *8*, 641, doi:10.3390/cells8070641 175

Olga Chernysheva, Irina Markina, Lev Demidov, Natalia Kupryshina, Svetlana Chulkova, Alexandra Palladina, Alina Antipova and Nikolai Tupitsyn
Bone Marrow Involvement in Melanoma. Potentials for Detection of Disseminated Tumor Cells and Characterization of Their Subsets by Flow Cytometry
Reprinted from: *Cells* **2019**, *8*, 627, doi:10.3390/cells8060627 188

Jeannette Huaman, Michelle Naidoo, Xingxing Zang and Olorunseun O. Ogunwobi
Fibronectin Regulation of Integrin B1 and SLUG in Circulating Tumor Cells
Reprinted from: *Cells* **2019**, *8*, 618, doi:10.3390/cells8060618 197

François-Clément Bidard, Nicolas Kiavue, Marc Ychou, Luc Cabel, Marc-Henri Stern, Jordan Madic, Adrien Saliou, Aurore Rampanou, Charles Decraene, Olivier Bouché, Michel Rivoire, François Ghiringhelli, Eric Francois, Rosine Guimbaud, Laurent Mineur, Faiza Khemissa-Akouz, Thibault Mazard, Driffa Moussata, Charlotte Proudhon, Jean-Yves Pierga, Trevor Stanbury, Simon Thézenas and Pascale Mariani
Circulating Tumor Cells and Circulating Tumor DNA Detection in Potentially Resectable Metastatic Colorectal Cancer: A Prospective Ancillary Study to the Unicancer Prodige-14 Trial
Reprinted from: *Cells* **2019**, *8*, 516, doi:10.3390/cells8060516 215

Fabienne Schochter, Thomas W. P. Friedl, Amelie deGregorio, Sabrina Krause, Jens Huober, Brigitte Rack and Wolfgang Janni
Are Circulating Tumor Cells (CTCs) Ready for Clinical Use in Breast Cancer? An Overview of Completed and Ongoing Trials Using CTCs for Clinical Treatment Decisions
Reprinted from: *Cells* **2019**, *8*, 1412, doi:10.3390/cells8111412 228

Carmen Garrido-Navas, Diego de Miguel-Pérez, Jose Exposito-Hernandez, Clara Bayarri, Victor Amezcua, Alba Ortigosa, Javier Valdivia, Rosa Guerrero, Jose Luis Garcia Puche, Jose Antonio Lorente and Maria José Serrano
Cooperative and Escaping Mechanisms between Circulating Tumor Cells and Blood Constituents
Reprinted from: *Cells* **2019**, *8*, 1382, doi:10.3390/cells8111382 . . . . **237**

Olga A. Sindeeva, Roman A. Verkhovskii, Mustafa Sarimollaoglu, Galina A. Afanaseva, Alexander S. Fedonnikov, Evgeny Yu. Osintsev, Elena N. Kurochkina, Dmitry A. Gorin, Sergey M. Deyev, Vladimir P. Zharov and Ekaterina I. Galanzha
New Frontiers in Diagnosis and Therapy of Circulating Tumor Markers in Cerebrospinal Fluid In Vitro and In Vivo
Reprinted from: *Cells* **2019**, *8*, 1195, doi:10.3390/cells8101195 . . . . **247**

Tala Tayoun, Vincent Faugeroux, Marianne Oulhen, Agathe Aberlenc, Patrycja Pawlikowska and Françoise Farace
CTC-Derived Models: A Window into the Seeding Capacity of Circulating Tumor Cells (CTCs)
Reprinted from: *Cells* **2019**, *8*, 1145, doi:10.3390/cells8101145 . . . . **262**

Elisabetta Rossi and Francesco Fabbri
CTCs 2020: Great Expectations or Unreasonable Dreams
Reprinted from: *Cells* **2019**, *8*, 989, doi:10.3390/cells8090989 . . . . **280**

Vera Kloten, Rita Lampignano, Thomas Krahn and Thomas Schlange
Circulating Tumor Cell PD-L1 Expression as Biomarker for Therapeutic Efficacy of Immune Checkpoint Inhibition in NSCLC
Reprinted from: *Cells* **2019**, *8*, 809, doi:10.3390/cells8080809 . . . . **294**

Simon Heeke, Baharia Mograbi, Catherine Alix-Panabières and Paul Hofman
Never Travel Alone: The Crosstalk of Circulating Tumor Cells and the Blood Microenvironment
Reprinted from: *Cells* **2019**, *8*, 714, doi:10.3390/cells8070714 . . . . **306**

Lucile Broncy and Patrizia Paterlini-Bréchot
Clinical Impact of Circulating Tumor Cells in Patients with Localized Prostate Cancer
Reprinted from: *Cells* **2019**, *8*, 676, doi:10.3390/cells8070676 . . . . **318**

Patrick C. Bailey and Stuart S. Martin
Insights on CTC Biology and Clinical Impact Emerging from Advances in Capture Technology
Reprinted from: *Cells* **2019**, *8*, 553, doi:10.3390/cells8070553 . . . . **333**

About the Special Issue Editor

Catherine Alix-Panabières received her Ph.D. degree in 1998 at the Institute of Virology, University Louis Pasteur, in Strasbourg in France. In 1999, she moved to Montpellier where she conducted postdoctoral research at the University Medical Centre. During the last decade, Dr. Alix-Panabières has focused on optimizing new techniques of enrichment, detection, and characterization of viable circulating tumor cells (CTCs) in patients with solid tumors. She is the expert for the EPISPOT technology that is used to detect viable CTCs in patients with breast, prostate, colon, head and neck cancer and melanoma. This technology has been recently improved to detect functional CTCs at the single-cell level (EPIDROP). In 2010, the term *'Liquid Biopsy'* was coined for the first time by Dr. Catherine Alix-Panabières and Professor Pantel (Trends Mol Med).

In 2010, she achieved a permanent position at the Hospital and the Faculty of Medicine of Montpellier (MCU-PH). As an associate professor, she became the new director of the Laboratory of Rare Human Circulating Cells (LCCRH) in the Department of Pathology and Onco-Biology.

In this unique platform, LCCRH, they isolate, detect, and characterize circulating tumor cells using combinations of many technologies. She has authored or co-authored over 85 scientific publications in this field during the last years and 10 book chapters, she is the inventor of three patents in the liquid biopsy field and she is part of several French national projects: for example, PANTHER (FUI project), STIC-METABREAST, and TACTIK (PHRC), as well as big European projects: CTC-SCAN (Transcan project), CANCER-ID (IMI project), PROLIPSY (Transcan project), and European Liquid Biopsy Academy (ELBA, Marie-Curie project).

It was a great honor for her to receive the Gallet et Breton Cancer Prize, the highest honor conferred by the French Academy of Medicine in November 2012 and, very recently, the 2017 AACR Award for the most cited scientific article in 2015 (Cayrefourcq et al. *Cancer Res*).

Editorial

"Circulating Tumor Cells: Finding Rare Events for a Huge Knowledge of Cancer Dissemination"

Catherine Alix-Panabières

Laboratory of Rare Human Circulating Cells (LCCRH), University Medical Centre of Montpellier, 641 Avenue du Doyen Gaston Giraud, 34093 Montpellier CEDEX 5, France; c-panabieres@chu-montpellier.fr; Tel.: +33-(0)4-1175-9931; Fax: +33-(0)4-1175-9933

Received: 14 January 2020; Accepted: 5 March 2020; Published: 9 March 2020

Circulating tumor cells (CTCs) as real-time liquid biopsy [1] can be used to obtain new insights into the biology of the metastatic cascade, and as a companion diagnostic to improve the stratification of therapies and to obtain new insights into therapy-induced selection of cancer cells. Combining different circulating biomarkers, such as CTCs, circulating tumor DNA (ctDNA) and extracellular vesicles (EVs) analysis, will provide different and complementary information. Technical and clinical assay validation in big cohorts of cancer patients is crucial and can be achieved in international consortia such as the European Liquid Biopsy Society (ELBS) [2].

This Special Issue, "Circulating Tumor Cells: Finding Rare Events for A Huge Knowledge of Cancer Dissemination", includes 23 articles written by experts in this field and covers multiple facets of CTCs in order to assemble a huge corpus of knowledge on cancer biology with emphasis on (i) technical challenges to enrich, detect, isolate and characterize CTCs at the single cell level, (ii) cancer biology with emphasis on metastasis including cancer stemness, epithelial-mesenchymal transition as well as immunomodulation of tumor cells, (iii) clinical studies on liquid biopsy, a new diagnostic concept introduced and coined for the first time in 2010 [1] for the analysis of CTCs and now extended to material (in particular DNA and EVs) released by tumor cells in the peripheral blood of cancer patients (Figure 1).

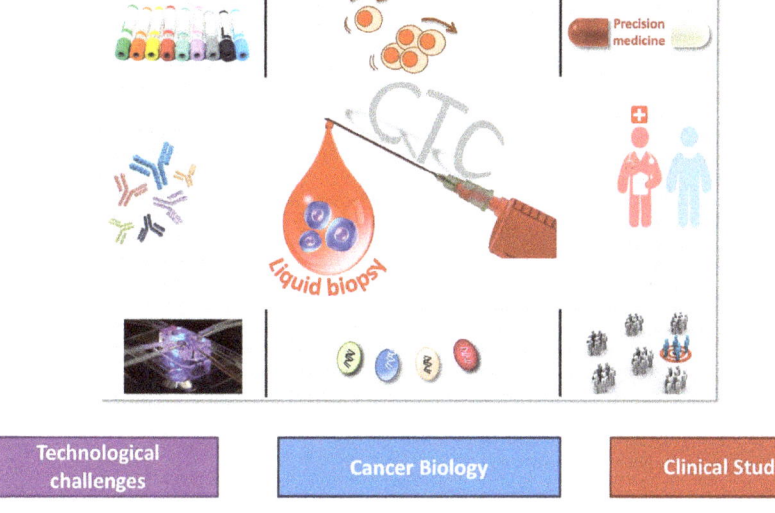

Figure 1. The different aspects of Liquid Biopsy: Technologies–Biology–Trials.

TECHNICAL CHALLENGES. Efficient enrichment of CTCs can be achieved by approaches that exploit differences between tumour cells and blood cells. Subsequently, enriched CTC populations might still contain among them hundreds to thousands of undesirable leukocytes, which requires the use of reliable methods to identify CTCs. Rossi et al. also highlighted the two major issues in the CTC field: rarity and heterogeneity [3]. Most current CTC assays use the same identification step as the FDA-approved CellSearch® system. However, using physical properties to enrich CTCs, Obermayr et al. showed that CTCs enriched by the Parsortix system in small-cell lung cancer patients can be assessed using epithelial and neuroendocrine cell lineage markers at the molecular level [4]. Moreover, Bailey et al. reported emerging technologies for the capture of CTC/microemboli showing important biological and functional information that can lead to important alterations in how therapies are administered [5]. Moreover, Weerakoon-Ratnayake et al. detailed a novel microfluidic technology (i.e., microtrap device) that can perform immunophenotyping and FISH on CTCs [6].

One of the greatest challenges in neuro-oncology is the theranostic of leptomeningeal metastasis, brain metastasis and brain tumors, which are associated with poor prognosis in patients. Although extracranial metastases in glioma patients are rarely observed, recent studies have shown the presence of CTCs in the bloodstream. Bang-Christensen et al. demonstrated how the recombinant malaria VAR2CSA protein (rVAR2) can be used for the capture and detection of glioma cell lines and identified a panel of proteoglycans, known to be essential for glioma progression [7]. Cerebrospinal fluid (CSF) is one of the promising diagnostic targets because CSF passes through the central nervous system, harvests tumor-related markers from brain tissue and then, delivers them into peripheral parts of the human body where CSF can be sampled using minimally invasive and routine clinical procedure. Sindeeva et al. outlined the advantages, limitations and clinical utility of emerging liquid biopsy in vitro and photoacoustic flow cytometry (PAFC) in vivo for assessment of CSF markers including CTCs, ctDNA, miRNA, proteins, exosomes and emboli [8].

Chernysheva et al. studied disseminated tumor cells (DTCs) as a prognostic factor in many non-hematopoietic tumors [9]. They evaluated the possibility of detecting subsets of melanoma DTCs in the bone marrow based on the expression of a cytoplasmic premelanocytic glycoprotein HMB-45 using flow cytometry.

In addition to CTCs, circulating EVs can be of important interest. For instance, large tumor-derived extracellular vesicles (tdEVs) detected in blood of metastatic prostate, breast, colorectal, and non-small cell lung cancer patients are negatively associated with the overall survival of patients. Nanou et al. investigated whether, similarly to tdEVs, leukocyte-derived EVs (ldEVs) could also be detected in EpCAM-enriched blood [10].

CANCER BIOLOGY. The huge advantage of CTCs against other blood biomarkers is that the knowledge derived from a single-cell analysis of CTCs can be obtained at the DNA, RNA, and protein level [11].

As a first example, the expression of the androgen receptor splice variant 7 (ARV7) in circulating tumor cells (CTCs) has been associated with resistance towards novel androgen receptor (AR)-targeting therapies. A highly sensitive and specific qPCR-based assay was developed by Hille et al., allowing detection of ARV7 and keratin 19 transcripts from as low as a single ARV7$^+$/K19$^+$ cell [12]. Moreover, detection of AR and AR-V7 was also performed by Nimir et al. using a highly sensitive droplet digital PCR-based assay [13]. In that case, AR and AR-V7 RNA were detectable in CTCs, ctRNA and exosome samples. They could show that AR-V7 detected from CTCs could be done with a higher sensitivity and specificity compared to that detected from ctRNA and exosomes.

A second example is PD-L1 as one immune checkpoint regulator; it has become an exciting new therapeutic target leading to long lasting remissions in patients with advanced malignancies [14,15]. Kloten et al. highlighted the use of CTCs as a complementary diagnostic tool for PD-L1 expression analysis in advanced NSCLC patients [16].

Concerning (i) the stemness status of cancer cells, Strati et al. showed that the detection of TWIST1 overexpression and stem-cell (CD24, CD44, ALDH1) transcripts in EpCAM$^+$ CTCs provides prognostic information in early-stage breast cancer patients [17]; and (ii) the capacity to disseminate, Huaman et al. studied hepatocellular carcinoma and castration-resistant prostate cancer and showed that CTCs exhibit distinct characteristics from primary tumor-derived cells: they highlighted an enhanced migration in part through fibronectin regulation of integrin B1 and SLUG [18].

While CTCs have long been considered to be isolated cells circulating in the bloodstream, recent research demonstrated the close interaction of CTCs with the blood microenvironment. CTCs need to establish close interaction not only with platelets and neutrophils, but also with macrophages and endothelial cells to resist the physical stress in the bloodstream. Heeke et al. discussed the recent research on the crosstalk between CTCs and the blood microenvironment and outlined currently investigated treatment strategies [19] and Garrido-Navas et al. reported the findings regarding active interactions between CTCs and platelets, myeloid cells, macrophages, neutrophils, and other hematopoietic cells that aid CTCs to evade the immune system and enable metastasis [20]. In addition, Cleris et al. detected and analyzed the morphology of CTCs and could show that orthotopic xenografts of breast cancer cell lines offer valid models of hematogenous dissemination and a possible experimental setting to study CTC–blood microenvironment interactions [21].

As microsatellite instability (MSI) in colorectal cancer is a marker of immunogenicity and associated with an increased abundance of tumour infiltrating lymphocytes (TILS), Toh et al. compared for the first time the MSI status with the prevalence of CTCs in the peri-operative colorectal surgery setting [22].

In cancer biology, a crucial point is to identify the CTCs able to initiate metastases. Different methods have been developed to expand CTCs in vitro and in vivo with the aim of characterizing functional metastasis-initiator CTCs with stemness traits, and to obtain new diagnostics and therapeutic tools [23]. Tayoun et al. evaluated CTC-derived models generated in different types of cancer and shed a light on challenges and key findings associated with these novel assays [24].

CLINICAL STUDIES. Rossi et al. discussed how CTCs could drastically improve tumor companion diagnostics, personalized treatment strategies, overall patient's management, and reduce healthcare costs. Broncy et al. highlighted the clinical impact of CTCs in patients with localized prostate cancer [25] and Schochter et al. summarized the completed and ongoing clinical trials using CTC number or phenotype for treatment decisions in breast cancer [26].

Concerning colorectal cancer, the management of patients and potentially resectable liver metastases requires quick assessment of mutational status and of response to pre-operative systemic therapy. In a prospective phase II trial (NCT01442935), Bidard et al. investigated the clinical validity of CTCs and ctDNA [27]. They concluded that ctDNA detection could help to select patients eligible for liver metastases resection. In addition, Troncarelli Flores et al. reported that molecular and kinetic analyses of CTCs are predictive markers of treatment response in locally advanced rectal cancer patients [28].

Finally, metastatic melanoma is one of the most aggressive and drug-resistant cancers with very poor overall survival. Circulating melanoma cells (CMCs) were first described in 1991 and here, Cayrefourcq et al. developed a new EPISPOT assay to detect viable CMCs based on their secretion of the S100 protein using the functional S100-EPISPOT assay [29]. They showed that the S100-EPISPOT sensitivity was significantly higher than that of the CellSearch® system. It will be interesting in the future to determine whether this functional test could be used in patients with non-metastatic melanoma for the early detection of tumor relapse and for monitoring the treatment response.

References

1. Pantel, K.; Alix-Panabieres, C. Circulating tumour cells in cancer patients: Challenges and perspectives. *Trends Mol. Med.* **2010**, *16*, 398–406. [CrossRef] [PubMed]
2. Pantel, K.; Alix-Panabieres, C. Liquid biopsy and minimal residual disease—Latest advances and implications for cure. *Nat. Rev. Clin. Oncol.* **2019**, *16*, 409–424. [CrossRef] [PubMed]
3. Rossi, E.; Fabbri, F. CTCs 2020: Great Expectations or Unreasonable Dreams. *Cells* **2019**, *8*, 989. [CrossRef] [PubMed]

4. Obermayr, E.; Agreiter, C.; Schuster, E.; Fabikan, H.; Weinlinger, C.; Baluchova, K.; Hamilton, G.; Hochmair, M.; Zeillinger, R. Molecular Characterization of Circulating Tumor Cells Enriched by A Microfluidic Platform in Patients with Small-Cell Lung Cancer. *Cells* **2019**, *8*, 880. [CrossRef]
5. Bailey, P.C.; Martin, S.S. Insights on CTC Biology and Clinical Impact Emerging from Advances in Capture Technology. *Cells* **2019**, *8*, 553. [CrossRef]
6. M Weerakoon-Ratnayake, K.; Vaidyanathan, S.; Larky, N.; Dathathreya, K.; Hu, M.; Jose, J.; Mog, S.; August, K.; K Godwin, A.; L Hupert, M.; et al. Microfluidic Device for On-Chip Immunophenotyping and Cytogenetic Analysis of Rare Biological Cells. *Cells* **2020**, *9*, 519. [CrossRef]
7. Bang-Christensen, S.R.; Pedersen, R.S.; Pereira, M.A.; Clausen, T.M.; Løppke, C.; Sand, N.T.; Ahrens, T.D.; Jørgensen, A.M.; Lim, Y.C.; Goksøyr, L.; et al. Capture and Detection of Circulating Glioma Cells Using the Recombinant VAR2CSA Malaria Protein. *Cells* **2019**, *8*, 998. [CrossRef]
8. Sindeeva, O.A.; Verkhovskii, R.A.; Sarimollaoglu, M.; Afanaseva, G.A.; Fedonnikov, A.S.; Osintsev, E.Y.; Kurochkina, E.N.; Gorin, D.A.; Deyev, S.M.; Zharov, V.P.; et al. New Frontiers in Diagnosis and Therapy of Circulating Tumor Markers in Cerebrospinal Fluid In Vitro and In Vivo. *Cells* **2019**, *8*, 1195. [CrossRef]
9. Chernysheva, O.; Markina, I.; Demidov, L.; Kupryshina, N.; Chulkova, S.; Palladina, A.; Antipova, A.; Tupitsyn, N. Bone Marrow Involvement in Melanoma. Potentials for Detection of Disseminated Tumor Cells and Characterization of Their Subsets by Flow Cytometry. *Cells* **2019**, *8*, 627. [CrossRef]
10. Nanou, A.; Zeune, L.L.; Terstappen, L. Leukocyte-Derived Extracellular Vesicles in Blood with and without EpCAM Enrichment. *Cells* **2019**, *8*, 937. [CrossRef]
11. Alix-Panabieres, C.; Pantel, K. Characterization of single circulating tumor cells. *FEBS Lett.* **2017**, *591*, 2241–2250. [CrossRef] [PubMed]
12. Hille, C.; Gorges, T.M.; Riethdorf, S.; Mazel, M.; Steuber, T.; Amsberg, G.V.; König, F.; Peine, S.; Alix-Panabières, C.; Pantel, K. Detection of Androgen Receptor Variant 7 (ARV7) mRNA Levels in EpCAM-Enriched CTC Fractions for Monitoring Response to Androgen Targeting Therapies in Prostate Cancer. *Cells* **2019**, *8*, 1067. [CrossRef] [PubMed]
13. Nimir, M.; Ma, Y.; Jeffreys, S.A.; Opperman, T.; Young, F.; Khan, T.; Ding, P.; Chua, W.; Balakrishnar, B.; Cooper, A.; et al. Detection of AR-V7 in Liquid Biopsies of Castrate Resistant Prostate Cancer Patients: A Comparison of AR-V7 Analysis in Circulating Tumor Cells, Circulating Tumor RNA and Exosomes. *Cells* **2019**, *8*, 688. [CrossRef] [PubMed]
14. Hofman, P.; Heeke, S.; Alix-Panabieres, C.; Pantel, K. Liquid biopsy in the era of immuno-oncology: Is it ready for prime-time use for cancer patients? *Ann. Oncol.* **2019**, *30*, 1448–1459. [CrossRef]
15. Strati, A.; Koutsodontis, G.; Papaxoinis, G.; Angelidis, I.; Zavridou, M.; Economopoulou, P.; Kotsantis, I.; Avgeris, M.; Mazel, M.; Perisanidis, C.; et al. Prognostic significance of PD-L1 expression on circulating tumor cells in patients with head and neck squamous cell carcinoma. *Ann. Oncol.* **2017**, *28*, 1923–1933. [CrossRef]
16. Kloten, V.; Lampignano, R.; Krahn, T.; Schlange, T. Circulating Tumor Cell PD-L1 Expression as Biomarker for Therapeutic Efficacy of Immune Checkpoint Inhibition in NSCLC. *Cells* **2019**, *8*, 809. [CrossRef]
17. Strati, A.; Nikolaou, M.; Georgoulias, V.; Lianidou, E.S. Prognostic Significance of TWIST1, CD24, CD44, and ALDH1 Transcript Quantification in EpCAM-Positive Circulating Tumor Cells from Early Stage Breast Cancer Patients. *Cells* **2019**, *8*, 652. [CrossRef]
18. Huaman, J.; Naidoo, M.; Zang, X.; Ogunwobi, O.O. Fibronectin Regulation of Integrin B1 and SLUG in Circulating Tumor Cells. *Cells* **2019**, *8*, 618. [CrossRef]
19. Heeke, S.; Mograbi, B.; Alix-Panabieres, C.; Hofman, P. Never Travel Alone: The Crosstalk of Circulating Tumor Cells and the Blood Microenvironment. *Cells* **2019**, *8*, 714. [CrossRef]
20. Garrido-Navas, C.; de Miguel-Pérez, D.; Exposito-Hernandez, J.; Bayarri, C.; Amezcua, V.; Ortigosa, A.; Valdivia, J.; Guerrero, R.; Puche, J.L.G.; Lorente, J.A.; et al. Cooperative and Escaping Mechanisms between Circulating Tumor Cells and Blood Constituents. *Cells* **2019**, *8*, 1382. [CrossRef]
21. Cleris, L.; Daidone, M.G.; Fina, E.; Cappelletti, V. The Detection and Morphological Analysis of Circulating Tumor and Host Cells in Breast Cancer Xenograft Models. *Cells* **2019**, *8*, 683. [CrossRef]
22. Toh, J.W.T.; Lim, S.H.; MacKenzie, S.; de Souza, P.; Bokey, L.; Chapuis, P.; Spring, K.J. Association Between Microsatellite Instability Status and Peri-Operative Release of Circulating Tumour Cells in Colorectal Cancer. *Cells* **2020**, *9*, 425. [CrossRef] [PubMed]

23. Cortes-Hernandez, L.E.; Eslami, S.Z.; Alix-Panabieres, C. Circulating tumor cell as the functional aspect of liquid biopsy to understand the metastatic cascade in solid cancer. *Mol. Asp. Med.* **2019**, 100816. [CrossRef] [PubMed]
24. Tayoun, T.; Faugeroux, V.; Oulhen, M.; Aberlenc, A.; Pawlikowska, P. Farace FCTC-Derived Models: A Window into the Seeding Capacity of Circulating Tumor Cells (CTCs). *Cells* **2019**, *8*, 1145. [CrossRef] [PubMed]
25. Broncy, L.; Paterlini-Brechot, P. Clinical Impact of Circulating Tumor Cells in Patients with Localized Prostate Cancer. *Cells* **2019**, *8*, 676. [CrossRef] [PubMed]
26. Schochter, F.; Friedl, T.W.P.; deGregorio, A.; Krause, S.; Huober, J.; Rack, B.; Janni, W. Are Circulating Tumor Cells (CTCs) Ready for Clinical Use in Breast Cancer? An Overview of Completed and Ongoing Trials Using CTCs for Clinical Treatment Decisions. *Cells* **2019**, *8*, 1412. [CrossRef] [PubMed]
27. Bidard, F.C.; Kiavue, N.; Ychou, M.; Cabel, L.; Stern, M.H.; Madic, J.; Saliou, A.; Rampanou, A.; Decraene, C.; Bouché, O.; et al. Circulating Tumor Cells and Circulating Tumor DNA Detection in Potentially Resectable Metastatic Colorectal Cancer: A Prospective Ancillary Study to the Unicancer Prodige-14 Trial. *Cells* **2019**, *8*, 516. [CrossRef]
28. Troncarelli Flores, B.C.; Souza E Silva, V.; Ali Abdallah, E.; Mello, C.A.L.; Gobo Silva, M.L.; Gomes Mendes, G.; Camila Braun, A.; Aguiar, S., Jr.; Thomé Domingos Chinen, L. Molecular and Kinetic Analyses of Circulating Tumor Cells as Predictive Markers of Treatment Response in Locally Advanced Rectal Cancer Patients. *Cells* **2019**, *8*, 641. [CrossRef]
29. Cayrefourcq, L.; de Roeck, A.; Garcia, C.; Stoebner, P.-E.; Fichel, F.; Garima, F.; Perriard, F.; Daures, J.-P.; Meunier, L.; Alix-Panabières, C. S100-EPISPOT: A New Tool to Detect Viable Circulating Melanoma Cells. *Cells* **2019**, *8*, 755. [CrossRef]

© 2020 by the author. Licensee MDPI, Basel, Switzerland. This article is an open access article distributed under the terms and conditions of the Creative Commons Attribution (CC BY) license (http://creativecommons.org/licenses/by/4.0/).

Article

Microfluidic Device for On-Chip Immunophenotyping and Cytogenetic Analysis of Rare Biological Cells

Kumuditha M. Weerakoon-Ratnayake [1,2], Swarnagowri Vaidyanathan [2,3], Nicholas Larkey [2,4], Kavya Dathathreya [1,2], Mengjia Hu [2,4], Jilsha Jose [2], Shalee Mog [1,2], Keith August [5], Andrew K. Godwin [4], Mateusz L. Hupert [6,*], Malgorzata A. Witek [1,2,*] and Steven A. Soper [1,2,4,6,7,*]

[1] Department of Chemistry, The University of Kansas, Lawrence, KS 66047, USA; kratnayake@ku.edu (K.M.W.-R.); kavya87.atreya@gmail.com (K.D.); shaleemog@ku.edu (S.M.)
[2] Center of BioModular Multiscale Systems for Precision Medicine, Lawrence, KS 66045, USA; swarna_vaidy@ku.edu (S.V.); larkeyn@ku.edu (N.L.); m367h922@kumc.edu (M.H.); jiljose97@gmail.com (J.J.)
[3] Bioengineering, The University of Kansas, Lawrence, KS 66045, USA
[4] Department of Pathology & Laboratory Medicine, University of Kansas Medical Center, Kansas City, KS 66160, USA; agodwin@kumc.edu
[5] Children's Mercy Hospital, Kansas City, MO 64108, USA; kjaugust@cmh.edu
[6] Biofluidica Inc., BioFluidica Research Laboratory, Lawrence, KS 66047, USA
[7] Department of Mechanical Engineering, The University of Kansas, Lawrence, KS 66045, USA
* Correspondence: matt@biofluidica.com (M.L.H.); mwitek@ku.edu (M.A.W.); ssoper@ku.edu (S.A.S.); Tel.: +1-785-864-3072 (S.A.S.)

Received: 19 December 2019; Accepted: 18 February 2020; Published: 24 February 2020

Abstract: The role of circulating plasma cells (CPCs) and circulating leukemic cells (CLCs) as biomarkers for several blood cancers, such as multiple myeloma and leukemia, respectively, have recently been reported. These markers can be attractive due to the minimally invasive nature of their acquisition through a blood draw (i.e., liquid biopsy), negating the need for painful bone marrow biopsies. CPCs or CLCs can be used for cellular/molecular analyses as well, such as immunophenotyping or fluorescence in situ hybridization (FISH). FISH, which is typically carried out on slides involving complex workflows, becomes problematic when operating on CLCs or CPCs due to their relatively modest numbers. Here, we present a microfluidic device for characterizing CPCs and CLCs using immunofluorescence or FISH that have been enriched from peripheral blood using a different microfluidic device. The microfluidic possessed an array of cross-channels (2–4 µm in depth and width) that interconnected a series of input and output fluidic channels. Placing a cover plate over the device formed microtraps, the size of which was defined by the width and depth of the cross-channels. This microfluidic chip allowed for automation of immunofluorescence and FISH, requiring the use of small volumes of reagents, such as antibodies and probes, as compared to slide-based immunophenotyping and FISH. In addition, the device could secure FISH results in <4 h compared to 2–3 days for conventional FISH.

Keywords: microfluidics; immunophenotyping; fish; liquid biopsy; circulating leukemia cells; circulating plasma cells

---

1. Introduction

Molecular diagnostics are growing immensely due in part to the Precision Medicine Initiative (www.whitehouse.gov/precision-medicine), which seeks to match appropriate therapies to the molecular characteristics of a patient's disease. Unfortunately, the majority of molecular diagnostic tests are expensive, involve slow turnaround times from centralized laboratories, and require highly specialized equipment with seasoned technicians to carry out the assay. In addition, acquisition of the molecular biomarkers requires a solid tissue or bone marrow biopsy, which can be an invasive procedure, especially for anatomically inaccessible organs. For example, bone marrow biopsies are typically required to monitor leukemia or multiple myeloma status, which not only complicates sample acquisition but limits the frequency of testing.

Liquid biopsies are generating a significant amount of interest in the medical community owing to the minimally invasive nature of acquiring biomarkers and the fact that they can enable precision decisions on managing a variety of diseases [1,2]. Liquid biopsy markers include but are not limited to circulating tumor cells (CTCs), cell-free DNA (cfDNA), and extracellular vesicles (EVs). As an example of the utility of liquid biopsy analysis for some blood cancers, we have shown that circulating plasma cells (CPCs) can be used to stage patients diagnosed with multiple myeloma [3]. Circulating leukemia cells (CLCs) can be used to determine relapse from minimum residual disease (MRD) in patients with acute myeloid leukemia (AML) or acute lymphoblastic leukemia (ALL) [4], all of which typically require a highly painful and invasive bone marrow biopsy. The challenge with using CLCs or CPCs is that their abundance in blood is lower than what is found in the bone marrow, requiring highly sensitive assays to analyze their molecular content following enrichment to remove interfering white and red blood cells.

A cytogenetic method called fluorescence in-situ hybridization (FISH), which was discovered in the early 1980s [5], can be used to detect chromosomal modifications [6–8]. FISH identifies abnormalities in chromosomes using fluorescent DNA probes that hybridize to a specific gene region. When properly hybridized to its complementary sequence, FISH allows for the visualization of chromosomal aberrations, such as deletions, fusions, balanced translocations, etc. [9]. For example, cytogenetic abnormalities are found in most cases of multiple myeloma, in which IGH translocations initiate events associated with tumorigenesis and disease progression [10]. The progression of multiple myeloma was discovered in clinical studies where investigators found frequent chromosomal aberrations, such as 13q14 deletions (del13q14), 1q21 gains (amp1q21), and monosomy 13 and 17p13 deletions (del17p13) [10,11]. B-type acute lymphoblastic leukemia (B-ALL), which is a common childhood malignancy, is prognosed by BCR/ABL [t(9;22)], MLL [t(4;1)], and TEL/AML1 [t12;21] [12] aberrations.

While FISH assays are widely used in clinical settings, the workflow requires labor-intensive and time-consuming protocols. Conventional slide-based methods for FISH utilize workflows that necessitate the need for highly trained professionals and relatively high volumes of costly FISH probes; the full assay may require 2-3 days of processing. Therefore, it is critical to develop alternative methods and platforms to undertake FISH that address the aforementioned limitations [9,13].

Microfluidic FISH assays can address many of the limitations associated with slide-based FISH, such as providing process automation and reducing reagent requirements and processing time [9]. Even though there are microfluidic assays that have been developed over the past years for genetic techniques, such as PCR and DNA microarrays, less effort has been devoted toward realizing the implementation of FISH assays using microfluidic devices [14–16]. A summary of microfluidic devices for performing FISH are summarized in Table S1 along with their operational characteristics.

The first microfluidic for FISH was developed by Sieben and coworkers in 2007 [17]. The study demonstrated the ability to detect a FISH signal with 10-fold higher throughput and 1/10th reagent consumption compared to slide-based FISH. Modification of the microfluidic substrate to achieve cell adherence is a common protocol to allow for processing the cells and imaging. Therefore, in this report, TiO_2-modified glass slides were used for cell adherence to identify FISH signals, with the PDMS fluidic network used to introduce FISH reagents to the target cells. The PDMS was removed from the TiO_2-modified glass slide, and imaging was undertaken using a 100× objective [18]. Liu et al. [19] performed FISH on centromere-sized cell arrays modified with 3-aminopropyltriethoxysilane (APTES) or polyethylene glycol (PEG)-coated glass slides with overnight hybridization. Wang et al. [20] introduced an APTES-coated glass slide with a PDMS microfluidic for stretching chromosomal DNA from a single cell to perform FISH. In a recent study, a microfluidic consisting of a Pyrex-Si stack was generated and used for FISH to provide breast cancer prognosis [21–24]. This report reduced FISH reagent consumption by 70% and hybridization time to 2 h.

Mayer et al. [24] used a microfluidic device to investigate HER2 amplification and immunohistochemistry of breast cancer patients. In a further extension of this study, the authors reported short incubation microfluidic-assisted FISH [23]. The researchers were able to reduce FISH hybridization time to 15 min for cell lines and 35 min for human tissue slides. Microfluidic channels etched into a glass slide, called FISHing lines, allowed the processing of 10 samples on a glass slide with 0.2 μL of FISH probe. MRD screening using the BCR/ABL fusion gene for chronic myeloid leukemia was performed by Mughal et al. [25].

Researchers have also reported using plastics to make the microfluidic for FISH to reduce the fabrication cost. Kwasny et al. [26] described the use of cyclic olefin copolymer (COC) devices sealed with glass or COC cover plates that were surface modified to allow for stretching chromosomes. Perez-Torella and coworkers reported a COC chamber, which was capable of delivering FISH reagents to cells [27]. The chamber was modified with 2-hydroxyethyl cellulose (HEC) to allow for cell adherence to the chamber walls. Micro-FISH devices fabricated with CO_2 laser ablation in a plastic were reported, which resulted in a 20-fold reduction in the sample volume [28].

Our group has developed a highly sensitive microfluidic for rare cell isolation. The isolated cells were immunostained directly within the selection chip, released into a 96-well plate, and visualized/enumerated using fluorescence microscopy. In addition, following cell selection and release from the isolation chip, the cells could be subjected to FISH using conventional slide-based approaches [3,4]. Unfortunately, the workflow required extensive manual handling of cells.

To address the workflow challenge, we developed a microfluidic device named "microtrap" for both immunophenotyping and FISH analysis of biological cells enriched from clinical samples, which we report herein. Cells isolated on the isolation chip could be transferred to the microtrap device for immunophenotyping and/or FISH in an automated fashion with reduced processing time.

The microtrap device consisted of an array of 80,000 containment pores generated by cross-channels (2–4 μm in width and depth) connecting a network of interleaving fluidic channels. Unique to this device is that it was fabricated with containment pores or microtraps that arrayed single cells in a 2-D format and within a common imaging plane. The device does not rely on modification of the surface to retain cells; rather, it physically entraps the cells. Additionally, imaging of the cells in the microtrap device on different z-planes demonstrated better spatial resolution of the hybridization probes to designate the detection of a genetic abnormality. This feature cannot be achieved in Flow-FISH analysis [29,30]. The microtrap device provided reduced processing time (18 → 3 h) and lower amounts of FISH probes compared to slide-based processing. The microtrap device integration to a cell selection microfluidic allowed for automated cell processing as well as minimized the amount of operator handling of the enriched cells, simplifying the workflow and reducing cell loss.

We demonstrate the utility of the device for the processing of non-adherent cells, such as CPCs and CLCs, with the device being able to perform immunophenotyping and FISH from liquid biopsy markers. In addition, we demonstrate that we can affinity-enrich B-ALL CLCs from a pediatric patient blood sample using a cell enrichment microfluidic decorated with anti-CD19 antibodies and perform FISH and immunophenotyping on the enriched cells using the device reported herein.

2. Materials and Methods

2.1. Design and Fabrication of the Microtrap Device

The design of the microtrap device is shown in Figure 1. There are two basic renditions, with each differing in terms of the number of microtraps the device possessed. Design 1—single bed with 7200 microtraps; and Design 2—8-bed device with 80,000 microtraps. The microtrap size could vary depending on the width and depth of the cross-channels and was as small as 4 μm (width) × 2 μm (depth) to accommodate the containment of smaller cells, such as CPCs and CLCs (6–16 µm in diameter), compared to the larger-sized CTCs [31]. The 8-bed device (shown in Figure 1A,B) consisted of a significantly increased number of microtraps as compared to our previous design due to the high numbers of cells that are enriched for leukemia and multiple myeloma diseases compared to CTCs' affinity enriched from epithelial cancers [3,4,31].

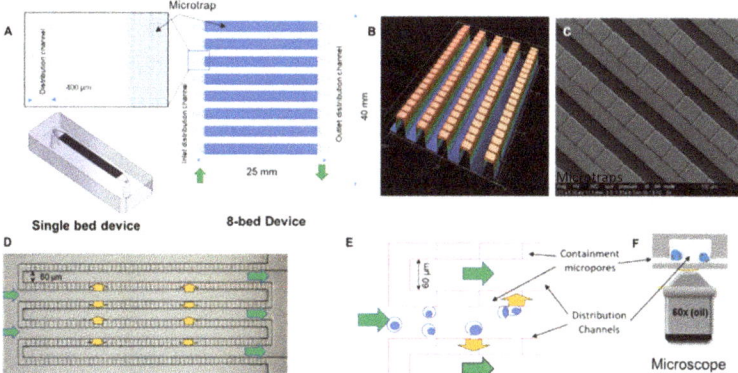

Figure 1. Microfluidic device for performing automated immunophenotyping and FISH. (**A**) Design of the microfluidic network composed of a single bed with 7200 microtraps and the 8-bed device containing 10,000 microtraps in each bed for a total of 80,000 traps per device. Microtrap size: $4 \times 2 \times 50$ µm (w × d × l). (**B**) Profilometer scan of the microtrap chip replicated in PDMS from a 3-level SU-8 relief and a Si master showing microchannel depth varying between input/output distribution channels, interleaving channels, and cross-channels. (**C**) Cross-channels and the deeper interleaving channels are shown in the SEM image. (**D**) Optical microscope image of a lithographically patterned 2-level SU-8 relief for preparing a single bed microtrap device. The arrows show the fluid path. (**E**) Schematic showing operation of the microtrap chip. Cells in solution (green arrows) are contained at the entrances of the microtraps, letting the fluid pass (yellow arrows) into the outlet channels of the interleaving network. (**F**) Schematic showing the 3-dimensionality of cells captured in the microtrap chip and imaging using a high magnification (60× or 100×) objective through a thin cover plate.

The lithography and fabrication steps of both devices are discussed in detail in the Supplementary Materials. The architecture of the single bed device (2-step lithography to create two levels) consisted of two independent networks of interleaving channels that were 60×40 µm^2 (W × D) and interconnected using an array of cross-channels orthogonally placed in between the interleaving channels (see Figure S1). The volume of this device was ~1 µL. For the 8-bed device, 3-level lithography was used to allow generation of a deep layer for the distribution channels, interleaving channels, and cross-channels (see Figure S2). Profilometry scans of the 8-bed chip showed the 3 levels for the SU-8 relief (Figure 1B). Figure 1C shows an SEM of the lithographically patterned 2-level SU-8 relief of a single bed device, with a magnified section showing the replica PDMS device with interleaving channels (30 µm) and shallow cross-channels (2–4 µm). In the 8-bed device, the 3 rd lithography layer was used to fabricate a deeper distribution channel network to distribute fluid to all 8 beds equally [32]. The distribution channels of the 8-bed device were 400×150 µm (W × D). The internal volume of this device, including all fluidic channels, was ~10 µL.

The fluidic operation of the device is shown in Figure 1D,E, where the flow from the interleaving channel input flows across (90° angle) the interconnecting cross-channels and then to the output channels of the interleaving network. The cross-channel, when a cover plate was bonded to the chip substrate, generated a microtrap structure where cells were contained but still allowed fluid to flow through it.

The optimal device design was chosen from testing 14 different designs for which we determined the containment efficiency of CPCs and CLCs. The 14 devices with different pore widths (4, 6, and 8 µm) and different depths (2 and 4 µm) were evaluated in terms of their ability to trap CPCs and CLCs (data not shown). The schematic in Figure 1D shows how the cells were contained at the entrance of the microtraps and high magnification imaging of the trapped cell as shown in Figure 1E.

An SU-8 relief was used to prepare PDMS trapping devices that were bonded to glass cover plates to allow for high resolution fluorescence imaging. Reservoirs were formed in the PDMS device using a sharp biopsy puncture. PDMS devices and glass cover plates (No.1 coverslips with a thickness of 0.13–0.16 mm) were cleaned with IPA, then washed with water, and air dried. We used thin glass coverslips to facilitate the use of high NA objectives to accommodate their short working distances and high magnification necessary for FISH. PDMS devices were bonded to the glass coverslips by treating the surfaces with O_2 plasma (50 W for 1 min). After plasma treatment, slight pressure was applied starting from one edge to the other to avoid the trapping of air bubbles. Peak tubing was sealed with epoxy glue to the reservoirs and a syringe pump was used to provide continuous fluid flow through the device.

2.2. Sample loading

The microtrap device was flooded with a continuous flow of 25 µL/min of PBS buffer (pH 7.4, 100 mM) followed by flowing 0.5% BSA/PBS solution through the device using a PHD2000 syringe pump (Harvard Apparatus, Holliston, MA, USA). Forty µL/min flow rates were used to remove air bubbles from the device. Once the device was fully wetted with the 0.5% BSA/PBS solution, cells were introduced into the device for either immunophenotyping or FISH at a volume flow rate of 10 µL/min.

2.3. On-Chip immunostaining

For immunostaining, live cells (RPMI-8226 or SUP-B15) were loaded as described above onto the microtrap device. After loading target cells onto the device, 2% paraformaldehyde (PFA) was injected at 10 µL/min for 2 min and allowed to incubate for 15 min. After incubation, the device was washed with PBS buffer for 5 min. Fixed cells were then treated with human Fc blocker (IgG1) for 15 min to block any Fc receptors on the cell surface followed by incubation with monoclonal antibodies for 30 min. For RPMI-8226 cells, anti-CD138 FITC (MI15, 5.0 µg/mL), anti-CD56-PE (MEM-188 clone, 20 µg/mL), and anti-CD38-APC (HIT2 clone, 2.5 µg/mL) monoclonal antibodies were used. Cells were then washed with PBS and permeabilized with 0.1% Triton X-100 for 5 min followed by counter staining with DAPI for 2 min.

2.4. Sample Preparation for FISH

RPMI-8226 cells were cultured as a model CPC cell line for multiple myeloma and similarly, a SUP-B15 cell line was cultured as a model CLC for B-ALL (see Supplementary Materials for more information). Once the cells were removed from the culture media, they were washed with 1× PBS twice. Following the wash, the cells were re-suspended in 0.075 M KCl hypotonic solution plus 100 µL of Colcemid to swell the cells. Colcemid helps the chromosomes to stretch, thus enhancing the clarity and the resolution of the fluorescent probes used for FISH. Cells were fixed using Carnoy's fixative (methanol: glacial acetic acid = 3:1 (v/v)). Carnoy's solution is a light fixative (no crosslinking) as opposed to 2% PFA. During fixation, Carnoy's fixative was added dropwise to the cells and mixed with gentle agitation. The solution was then centrifuged at 1200 rpm for 10 min. This step was repeated 3 times and the fixed cells were stored at −20 °C until being used for experiments.

2.5. On-Chip FISH

Before using fixed cell samples prepared as described above, they were mixed with fresh Carnoy's solution, and introduced into the microtrap device at 10 µL/min. In the case of live cell samples, cells were fixed on-chip before processing for FISH. Once the live cells were injected and contained at the microtraps in the device, 0.056 M KCl hypotonic solution was injected and cells were incubated for 10 min. After KCl treatment, cells were fixed by injecting Carnoy's fixative and incubated for 30 min, replacing the solution with fresh Carnoy's every 10 min. Then, all of the Carnoy's solution was removed by washing with PBS.

Next, the chip was washed with 2X SSC for 2 min at room temperature followed by a series of ethanol washes (70%, 85%, and 100% EtOH) injected for 1–2 min each and dried with 100% EtOH for 5 min. After the EtOH wash, the chip was dried completely by heating and evaporating all of the EtOH. One µL and 10 µL of 5× diluted FISH probe mix was introduced into the single and 8-bed device, respectively, while applying light vacuum to the outlet tubing. Once the device was filled with FISH probes, the inlet and outlet were sealed with rubber cement. The chip was heated to 75 °C for 5 min and incubated in a hybridization oven (Bambino II™ Hybridization Oven) (Boekel Scientific, Feasterville, PA, USA) for 2 h at 37 °C. Afterwards, the rubber cement was removed, and the chip was kept at 72 °C (±1 °C) while washing for 2 min with 0.4X SSC (pH 7.0). The temperature of this step is critical as it will remove the remaining free probes and keep hybridized ones associated to the target DNA to improve imaging of the FISH signal. Next, the device was washed with 2 × SSC + 0.05%, and Tween-20 (pH 7.0) for 1 min at room temperature. Finally, the device was washed with EtOH for 5 min and dried completely before applying 2 µL of 4-diamidino-2-phenylindole (DAPI II) for nuclear staining. After these steps, the chip could be stored at 4 °C until imaging.

2.6. Imaging On-chip and Image Processing

All imaging was performed using a Keyence BZ-X710 microscope (Keyence Cooperation of America, Itasca, IL, USA) equipped with BZ-X filters; DAPI (Ex: 360/40 nm, Em: 460/50 nm, dichroic mirror wavelength (DMW): 400 nm), GFP (Ex: 470/40, Em: 525/50 nm, DMW: 495 nm), TRITC (Ex: 545/25 nm, Em: 605/70 nm, DMW: 565 nm), and Cy5 (Ex: 620/60 nm, Em: 700/75 nm, DMW: 660 nm). The microscope was equipped with a Nikon objectives (Nikon Instruments, Melville, NY, USA) CFI Plan Apo 4× (λ 4×, NA 0.20 WD 20.00 mm), 10× (λ 10×, NA 0.45 WD 4.00 mm), 20× (λ 20×, NA 0.75 WD 1.00 mm), 40× (λ 40×, NA 0.95 WD 0.21 mm), and 60× (Apo VC 60× NA 1.40 NA WD 0.13 mm) objectives. For immunophenotyping, exposure times of 50 ms for DAPI, 500 ms for FITC, and 1500 ms for TRITC/Cy3 and Cy5/APC were used at 10×, 20×, and 40× magnification. All images for FISH experiments were acquired in high-resolution mode. Cell nuclei were imaged with the 60× objective using DAPI (blue) filters (200 ms) and the FISH signals were acquired using FITC/GFP (green, 1500 ms) and (Cy5/APC)/Cy3 (red, 2500/3000 ms) channels. Due to the cell's 3-D profile while they were contained at the microtrap inlets, it was necessary to do imaging across the z-axis. For the imaging of slides, we used $\Delta z = 2$ µm for 5 different planes while for on-chip FISH imaging, cells were imaged at 10 different planes along the z-axis with $\Delta z = 1$ µm. All images were processed using BZ-X Analyzer (Keyence Cooperation) and FIJI software (NIH) [33].

2.7. Patient Sample Processing for FISH

CLCs from B-ALL patients were captured using the CTC isolation device described in our earlier publications (see Supplementary Materials for more information), which was also used for acute myeloid leukemia (AML) and multiple myeloma work [3,4]. Healthy donor blood sample was obtained from the University of Kansas Medical Center (KUMC) IRB-approved Biospecimen Repository Core Facility. Blood samples from a patient diagnosed with B-ALL was collected according to an approved Children's Mercy Hospital Institutional Review Board procedure. Written informed consent was obtained from the patient included in the study before enrollment. Peripheral blood samples (5 mL) were drawn by venipuncture into Vacuette® K3EDTA (Greiner Bio-one, Monroe, NC, USA) tubes. Following affinity enrichment using the microfluidic enrichment chip, cells were released from the capture device's surface using USER enzyme that cleaved a single-stranded bifunctional linker containing a uracil residue [34]. The released cells were collected into a microfuge tube and centrifuged to prepare for immunophenotyping or FISH. For FISH, once the cells were spun down, supernatant was removed and pre-heated (37 °C) in 0.056 M KCl and incubated for 10 min. The mixture was centrifuged again and after removing the supernatant, ice-cold Carnoy's fixative (if fixation was done before introduction to the microtrap device) was added to the cells and centrifuged again. This step was repeated 3 more times and cells were stored in Carnoy's fixative at 20 °C until further use. CLCs were spun down and resuspended in fresh Carnoy's solution before use. These samples were infused into the microtrap device and processed according to the on-chip FISH procedure as described above.

3. Results

3.1. Microtrap Device for Immunophenotyping and Cytogenetic Analysis

The large number of traps in the microtrap device needed to be contained for analysis of CLCs' and CPCs' affinity selected from a blood sample (see Table 1). Contrary to CTCs, which use, for example, EpCAM as the enrichment antigen, the number of non-diseased cells selected by antibodies targeting leukemic associated and multiple myeloma antigens is high because non-diseased cells can also express the enrichment antigen (i.e., CD19, CD34, CD117, CD33, or CD138). Following CLC or CPC affinity selection and release from the selection chips [3], cells were trapped without surface modifications at containment pores of the microtrap device, which were arranged in a 2-D format to make it easier for imaging single cells. This simplified imaging was compared to cells stochastically arranged on glass slides or wells possessing the appropriate surface chemistry. In addition, we will show that we can interface the microtrap device for immunophenotyping and FISH of CLCs and CPCs to a device used for their enrichment from the blood of patients.

Table 1. Number of CLCs and CPCs compared to CTCs detected in cancer patients.

Cancer Cells (Antigen Used for Selection)	Target Cells (mL^{-1})	Non-Aberrant Cells (mL^{-1})	Cell Diameter (μm)	Reference
CLCs (AML, CD33, CD34, CD117)	11–2684	10–2450	11–16	[4]
CLCs (B-ALL, CD19)	40–840	400–2050	6–12	this work
CPCs (CD138)	10–5900	43–1875	12–16	[3]
CTCs in epithelial tumors (EpCAM)	1–800	3–10	10–23	[35]

Note: The references used here were taken from a single type of cell selection chip (sinusoidal device) so that comparisons could be made as to the numbers of the CLCs, CPCs, and CTCs secured from liquid biopsies.

Considering the size of the CLCs and CPCs with an average diameter of 6–16 μm, we used 4×2 μm^2 microtraps to maximize the containment efficiency of CLCs and CPCs. Table 1 shows the number of CLCs for both AML and B-ALL as well as CPCs enriched using the appropriate marker(s). The numbers of CTCs enriched from different epithelial cancers (i.e., pancreatic, breast, prostate, colorectal, and ovarian) is shown as well in Table 1. Not only are the number of aberrant cells enriched higher for the leukemia and multiple myeloma diseases compared to epithelial cancers, but the number of non-diseased cells enriched is higher as well compared to epithelial cancers due to the fact that non-diseased blood cells can carry the same antigens as CLCs and CPCs, whereas for CTCs the blood cells do not express EpCAM. Due to the higher number of cells anticipated for the leukemic and multiple myeloma diseases, we had to build devices with high numbers of microtraps.

3.2. Microfluidic Containment Device Operation

The microtrap device (both single and 8-bed devices) used two independent networks of interleaving channels that were interconnected using smaller cross-channels positioned orthogonally to the interleaving channels. The cross-channel, when sealed with a cover plate, generated a pore structure whose dimensions were determined by the size of the cross-channels (Figure 2A) and allowed fluid flow between the input and output interleaving channels. During operation, CPCs/CLCs are released from the isolation chip and directed to the microtrap device, where they become physically trapped with the efficiency of trapping dependent on the size of the microtrap with respect to the cell of interest. Trapped cells could then be immunophenotyped or subjected to FISH followed by imaging with a fluorescence microscope to read out the appropriate signals. Microtrap beds were connected in parallel to provide sufficient numbers of equally accessible pores to retain a large number of cells (see Table 1). If the eight beds were connected in series, a majority, if not all, of the cells would be retained within the first bed, generating cell pileup and crowding, thus making imaging difficult. Conversely, placing the beds in parallel (see Figure 1A) allowed access to all beds from a common input, and thus, dispersed the cells with equal probability at the 80,000 containment pores associated with the 8-bed device.

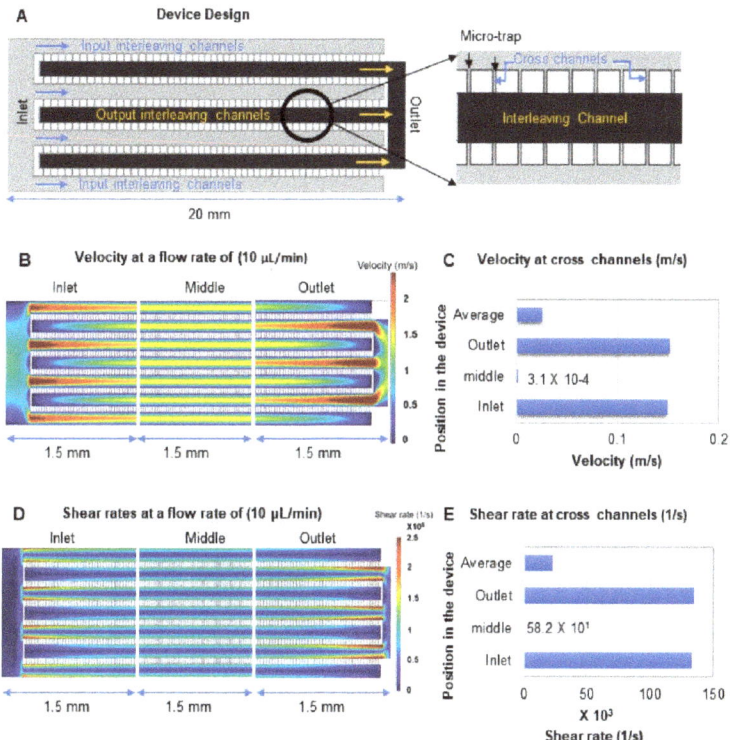

Figure 2. Simulations of the microtrap device. (A) 2-D CAD design of the microtrap device used for COMSOL simulations showing the interleaving network for the flow of fluid, and the cross-channels, which produce the microtraps when a cover plate is sealed to the device. The magnified image of the microtrap area is shown on the right with a single interleaving output channel (red) and two interleaving input channels (gray). (B) The simulated linear fluid velocity throughout the microtrap chip. The simulation shows three sections of the device: (i) input section; (ii) middle section; and (iii) outlet section. Flow was simulated across the interleaving input/output channels and the cross-channels. The dashed box shown here is the region of the device that was simulated in Figure S3 (see Supplementary Materials). (C) Bar graph representing the mean velocities expressed in m/s observed for the cross-channels at different sections of the device and at a 10 µL/min volume flow rate. The sections labeled here correspond to the sections of the device simulated in (B). (D) Simulated shear rate at three different sections of the device, inlet, middle, and outlet sections. (E) Bar graphs representing the mean shear rates across the cross-channels at different sections of the device at a volume flow rate of 10 µL/min. The sections of the device listed here correspond to those sections shown in (D).

Even though earlier studies from our group showed that live cells with a cell diameter of ~16 μm (CTCs) were successfully contained by microtraps with dimensions of 8 × 4 μm² [31], the current study required containment of smaller cells and cell numbers that were higher (see Table 1). RPMI-8226 cells (multiple myeloma model) have a size range of 6–16 μm with an average diameter of ~13 μm whereas SUP-B15 cells (B-ALL model cell line) have an average size of ~10 μm with a size range of 8–12 μm. Thus, when devices with microtraps of 8 × 4 μm² (width × depth) were used, we observed that the containment efficiency was <50% (data not shown). However, devices with microtraps of 4 × 2 μm² produced a containment efficiency >90% for these cells (Figures S1 and S3B).

3.3. Device Design and COMSOL Simulations

Finite element analysis (COMSOL) was performed on the microtrap device to deduce the projected linear velocity through the fluidic network and the corresponding shear rates to help determine the containment efficiency of the microfluidic device to physically trap live cells without damaging them. Laminar flow was validated across the entire device for the following flow rates: 1, 3, 5, and 10 μL/min with an aqueous fluid (Figure S3). The rationale behind choosing these flow rates was to test the optimal flow rate for effective containment of live cells without damage but having sufficient pressure to fill the device without generating air bubbles. The cells used for our studies were human cancer cell lines having a diffusion coefficient of 5×10^{-14} m²/s [36].

The flow within the microtrap device was driven hydrodynamically, hence, a parabolic flow profile existed with higher velocity in the center of each individual channel as compared to the channel walls (no-slip condition; see Figure 2B) [37]. Figure 2A shows a CAD drawing of the device with 4-μm-wide microtraps and a depth of 2 μm. This device was found to provide 90% containment efficiency for unfixed RPMI-8226 cells as shown in Figure S3A. For larger epithelial cells (i.e., SKBR3), the containment efficiency was 96%. Cells were evenly distributed throughout the microtrap device as well (Figure 3; Figure 4). In the experiments evaluating trapping efficiency, the cells were DAPI stained and counted using a microscope to verify the number of cells captured. Cells that were not retained were collected at the outlet of the microtrap device into a flat-bottomed plate and enumerated and inspected for damage. The containment efficiency was determined from the ratio of cells in the microtrap to the total number of cells introduced (i.e., cells trapped and cells passing through the microtrap device).

The average velocity at the inlet and outlet interleaving channels toward the input end of the device was 2.5 m/s and in the interleaving channels in the center of the device it was 1.5 m/s at a volumetric flow rate of 10 μL/min. Even at these relatively high velocities, the flow was still laminar (see Figure S3B,C). Accounting for differences in the velocity between the interleaving and cross channels, the average velocity was calculated for the cross-channels at different sections of the device at a 10 μL/min volume flow rate and is plotted in Figure 2C.

The average linear velocity in the cross-channels was 0.02 m/s at the 10 μL/min volumetric flow rate. To measure the pressure drop across the device, the relative pressure at the outlet was defined in absolute terms ($p_A = p + p_{ref}$, where p_A is the absolute pressure, p is the relative pressure, and p_{ref} is the reference pressure, which was set to 1 atm (101 kPa) [3]. A gradual drop in pressure across the length of the device was noted, with this drop being ~14 kPa (16 and 2 kPa at the inlet and outlet, respectively, at 10 μL/min). The calculated shear rates at different volumetric flow rates were used to determine the shear stress in the microtrap device [38]. According to Newton's law, shear stress is the shear rate times the viscosity:

$$\text{Shear stress (dynes/(cm}^2\text{))} = \text{Shear rate (1/s)} \times T, \tag{1}$$

where T is the dynamic viscosity (T for water is 8.90×10^{-3} dynes*s/cm² at 25 °C).

We calculated the average shear stress on the cells experienced in the microtrap device through the entire device at different flow rates. At a flowrate of 1 μL/min, the shear rate calculated was 6042 s⁻¹, which corresponds to a shear stress of 54 dynes/cm² and is 10 times higher at 10 μL/min (Table 2). Moreover, higher shear rates were observed in the inlet and outlet of the device, where cells have potentially the highest probability of being damaged when flowing near the wall of the device as opposed to the center of the channel or the center area of the device where lower shear stress is observed (Figure 2D,E). Shear rate distributions across a section of the device can be found in Figure S3D,E.

Table 2. Average shear rate and calculated shear stress on cells at each microtrap for the flow rates listed.

Flow Rate (µL/min)	Shear Rate (1/s)	Shear Stress (dynes/cm^2)
1	6042	53.8
3	18,206	162.0
5	30,454	271.0
10	63,750	567.4

The shear stress experienced by cells in physiological conditions as they travel through capillaries and arterioles ranges between 40 and 55 dynes/cm^2 [39–42]. Interestingly, even though the shear stress at a flow rate of 10 µL/min was 10 times higher than the average shear stress of cells traveling through arterioles, we did not observe obvious damage of RPMI-8226 or SUP-B15 cells when contained at the entrance of any microtrap within the device. Cells tolerate transiently high shear stress of ~3000 dyn/cm^2 or higher [43,44]. In our microtrap device, the average transit time of the cell traveling through the chip is 210 ms if it is not retained by the microtrap and only 1 min when cells are retained within the microtrap device. Because the cells experience high shear stress briefly, they remain intact during the transport in high shear stress environments [45].

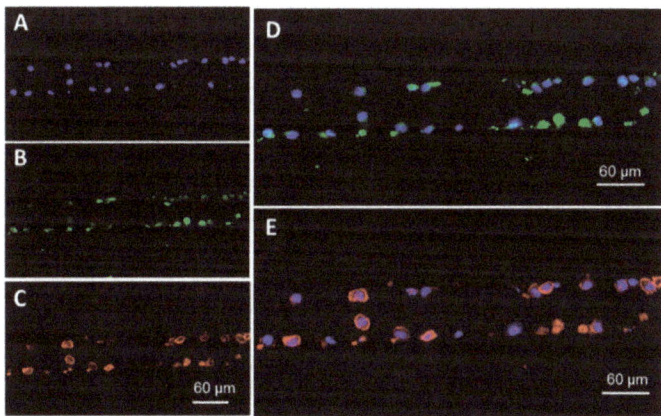

Figure 3. On-chip immunophenotyping of RPMI-8226 cells. (**A**) DAPI-labeled RPMI-8226 cell nucleus aligned at the entrance of the microtraps formed by the cross-channels and the cover plate assembled to the device. (**B**) CD138 expression of the RPMI-8226 cells and (**C**) CD38 expression for the same cells. (**D**) Composite image of CD138 expression (FITC channel) with the cell nucleus (DAPI channel of the microscope). (**E**) Composite image of CD38 expression (APC channel) with the cell nucleus that was DAPI stained. Exposure times were DAPI 50 ms, FITC 500 ms, and APC 1500 ms with 20× magnification. All images were collected using the Keyence fluorescence microscope. Shown in this fluorescence image are cells aligned along one interleaving input channel with cross-channels on either side of that channel.

As a control for on-chip immunophenotyping using the microtrap device, we performed immunophenotyping on a slide (see Figures S4 and S5) to compare with our on-chip results. RPMI-8226 cells were labeled with FITC-anti-CD138 antibodies and APC-anti-CD38 antibodies. RPMI-8226 cells express these markers [3,46,47]. See Figures S4 and S5 for the flow cytometry results for RPMI-8226 and on-slide immunophenotyping data, respectively.

Figure 3A shows DAPI-stained cells positioned at the microtraps in the single bed device. Here, we injected live cells followed by introduction of a fixative (2% PFA) to demonstrate the ability to fix cells on-chip. It was evident that our device could contain live cells without damaging their integrity. Trapped cells were stained with anti-CD138-FITC and anti-CD38-APC human antibodies (Figure 3B,C, respectively). Figure 3D,E show the composite images of the cell nucleus (DAPI channel) with the corresponding FITC and APC fluorescence emission signals.

Imaging of retained cells in the device for immunophenotyping was rapid. The microtrap device could be imaged at 20× magnification for all three colors (DAPI 50 ms, FITC 500 ms and APC 1500 ms) in <2 min. An advantage of using the microtrap device is the fact that single cells are positioned at the micropore entrance in a 2–D format, making them easy to locate. For immunophenotyping using the 8-bed device, >98% of cells were imaged in one plane without requiring z-stacking using a 20× microscope objective.

3.4. Microchip Processing and Imaging of a Large Number of Single Cells

Results for isolating CTCs, CLCs, and CPCs are summarized in Table 1, and show a high number of cells enriched from 1 mL of a patient blood sample using the sinusoidal microfluidic enrichment chip when analyzing CPCs and CLCs due to the fact that the enrichment antibody also selects non-diseased cells as opposed to CTCs, where the enriched fraction possesses only a few non-diseased cells. As such, thousands of cells may be required to be analyzed via immunophenotyping or FISH [3] to identify cancer cells (i.e., CLCs and CPCs). To facilitate the analysis of a vast number of cells, we designed the microtrap device with eight beds capable of entrapping enriched cells, subject them to staining, and present them in a 2-D array format for microscopic evaluation.

The 8-bed microtrap device possessed 80,000 containment pores patterned in PDMS from SU-8 reliefs. Three-level reliefs (i.e., three different heights of microstructures; see Supplementary Materials for fabrication description) were required to reduce the pressure in the chip and achieve well-balanced flow through the entire fluidic network (Figures 1 and 3). Because CLCs and CPCs have a diameter ranging from 6–16 µm (see Table 1), to ensure maximum containment efficiency by the microtraps, we used 4×2 µm^2 cross-sections for the containment microtraps (see Figure S3A). Fluorescence microscope images of cells contained at the microtraps and immunostained are presented in Figure 4.

Figure 4A shows a brightfield image of a single bed in the 8-bed device. Merged images of cells aligned at micropore entrances were imaged with DAPI (50 ms acquisition time) and anti-CD38-APC antibodies (1500 ms acquisition time) to identify CD38 on the RPMI-8226 cell surfaces; see Figure 4B. Figure 4C shows a single bed device containing DAPI-stained cells. DAPI-stained cells in two consecutive beds from the 8-bed device are imaged and presented in Figure 4D. As can be seen from Figure 4C,D, the contained cells are fairly well distributed throughout the microtrap 2-D array in spite of the decrease in the linear velocity seen down the length of the interleaving input channels (see Figure 2B). We noticed no loss of cell integrity at the microtraps even in the region of the input/output ends of the microtrap array where the shear stress was high (Figure 3D). Finally, when the flow was stopped, the cells remained at their trapped location.

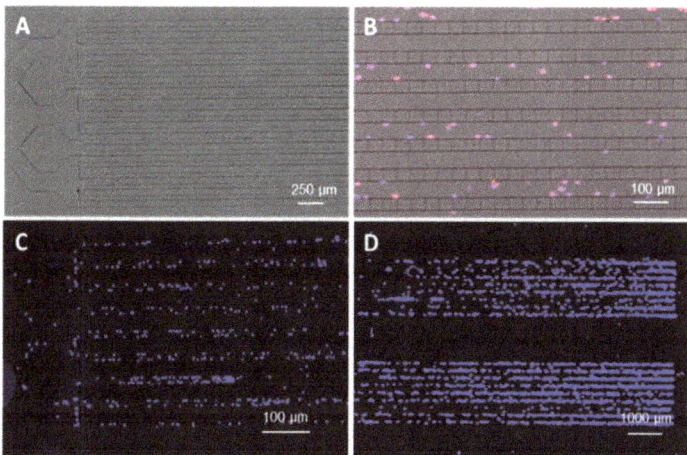

Figure 4. (**A**) Brightfield image of the bifurcated entrance channels of the microtrap device. RPMI-8226 cells were injected into the device at 10 µL/min and contained at the entrance of the microtraps. Cell images were processed according to the procedure listed in the materials and methods section of this manuscript and labeled with DAPI (nuclear stain) and CD38-APC markers. (**B**) Brightfield image merged with DAPI and APC channels showing the presence of the cell nucleus and CD38 on the cell surface aligned mainly at the microtrap entrances. (**C**) Entrance of the single bed device imaged using DAPI. RPMI-8226 cells were trapped inside the device at the entrance to the microtraps. (**D**) Two consecutive beds of the 8-bed device imaged with the DAPI channel of the microscope for stained RPMI-8226 cells.

3.5. On-Chip FISH

The microtrap device can be used for immunophenotyping and cytogenetic analysis, such as FISH. FISH determines aberrations in a metaphase chromosome or chromosomes buried in interphase nuclei from a fixed cytogenetic sample. The procedure for FISH processing on-chip is detailed in the materials and methods section as well as the Supplementary Materials. FISH experiments were carried out on RPMI-8226 and SUP-B15 cells with Cytocell FISH probes. The conventional workflow using microscope slides for FISH is a tedious and time-consuming process (see Figure S6), which requires 2–3 days including overnight hybridization of FISH probes. Figure 5 shows a step-by-step workflow for the on-chip FISH procedure, which required ~4 h of processing time and 2 µL of stock FISH probes for the 8-bed microtrap device, producing a 5-fold reduction in FISH probe volume compared to the slide-based FISH assay.

Figure 6A (i–iv) shows FISH signals from RPMI 8226-cells processed on-chip. FISH signals present in both the red and green channels of the fluorescence microscope in all of the cells were seen except for the image shown in Figure 6A (ii), where only one green signal was present due to deletion of the target gene region. In some of the cells, only one set (1 red and 1 green) of signal was present. One reason for losing some FISH signals is that the cells possess a 3-D structure (Figure 1E) even after entrapment at the microtrap and the fact that a high numerical aperture (NA) microscope was used with a short focal length; the images shown in Figure 6A were processed using only a single imaging plane (z-axis). This issue was addressed by using z-stacking of the imaging planes over a range equal to the average cell diameter.

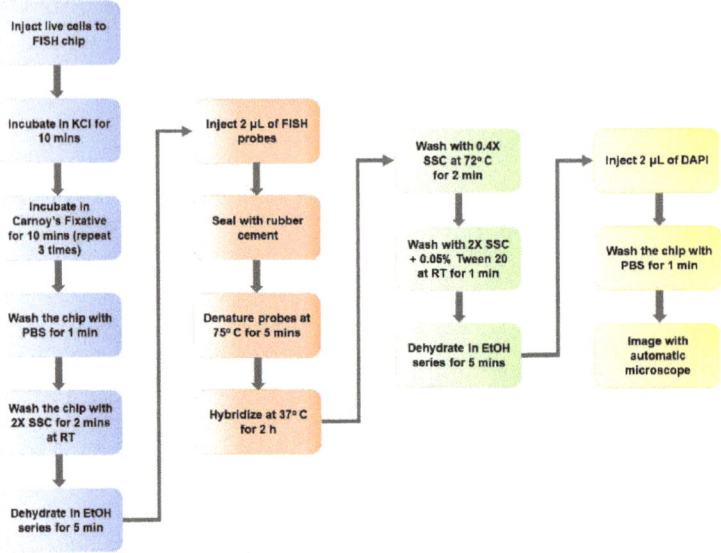

Figure 5. Workflow of FISH using the microtrap device. The workload was reduced from 2 days (slide method) to 4 h using the microtrap device primarily due to the hybridization time reduced from overnight to 2 h. The probe volume required for the assay was also reduced from 10 to 2 µL as well as using the microtrap device. Live cells were injected into the microtrap device at a flow rate of 10 µL/min and the washing steps were done at 5 µL/min to reduce the shear stress on the fixed cells contained within the microfluidic device.

Figure 6B shows a set of SUP-B15 cells processed using the microtrap chip for the TEL/AML1 FISH probes imaged in one image plane (i.e., no z-stacking). For TEL/AML1, probe TEL (ETV6—Erythroblastosis Variant Gene 6 translocation, ETS) refers to a region in chromosome 12 p-arm (12p13.2), and AML1 (or RUNX1—Runt-Related Transcription Factor 1) refers to the region in the q-arm of chromosome 21 (21q22.12). In a normal cell, there should be two red and two green signals and in a diseased cell, two yellow fusion signals are expected due to translocation of the TEL and AML1 genes. All of the images (Figure 6B) showed distinct red and green signals, with no clear indication of a yellow fusion signal to identify any cell as positive for the t(12;21) translocation.

Figure 6C,D shows two examples of SUP-B15 cells processed with BCR/ABL1 FISH probes for Ph t(9;22) (q34.12; q11.23) imaged with z-stacking (1-µm increments along the z-axis over ~15 µm). Ph t(9;22) (q34.12; q11.23) consists of two gene regions, with one from chromosome 9 corresponding to ABL1 gene (red labeled) and the other for the BCR (breakpoint cluster region) gene in chromosome 22 (green labeled). In a cell without a chromosomal fusion aberration, there are two green and two red signals. If there are yellow fusion signals detected, the cell can be identified as Ph(+). Imaging of 1-µm z-planes over a 15-µm range covered the entire cell as noted in Figure 6C,D. Those images showed two or more signals present within the cells entrapped by the microtrap device. Figure 6C shows two distinct red and green signals (confirmation as a cell not possessing the fusion product) and Figure 6D shows one yellow fusion signal (second signal not visible or merged with the first one) and one red and green signal, confirming it as a B-ALL cell that is Ph(+).

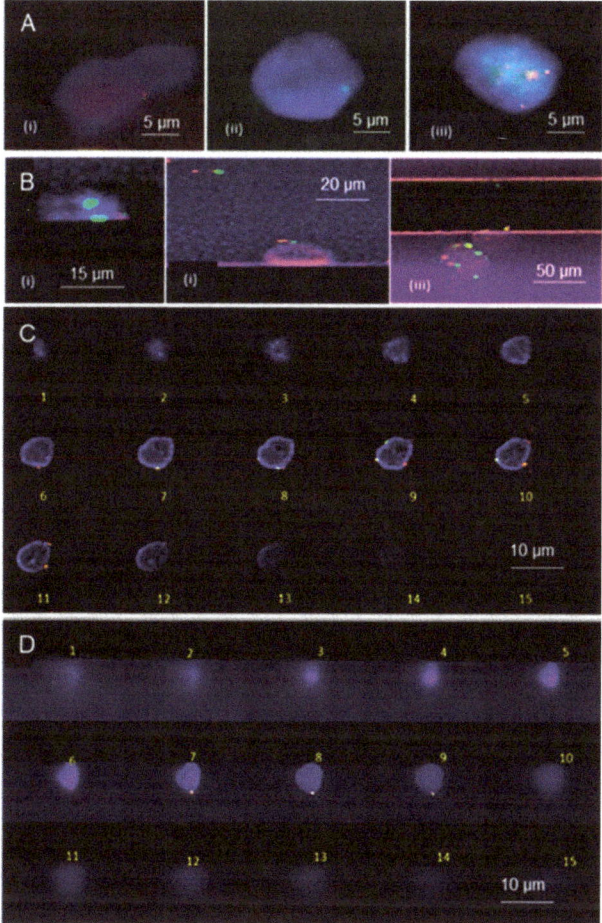

Figure 6. FISH-on-chip analysis of RPMI-8226 and SUP-B15 cells. (**A**) RPMI-8226 cells after FISH processing using the microtrap chip with the D13S319 plus deletion probe; (i) a cell that shows one green and one red FISH signal; (ii) a cell with only one green signal; (iii) a cell with 2 green and 2 red signals. (**B**) FISH analysis of SUP-B15 cells processed with the TEL/AML1 translocation, dual fusion probes showing the TEL (ETV6, 12p13.2) region in red, and AML1 (RUNX1, 21q22.12) region in green. (i) Two cells contained at the entrance of two different microtraps, but the FISH probes were visible in only one cell with two green signals; (ii) shows one red and two green signals with no clear yellow signals present; (iii) two cells that show distinct red and green signals, one cell captured at the entrance of microtrap shows one red, and one green signal with a possible yellow fusion signal. Both (**A**,**B**) were imaged in one single z-plane without z-stacking. (**C**,**D**) show z-stacking planes of 15 different image planes for FISH images from SUP-B15 cells captured at the microtrap and FISH processed with BCR/ABL plus translocation, dual fusion probe. (**C**) SUP-B15 cell with two green and two red signals. (**D**) SUP-B15 cell with one yellow fusion signal (second yellow signal not visible) and one red and green signal. Each image shows 15 separate images through the 15 µm distance range taken at 1-µm imaging intervals. FISH probes were specific to the BCR/ABL gene region, Philadelphia (Ph) chromosome tagging. All images were acquired using a Nikon 60× oil objective with DAPI—200 ms, FITC—1500 ms, TRITC—2500 ms integration times. The average SNR was 59 for the green probe and 68 for the red probe.

3.6. Measurement of MRD Status in Pediatric B-ALL Patients Using CLCs

CLCs and normal B-cells were affinity selected from a patient's peripheral blood with anti-human CD19 antibodies attached to the surface of a CLC enrichment (sinusoidal) microchip via a single-stranded oligonucleotide cleavable linker containing a dU residue [34]. Blood was collected from a pediatric patient (1–18 years) diagnosed with B-ALL undergoing induction and consolidation chemotherapy. Released cells following enrichment were immunophenotyped to distinguish CLCs from normal B-cells using the microtrap device; cells that demonstrated the expression of terminal deoxynucleotidyl transferase (TdT) in the nucleus were classified as CLCs. Additionally, the staining cocktail contained CD19/CD34/CD10 fluorescently labeled monoclonal antibodies to provide additional phenotypic data (Figure 7A).

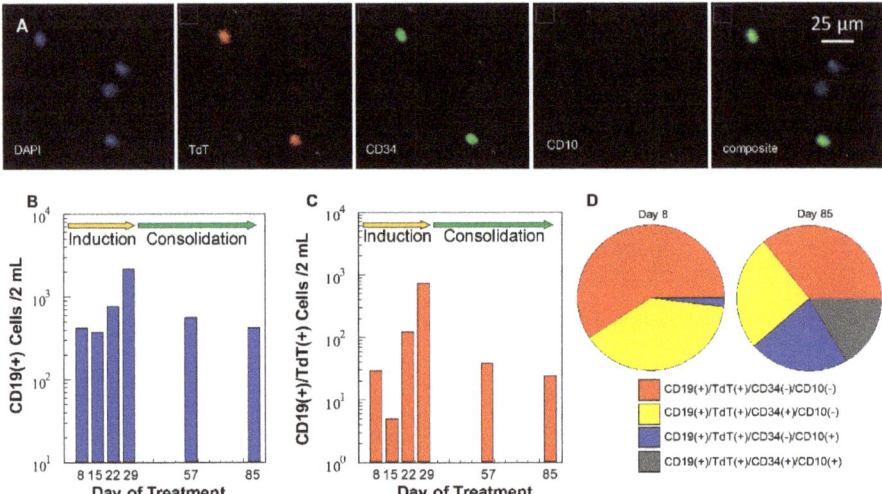

Figure 7. (**A**) Immunophenotyping of cells enriched from peripheral blood of a B-ALL patient by targeting cells with that express the CD19 antigen. The cells were stained using DAPI (nucleus), and monoclonal antibodies directed against TdT (FITC), CD34 (Cy3), and CD10 (Cy5). The images were acquired using a 40× microscope objective. The CLCs shown were DAPI(+)/CD34(+) and TdT(+), but CD10(−). (**B**) Microfluidic monitoring of a B-ALL patient from day 8 to 85 of chemotherapy. Total cell count represents all DAPI(+)/CD19(+) cells selected. (**C**) Number of CLCs identified as DAPI(+)/CD19(+)/TdT(+)/CD34(±)/CD10(±). (**D**) Change in phenotype among CLCs for this patient for days 8 and 85.

We analyzed the blood of a pediatric B-ALL patient to determine MRD status during chemotherapy on days 8, 15, 22, 29, 57, and 85 (Figure 7B–D). The clinical specificity was determined based on a threshold value established from the analysis of healthy donors as negative controls (average CD19 expressing cells was 68 cells/mL of peripheral blood). Grounded on that, we classified this patient as MRD(−) upon completion of induction and consolidation therapy on day 85. In this particular patient, on day 85, we observed a new phenotypic population of cells (Figure 7D) not observed during the first two analyses, which were secured on days 8 and 15 during induction therapy. Cells with the CD19(+)/TdT(+)/CD34(+)/CD10(+) phenotype began to appear in the blood on day 22. Although leukemic cell phenotype changes are common in B-ALL due to the effects of steroids as part of chemotherapy (i.e., loss of CD34), it is likely that the aforementioned cells represent normal immature lymphoid precursors whose morphology and immunophenotype are similar to the CLCs found in B-ALL. MRD status of this patient was determined to be positive only once, which was on day 29 of treatment when the level of enriched cells classified as CLCs (361 cells/mL) were above the threshold value.

To confirm the chromosomal status of the CD19-expressing cells on day 29 of treatment, we tested the enriched cells for chromosomal aberrations via FISH. Figure 8 shows a stitched image of cells contained on the microtrap device from a B-ALL patient. The sample was processed using TEL/AML1 FISH probes, which were able to identify the t(12;21) translocation. The TEL/AML1 fusion FISH probes identifies the most common rearrangements in childhood B-ALL, which is seen in around 17% of patients [48]. Figure 8a shows a single cell with two green FISH signals and Figure 8b shows two green FISH signals, with one signal showing a yellow signal (see arrow). Figure 8c shows a single cell with discrete red, green, and yellow signals. There was no visible evidence of a second yellow signal for confirmation of both t(12;21) translocations. Figure 8d shows a cell with the same observation of one red, one green, and one yellow signal, which were closely packed together. In Figure 8e, there are two green and one red signal visible. Figure 8a,b,e shows a lack of a red signal from the TEL gene. It has been observed that in B-ALL patients, there is the possibility for deletion of one TEL allele [49–51].

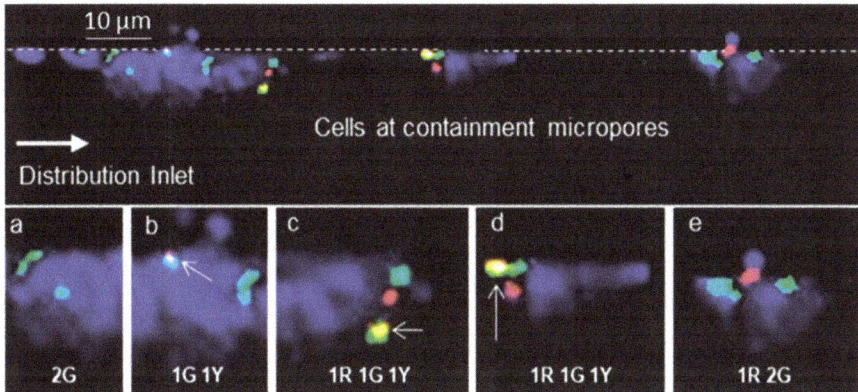

Figure 8. On-chip FISH processed B-cells isolated from a diagnosed B-ALL patient. TEL/AML1 FISH probes were used for the chromosomal aberration of t (12;21) translocation. Cells were imaged at the microtraps. Zoomed images show (**a**) single cell with 2 green FISH signals; (**b**) single cell with one green and one yellow signal; (**c**) a single cell with one red, one green, and one yellow FISH signal; (**d**) single cell with one red, one green, and one yellow FISH signals close to each other in the cell; and (**e**) single cell with one red and two green signals. In all cases, the images were collected using a 60× objective with z-stacking.

4. Discussion

FISH testing constitutes important and independent prognostic factors and is considered obligatory for analyzing patient outcome [52]. Of the current ~117 human genetic tests approved by the US Food and Drug Administration, 18 of these are FISH-based assays and most are directed toward hematological diseases, such as AML, multiple myeloma, and ALL.

AML arises from mutations occurring in progenitor cells of the myeloid lineage, which results in the inability of these cells to differentiate into functional blood cells. AML is the most common adult leukemia, with >21,000 new cases in the US in 2018, with a 5-year survival rate of 25%. The primary cause of death for AML patients is due to disease relapse [53]. The WHO currently categorizes patients into four groups [54]. For example, one category is patients with recurrent genetic abnormalities, which can consist of seven different chromosomal aberrations (typically balanced translocations or inversions, inv). Some of these aberrations are t(8;21)(q22;q22) associated with the *RUNX1/RUNX1T1* genes in chromosome 8, inv(16)(p13.1q22)/t(16;16)(p13.1;q22) occurring in the *CBFB/MYH11* genes of chromosome 16, and inv(3)(q21q26)/t(3;3)(q21;q26) of the *RPN1/EVI1* genes in chromosome 3. While AML MRD is typically managed using bone marrow biopsies, we have shown that CLCs can be used to determine recurrence from MRD in AML. The CLCs were enriched from blood samples using three sinusoidal microfluidic devices, with each one targeting a specific AML-associated antigen, CD117, CD34, and CD33 [54].

Multiple myeloma is associated with the abnormal expansion of terminally differentiated B clonal plasma cells in the bone marrow that produces an abnormal monoclonal paraprotein [55,56]. Multiple myeloma has three clinically defined stages: (i) MGUS (monoclonal gammopathy of undetermined significance), which is the asymptomatic stage; (ii) SMM (smoldering multiple myeloma) an intermediate phase; and (iii) the symptomatic stage referred to as active multiple myeloma [57]. In most cases, bone marrow biopsies are used to manage multiple myeloma. However, we and others have shown that CPCs can be used to manage this disease, which used a minimally invasive liquid biopsy [3,4,31]. In our study, we used a microfluidic device containing an array of sinusoidal microchannels with anti-CD138 monoclonal antibodies used to enrich CPCs from multiple myeloma patients [3]. It has been reported that in 16–50% of all multiple myeloma cases, chromosome 13q aberrations are present [58,59]. More than 90% of reported cases show the chromosomal aberration specifically in the 13q14 region [60]. We were able to perform FISH in the CPCs to detect the presence of chromosome 13q deletions using a slide-based FISH method (see Figures S6 and S7).

The FISH probes used for the RPMI-8226 cells, a model of multiple myeloma, identifies the DLEU region of chromosome 13 covering the 13q14 gene and used a red (APC channel) fluorescent probe. The control gene, 13qter located at the end of chromosome 13, was labeled with a green fluorescent probe (FITC channel). In a normal cell, there are two green signals and two red signals. However, due to the polyploidy nature in some cells, there may be multiple chromosomes (>2). In CPCs, it is expected that one or both DLEU regions (DLEU1 and DLEU2) may be deleted [61].

Figures 3 and 6 show immunophenotyping and FISH processing of RPMI-8226 cells using our microtrap device. As expected, the data seen in Figures S4 and S5 in the Supplementary Materials and our previous studies [3] confirmed the expression of CD138 and CD38 proteins for RPMI-8226 cells. We detected the presence of chromosome 13 as a green FISH signal corresponding to the 13qter gene (100 kb), which was present in all images, as shown in Figure 6A. Figure 6A (i) and (ii) shows deletion of the red signal corresponding to gene regions covering DLEU1, DLEU2, D13S319, D13S272, and RH47934 (156 kb) as expected for the RPMI-8226 cell line, as well. Most of the RPMI-8226 cells contained both red and green signals (Figure S7 in Supplementary Materials) lacking deletion, which is consistent with the karyotype data for this cell line.

B-ALL is the most common cancer diagnosed in children, representing ~30% of cancer diagnoses [62]. Despite significant improvements in the overall survival of children with B-ALL, there is a group of patients that experience relapse and ultimately die from their disease [63]. In fact, the likelihood of relapse is 80% for patients who have MRD at the end of induction therapy, indicative of active disease [64]. Monitoring of MRD, therefore, is considered a powerful predictor of outcome in B-ALL.

Cytogenetic abnormalities detected at diagnosis or generated during chemotherapy constitute important prognostic factors [52]. In B-ALL patients, 25–30% of patients have hyperdiploidy, 25% have t(12;21), 3–5% have t(9;22), 10% have MLL translocations, and 2% have iAMP21 chromosomal abnormalities. Once an aberration is detected, it can aid in the determination of the treatment regimen [65,66]. As another example, the detection of specific chromosome aberrations, such as t(9;22)(q34;q11.2) for BCR-ABL1, which results in the formation of the Philadelphia (Ph) chromosome, or t(12;21) aberrations of TEL/AML1 gene translocations are used to assign B-ALL patients to specific targeted therapies [67].

For the SUP-B15 cell line, which is a model for B-ALL [68], there are a few targeted gene variations that are typically evaluated using FISH [50,69,70]. MLL break-apart probes are used to detect the breakage of the MLL gene, which is frequently found in infant B-ALL [71–73]. BCR/ABL1 probes are used to detect the formation of the Philadelphia (Ph) chromosome produced by the fusion of two genes from chromosome 9 and 22, which is one the most important prognostic indicators for several hematological disorders, including B-ALL (see Figures S7, S8, and S9 for on-slide FISH analysis for some of these chromosomal abnormalities) [74–76].

In this study, the SUP-B15 cells were tested for TEL/AML1 translocations and BCR/ABL1 (Ph chromosome) using the microtrap device for FISH. For the SUP-B15 cell line, it is expected to see two distinct red and two green signals with the TEL/AML1 FISH probes. The TEL (ETV6) gene region marked in red corresponds to the 12p13.2 in chromosome 12 covering 168 kb of D12S1898 region and the green marker covers AML1 (RUNX1) in chromosome 21q22.12, a 167 kb gene region, including the CLIC6 gene [50]. On-chip FISH results (Figure 6B) showed distinct red and green signals corresponding to the presence of chromosome 12 and 21. Lack of a yellow signal confirmed t(12;21)(p13.2;q22.12) translocations were not present in the SUP-B15 cell line, which agrees with the karyotype as noted in the literature. The Philadelphia chromosome results from translocations of the ABL1 gene (9q34.11-q34.12, red) in chromosome 9 and the BCR gene (22q11.22-q11.23, green) in chromosome 22. The BCR probe region covers the GNAZ and RAB36 genes in a 169 kb region plus 148 kb region in telemetric BCR. The ABL1 probe covers a 346 kb region in the middle of the FUBP3 gene. As for the karyotype data of SUP-B15, we expected to see >90% of cells possessing the Ph chromosome (yellow fusion signal present). No yellow signal would be considered a cell with no Ph chromosome.

Figure 6C showed that SUP-B15 cells processed on-chip with BCR/ABL1 genes expressed two distinct red and green signals. This confirmed that there was no Ph chromosome present while in Figure 6D it showed the presence of one yellow fusion signal, confirming Ph(+) in that cell. In Figure 6D, we did not identify the fusion signal in this cell. However, the patient sample processed for FISH on-chip showed improved FISH signals as seen in Figure 8.

The microtrap FISH assay resulted in an SNR of 59 for the green signal, and 68 for the red signal. In the case of the slide-based assays, the SNR for the green and red signals were 64 and 63, respectively (see Figure S9), indicating that the ability to detect single molecules associated with the fluorescent reporter attached to each FISH probe was clearly visible using the microtrap device. The challenge is that in the FISH experiments, we used a high numerical objective with a small focal distance and as seen in Figure 1E, z-stacking was necessary to cover the genetic material housed within the nucleus. This may have been the reason that some signals were missed. This can be obviated by using a high numerical objective with a larger focal distance to better cover the entire nuclear region when the cells are located at the pore entrance.

Most FISH-based assays are predicated on the use of bone marrow, which is enriched in diseased cells compared to blood. For example, in the case of multiple myeloma, CPCs in peripheral blood are reported to be >100-fold lower than in bone marrow [77,78]. If disease relapse and chromosomal defects could be detected from peripheral blood, painful bone marrow biopsies could be avoided, and physicians could obtain information in near real time and potentially implement changes in treatment to affect better outcomes for patients with hematological diseases.

To obviate the need for a bone marrow biopsy, we used a liquid biopsy secured from a B-ALL patient using an affinity microfluidic chip and performed immunophenotyping and FISH on those enriched cells using our microtrap device. Similar to our previous work on the isolation of leukemic cells from blood of patients diagnosed with AML, a sinusoidal microfluidic chip with positive affinity selection was used [3,4], but in this case the affinity selection used a different antibody (anti-CD19 monoclonal antibodies) to enrich B-cells. While the enrichment of the CLCs and CPCs in our previous work was accomplished using a microfluidic chip, the downstream analysis was done off-chip, including immunophenotyping and FISH. Standard FISH workflow demands highly trained and experienced personnel (see Figure S6), making it difficult to implement in clinical laboratories not possessing the specialized facilities and trained personnel. Additionally, blood cells collected from a bone marrow biopsy are used for cytogenetic analysis, requiring an invasive procedure [64].

Using our microfluidic assay, a blood sample was subjected to affinity enrichment with high efficiency in terms of recovery and purity, and thus, the entire leukocyte population of peripheral blood did not require cytogenetic interrogation. The only cell population interrogated was those that expressed the target antigen, which in the case of B-ALL was CD19-expressing cells. Additionally, enriched cells were distributed in an array-like format as determined by the position of the microtraps of the microtrap device (see Figure 4C,D), which made them easier to image as opposed to stochastically distributed on a properly functionalized surface to induce cell adhesion to the surface. In addition, the 8-bed version of the microtrap device possessed 80,000 pores for retaining cells. While the data displayed in Table 1 show the total number of cells affinity selected (aberrant and non-aberrant) were 5314 for AML, 2840 for ALL, and 7775 for multiple myeloma, these numbers were based on a per mL sample volume. In some cases, it may be necessary to use larger input volumes, such as 10 mL, to search for rare CLCs and CPCs to find cells in the correct phase to elicit proper FISH signals. In these cases, the full advantage of the large dynamic range of the 8-bed device can be realized.

To reduce the workflow for FISH, microfluidics has been suggested by several groups, with the processing time reduced from several days using conventional slide-based FISH to several hours using FISH-on-chip platforms [14,26–28]. In addition, microfluidics has also resulted in a reduction in the use of expensive FISH probes. However, the reported platforms (see Table S1) required the use of special surface coatings to allow for cells to adhere to the surface of the device. Because of the stochastic nature of the attachment to the surface of the chip, this can create cell aggregates that made it difficult to image single cells under high magnification to determine the chromosomal status of the cells. Our device obviated the need of surface coatings and ordered the cells in a 2-D format to reduce device preparation steps and simplify single cell imaging, respectively. The microfluidic was comprised of an array of microtraps that were easily formed via a replication step in PDMS from a relief prepared by lithography. The relief also possessed the fluidic network.

We showed in this work that the microtrap device could be coupled to a rare cell enrichment chip to allow processing of circulating cells, such as CPCs or CLCs, with the ability to perform immunophenotyping and FISH of the enriched cells directly from blood samples. Moreover, experiments showed that our microtrap device was capable of containing live cells with >90% efficiency with sufficient traps to process CLCs and CPCs enriched from blood. The microtrap device was operated at a 10 µL/min volume flow rate to facilitate proper filling of the device without air bubbles. At this flow rate, even though cells experienced ~570 dynes/cm^2 shear stress, no obvious cell damage was observed for cells contained by the microtraps. Cell physical survival was attributed to the cell membrane's ability to handle relatively high shear stress for a brief time.

Using our 8-bed device with 80,000 microtraps, we could process thousands of cells for molecular profiling following enrichment. FISH results were achieved in <4 h by reducing the hybridization time from overnight to 2 h. Also, automated imaging was demonstrated for phenotyping in 2 min and FISH imaging in < 5 min, reducing the workflow for FISH compared to conventional slide-based assays, which requires 2–3 days of processing time, most of which is done manually. The FISH-on-chip provides full process automation.

When the cells were physically retained at the microtraps, they did possess a 3-D structure, requiring z-stacking to cover all of the FISH probes present inside the cell nucleus. Unlike immunophenotyping, where it was possible to use a single focusing plane because a lower magnification and associated longer focal length was required, FISH required the use of a high numerical aperture objective with a smaller focal length to capture high resolution images along several focal planes to image all of the FISH probes present in the cell nucleus. Even though z-stacking was necessary for FISH imaging, it was possible to automate this process. We set a common upper and lower threshold point in the device and selected the points where the cells were present at the arrayed microtraps and proceeded with automated imaging with pre-set exposure times for different filters (DAPI 200 ms, FITC 1500 ms, Cy3/Cy5 2500 ms). Processing of the captured images using BZ-X Analyzer and FIJI is detailed in the Supplementary Materials.

As opposed to our previous version of this device, which possessed larger microtraps [31], this device was designed to accommodate smaller CLCs and CPCs and the higher number of cells to analyze (see Table 1). For example, the CLCs (i.e., B-ALL cells) are smaller than CTCs and, as such, required the use of a smaller containment pore (4×2 µm^2 compared to 8×6 µm^2 in our previous report) [31]. Because the affinity selection process for CLCs and CPCs results in the enrichment of a much larger number of cells due to the fact that the even non-diseased cells express the capture antigen, a larger number of containment pores were required (80,000 herein compared to 5000 in our previous device) [31]. In addition, our previous report only performed immunophenotyping and did not carry out FISH on the chip, as was demonstrated here.

In the current rendition, the microtrap device was made from PDMS by casting it against a relief. However, the same device architecture can be made from a thermoplastic, such as cyclic olefin copolymer (COC), which has some decisive advantages compared to PDMS [79]. For example, because COC is a thermoplastic, it can be injection molded to allow production of devices at high rates and at significantly lower chip cost compared to PDMS [79]. In fact, the entire device can be injection molded in a single cycle, with the only requirement being cover plate bonding as a finishing step. Also, COC has excellent optical properties that allow for high sensitivity imaging using the spectral range typically employed for FISH [27]. COC can be UV/O_3 activated to change its wettability to allow efficient filling with aqueous solutions without creating air bubbles and does not show the typical rapid hydrophobic recovery as seen with PDMS [80]. This will allow for the generation of a low-cost disposable appropriate for in vitro diagnostics. When the microtrap device is physically integrated to the cell enrichment device via a fluidic motherboard, fully automated processing of liquid biopsy samples can be envisioned to enable clinical use.

5. Conclusions

The ability of our microtrap device was demonstrated using CPCs and CLCs for immunophenotyping and FISH analyses of relatively small cells (D_{avg} ~12 µm) and in high numbers. The same device could be used to identify expression patterns of proteins and detect targeted chromosomal aberrations in single cells. Using the 8-bed device with 80,000 containment microtraps, we could process thousands of live or fixed cells enriched using a cell isolation microchip. FISH results were achieved in <4 h by primarily reducing the hybridization time from overnight to 2 h, and lowering the volume of the FISH probes required for analysis. The microtrap device was used for automated imaging for phenotypic identification of cells in 2 min and for FISH in <5 min for all fluorescence channels without the need to scan a relatively large area. Moreover, we were able to enrich B-cells from an ALL patient and process those cells to identify chromosomal aberrations. In future work, the use of a thermoplastic, such as COC, instead of PDMS will be undertaken to produce a low-cost, disposable device appropriate for use in clinical applications.

Supplementary Materials: The following are available online at http://www.mdpi.com/2073-4409/9/2/519/s1, Table S1. Summary of FISH-based microfluidic technologies. Figure S1: metrology of microtrap device, Figure S2: fabrication steps for 8-bed microtrap device, Figure S3: COMSOL simulations of microtrap device, Figure S4: flow cytometry results of RPMI-8226 cells, Figure S5: immunophenotyping of RPMI-8226 cells, Figure S6: chart of FISH workflow using slide-based method, Figure S7: slide-based FISH results for RPMI-8226 cells, Figure S8: slide-based FISH results for SUP-B15 cells, Figure S9: slide-based FISH results for SUB-B15 cells.

Author Contributions: Conceptualization, K.A., M.L.H., M.A.W., S.A.S.; methodology, K.A., A.K.G., M.L.H., M.A.W., S.A.S.; formal analysis, K.A., M.A.W., S.A.S.; investigation, K.M.W.-R., S.V., N.L., K.D., M.H., J.J., S.M., M.L.H., M.A.W.; writing—original draft preparation, K.M.W.-R., S.V.; writing—review and editing, K.A., M.A.W., S.A.S.; supervision, M.A.W., S.A.S.; project administration, K.A., S.A.S.; funding acquisition, S.A.S., M.A.W. All authors have read and agreed to the published version of the manuscript.

Funding: The authors thank the NIH for financial support of this work via NIBIB: P41-EB020594; NIGMS: P20-GM130423; and NCI: P30-CA168524. The authors also acknowledge financial support from the Midwest Cancer Alliance, NCI: R44-CA224848, and R33-CA235597.

Acknowledgments: The authors also acknowledge the KU Nanofabrication Facility, and the University of Kansas Cancer Center's Biospecimen Repository Core (NCI: P30-CA168524) for sample collection. We thank Lindsey Roe for help in editing the manuscript.

Conflicts of Interest: S.A.S. and M.L.H. hold equity shares in BioFluidica, Inc., a company that holds commercialization rights to the cell isolation technology described herein. M.A.W declares COI as a spouse of a BioFluidica, Inc. employee.

References

1. Macías, M.; Alegre, E.; Díaz-Lagares, A.; Patiño-García, A.; Pérez-Gracia, J.L.; Sanmamed, M.F.; López-López, R.; Varo, N.; Gonzalez, A. Liquid Biopsy: From Basic Research to Clinical Practice. In *Advances in Virus Research*; Elsevier BV: Bethesda, MD, USA, 2018; pp. 73–119.
2. Jeffrey, S.S.; Toner, M. Liquid biopsy: A perspective for probing blood for cancer. *Lab Chip* **2019**, *19*, 548–549. [CrossRef]
3. Kamande, J.W.; Lindell, M.A.M.; Witek, M.A.; Voorhees, P.M.; Soper, S.A. Isolation of circulating plasma cells from blood of patients diagnosed with clonal plasma cell disorders using cell selection microfluidics. *Integr. Biol.* **2018**, *10*, 82–91. [CrossRef]
4. Jackson, J.M.; Taylor, J.B.; Witek, M.A.; Hunsucker, S.; Waugh, J.P.; Fedoriw, Y.; Shea, T.C.; Soper, S.A.; Armistead, P.M. Microfluidics for the detection of minimal residual disease in acute myeloid leukemia patients using circulating leukemic cells selected from blood. *Analyst* **2016**, *141*, 640–651. [CrossRef]
5. Bauman, J.; Wiegant, J.; Borst, P.; Van Duijn, P. A new method for fluorescence microscopical localization of specific DNA sequences by in situ hybridization of fluorochrome-labelled RNA. *Exp. Cell Res.* **1980**, *128*, 485–490. [CrossRef]
6. Langer-Safer, P.R.; Levine, M.; Ward, D.C. Immunological method for mapping genes on Drosophila polytene chromosomes. *Proc. Natl. Acad. Sci. USA* **1982**, *79*, 4381–4385. [CrossRef]
7. Andreeff, M.; Pinkel, D. *Introduction to Fluorescence in Situ Hybridization: Principles and Clinical Applications*; Wiley-Liss: New York, NY, USA, 1999.
8. Ye, C.; Stevens, J.; Liu, G.; Ye, K.; Yang, F.; Bremer, S.; Heng, H. Combined multicolor-FISH and immunostaining. *Cytogenet. Genome Res.* **2006**, *114*, 227–234. [CrossRef]
9. Sato, K. Microdevice in Cellular Pathology: Microfluidic Platforms for Fluorescence in situ Hybridization and Analysis of Circulating Tumor Cells. *Anal. Sci.* **2015**, *31*, 867–873. [CrossRef]
10. Dowd, A.; Homeida, S.; Elkarem, H.A. Detection of chromosome 13 (13q14) deletion among Sudanese patients with multiple myeloma using a molecular genetics fluorescent in situ hybridization technique (FISH). *Malays. J. Pathol.* **2015**, *37*, 95–100.
11. Hu, Y.; Chen, L.; Sun, C.; She, X.-M.; Ai, L.; Qin, Y. Clinical significance of chromosomal abnormalities detected by interphase fluorescence in situ hybridization in newly diagnosed multiple myeloma patients. *Chin. Med. J.* **2011**, *124*, 2981–2985.
12. Conter, V.; Bartram, C.R.; Valsecchi, M.G.; Schrauder, A.; Panzer-Grümayer, R.; Moricke, A.; Aricò, M.; Zimmermann, M.; Mann, G.; De Rossi, G.; et al. Molecular response to treatment redefines all prognostic factors in children and adolescents with B-cell precursor acute lymphoblastic leukemia: Results in 3184 patients of the AIEOP-BFM ALL 2000 study. *Blood* **2010**, *115*, 3206–3214. [CrossRef]
13. Tai, C.-H.; Ho, C.-L.; Chen, Y.-L.; Chen, W.L.; Lee, G.-B. A novel integrated microfluidic platform to perform fluorescence in situ hybridization for chromosomal analysis. *Microfluid. Nanofluidics* **2013**, *15*, 745–752. [CrossRef]
14. Sieben, V.J.; Debes-Marun, C.S.; Pilarski, L.M.; Backhouse, C.J. An integrated microfluidic chip for chromosome enumeration using fluorescence in situ hybridization. *Lab Chip* **2008**, *8*, 2151. [CrossRef]
15. Chen, L.; Manz, A.; Day, P.J.R. Total nucleic acid analysis integrated on microfluidic devices. *Lab Chip* **2007**, *7*, 1413. [CrossRef]
16. Packard, M.M.; Shusteff, M.; Alocilja, E. Microfluidic-Based Amplification-Free Bacterial DNA Detection by Dielectrophoretic Concentration and Fluorescent Resonance Energy Transfer Assisted in Situ Hybridization (FRET-ISH). *Biosensors* **2012**, *2*, 405–416. [CrossRef]
17. Sieben, V.; Marun, C.D.; Pilarski, P.; Kaigala, G.; Pilarski, L.; Backhouse, C. FISH and chips: Chromosomal analysis on microfluidic platforms. *IET Nanobiotechnol.* **2007**, *1*, 27. [CrossRef]
18. Zanardi, A.; Bandiera, D.; Bertolini, F.; Corsini, C.; Gregato, G.; Milani, P.; Barborini, E.; Carbone, R. Miniaturized FISH for screening of onco-hematological malignancies. *Biotechniques* **2010**, *49*, 497–504. [CrossRef]
19. Liu, Y.; Kirkland, B.; Shirley, J.; Wang, Z.; Zhang, P.; Stembridge, J.; Wong, W.; Takebayashi, S.-I.; Gilbert, D.M.; Lenhert, S.; et al. Development of a single-cell array for large-scale DNA fluorescence in situ hybridization. *Lab Chip* **2013**, *13*, 1316–1324. [CrossRef]

20. Wang, X.; Takebayashi, S.-I.; Bernardin, E.; Gilbert, D.M.; Chella, R.; Guan, J. Microfluidic extraction and stretching of chromosomal DNA from single cell nuclei for DNA fluorescence in situ hybridization. *Biomed. Microdevices* **2012**, *14*, 443–451. [CrossRef]
21. Ciftlik, A.T.; Lehr, H.-A.; Gijs, M.A.M. Microfluidic processor allows rapid HER2 immunohistochemistry of breast carcinomas and significantly reduces ambiguous (2+) read-outs. *Proc. Natl. Acad. Sci. USA* **2013**, *110*, 5363–5368. [CrossRef]
22. Nguyen, H.T.; Trouillon, R.; Matsuoka, S.; Fiche, M.; De Leval, L.; Bisig, B.; Gijs, M.A. Microfluidics-assisted fluorescence in situ hybridization for advantageous human epidermal growth factor receptor 2 assessment in breast cancer. *Lab. Investig.* **2016**, *97*, 93–103. [CrossRef]
23. Nguyen, H.T.; Dupont, L.N.; Cuttaz, E.A.; Jean, A.M.; Trouillon, R.; Gijs, M.A.M. Breast cancer HER2 analysis by extra-short incubation microfluidics-assisted fluorescence in situ hybridization (ESIMA FISH). *Microelectron. Eng.* **2018**, *189*, 33–38. [CrossRef]
24. Mayer, J.A.; Pham, T.; Wong, K.L.; Scoggin, J.; Sales, E.V.; Clarin, T.; Pircher, T.J.; Mikolajczyk, S.D.; Cotter, P.D.; Bischoff, F.Z. Fish-Based Determination of Her2 Status in Circulating Tumor Cells Isolated with the Microfluidic Cee™ Platform. *Cancer Genet.* **2011**, *204*, 589–595. [CrossRef]
25. Mughal, F.; Baldock, S.J.; Karimiani, E.G.; Telford, N.; Goddard, N.J.; Day, P.J. Microfluidic channel-assisted screening of hematopoietic malignancies. *Genes Chromosom. Cancer* **2013**, *53*, 255–263. [CrossRef]
26. Kwasny, D.; Mednova, O.; Vedarethinam, I.; Dimaki, M.; Silahtaroglu, A.; Tümer, Z.; Almdal, K.; Svendsen, W.E. A Semi-Closed Device for Chromosome Spreading for Cytogenetic Analysis. *Micromachines* **2014**, *5*, 158–170. [CrossRef]
27. Perez-Toralla, K.; Guneri, E.T.; Champ, J.; Mottet, G.; Bidard, F.-C.; Pierga, J.-Y.; Klijanienko, J.; Draskovic, I.; Malaquin, L.; Viovy, J.-L.; et al. FISH in chips: Turning microfluidic fluorescence in situ hybridization into a quantitative and clinically reliable molecular diagnosis tool. *Lab Chip* **2015**, *15*, 811–822. [CrossRef]
28. Vedarethinam, I.; Shah, P.; Dimaki, M.; Tümer, Z.; Tommerup, N.; Svendsen, W.E. Metaphase FISH on a Chip: Miniaturized Microfluidic Device for Fluorescence in situ Hybridization. *Sensors* **2010**, *10*, 9831–9846. [CrossRef]
29. Baerlocher, G.M.; Vulto, I.; De Jong, G.; Lansdorp, P.M. Flow cytometry and FISH to measure the average length of telomeres (flow FISH). *Nat. Protoc.* **2006**, *1*, 2365–2376. [CrossRef]
30. Arrigucci, R.; Bushkin, Y.; Radford, F.; Lakehal, K.; Vir, P.; Pine, R.; Martin, D.; Sugarman, J.; Zhao, Y.; Yap, G.S.; et al. FISH-Flow, a protocol for the concurrent detection of mRNA and protein in single cells using fluorescence in situ hybridization and flow cytometry. *Nat. Protoc.* **2017**, *12*, 1245–1260. [CrossRef]
31. Kamande, J.W.; Hupert, M.L.; Witek, M.A.; Wang, H.; Torphy, R.J.; Dharmasiri, U.; Njoroge, S.K.; Jackson, J.M.; Aufforth, R.D.; Snavely, A.; et al. Modular Microsystem for the Isolation, Enumeration, and Phenotyping of Circulating Tumor Cells in Patients with Pancreatic Cancer. *Anal. Chem.* **2013**, *85*, 9092–9100. [CrossRef]
32. Hupert, M.L.; Jackson, J.M.; Wang, H.; Witek, M.A.; Kamande, J.; Milowsky, M.I.; Whang, Y.E.; Soper, S.A. Arrays of High-Aspect Ratio Microchannels for High-Throughput Isolation of Circulating Tumor Cells (CTCs). *Microsyst. Technol.* **2013**, *20*, 1815–1825. [CrossRef]
33. Schindelin, J.; Arganda-Carreras, I.; Frise, E.; Kaynig, V.; Longair, M.; Pietzsch, T.; Preibisch, S.; Rueden, C.; Saalfeld, S.; Schmid, B.; et al. Fiji: An Open-Source Platform for Biological-Image Analysis. *Nat. Methods* **2012**, *9*, 676–682. [CrossRef]
34. Nair, S.V.; Witek, M.A.; Jackson, J.M.; Lindell, M.A.M.; Hunsucker, S.; Sapp, T.; Perry, C.E.; Hupert, M.L.; Bae-Jump, V.; Gehrig, P.A.; et al. Enzymatic cleavage of uracil-containing single-stranded DNA linkers for the efficient release of affinity-selected circulating tumor cells. *Chem. Commun.* **2015**, *51*, 3266–3269. [CrossRef]
35. Witek, M.A.; Aufforth, R.D.; Wang, H.; Kamande, J.W.; Jackson, J.M.; Pullagurla, S.R.; Hupert, M.L.; Usary, J.; Wysham, W.Z.; Hilliard, D.; et al. Discrete microfluidics for the isolation of circulating tumor cell subpopulations targeting fibroblast activation protein alpha and epithelial cell adhesion molecule. *NPJ Precis. Oncol.* **2017**, *1*, 24. [CrossRef]
36. Bray, D. *Cell Movements: From Molecules to Motility*, 2nd ed.; Garland Science: New York, NY, USA, 1992; pp. 1–386. [CrossRef]
37. Lauga, E.; Brenner, M.P.; Stone, H. Microfluidics: The No-Slip Boundary Condition. In *Springer Handbook of Experimental Fluid Mechanics*; Springer Science and Business Media LLC: New York, NY, USA, 2007; pp. 1219–1240.

38. Rosser, J.; Thomas, D. Bioreactor processes for maturation of 3D bioprinted tissue. In *3D Bioprinting for Reconstructive Surgery*; Elsevier BV: Duxford, UK, 2018; pp. 191–215.
39. Lipowsky, H.H.; Kovalcheck, S.; Zweifach, B.W. The distribution of blood rheological parameters in the microvasculature of cat mesentery. *Circ. Res.* **1978**, *43*, 738–749. [CrossRef]
40. Lipowsky, H.H.; Usami, S.; Chien, S. In vivo measurements of "apparent viscosity" and microvessel hematocrit in the mesentery of the cat. *Microvasc. Res.* **1980**, *19*, 297–319. [CrossRef]
41. Varma, S.; Voldman, J. A cell-based sensor of fluid shear stress for microfluidics. *Lab Chip* **2015**, *15*, 1563–1573. [CrossRef]
42. Papaioannou, T.G.; Stefanadis, C. Vascular wall shear stress: Basic principles and methods. *Hell. J. Cardiol.* **2005**, *46*, 9–15.
43. Malek, A.M.; Alper, S.L.; Izumo, S. Hemodynamic shear stress and its role in atherosclerosis. *JAMA* **1999**, *282*, 2035–2042. [CrossRef]
44. Strony, J.; Beaudoin, A.; Brands, D.; Adelman, B. Analysis of shear stress and hemodynamic factors in a model of coronary artery stenosis and thrombosis. *Am. J. Physiol. Circ. Physiol.* **1993**, *265*, H1787–H1796. [CrossRef]
45. Barnes, J.M.; Nauseef, J.T.; Henry, M. Resistance to Fluid Shear Stress Is a Conserved Biophysical Property of Malignant Cells. *PLoS ONE* **2012**, *7*, e50973. [CrossRef]
46. Pellat-Deceunynck, C.; Barillé, S.; Puthier, D.; Rapp, M.J.; Harousseau, J.L.; Bataille, R.; Amiot, M. Adhesion molecules on human myeloma cells: Significant changes in expression related to malignancy, tumor spreading, and immortalization. *Cancer Res.* **1995**, *55*, 3647–3653. [PubMed]
47. Sahara, N.; Takeshita, A.; Shigeno, K.; Fujisawa, S.; Takeshita, K.; Naito, K.; Ihara, M.; Ono, T.; Tamashima, S.; Nara, K.; et al. Clinicopathological and prognostic characteristics of CD56-negative multiple myeloma. *Br. J. Haematol.* **2002**, *117*, 882–885. [CrossRef] [PubMed]
48. Jamil, A.; Theil, K.S.; Kahwash, S.; Ruymann, F.B.; Klopfenstein, K.J. TEL/AML-1 fusion gene. Its frequency and prognostic significance in childhood acute lymphoblastic leukemia. *Cancer Genet. Cytogenet.* **2000**, *122*, 73–78. [CrossRef]
49. Mikhail, F.M.; Serry, K.A.; Hatem, N.; Mourad, Z.I.; Farawela, H.M.; El Kaffash, D.M.; Coignet, L.; Nucifora, G. AML1 gene over-expression in childhood acute lymphoblastic leukemia. *Leukemia* **2002**, *16*, 658–668. [CrossRef]
50. Raynaud, S.; Cave, H.; Baens, M.; Bastard, C.; Cacheux, V.; Grosgeorge, J.; Guidal-Giroux, C.; Guo, C.; Vilmer, E.; Marynen, P.; et al. The 12;21 translocation involving TEL and deletion of the other TEL allele: Two frequently associated alterations found in childhood acute lymphoblastic leukemia. *Blood* **1996**, *87*, 2891–2899. [CrossRef]
51. Chung, H.Y.; Kim, K.-H.; Jun, K.R.; Jang, S.; Park, C.-J.; Chi, H.-S.; Im, H.J.; Seo, J.J.; Seo, E.-J. Prognostic Significance of TEL/AML1 Rearrangement and Its Additional Genetic Changes in Korean Childhood Precursor B-Acute Lymphoblastic Leukemia. *Korean J. Lab. Med.* **2010**, *30*, 1. [CrossRef]
52. Mrózek, K.; Harper, D.P.; Aplan, P.D. Cytogenetics and Molecular Genetics of Acute Lymphoblastic Leukemia. *Hematol. Clin. N. Am.* **2009**, *23*, 991–1010. [CrossRef]
53. Jaso, J.M.; Wang, S.; Jorgensen, J.L.; Lin, P. Multi-color flow cytometric immunophenotyping for detection of minimal residual disease in AML: Past, present and future. *Bone Marrow Transplant.* **2014**, *49*, 1129–1138. [CrossRef]
54. Morrissette, J.J.; Bagg, A. Acute Myeloid Leukemia: Conventional Cytogenetics, FISH, and Moleculocentric Methodologies. *Clin. Lab. Med.* **2011**, *31*, 659–686. [CrossRef]
55. Tschumper, R.C.; Asmann, Y.W.; Hossain, A.; Huddleston, P.; Wu, X.; Dispenzieri, A.; Eckloff, B.W.; Jelinek, D.F. Comprehensive Assessment of Potential Multiple Myeloma Immunoglobulin Heavy Chain V-D-J Intraclonal Variation Using Massively Parallel Pyrosequencing. *Oncotarget* **2012**, *3*, 502–513. [CrossRef]
56. Hideshima, T.; Anderson, K.C. Molecular mechanisms of novel therapeutic approaches for multiple myeloma. *Nat. Rev. Cancer* **2002**, *2*, 927–937. [CrossRef] [PubMed]
57. Lee, H.C.; Wang, H.; Baladandayuthapani, V.; Lin, H.; He, J.; Jones, R.J.; Kuiatse, I.; Gu, N.; Wang, Z.; Ma, W.; et al. RNA Polymerase I Inhibition with CX-5461 as a Novel Therapeutic Strategy to Target MYC in Multiple Myeloma. *Br. J. Haematol.* **2017**, *177*, 80–94. [CrossRef] [PubMed]

58. Fonseca, R.; Oken, M.M.; Harrington, D.; Bailey, R.J.; A Van Wier, S.; Henderson, K.J.; E Kay, N.; Van Ness, B.; Greipp, P.R.; Dewald, G.W. Deletions of chromosome 13 in multiple myeloma identified by interphase FISH usually denote large deletions of the q arm or monosomy. *Leukemia* **2001**, *15*, 981–986. [CrossRef] [PubMed]
59. Bullrich, F.; Fujii, H.; Calin, G.A.; Mabuchi, H.; Negrini, M.; Pekarsky, Y.; Rassenti, L.; Alder, H.; Reed, J.C.; Keating, M.J.; et al. Characterization of the 13q14 tumor suppressor locus in CLL: Identification of ALT1, an alternative splice variant of the LEU2 gene. *Cancer Res.* **2001**, *61*, 6640. [PubMed]
60. Shaughnessy, J.; Erming, T.; Jeffrey, S.; Klaus, B.; Reid, L.; Ashraf, B.; Christopher, M.; Guido, T.; Joshua, E.; Bart, B. High Incidence of Chromosome 13 Deletion in Multiple Myeloma Detected by Multiprobe Interphase Fish. *Blood* **2000**, *4*, 1505–1511. [CrossRef]
61. Bullrich, F.; Veronese, M.; Kitada, S.; Jurlander, J.; Caligiuri, M.; Reed, J.; Croce, C. Minimal region of loss at 13q14 in B-cell chronic lymphocytic leukemia. *Blood* **1996**, *88*, 3109–3115. [CrossRef]
62. Childhood Acute Lymphoblastic Leukemia Treatment (PDQ®): Health Professional Version. Available online: https://www.ncbi.nlm.nih.gov/pubmed/26389206 (accessed on 21 February 2020).
63. Seibel, N.L. Treatment of Acute Lymphoblastic Leukemia in Children and Adolescents: Peaks and Pitfalls. *Hematology* **2008**, *2008*, 374–380. [CrossRef]
64. Coustan-Smith, E.; Sancho, J.; Hancock, M.L.; Razzouk, B.I.; Ribeiro, R.C.; Rivera, G.K.; Rubnitz, J.; Sandlund, J.T.; Pui, C.-H.; Campana, D. Use of peripheral blood instead of bone marrow to monitor residual disease in children with acute lymphoblastic leukemia. *Blood* **2002**, *100*, 2399–2402. [CrossRef]
65. Blunck, C.B.; Terra-Granado, E.; Noronha, E.P.; Wajnberg, G.; Passetti, F.; Pombo-De-Oliveira, M.S.; Emerenciano, M. CD9 predicts ETV6-RUNX1 in childhood B-cell precursor acute lymphoblastic leukemia. *Hematol. Transfus. Cell Ther.* **2019**, *41*, 205–211. [CrossRef]
66. Van Delft, F.W.; Horsley, S.; Colman, S.; Anderson, K.; Bateman, C.; Kempski, H.; Zuna, J.; Eckert, C.; Saha, V.; Kearney, L.; et al. Clonal origins of relapse in ETV6-RUNX1 acute lymphoblastic leukemia. *Blood* **2011**, *117*, 6247–6254. [CrossRef]
67. Marshall, G.M.; Pozza, L.D.; Sutton, R.; Ng, A.; A De Groot-Kruseman, H.; Van Der Velden, V.H.; Venn, N.C.; Berg, H.V.D.; De Bont, E.S.J.M.; Egeler, R.M.; et al. High-risk childhood acute lymphoblastic leukemia in first remission treated with novel intensive chemotherapy and allogeneic transplantation. *Leukemia* **2013**, *27*, 1497–1503. [CrossRef] [PubMed]
68. Thomas, E.; Young, M.; Wimalendra, M.; Tunstall, O. PO-0168 Leukaemia Cutis: An Unusual Paediatric Presentation Of Acute Lymphoblastic Leukaemia. *Arch. Dis. Child.* **2014**, *99*, 1030–1043. [CrossRef]
69. Borkhardt, A.; Cazzaniga, G.; Viehmann, S.; Valsecchi, M.G.; Ludwig, W.D.; Mangioni, S.; Schrappe, M.; Riehm, H.; Lampert, F.; Basso, G.; et al. Incidence and Clinical Relevance of Tel/Aml1 Fusion Genes in Children with Acute Lymphoblastic Leukemia Enrolled in the German and Italian Multicenter Therapy Trials. *Blood* **1997**, *2*, 571–577. [CrossRef]
70. Mosad, E.; Hamed, H.; Bakry, R.; Ezz-Eldin, A.M.; Khalifa, N.M. Persistence of TEL-AML1 fusion gene as minimal residual disease has no additive prognostic value in CD 10 positive B-acute lymphoblastic leukemia: A FISH study. *J. Hematol. Oncol.* **2008**, *1*, 17. [CrossRef]
71. Rubnitz, J.E.; Link, M.P.; Shuster, J.J.; Carroll, A.J.; Hakami, N.; Frankel, L.S.; Pullen, D.J.; Cleary, M.L. Frequency and Prognostic Significance of Hrx Rearrangements in Infant Acute Lymphoblastic Leukemia: A Pediatric Oncology Group Study. *Blood* **1994**, *2*, 570–573. [CrossRef]
72. Secker-Walker, L.M.; Moorman, A.V.; Bain, B.J.; Mehta, A.B. Secondary acute leukemia and myelodysplastic syndrome with 11q23 abnormalities. *Leukemia* **1998**, *12*, 840–844. [CrossRef]
73. Rowley, J.D. The Critical Role of Chromosome Translocation in Human Leukemias. *Annu. Rev. Genet.* **1998**, *32*, 495–519. [CrossRef]
74. Naumovski, L.; Morgan, R.; Hecht, F.; Link, M.P.; E Glader, B.; Smith, S.D. Philadelphia chromosome-positive acute lymphoblastic leukemia cell lines without classical breakpoint cluster region rearrangement. *Cancer Res.* **1988**, *48*, 2876.
75. Harata, M.; Soda, Y.; Tani, K.; Ooi, J.; Takizawa, T.; Chen, M.; Bai, Y.; Izawa, K.; Kobayashi, S.; Tomonari, A.; et al. CD19-targeting liposomes containing imatinib efficiently kill Philadelphia chromosome-positive acute lymphoblastic leukemia cells. *Blood* **2004**, *104*, 1442–1449. [CrossRef]
76. Groffen, J.; Stephenson, J.R.; Heisterkamp, N.; De Klein, A.; Bartram, C.R.; Grosveld, G. Philadelphia chromosomal breakpoints are clustered within a limited region, bcr, on chromosome 22. *Cell* **1984**, *36*, 93–99. [CrossRef]

77. Vij, R.; Mazumder, A.; Klinger, M.; O'Dea, D.; Paasch, J.; Martin, T.; Weng, L.; Park, J.; Fiala, M.; Faham, M.; et al. Deep Sequencing Reveals Myeloma Cells in Peripheral Blood in Majority of Multiple Myeloma Patients. *Clin. Lymphoma Myeloma Leuk.* **2014**, *14*, 131–1390. [CrossRef]
78. Rajkumar, S.V.; Rajkumar, S.V.; Greipp, P.R.; Witzig, T.E. Cell proliferation of myeloma plasma cells: Comparison of the blood and marrow compartments. *Am. J. Hematol.* **2004**, *77*, 7–11.
79. Aghvami, S.A.; Opathalage, A.; Zhang, Z.; Ludwig, M.; Heymann, M.; Norton, M.; Wilkins, N.; Fraden, S. Rapid prototyping of cyclic olefin copolymer (COC) microfluidic devices. *Sens. Actuators B Chem.* **2017**, *247*, 940–949. [CrossRef]
80. Jackson, J.M.; Witek, M.A.; Hupert, M.L.; Brady, C.; Pullagurla, S.; Kamande, J.; Aufforth, R.D.; Tignanelli, C.; Torphy, R.J.; Yeh, J.J.; et al. UV activation of polymeric high aspect ratio microstructures: Ramifications in antibody surface loading for circulating tumor cell selection. *Lab Chip* **2014**, *14*, 106–117. [CrossRef]

© 2020 by the authors. Licensee MDPI, Basel, Switzerland. This article is an open access article distributed under the terms and conditions of the Creative Commons Attribution (CC BY) license (http://creativecommons.org/licenses/by/4.0/).

Article

Association between Microsatellite Instability Status and Peri-Operative Release of Circulating Tumour Cells in Colorectal Cancer

James W. T. Toh [1,2,3,*], Stephanie H. Lim [1], Scott MacKenzie [4], Paul de Souza [1,4], Les Bokey [4], Pierre Chapuis [3] and Kevin J. Spring [1,4,*]

1. Medical Oncology, Ingham Institute of Applied Research, School of Medicine, Western Sydney University and SWS Clinical School, UNSW Sydney 2170, NSW, Australia; stephanie.lim@health.nsw.gov.au (S.H.L.); P.DeSouza@westernsydney.edu.au (P.d.S.)
2. Division of Colorectal Surgery, Department of Surgery, Westmead Hospital, Sydney 2145, Australia
3. Department of Colorectal Surgery, Concord Hospital and Discipline of Surgery, Sydney Medical School, University of Sydney, Sydney 2137, Australia; pierre.chapuis@sydney.edu.au
4. Liverpool Clinical School, Western Sydney University, Sydney 2170, Australia; S.Mackenzie@westernsydney.edu.au (S.M.); L.Bokey@westernsydney.edu.au (L.B.)
* Correspondence: james.toh@usyd.edu.au (J.W.T.T.); k.spring@westernsydney.edu.au (K.J.S.); Tel.: +61-2-8738-9032 (K.J.S.)

Received: 8 January 2020; Accepted: 11 February 2020; Published: 12 February 2020

Abstract: Microsatellite instability (MSI) in colorectal cancer (CRC) is a marker of immunogenicity and is associated with an increased abundance of tumour infiltrating lymphocytes (TILs). In this subgroup of colorectal cancer, it is unknown if these characteristics translate into a measurable difference in circulating tumour cell (CTC) release into peripheral circulation. This is the first study to compare MSI status with the prevalence of circulating CTCs in the peri-operative colorectal surgery setting. For this purpose, 20 patients who underwent CRC surgery with curative intent were enrolled in the study, and peripheral venous blood was collected at pre- (t1), intra- (t2), immediately post-operative (t3), and 14–16 h post-operative (t4) time points. Of these, one patient was excluded due to insufficient blood sample. CTCs were isolated from 19 patients using the Isoflux^{TM} system, and the data were analysed using the STATA statistical package. CTC number was presented as the mean values, and comparisons were made using the Student t-test. There was a trend toward increased CTC presence in the MSI-high (H) CRC group, but this was not statistically significant. In addition, a Poisson regression was performed adjusting for stage (I-IV). This demonstrated no significant difference between the two MSI groups for pre-operative time point t1. However, time points t2, t3, and t4 were associated with increased CTC presence for MSI-H CRCs. In conclusion, there was a trend toward increased CTC release pre-, intra-, and post-operatively in MSI-H CRCs, but this was only statistically significant intra-operatively. When adjusting for stage, MSI-H was associated with an increase in CTC numbers intra-operatively and post-operatively, but not pre-operatively.

Keywords: circulating tumour cells; colorectal cancer; colorectal surgery; microsatellite instability

1. Introduction

Biomarkers in colorectal cancer (CRC) have had limited success in clinical application to date, but microsatellite instability (MSI) status is emerging as a biomarker of clinical relevance. It is known that CRC exhibiting high level MSI (MSI-H) is associated with increased tumour infiltrating lymphocytes (TILs) and is a marker of immunogenicity [1–3]. MSI-H CRCs are less likely to disseminate due to TILs as a protective factor, yet a double-edged sword exists in that MSI-H CRCs have more mutations and are associated with more adverse pathological features. However, what is not known is whether the abundance of TILs decreases the risk of tumour dissemination by reducing the release of circulating tumour cells (CTCs), which have metastatic potential.

Microsatellites are short tracts of repetitive sequence (1–6 base pairs or more that are generally repeated between 5 and 50 times) found disseminated throughout the genome. Due to the repetitive nature of microsatellites, these regions are prone to change (instability) during replication. In MSI-H CRC, the resultant microsatellite alterations result in frameshifts that truncate proteins and may lead to inactivation of affected coding regions. Usually, microsatellite alterations are sensed by mismatch repair (MMR) genes that act like spellcheckers or DNA damage sensors, which detect mutations and signal for repair or apoptosis. When there is a loss of DNA damage sensors, either through genetic or epigenetic inactivation of MMR genes, this leads to loss of appropriate signalling and an accumulation of genetic mutations. In clinical practice, MSI-H occurs in 10–15% of colorectal cancers and is defined by IHC staining demonstrating MMR deficiency (MMRD).

The serrated neoplastic pathway is one of the two sporadic pathways that result in MSI-H cancers, the other classical pathway being the adenoma-carcinoma pathway involving chromosomal instability (CIN) and that results in microsatellite stable (MSS) cancers. Interestingly, it appears that MSI-H colorectal cancers are less likely to progress to stage IV disease compared to their MSS counterpart [4,5]. However, it is unclear if the biology of this observation is associated with decreased release of CTC. The hypothesis we sought to test was that the abundance of TILs in MSI-H CRCs may reduce the release of CTCs and, by doing so, protect against the risk of dissemination.

CTCs were first identified by Dr. Thomas Ashworth in 1869 [6]. Under the microscope, it was observed that "cells identical with those of the cancer itself" were present in the blood of a man with metastatic cancer. Since then, CTCs have been shown to be both a predictive and prognostic biomarker, but have remained in the research domain rather than clinical application due to cost. Evaluating circulating tumour DNA (ctDNA) has also shown great utility as a diagnostic approach for cancer management. Instead of identifying cancer cells (CTCs) in the bloodstream, identifying ctDNA depends on DNA released into the bloodstream from the tumour cell nucleus as it dies and is replaced by new cancer cells. A recent study by Tie et al. investigated ctDNA in stage II colon cancer to detect patients at high risk of recurrence. In that study, they also assessed the association between post-operative ctDNA status and conventional high-risk clinicopathological factors, but were not able to show an association, albeit that the majority of patients in the study were ctDNA negative [7].

With the call for universal MSI testing in CRC, it is important to understand the immunobiology of MSI to understand its clinical implications and its role in guiding prognosis and adjuvant therapy. It is known that MSI is associated with TILs [8]. However, it is not yet known if MSI status affects the release of CTC. It is not clear if patients with abundant TILs have a reduction of CTC count and whether this is stage dependent. This pilot study is the first to investigate CTC count in elective colorectal surgery and to analyse possible differences in the pre-operative, intra-operative, and post-operative stage of treatment and correlate this with MSI status.

2. Materials and Methods

2.1. Patients and Blood Samples

Twenty patients undergoing elective laparoscopic or open colorectal surgery at either Liverpool or Westmead Hospitals were enrolled in the study approved by the South Western Sydney Local Health District Ethics (Ref: HREC/13/LPOOL/158). All patients gave informed written consent for blood collection and CTC analysis. Peripheral venous blood was collected at four time points: (t1) pre-operative blood collection in the anaesthetic bay of operating theatres; (t2) intra-operatively after mobilisation of bowel was completed; (t3) at time of completion of surgery; and (t4) fourteen-sixteen hours post-operatively. One patient had insufficient blood volume collected and was excluded from analysis.

2.2. CTC Enrichment and Enumeration

Quantification of CTCs was performed using the IsoFlux^{TM} instrument (Fluxion Biosciences Inc, Alameda, CA, USA). Peripheral venous blood was collected into 9 mL anti-coagulant K_2EDTA tubes (Vacuette 455036) and processed within 24 h of collection in accordance with the Isoflux protocol using the CTC enrichment (910-0091) and enumeration (910-0093) kits supplied by the manufacturer. Briefly, immuno-magnetic EpCAM linked beads was used to capture CTCs, and after processing through the Isoflux instrument, CTCs were identified by immune staining using anti-cytokeratin (CK-7, -8, -18, and -19), Hoechst 33342 dye, and anti-CD45. After transferring each sample to 24 well SensoPlates^{TM} (Cat. No. 662892, Geriner Bio-One, GmbH, Kremsminster, Austria) and applying coverslips, each sample was scanned and visualized using a 10× objective on a fluorescence Olympus IX71 inverted microscope. Putative CTCs were defined as CK^+, $DAPI^+$, and $CD45^-$, nucleated and morphologically intact cells.

2.3. Clinical and Histopathological Data

Carcinoembryonic antigen (CEA), tumour infiltrating lymphocytes (TILs), microsatellite instability (MSI), and BRAF status were recorded. TILs were reported by the pathologist as present when there were more than 5 intraepithelial lymphocytes/100 epithelial cells (assessed on minimum three high power (×400) fields) [8]. MSI and BRAF status was tested by immunohistochemistry.

Data on patient demographics, histopathological features of the tumour, and CTC count at four time periods were collected. Certified pathologists examined the tissue biopsy specimens post-operatively and provided the histopathology diagnosis. All patients had follow-up at one year with disease-free survival (DFS) being the main outcome.

2.4. Statistical Analysis

Analysis was performed on STATA (Stata MP, Version 15; StataCorp LP). The Student *t*-test was used to compare between groups, and a Poisson regression was used to adjust for stage. The Student *t*-test was used instead of the Wilcoxon–Mann–Whitney U-test and the Kruskal–Wallis tests as these non-parametric tests are better used to compare medians, whereas the Student *t*-test provides a better assessment of means.

3. Results

3.1. Clinical and Surgical Characteristics

In total, 80 samples from 20 patients with colorectal cancer who underwent elective open or laparoscopic colorectal surgery with curative intent were recruited for this study. However, four samples from one patient had insufficient blood volume collected, and this patient was excluded from analysis. CTC isolation and enrichment were performed using the IsoFlux^{TM} system. Of the nineteen patients who had CTCs enumerated, two patients had high-grade dysplasia without malignancy (these were 30 × 33 × 23 mm and 57 × 50 × 55 mm villous adenomas). Of the remaining 17 patients, 3 (17.6%) were stage I, 6 (35.3%) stage II, 7 (41.2%) stage III, and 1 (5.9%) stage IV. Nine (52.9%) were right-sided (caecal (*n* = 4), ascending colon (*n* = 4), and transverse colon (*n* = 1)), the rest (41.2%) were left-sided (rectum (*n* = 4), rectosigmoid (*n* = 1), sigmoid (*n* = 3)). The histopathology of all seventeen patients was adenocarcinoma. Further, two of the four patients with rectal cancer had neoadjuvant chemoradiotherapy. Patient demographic, clinical, and surgical characteristics are summarized in Table 1, and the CTC yield for each patient at the different time points are in Table S1 (Supplementary Material).

Table 1. Patient demographic, clinical, and surgical characteristics and circulating tumour cell (CTC) number. MSI-H, microsatellite instability-high; MSS, microsatellite stable.

Patient Characteristics	Microsatellite Status	
	MSI-H	MSS
Patient number	4	13
Age	85.5 (54–86)	66 (44–86)
Female:male	3:1	6:7
Right colon	4 (100%)	5 (38.5%)
Left colon	0	3 (23.1%)
Rectal/rectosigmoid	0	5 (38.5%)
Grade		
High	3 (75%)	1 (8.3%)
Moderate	1 (25%)	10 (83.3%)
Low	0	1 (8.3%)
BRAF mutant:wild-type	3:1	N/A
Stage		
I	0	3 (23.1%)
II	2 (50%)	4 (30.8%)
III	2 (50%)	5 (38.5%)
IV	0	1 (7.7%)
CTC number		
t1	10.5 (0–29)	1 (0–61)
t2	52 (44–189)	1 (0–74)
t3	23 (1–83)	1 (0–17)
t4	34 (6–65)	1 (0–12)

Continuous data shown as the mean with the range; count data presented as the frequencies and percentages. Three patients are not included in the Table as the histopathology was villous adenoma, with high-grade dysplasia for two patients, and one patient had insufficient samples.

3.2. CTC Yield in All Patients

First, we looked at the number of CTCs enumerated for all 19 patients (Table 1 and supplementary Table S1. CTC number was presented as the mean values and comparisons made using the Student t-test. CTCs were enumerated for 19 patients: microsatellite stable (MSS) ($n = 15$); microsatellite unstable (MSI-H) ($n = 4$), respectively. A Student t-test was used to test the difference in CTC number between MSS and MSI-H CRCs, respectively, at the four different time points: t1 (8.2 vs. 12.5, $p = 0.6191$); t2 (23.7 vs. 37.8, $p = 0.5893$); t3 (9.3 vs. 12.3, $p = 0.7798$); and t4 (8.1 vs. 18.8, $p = 0.3696$). It was apparent that at each of these time points, there was no significant difference between MSS and MSI-H patient groups.

3.3. CTC Yield in Cancer Patients Only

Excluding the two patients with villous adenoma and high-grade dysplasia, there was a trend towards higher CTC number for MSI-H CRC, but this was not statistically significant between the two groups: t1 (7.9 vs. 12.5, $p = 0.6191$); t2 (22.2 vs. 37.8, $p = 0.5893$); t3 (8.7 vs. 12.3, $p = 0.7798$); t4 (8.3 vs. 18.8, $p = 0.3696$). In addition, a Poisson regression was performed adjusting for stage (I-IV). This demonstrated no significant difference between the two MSI groups for t1. However, t2, t3, and t4 were all associated with an increase in CTC number for MSI-H CRCs.

3.4. MSI Status and CTC Number by Stage of CRC

For this analysis, there were no MSI-H patients in the stage I and IV groups; however, there were two MSI-H (caecum and ascending colon) patients in the stage II group; whereas there were four stage II MSS CRC (caecum, rectosigmoid and two rectum) patients. For the stage II patients, there was a significant spike at the t2 timepoint for the MSI-H group (Figure 1; Panel A and Panel B). However, the sample size was too small to perform reliable statistical analysis. For stage III patients, there appeared to be a trend towards higher CTC count for the MSI-H group, but this was not statistically significant: t1 (13.2 vs. 14.5, $p = 0.9540$); t2 (15.2 vs. 23.5, $p = 0.7700$); t3 (3.6 vs. 20, $p = 0.1981$); t4 (3.6 vs. 29, $p = 0.1589$); Figure 1; C and D. When combining data for all stage I-III patients, there was a statistically significant spike in t2 count in the MSI-H group. There was also a trend toward higher CTC number for t1, t3, and t4 in the MSI-H group, but this was not statistically significant (Figure 1; E and F): t1 (7.5 vs. 12.5, $p = 0.6027$); t2 (8.25 vs. 37.75, $p = 0.0328$); t3 (2.5 vs. 12.25, $p = 0.0878$); t4 (3.67 vs. 18.75, $p = 0.0604$).

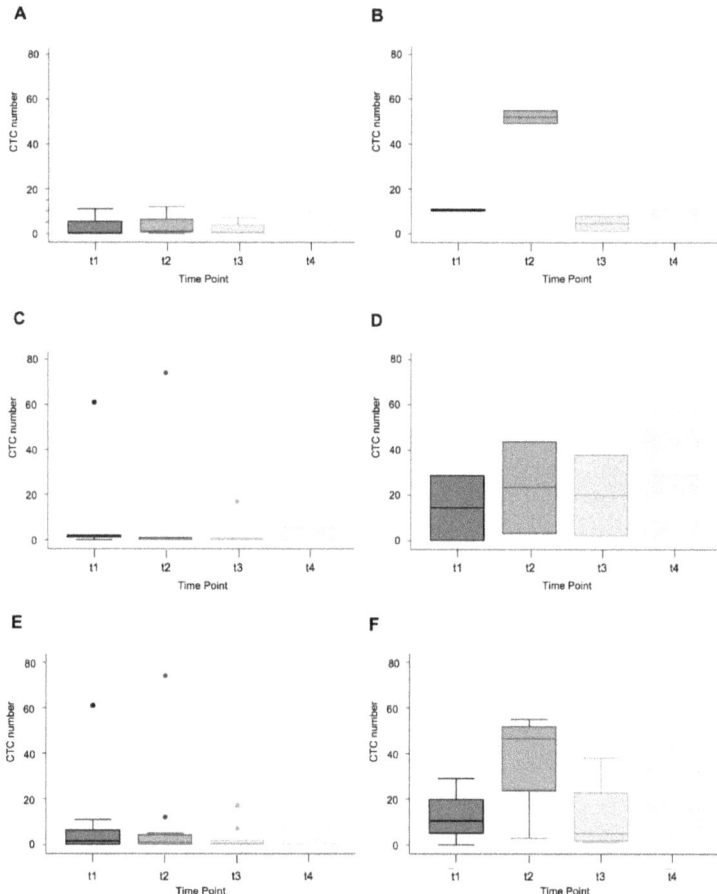

Figure 1. CTC number for stage II (Panels **A** and **B**), III (**C** and **D**) and I-III (Panels **E** and **F**) MSS (left) vs. MSI-H (right) colorectal cancer at different sample time points: t1 (pre-operative), t2 (intra-operative), t3 (immediate post-operative), t4 (14–16 h post-operative).

There was only one patient with stage IV colon cancer. This patient had a right-sided cancer that was MSS and had high CTCs that were persistently elevated with 13 CTCs detected at the pre-operative time point (t1), which increased to 189 during surgery and then remained high post-operatively with 83 and 65 CTCs detected, respectively (Figure 2).

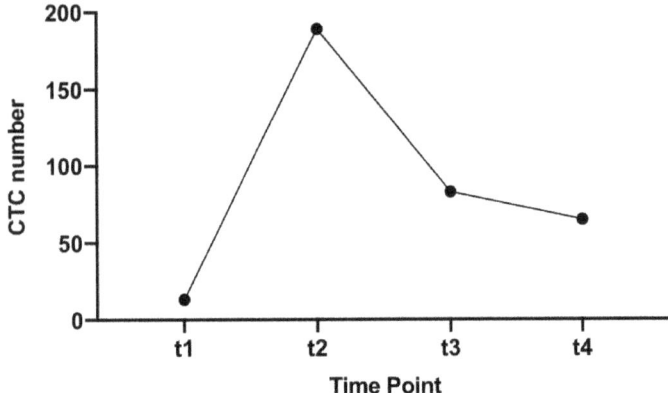

Figure 2. CTC number at different time points for patient S16 (stage IV, MSS CRC).

3.5. MSI Status and CTC by Side (Right vs. Left) Stage I-III Colon Cancer

There were no left-sided MSI-H CRCs in this study, so it was not possible to compare between MSI-H and MSS for left-sided colon and rectal cancer. However, there were four MSI-H and five MSS right-sided colon cancers. One MSS was excluded from analysis as it was a stage IV CRC. There was no statistically significant difference in CTC number for right-sided colon cancer by MSI status, but there was a trend for increased CTC number at t2, t3, and t4 time points (t1 (20 vs. 12.5, $p = 0.6375$); t2 (22.75 vs. 37.75, $p = 0.5001$); t3 (6.5 vs. 12.25, $p = 0.5677$); t4 (6 vs. 18.75, $p = 0.0942$); Figure 3, Panels A and B).

There were eight left-sided CRCs in this study, of which none exhibited MSI. Two patients received neoadjuvant therapy, whereas six patients did not receive neoadjuvant therapy. There were four rectal cancers, one rectosigmoid and three sigmoid cancers. The CTC number overall was low in this subgroup; however, it appeared there was a difference between those who received neoadjuvant therapy and those who did not (Figure 3, Panels C and D). Reliable statistical analysis was not performed due to the small sample size.

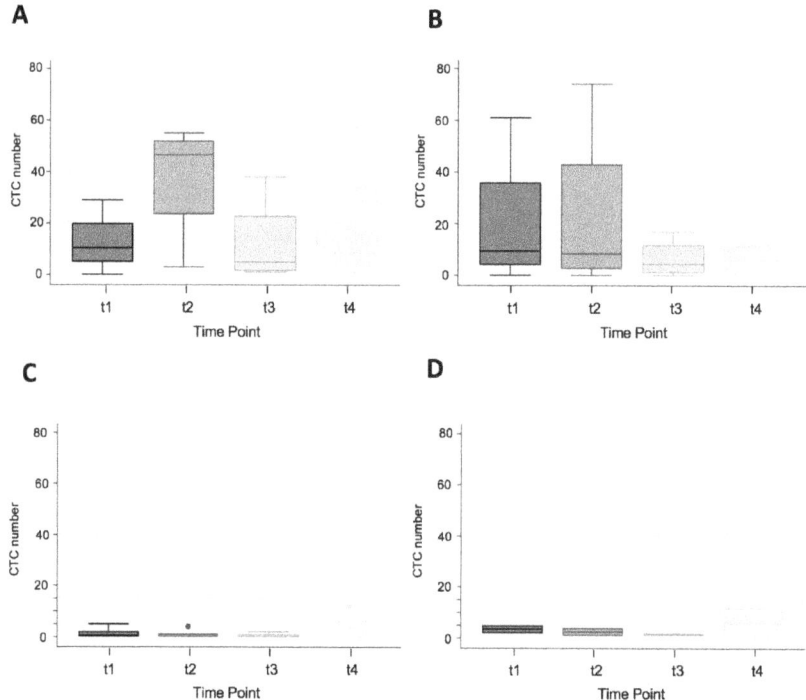

Figure 3. CTC number and MSI status for right-sided stage I-III colon cancer, (Panel **A**) MSS, (Panel **B**) MSI-H and CTC number for left-sided colorectal cancer, no neoadjuvant (Panel **C**), and neoadjuvant therapy (Panel **D**) at sample time points t1 (pre-operative), t2 (Intra-operative), t3 (immediate post-operative), t4 (14–16 h post-operative).

3.6. Poisson Regression Model with Post-Estimation Marginal Fundamental Analysis

A Poisson regression model was run with MSI and stage as independent variables. A post-estimation marginal means and marginal effects fundamental analysis was performed. This showed no difference in CTC number at the t1 time point, but a statistically significant difference in CTC number at the t2, t3, and t4 time points with stage as a covariate: t1: 7.96 (6.42–9.50) vs. 12.29 (8.88–15.71); t2: (20.99 (18.55–23.43) vs. 46.01 (38.18–53.83); t3: 7.72 (6.28–9.16) vs. 20.69 (13.88–27.49); t4: 8.03 (6.51–9.54) vs. 21.91 (16.62–27.18). In addition, a Poisson regression model was run for stage of CRC, and a post-estimation marginal fundamental analysis was performed. This showed a significant difference in CTC number at the t1, t2, t3, and t4 time points between stages independent of MSI status (Figure 4). Preoperatively, the CTC number at the t1 time point for stage I was 3.99 (2.48–5.49); for stage II 6.78 (5.33–8.22); for stage III 11.53 (9.58–13.48); for stage IV 19.61 (13.63–25.59).

3.7. Non-Parametric Wilcoxon–Mann–Whitney U-test

When comparing the median CTC number between MSS and MSI-H CRCs, there was again a trend toward increased CTC number with MSI-H CRC, but this was only statistically significant at the t2 time point when comparing stage I-III MSS and MSI-H CRC.

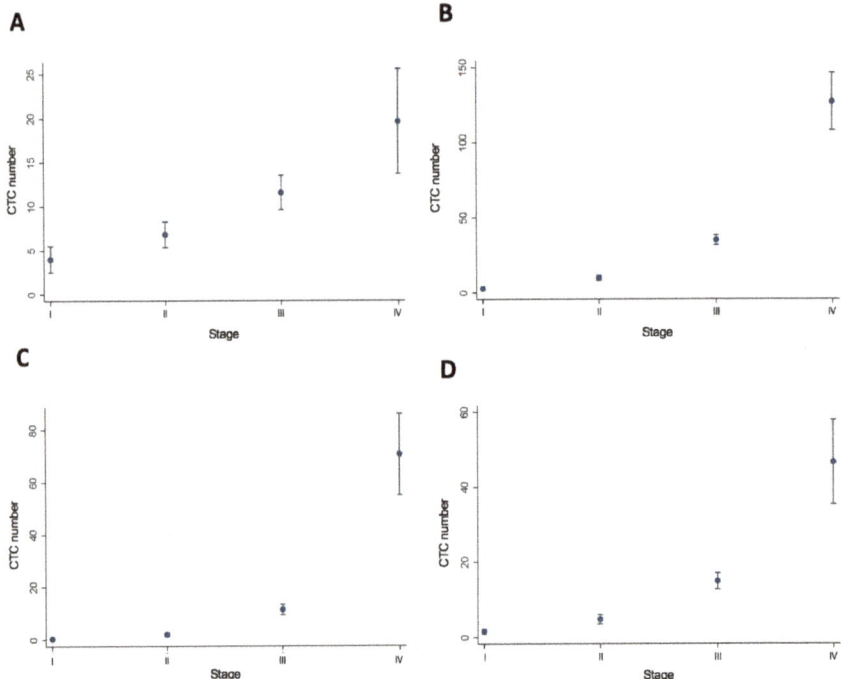

Figure 4. Poisson regression model with post-estimation marginal fundamental analysis of stage and CTC number at sample time points t1 and t2 (Panels **A** and **B**) and t3 and t4 (Panels **C** and **D**).

4. Discussion

There is an abundance of clinical data reporting on the association between MSI status and prognosis in CRC showing that MSI-H may be associated with better prognosis [3,4]. The evidence in the literature shows that CTCs may be important in prognostication of colon cancer [5] in predicting dissemination [6], overall and disease-free survival [7], and lymph node involvement [8,9]. Higher levels of CTCs have been associated with worse outcome and may predict for poor disease-free survival [10–12]. Most existing studies assessing the relationship between MSI and prognosis in CRC have been clinical studies or histopathological studies. We took a fundamentally different approach to instead examine the immune-biological characteristics of CRC based on MSI status.

Our study confirmed that CTC measurements correlated with dissemination and stage of disease. There was a statistically significant difference in CTC number with stage (I-IV) and across all time points (preoperatively, intraoperatively, and postoperatively, as shown in Figure 4). However, notably, our study showed that MSI-H CRCs (which have been reported to be immunogenic and associated with enhanced survival) [3,4] were associated with increased peri-operative release of CTCs (Figure 1). Further, both the mean and median CTC count at all time points were higher in the MSI-H group compared to the MSS group. Our hypothesis that increased TILs would decrease peri-operative release of CTCs in MSI-H CRCs was not supported by the data.

While most of the analyses performed were not statistically significant, overall, there was a trend towards increased CTC number at all time points (t1, t2, t3, and t4) in the MSI-H group. When analysing stage II CRC and stage I-III CRC, there was a statistically significant increase in the CTC number for the t2 time point: 3.5 vs. 52, $p = 0.0005$ and 8.25 vs. 37.75, $p = 0.0328$, respectively, in the MSI-H group. For all other comparisons, there was increased CTC number in the MSI-H group, but it was not statistically significant. A Poisson regression was performed to adjust for stage (I-IV). This demonstrated no significant difference between the two MSI groups for the t1 time point. However, the t2, t3, and t4 time points were all associated with increased CTC number for MSI-H CRCs.

The literature on CTCs has shown that a measurement of more than three CTCs per 7.5 mL peripheral blood may be associated with poor survival, although some studies suggest 1–2 CTCs per 7.5 mL may also be associated with a worse outcome [13]. Furthermore, higher post-operative CTC numbers may be associated with a higher risk of recurrence [14], whereas improvements in post-operative CTC number from pre-operative baselines has been associated with better survival [15]. In this study, the median CTC number for the t1, t2, t3, and t4 time points for MSS CRC was one, whereas the median for MSI-H CRC was significantly >3 in all the corresponding time points. However, with data in the literature showing enhanced survival in MSI-H CRC, the high CTC numbers for MSI-H CRCs and low CTC numbers for MSS CRCs found in this study did not correlate with the clinicopathological data reporting on survival existing in the literature. On the other hand, this study may have shown that the cut-off for prognostication of CRC by CTC measurement may actually be influenced by MSI status and that there should not be a single cut-off, but the benchmark may depend on characteristics such as MSI. One hypothesis was that CTCs released into the bloodstream of patients with MSI-H CRCs usually remain microsatellite unstable [16], which maintains an enhanced immunogenic response from circulating lymphocytes in the bloodstream [17]. Thus, its presence may not represent the same risk of metastases as CTCs associated with MSS CRCs [18,19].

What about the protective effect of MSI? The immunogenicity of MSI is believed to be associated with the presence of TILs. Collinearity between MSI-H status and TILs has been shown in clinicopathological data, including the data from our own cohort study [20], and has been traditionally associated with a better prognosis [3,4]. It is believed that the survival and clinical benefits of MSI may be due to its immunogenicity, with MSI associated with increased TILs [2,21–24] and TILs associated with a better prognosis [2,25–34]. MSI-H CRCs may also be associated not only with an abundance of intra-tumoral TILs, but also with a higher density of associated cytotoxic, helper, and regulatory T lymphocytes in peripheral blood [17], as well as increased activity in the bone marrow [35]. Studies have shown that both the immunogenic TILs response at the tumour site, as well as the circulatory system are believed to be associated with a decreased risk of lymph node metastases [19,36] and distant metastases [18] in MSI CRCs. The assumption, hence, would be that MSI-H CRCs should be associated with decreased CTC dissemination into the blood.

However, this study showed that the immuno-biology of MSI-H CRCs is more complex than this. As is well established in the literature, MSI-H CRCs are also associated with poor differentiation [19,37], are larger and more likely to be mucinous [38–40], as well as being higher grade tumours with a greater mutational burden and the mucinous phenotype more likely to be associated with signet cells. It is also believed that MSI-H CRCs are associated with a high number of frameshift mutation peptides (FSPs) [41,42] when compared to MSS CRCs [43,44]. The MSI hyper-mutational state is also thought to be the reason for poor differentiation and other adverse pathological features of the tumour. The higher grade of tumour associated with hyper-mutational state may be the reason for the increased peri-operative release of CTCs. On Poisson regression analysis, our study showed no difference in pre-operative (t1) CTC numbers, but a statistically significant difference in intra-operative (t2 and t3) and post-operative (t4) release of CTCs. From the literature, immunogenic TILs may reduce the risk of dissemination, but this study showed that they did not do so by preventing the release of CTCs. We currently do not have long term survival and recurrence data, nor CTC measurements outside of the 24 h window peri-operatively, but a study looking at recurrence, survival, and CTC number at 6 months, 1 year, 3 year, and 5 years would be interesting and would examine the role that CTCs may play in the post-operative management of CRCs including whether this may be used to predict recurrence or survival accurately.

Notably, this pilot study showed results that warrant further investigations. The increased CTC positivity seen in MSI-H CRCs is a significant point of difference from the currently understood immune-biological mechanisms associated with MSI-H CRCs. In practice, this means that the benchmark cut-off points for CTC enumeration may be influenced by tumour characteristic, and future clinical applications of CTC in CRC management may need to take this into consideration. Another consideration is that the improved survival associated with immunogenic MSI-H CRCs may not be as profound as once believed, and the increased peri-operative release of CTCs in MSI-H CRCs may be revisited with future studies with a larger patient cohort. This could prove a useful follow up of the suggestions made by the European Society of Medical Oncology (ESMO) guidelines that patients with stage II colon cancer with MSI are at very low risk of recurrence and unlikely to benefit from chemotherapy.

In this study, the overall patient number ($n = 20$) was low and a limitation for insightful statistical analysis. There was significant heterogeneity within the study with two patients receiving neoadjuvant chemoradiotherapy, no direct comparisons available for stage I and IV (no MSI-H CRC in these subgroups), and the inclusion of right colon, left colon, and rectal cancers. The incidence of MSI in CRC is 10–15%. In our cohort, there were only four MSI-H CRCs, of which two were stage II and two were stage III, and the majority of patients were MSS CRCs, being a limitation of our data. From a tumour biology perspective, the main implication of our findings is that immunogenic MSI-H CRCs did not protect against release of CTCs, and the protective effect against metastatic disease was not by reducing CTCs. Further, the main clinical implication of this study involves the utility of CTCs for monitoring and surveillance with potentially differential baseline levels of CTCs associated with the two subtypes of CRC.

5. Conclusions

There was a trend toward increased CTC release pre-, intra-, and post-operatively in MSI-H CRCs, but this was only statistically significant intra-operatively. When adjusting for stage, MSI-H was associated with an increase in CTC number intra-operatively and post-operatively, but not pre-operatively. This dataset was limited, and further studies are required. Finally, these data suggested that immunogenic MSI-H CRCs did not suppress CTC release and that different reference ranges may be required for CTC enumeration of MSI-H and MSS CRC.

Supplementary Materials: The following is available online at http://www.mdpi.com/2073-4409/9/2/425/s1, Table S1: Patient, MSI status, stage, and CTC number at different time points.

Author Contributions: Conceptualization, J.W.T.T., K.J.S., and S.H.L.; formal analysis, J.W.T.T., K.J.S., and S.H.L.; investigation, J.W.T.T., S.M., S.H.L. and K.J.S.; resources, K.J.S. and P.d.S.; data curation, J.W.T.T., S.M. and S.H.L.; methodology, J.W.T.T., S.H.L. and K.J.S.; writing, original draft preparation, J.W.T.T. and K.J.S.; writing, review and editing, K.J.S., P.C., L.B. and P.d.S.; project administration, K.J.S.; funding acquisition, K.J.S. and P.d.S. All authors have read and agreed to the published version of the manuscript.

Funding: This work was supported through the Cancer Institute NSW (CINSW) funded Centre for Oncology Education and Research Translation (CONCERT). K.J.S. received partial funding as the CONCERT Centre Fellow.

Conflicts of Interest: The authors declare no conflict of interest.

References

1. Tie, J.; Wang, Y.; Tomasetti, C.; Li, L.; Springer, S.; Kinde, I.; Silliman, N.; Tacey, M.; Wong, H.L.; Christie, M.; et al. Circulating tumor DNA analysis detects minimal residual disease and predicts recurrence in patients with stage II colon cancer. *Sci. Transl. Med.* **2016**, *8*, 346ra392. [CrossRef]
2. Michael-Robinson, J.M.; Biemer-Huttmann, A.; Purdie, D.M.; Walsh, M.D.; Simms, L.A.; Biden, K.G.; Young, J.P.; Leggett, B.A.; Jass, J.R.; Radford-Smith, G.L. Tumour infiltrating lymphocytes and apoptosis are independent features in colorectal cancer stratified according to microsatellite instability status. *Gut* **2001**, *48*, 360–366. [CrossRef] [PubMed]
3. Popat, S.; Hubner, R.; Houlston, R.S. Systematic review of microsatellite instability and colorectal cancer prognosis. *J. Clin. Oncol.* **2005**, *23*, 609–618. [CrossRef]

4. Guastadisegni, C.; Colafranceschi, M.; Ottini, L.; Dogliotti, E. Microsatellite instability as a marker of prognosis and response to therapy: A meta-analysis of colorectal cancer survival data. *Eur. J. Cancer* **2010**, *46*, 2788–2798. [CrossRef]
5. Steinert, G.; Scholch, S.; Koch, M.; Weitz, J. Biology and significance of circulating and disseminated tumour cells in colorectal cancer. *Langenbeck's Arch. Surg.* **2012**, *397*, 535–542. [CrossRef]
6. Hiraiwa, K.; Takeuchi, H.; Hasegawa, H.; Saikawa, Y.; Suda, K.; Ando, T.; Kumagai, K.; Irino, T.; Yoshikawa, T.; Matsuda, S.; et al. Clinical significance of circulating tumor cells in blood from patients with gastrointestinal cancers. *Ann. Surg. Oncol.* **2008**, *15*, 3092–3100. [CrossRef]
7. Liu, Y.; Qian, J.; Feng, J.G.; Ju, H.X.; Zhu, Y.P.; Feng, H.Y.; Li, D.C. Detection of circulating tumor cells in peripheral blood of colorectal cancer patients without distant organ metastases. *Cell. Oncol.* **2013**, *36*, 43–53. [CrossRef]
8. Gazzaniga, P.; Gianni, W.; Raimondi, C.; Gradilone, A.; Russo, G.L.; Longo, F.; Gandini, O.; Tomao, S.; Frati, L. Circulating tumor cells in high-risk nonmetastatic colorectal cancer. *Tumour Biol.* **2013**, *34*, 2507–2509. [CrossRef]
9. Katsuno, H.; Zacharakis, E.; Aziz, O.; Rao, C.; Deeba, S.; Paraskeva, P.; Ziprin, P.; Athanasiou, T.; Darzi, A. Does the presence of circulating tumor cells in the venous drainage of curative colorectal cancer resections determine prognosis? A meta-analysis. *Ann. Surg. Oncol.* **2008**, *15*, 3083–3091. [CrossRef]
10. Groot Koerkamp, B.; Rahbari, N.N.; Buchler, M.W.; Koch, M.; Weitz, J. Circulating tumor cells and prognosis of patients with resectable colorectal liver metastases or widespread metastatic colorectal cancer: A meta-analysis. *Ann. Surg. Oncol.* **2013**, *20*, 2156–2165. [CrossRef]
11. Rahbari, N.N.; Aigner, M.; Thorlund, K.; Mollberg, N.; Motschall, E.; Jensen, K.; Diener, M.K.; Büchler, M.W.; Koch, M.; Weitz, J. Meta-analysis shows that detection of circulating tumor cells indicates poor prognosis in patients with colorectal cancer. *Gastroenterology.* **2010**, *138*, 1714–1726. [CrossRef]
12. Thorsteinsson, M.; Jess, P. The clinical significance of circulating tumor cells in non-metastatic colorectal cancer—a review. *Eur J. Surg Oncol.* **2011**, *37*, 459–465. [CrossRef]
13. Gazzaniga, P.; Raimondi, C.; Gradilone, A.; Biondi Zoccai, G.; Nicolazzo, C.; Gandini, O.; Longo, F.; Tomao, S.; Lo Russo, G.; Seminara, P.; et al. Circulating tumor cells in metastatic colorectal cancer: Do we need an alternative cutoff? *J. Cancer Res. Clin. Oncol.* **2013**, *139*, 1411–1416. [CrossRef]
14. Peach, G.; Kim, C.; Zacharakis, E.; Purkayastha, S.; Ziprin, P. Prognostic significance of circulating tumour cells following surgical resection of colorectal cancers: A systematic review. *Br. J. Cancer* **2010**, *102*, 1327–1334. [CrossRef]
15. Yalcin, S.; Kilickap, S.; Portakal, O.; Arslan, C.; Hascelik, G.; Kutluk, T. Determination of circulating tumor cells for detection of colorectal cancer progression or recurrence. *Hepato-gastroenterology* **2010**, *57*, 1395–1398.
16. Steinert, G.; Scholch, S.; Niemietz, T.; Iwata, N.; García, S.A.; Behrens, B.; Voigt, A.; Kloor, M.; Benner, A.; Bork, U. Immune escape and survival mechanisms in circulating tumor cells of colorectal cancer. *Cancer Res.* **2014**, *74*, 1694–1704. [CrossRef]
17. Schwitalle, Y.; Kloor, M.; Eiermann, S.; Linnebacher, M.; Kienle, P.; Knaebel, H.P.; Tariverdian, M.; Benner, A.; Doeberitz, M.v.K. Immune response against frameshift-induced neopeptides in HNPCC patients and healthy HNPCC mutation carriers. *Gastroenterology* **2008**, *134*, 988–997. [CrossRef]
18. Buckowitz, A.; Knaebel, H.P.; Benner, A.; Bläker, H.; Gebert, J.; Kienle, P.; Doeberitz, M.v.K.; Kloor, M. Microsatellite instability in colorectal cancer is associated with local lymphocyte infiltration and low frequency of distant metastases. *Br. J. Cancer* **2005**, *92*, 1746–1753. [CrossRef]
19. Kazama, Y.; Watanabe, T.; Kanazawa, T.; Tanaka, J.; Tanaka, T.; Nagawa, H. Microsatellite instability in poorly differentiated adenocarcinomas of the colon and rectum: Relationship to clinicopathological features. *J. Clin. Pathol.* **2007**, *60*, 701–704. [CrossRef]
20. Toh, J.; Chapuis, P.H.; Bokey, L.; Chan, C.; Spring, K.J.; Dent, O.F. Competing risks analysis of microsatellite instability as a prognostic factor in colorectal cancer. *Br. J. Surg.* **2017**, *104*, 1250–1259. [CrossRef]
21. Kim, J.H.; Kang, G.H. Molecular and prognostic heterogeneity of microsatellite-unstable colorectal cancer. *World J. Gastroenterol. WJG* **2014**, *20*, 4230–4243. [CrossRef] [PubMed]
22. Greenson, J.K.; Bonner, J.D.; Ben-Yzhak, O.; Cohen, H.I.; Miselevich, I.; Resnick, M.B.; Trougouboff, P.; Tomsho, L.; Kim, E.; Low, M.; et al. Phenotype of microsatellite unstable colorectal carcinomas: Well-differentiated and focally mucinous tumors and the absence of dirty necrosis correlate with microsatellite instability. *Am. J. Surg. Pathol.* **2003**, *27*, 563–570. [CrossRef] [PubMed]

23. Smyrk, T.C.; Watson, P.; Kaul, K.; Lynch, H.T. Tumor-infiltrating lymphocytes are a marker for microsatellite instability in colorectal carcinoma. *Cancer* **2001**, *91*, 2417–2422. [CrossRef]
24. Tougeron, D.; Fauquembergue, E.; Rouquette, A.; Pessot, F.L.; Sesboüé, R.; Laurent, M.; Berthet, P.; Mauillon, J.; Fiore, F.D.; Sabourin, J.C. Tumor-infiltrating lymphocytes in colorectal cancers with microsatellite instability are correlated with the number and spectrum of frameshift mutations. *Mod. Pathol.* **2009**, *22*, 1186–1195. [CrossRef] [PubMed]
25. Pages, F.; Berger, A.; Camus, M.; Sanchez-Cabo, F.; Costes, A.; Molidor, R.; Mlecnik, B.; Kirilovsky, A.; Nilsson, M.; Damotte, D.; et al. Effector memory T cells, early metastasis, and survival in colorectal cancer. *New Engl. J. Med.* **2005**, *353*, 2654–2666. [CrossRef]
26. Mlecnik, B.; Tosolini, M.; Kirilovsky, A.; Berger, A.; Bindea, G.; Meatchi, T.; Bruneval, P.; Trajanoski, Z.; Fridman, W.H.; Page's, F.; et al. Histopathologic-based prognostic factors of colorectal cancers are associated with the state of the local immune reaction. *J. Clin. Oncol.* **2011**, *29*, 610–618. [CrossRef]
27. Ling, A.; Edin, S.; Wikberg, M.L.; Oberg, A.; Palmqvist, R. The intratumoural subsite and relation of CD8(+) and FOXP3(+) T lymphocytes in colorectal cancer provide important prognostic clues. *Br. J. Cancer* **2014**, *110*, 2551–2559. [CrossRef]
28. Svennevig, J.L.; Lunde, O.C.; Holter, J.; Bjorgsvik, D. Lymphoid infiltration and prognosis in colorectal carcinoma. *Br. J. Cancer* **1984**, *49*, 375–377. [CrossRef]
29. Jass, J.R. Lymphocytic infiltration and survival in rectal cancer. *J. Clin. Pathol.* **1986**, *39*, 585–589. [CrossRef]
30. Shunyakov, L.; Ryan, C.K.; Sahasrabudhe, D.M.; Khorana, A.A. The influence of host response on colorectal cancer prognosis. *Clin. Colorectal Cancer* **2004**, *4*, 38–45. [CrossRef]
31. Dahlin, A.M.; Henriksson, M.L.; Van Guelpen, B.; Stenling, R.; Öberg, A.; Rutegård, J.; Palmqvist, R. Colorectal cancer prognosis depends on T-cell infiltration and molecular characteristics of the tumor. *Mod. Pathol.* **2011**, *24*, 671–682. [CrossRef] [PubMed]
32. Chiba, T.; Ohtani, H.; Mizoi, T.; Naito, Y.; Nagura, H.; Ohuchi, A.; Ohuchi, K.; Shiiba, K.; Kurokawa, Y.; Satomi, S. Intraepithelial CD8+ T-cell-count becomes a prognostic factor after a longer follow-up period in human colorectal carcinoma: Possible association with suppression of micrometastasis. *Br. J. Cancer* **2004**, *91*, 1711–1717. [CrossRef] [PubMed]
33. Zlobec, I.; Lugli, A.; Baker, K.; Roth, S.; Minoo, P.; Hayashi, S.; Terracciano, L.; Jass, J.R. Role of APAF-1, E-cadherin and peritumoral lymphocytic infiltration in tumour budding in colorectal cancer. *J. Pathol.* **2007**, *212*, 260–268. [CrossRef] [PubMed]
34. Prall, F.; Duhrkop, T.; Weirich, V.; Ostwald, C.; Lenz, P.; Nizze, H.; Barten, M. Prognostic role of CD8+ tumor-infiltrating lymphocytes in stage III colorectal cancer with and without microsatellite instability. *Hum. Pathol.* **2004**, *35*, 808–816. [CrossRef]
35. Koch, M.; Beckhove, P.; Op den Winkel, J.; Autenrieth, D.; Wagner, P.; Nummer, D.; Specht, S.; Antolovic, D.; Galindo, L. Schmitz-Winnenthal, F.H. Tumor infiltrating T lymphocytes in colorectal cancer: Tumor-selective activation and cytotoxic activity in situ. *Ann. Surg.* **2006**, *244*, 986–992. [CrossRef]
36. Lamberti, C.; Lundin, S.; Bogdanow, M.; Pagenstecher, C.; Friedrichs, N.; Büttner, R.; Sauerbruch, T. Microsatellite instability did not predict individual survival of unselected patients with colorectal cancer. *Int. J. Colorectal Dis.* **2007**, *22*, 145–152. [CrossRef]
37. Xiao, H.; Yoon, Y.S.; Hong, S.M.; Ae Roh, S.; Cho, D.H.; Yu, C.S.; Kim, J.C. Poorly differentiated colorectal cancers: Correlation of microsatellite instability with clinicopathologic features and survival. *Am. J. Clin. Pathol.* **2013**, *140*, 341–347. [CrossRef]
38. Thibodeau, S.N.; Bren, G.; Schaid, D. Microsatellite instability in cancer of the proximal colon. *Science* **1993**, *260*, 816–819. [CrossRef]
39. Yoon, Y.S.; Kim, J.; Hong, S.M.; Lee, J.L.; Kim, C.W.; Park, I.J.; Lim, S.B.; Yu, C.S.; Kim, J.C. Clinical implications of mucinous components correlated with microsatellite instability in patients with colorectal cancer. *Colorectal Dis.* **2015**, *17*, O161–O167. [CrossRef]
40. Karahan, B.; Argon, A.; Yildirim, M.; Vardar, E. Relationship between MLH-1, MSH-2, PMS-2,MSH-6 expression and clinicopathological features in colorectal cancer. *Int. J. Clin. Exp. Pathol.* **2015**, *8*, 4044–4053.
41. Maby, P.; Tougeron, D.; Hamieh, M.; Mlecnik, B.; Kora, H.; Bindea, G.; Angell, H.K.; Fredriksen, T.; Elie, N.; Fauquembergue, E.; et al. Correlation between density of CD8+ T cell infiltrates in microsatellite unstable colorectal cancers and frameshift mutations: A rationale for personalized immunotherapy. *Cancer Res.* **2015**. [CrossRef] [PubMed]

42. Saeterdal, I.; Bjorheim, J.; Lislerud, K.; Gjertsen, M.K.; Bukholm, I.K.; Olsen, O.C.; Nesland, J.M.; Eriksen, J.A.; Møller, M.; Lindblom, A.; et al. Frameshift-mutation-derived peptides as tumor-specific antigens in inherited and spontaneous colorectal cancer. *Proc. Natl. Acad. Sci. USA* **2001**, *98*, 13255–13260. [CrossRef] [PubMed]
43. Sun, Z.; Yu, X.; Wang, H.; Zhang, S.; Zhao, Z.; Xu, R. Clinical significance of mismatch repair gene expression in sporadic colorectal cancer. *Exp. Ther. Med.* **2014**, *8*, 1416–1422. [CrossRef] [PubMed]
44. Vogelstein, B.; Papadopoulos, N.; Velculescu, V.E.; Zhou, S.; Diaz, L.A., Jr.; Kinzler, K.W. Cancer genome landscapes. *Science* **2013**, *339*, 1546–1558. [CrossRef]

© 2020 by the authors. Licensee MDPI, Basel, Switzerland. This article is an open access article distributed under the terms and conditions of the Creative Commons Attribution (CC BY) license (http://creativecommons.org/licenses/by/4.0/).

Article

Detection of Androgen Receptor Variant 7 (*ARV7*) mRNA Levels in EpCAM-Enriched CTC Fractions for Monitoring Response to Androgen Targeting Therapies in Prostate Cancer

Claudia Hille [1], Tobias M. Gorges [1], Sabine Riethdorf [1], Martine Mazel [2], Thomas Steuber [3], Gunhild Von Amsberg [4], Frank König [5], Sven Peine [6], Catherine Alix-Panabières [2,†] and Klaus Pantel [1,*,†]

1. Department of Tumor Biology, University Medical Center Hamburg-Eppendorf, 20246 Hamburg, Germany; c.hille@uke.de (C.H.); t.gorges@uke.de (T.M.G.); s.riethdorf@uke.de (S.R.)
2. Laboratory of Rare Human Circulating Cells (LCCRH), University Medical Centre of Montpellier–UM EA2415, 34295 Montpellier, France; martine-mazel@chu-montpellier.fr (M.M.); c-panabieres@chu-montpellier.fr (C.A.-P.)
3. Martini Clinic, University Medical Center Hamburg-Eppendorf, 20246 Hamburg, Germany; steuber@uke.de
4. Department of Hematology and Oncology, University Medical Center Hamburg-Eppendorf, 20246 Hamburg, Germany; g.von-amsberg@uke.de
5. ATURO, Urology Practice, 14197 Berlin, Germany; frank.koenig@aturo.berlin
6. Department of Transfusion Medicine, University Medical Center Hamburg-Eppendorf, 20246 Hamburg, Germany; s.peine@uke.de
* Correspondence: pantel@uke.de; Tel.: 49-40-741053503; Fax: 49-40-7410-55379
† There authors contributed equally to this work.

Received: 15 August 2019; Accepted: 10 September 2019; Published: 11 September 2019

Abstract: Expression of the androgen receptor splice variant 7 (*ARV7*) in circulating tumor cells (CTCs) has been associated with resistance towards novel androgen receptor (AR)-targeting therapies. While a multitude of ARV7 detection approaches have been developed, the simultaneous enumeration of CTCs and assessment of *ARV7* status and the integration of validated technologies for CTC enrichment/detection into their workflow render interpretation of the results more difficult and/or require shipment to centralized labs. Here, we describe the establishment and technical validation of a novel *ARV7* detection method integrating the CellSearch® technology, the only FDA-cleared CTC-enrichment method for metastatic prostate cancer available so far. A highly sensitive and specific qPCR-based assay was developed, allowing detection of *ARV7* and *keratin 19* transcripts from as low as a single $ARV7^+/K19^+$ cell, even after 24 h of sample storage. Clinical feasibility was demonstrated on blood samples from 26 prostate cancer patients and assay sensitivity and specificity was corroborated. Our novel approach can now be included into prospective clinical trials aimed to assess the predictive values of CTC/ARV7 measurements in prostate cancer.

Keywords: prostate cancer; biomarkers; circulating tumor cells; androgen receptor; ARV7; abiraterone; enzalutamide

---

1. Introduction

Prostate cancer (PCa) remains the second most commonly diagnosed cancer among men worldwide with an estimated 1.3 million new cases each year [1]. In contrast to other cancer types such as pancreatic cancer, routine preventive medical screens for PCa are accessible to a broad spectrum of the public and have been widely accepted, leading to a drastic increase of newly diagnosed PCa cases. Tissue biopsies are invasive and can be associated with adverse effects for the patient [2]. Furthermore, routine tissue biopsy is challenging in metastatic PCa (mPCa). In recent years, minimally invasive liquid biopsies, focusing on the identification of circulating tumor cells (CTCs) and circulating nucleic acids (ctDNA, miRNA) from whole blood samples, have gained tremendous attention [3–7]. While the prognostic relevance of CTCs in PCa, especially in the metastatic setting, has been thoroughly shown in large clinical trials [8–11], predictive value of CTC analysis and their clinical utility are still being debated [12–17]. While a multitude of therapeutic approaches exist, aimed at treating PCa in various disease stages, a subset of patients develop aggressive PCa subtypes that defy current therapeutic options. Therefore, simple detection of PCa is not sufficient and robust biomarkers are urgently needed to discern aggressive subtypes from clinically well treatable cancers, preferably without exposing patients to unnecessary tissue biopsies.

With the advent of novel hormone therapies such as enzalutamide and abiraterone and the emergence of innate and acquired resistance towards these therapies, the androgen receptor splice variant 7 (ARV7) has become a leading target of CTC research in PCa [17–19]. Multiple studies indicate that *ARV7* mRNA and ARV7 protein expression in CTCs is associated with resistance towards novel hormone therapies [20–25] and that *ARV7* expressing patients benefit more from taxane-based therapy [25–27]. This implicates ARV7 as a possible treatment selection biomarker for PCa patients prior to receiving novel hormone therapy (e.g., enzalutamide, abiraterone). Additionally, the ARV7 status is subject to change during therapy regimens [25,28,29], underlining the benefit of sequential sampling which becomes possible through liquid biopsy. ARV7 could therefore also represent a biomarker to monitor treatment response and predict upcoming therapy resistance.

While many approaches have been developed to assess ARV7 either on protein or mRNA level [20,24,30], only very few of these approaches allow for parallel CTC enumeration and morphological characterization while giving information on ARV7 status for individual CTCs [24,31], a limitation recently highlighted [32]. Additionally, even fewer were designed to use the only FDA-cleared CTC enrichment and detection technology shown to have clinical prognostic relevance in prostate cancer, the CellSearch® system [33]. Here, we aimed to develop a protocol for *ARV7* detection using the CellSearch® technology. With our novel workflow we were able to detect *ARV7* mRNA in as low as one CTC in 7.5 mL of whole blood.

2. Materials and Methods

2.1. Cancer Cell Lines

The human prostate cancer cell lines 22Rv1 (ATCC® CRL-2505), VCaP (ATCC® CRL-2876), LNCaP (ATCC® CRL-1740) and PC3 (ATCC® CRL-1345) were obtained from the American Type Culture Collection (ATCC, Manassas, VA, USA) and cultured according to ATCC recommendations.

LNCaP and 22Rv1 cells were cultured in RPMI 1640 medium, while the VCaP and PC3 cells were maintained in Dulbecco's Modified Eagle Medium (DMEM). Media were additionally fortified with 10% fetal calf serum (FCS) (Gibco—Life Technologies, Darmstadt, Germany), 1% L-glutamine (Gibco—Life Technologies, Darmstadt, Germany) and 1% penicillin/streptomycin (Gibco—Life Technologies, Darmstadt, Germany), as recommended by ATCC. Cells were cultured in 25 cm^2 flasks at 37 °C in a humidified atmosphere containing 5% CO_2.

2.2. Blood Collection and Processing

Male healthy donor (HD) and patient blood samples were acquired in accordance to the World Medical Association Declaration of Helsinki and the guidelines for experimentation with humans by the Chambers of Physicians of the State of Hamburg ("Hamburger Ärztekammer"). All patients gave informed, written consent prior to blood collection (Ethics Approval: PV3779). Samples were drawn from 26 metastatic prostate cancer (mPCa) patients into standard 7.5 mL ethylenediaminetetraacetic acid (EDTA) vacutainers or CellSave® (Menarini-Silicon Biosystems, Florence, Italy) preservation tubes respectively. Each patient therefore provided a matched sample of EDTA-KE (Sarstedt, Rheinbach, Germany) and CellSave® blood for further analysis. CTCs from EDTA blood samples were enriched via the CellSearch® Profile Kit (Menarini-Silicon Biosystems, Florence, Italy) and further analyzed for *ARV7* expression as described below. Samples collected into CellSave® blood preservation tubes were processed via the CellSearch® CXC-Kit (FITC labelled pan-keratin) [34]. Phycoerythrin labelled androgen receptor CellTracks Anti-Androgen Receptor (Janssen Diagnostics) antibody (10 µg/mL) was used for full-length AR (AR-FL) detection in the fourth channel of the CellSearch® for 12/26 mPCa patients. All analyses were performed by trained CellSearch® analysist. CTCs were defined as keratin positive and CD45 negative cells with a nuclear DAPI staining.

2.3. Spiking of Healthy Donor Blood

For spiking experiments, cell line cells were washed once with 1 × PBS (Gibco-Life Technologies, Darmstadt, Germany) and treated with 0.25% trypsin-EDTA (Gibco-Life Technologies, Darmstadt, Germany) for 5 min at 37 °C prior to being resuspended in culture medium. The cell suspension was centrifuged at 190× g for 5 min after which the supernatant was discarded and the cells were again resuspended in fresh culture medium. The cells were spread to a petri dish filled with corresponding medium, manually counted and picked under a light microscope. Defined cell counts were added directly to healthy donor blood samples.

2.4. Immunocytochemical Stainings on Cell Culture Plates

Cells were seeded into 24-well plates at the rate of 50,000 cells/well, and maintained at 37 °C in a humidified atmosphere containing 5% CO_2 until reaching 80% confluence. Cells were then fixed and permeabilized using IntraPrep Permeabilization Reagent (A07803, Beckman Coulter, Brea, CA, USA), and blocked with 10% Goat serum for 1 h at room temperature. Cells were subsequently incubated with (i) primary antibodies Anti-AR (AR-V7 specific) antibody [EPR15656] (Abcam, Cambridge, United Kingdom) at a final concentration of 10 µg/mL, or (ii) Rabbit IgG, monoclonal [EPR25A]-Isotype Control (Abcam, Cambridge, United Kingdom) (our negative control) at a final concentration of 10 µg/mL. All wells were also incubated with the anti-PanCK^{PE} (Menarini-Silicon Biosystems, Florence, Italy). Following this first incubation, cells were washed with 1% goat serum in PBS, incubated with the FITC-conjugated secondary antibody (1:20 in PBS containing 10% Goat serum), and washed twice with 1% goat serum in PBS.

In parallel, the presence of the androgen receptor (AR) was tested using the anti-AR^{AF488} [D6F11] XP Rabbit antibody (0.5 µg/mL, Ozyme, Saint Cyr L'Ecole, France); in the control wells, the Rabbit [DA1E] IgG^{AF488} XP isotype (0.5 µg/mL, Ozyme, Saint Cyr L'Ecole, France) was used. Cell imaging was obtained under 20x magnification using a Fluorescent Axio Observer microscope (Carl Zeiss, Oberkochen, Germany).

2.5. Immunocytochemical Stainings on Cytospins

Cell suspensions of selected prostate cancer cell lines (22Rv1, LNCap, PC3) were spun down on glass slides (190× g, 5 min) and dried at room temperature (RT) over night. Cells were subsequently fixed and permeabilized using the respective CellSearch CXC Kit reagents (Menarini-Silicon Biosystems, Florence, Italy) and blocked with 10% AB-Serum (BioRad, Rüdigheim, Germany). Primary antibodies targeting ARV7, 4 µg/mL of clone AG10008 (unlabeled, Precision, Columbia, Maryland, USA) and 6 µg/mL EPR15656 (unlabeled, Abcam, Cambridge, United Kingdom) were tested. Secondary antibodies were applied and contained a DAPI nuclear counterstain. Secondary rabbit-anti mouse (Alexa 546, polyclonal, Thermo Fisher Scientific, Dreieich, Germany) and mouse-anti-rabbit (Alexa 546, polyclonal, Thermo Fisher Scientific, Dreieich, Germany) antibodies were used. Cytospins were covered in Prolong Gold Antifade Reagent (Thermo Fisher Scientific, Dreieich, Germany) and cover slipped for analysis. Slides were manually assessed using a fluorescence microscope (Axioplan 2, Carl Zeiss, Oberkochen, Germany).

2.6. Western Blots

Cell lines (22Rv1, VCaP, LNCaP, and PC3) were cultured to 70% confluency, harvested in urea lysis buffer (9.8 M Urea, 15 mM EDTA, 30 mM Tris) and homogenized by ultrasonic treatment. Protein concentration was measured with the Pierce BCA Protein Assay Kit (Pierce, Rockford, Illinois, USA). 40 µg of total protein was applied for Western Blot analysis for each respective cell line alongside pre-stained peqGold protein marker-V (VWR, Erlangen, Germany). Proteins were separated according to size using a Laemmli buffer system and 8% polyacrylamide separation gel. Two ARV7 antibodies, mouse-AG10008 (Precision, Columbia, MD, USA; 2 µg/mL) and rabbit-EPR15656 (Abcam, Cambridge, United Kingdom; 1.5 µg/mL) were applied in dilutions according to the supplier's instruction manual in 5% milk powder. Alpha-tubulin was used as a loading control (Cell Signaling Technology, Danvers, MA, USA). Species specific secondary antibodies (horseradish peroxidase conjugated, DAKO, Glostrup, Germany) were applied at 1:10.000 dilution in 5% milk powder. Protein bands were visualized using SignalFire™Plus ECL reagent (Cell Signaling Technology, Danvers, MA, USA) and X-ray films (CEA, Hamburg, Germany) according to the instruction manual.

2.7. RNA Extraction and cDNA Synthesis

For cell line characterization and PCR establishment RNA was isolated from prostate cancer cell lines grown in a T25 culture flask at 70% confluency using the NucleoSpin® RNA isolation kit (Macherey-Nagel, Düren, Germany) according to manufacturer's instructions. RNA concentration and purity were controlled via NanoDrop 1000 spectrophotometer (Thermo Fisher Scientific, Dreieich, Germany) following isolation. 0.5 µg of RNA per cell line were used for DNA synthesis with the First Strand cDNA Synthesis Kit (Thermo Fisher Scientific, Dreieich, Germany) according to manufacturer's instructions. cDNA Synthesis was carried out in a PeqSTAR 96 Universal Gradient thermocycler (VWR International, Darmstadt, Germany).

Following CTC enrichment via the CellSearch® Profile Kit (Menarini-Silicon Biosystems, Florence, Italy) samples were transferred to a fresh 1.5 mL tube (Sarstedt, Rheinbach, Germany). To do so, a 1000 µL pipette tip was first coated with a solution of 0.1 mg/mL of BSA/PBS to circumvent binding and sticking of CTCs to the pipette surface. All RNA work was performed using sterile, DNA/RNA-free, filtered Biosphere® plus pipette tipps (Sarstedt, Rheinbach, Germany). The Profile® sample tube was washed with 500 µL of 1x DPBS (cell culture use) (Thermo Fisher Scientific, Dreieich, Germany), which was also added to the sample. Subsequently the sample was placed in a magnetic rack (Magnetcellect; R&D systems, Minneapolis, MN, USA) for 10 min. The supernatant was discarded, and the sample was washed with 1000 µL of 1x DPBS, followed by another 10 min attached to the magnetic rack. This step was repeated with 500 µL of 1x DPBS prior to resuspension of the Profile® beads in 150µL of lysis buffer (Dynabeads mRNA DIRECT Kit; Thermo Fisher Scientific, Dreieich, Germany). Samples were immediately frozen at −80 °C. Sample lysates were stored for a maximum of 14 days prior to RNA isolation and cDNA synthesis.

For RNA extraction, the Dynabeads mRNA DIRECT Kit (Thermo Fisher Scientific, Dreieich, Germany) was applied according to manufacturer's instructions. Following the last wash step with Wash buffer B, supernatant was removed, and beads were resuspended in 14.75 µL of Nuclease-free H_2O (Qiagen, Hilden, Germany) and placed in a PCR cycler at 75 °C for 5 min to ensure elution of mRNA from the beads. Subsequently cDNA was synthesized using the Sensiscript Reverse Transcription Kit (Qiagen, Hilden, Germany) with Recombinant Rnasin® (Promega, Mannheim, Germany) as an added RNase inhibitor. Primer addition was not necessary as the contained dynabeads function as oligo-dT primers. RNase inhibitor was limited to 0.25 µL, leading to a total mastermix of 5.25 µL added to each RNA sample (total reaction volume of 20 µL). Following cDNA synthesis, beads were removed via magnet and supernatant was transferred to a fresh PCR tube for subsequent qPCR analysis.

2.8. Polymerase-Chain Reaction (PCR) Analysis

For *AR-FL* and *ARV7* primer evaluation, 10 ng of cDNA of each prostate cancer cell line was applied per PCR. The PCR reaction conditions for initial primer testing were adapted from the original Antonarakis et al. publication by the Johns Hopkins Group [20]. Reactions were run in a PeqSTAR 96 Universal Gradient thermocycler (VWR International, Darmstadt, Germany).

PCR primer pairs (Sigma Aldrich, Steinheim, Germany) chosen for PCR targeted *AR-FL* (fw-CAGCCTATTGCGAGAGAGCTG, rev-GAAAGGATCTTGGGCACTTGC, fragment size of 125 bp) [20] and *ARV7* (Antonarakis et al. [20]: fw-CCATCTTGTCGTCTTCGGAAATGTTA, rev-TTTGAATGAGGCAAGTCAGCCTTTCT, fragment size of 125 bp; Guo et al. [35]: fw-CTACTCCGGACCTTACGGGGACATGCG, rev-TGCCAACCCGGAATTTTTCTCCC, fragment size of 314 bp; Liu et al. [36]: fw- CAGGGATGACTCTGGGGAGAA, rev- GCCCTCTAGAGCCCTCATTT, fragment size of 112 bp; UKE: fw-AGAAAGGCTGACTTGCCTCA, rev- CGCCAGGTTTCTCCAGACTA, fragment size of 73 bp) gene sequences. Novel UKE primers were designed using the Primer 3 software [37]. Primers were aliquoted at stock concentrations of 100 µM with Nuclease-free H_2O (Qiagen, Hilden, Germany) and frozen at −20 °C. Final concentrations of 10 µM were applied to PCRs.

To visualize PCR products, they were mixed with DNA Gel loading dye (6x) (Thermo Fisher Scientific, Dreieich, Germany) and applied to 2% agarose gels containing GelRed® Nulceic Acid Gel Stain (Biotum, Fremont, CA, USA) at 1/µL per ml of agarose gel. The Quick-Load® 100 bp DNA Ladder (New England Biolabs, Frankfurt am Main, Germany) was used as a size standard. PCR fragments were visualized using the Gene Genius bioimaging system (Syngene, Bangalore, India).

2.9. Quantitative Polymerase-Chain Reaction (qPCR) Analysis

qPCRs were pipetted under a separate flow hood with sterile, DNA/RNA-free, filtered Biosphere® plus pipette tipps (Sarstedt, Rheinbach, Germany) and performed in a CFX96 Touch™ Real Time PCR Detection System (BioRad, Rüdigheim, Germany). Maxima SYBR-Green fluorescent dye (Thermo Fisher Scientific, Dreieich, Germany) was used for product detection. Amplification was performed under the following conditions: after an initial denaturation step (10 min at 95 °C), 40 amplification cycles were carried out, consisting of denaturation at 95 °C for 30 seconds, annealing at 60 °C for 30 s, and elongation for 30 s at 72 °C. A final elongation step at 72 °C (10 min) was followed by a melting curve analysis and storage of the samples at 4 °C. Data was summarized and converted into Excel files using the CFX Manager Software (BioRad, Rüdigheim, Germany). For qPCR analysis, two additional primer sets targeting *K19* (fw-CGAACCAAGTTTGAGACGGA; rev-GATCTGCATCTCCAGGTCGG; fragment size of 117 bp) and *Actin* (x) gene sequences were applied. Samples were applied in triplicates and average Cq values as well as standard deviations were calculated. Primers were aliquoted at stock concentrations of 100 µM with Nuclease-free H_2O (Qiagen, Hilden, Germany) and frozen at −20 °C. Final concentrations of 10 µM were applied to qPCRs.

Relative gene expression of *AR-FL* and *ARV7* in initial primer testing and cell line characterization was normalized from data sets by the comparative Cq method [38]. Briefly, the first amplification cycle showing significant increase of fluorescence signal over background level was defined as the cycle of quantification (Cq). Cq data of *AR-FL* and *ARV7* was normalized by subtracting the Cq value of *Actin* from the respective target gene for each cell line tested, generating a ΔCq value. Subsequently, the ΔΔCq values were calculated by subtracting the ΔCq of each specific gene calculated for the different cell lines (22Rv1, VCap, and LNCaP) from the ΔCq values calculated for gene expression in PC3 cells. Finally, ΔΔCq values were converted to log2 fold changes by applying $2^{-\Delta\Delta Cq}$. Ten nanograms of cDNA were applied per triplicate well.

Following RNA isolation and cDNA transcription from CellSearch® Profile Kit (Menarini-Silicon Biosystems, Florence, Italy) enriched samples, the 20 μL of cDNA mix was applied in triplicates for each gene (2 μL/well). No absolute quantification or normalization of genes was performed as levels of *Actin* gene expression is variable depending on background leucocyte cDNA co-amplified following CTC enrichment. Gene expression was confirmed when at least 2/3 triplicates showed detectable transcript levels under a Cq threshold of 35. Quality of the results was furthermore corroborated by melting curve analysis and subsequent visualization of amplified products on 2% agarose gels (see above).

3. Results

3.1. Test of Commercially Available ARV7 Antibodies for Fourth CellSearch® Channel

To allow assessment of ARV7 protein levels in parallel to CTC enumeration on a cell-specific level, we initially tested available ARV7 antibodies with the aim of adding them to the fourth channel of the CellSearch® system. Currently only few commercial antibodies are available, aimed at detecting ARV7 protein either via immunohistochemistry (IHC), immunocytochemistry (ICC) and/or western blot.

Three established prostate cancer cell lines were chosen for method establishment, each cell line representing a specific status of AR-full length (AR-FL) and ARV7 protein expression: 22Rv1 (AR-FL$^+$/AR-V7$^+$), LNCaP (AR-FL$^+$/AR-V7$^{+/-}$), and PC3 (AR-FL$^-$/AR-V7$^-$). First, the cell lines were characterized for AR-FL (Figure 1a), resulting in cell line specific nuclear ICC staining (22Rv1 and LNCaP) or absence of staining (PC3) for the full-length protein, seen in green. Next, we tested the anti-ARV7 antibody [EPR15656] described in literature to specifically stain nuclear ARV7 [25]. This antibody did not result in cell line specific staining results, as all three tested cell lines including the ARV7$^-$ PC3 cells showed green nuclear ARV7 staining (Figure 1b). Similar results were obtained using the antibody [EPR15656] as well as an additional commercially available ARV7 antibody on cell line cytospins (Figure S1). In western blot analysis the anti-ARV7 antibody [AG10008] by Precision showed cell line specific results, correctly detecting 22Rv1 and VCaP lysate as ARV7$^+$, LNCaP protein levels as below detection limit and identifying PC3 cells as ARV7$^-$ (Figure S1a). In contrast, the anti-ARV7 antibody [EPR15656], showed an unspecific western blot signal for PC3 cells (Figure S1a). In ICC both antibodies failed to correctly characterize the chosen prostate cancer cell lines, giving unspecific staining results (Figure S1b,c).

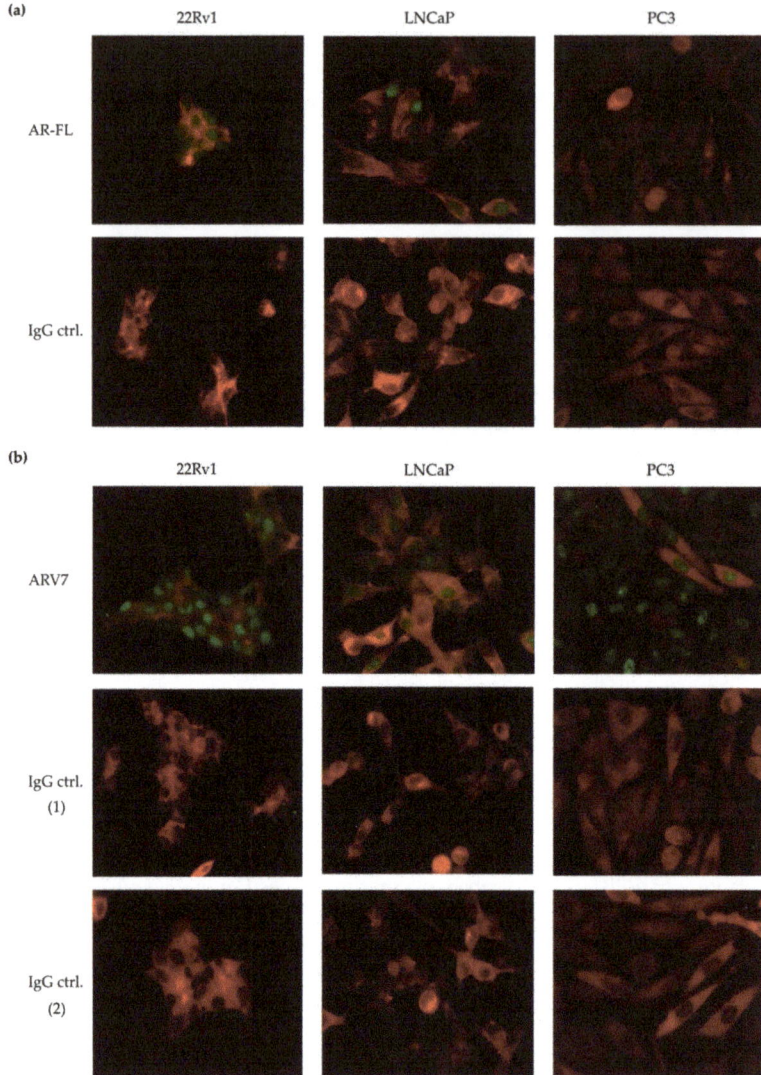

Figure 1. Immunocytochemical (ICC) staining of full-length androgen receptor (AR-FL) and ARV7 on three selected prostate cancer cell lines. Cells are stained for pan-keratin in red (anti-PanCK, CellSearch®, Menarini) in all images. (**a**) Upper panel: ICC staining performed using the anti-AR-FL antibody (7395S Ozyme) in green. AR-FL positive cells lines 22Rv1 and LNCaP show positive nuclear AR-FL staining, while PC3 cells remain unstained. Lower panel: ICC control staining using Rabbit [DA1E] IgG XP isotype (2975S Ozyme) in green showing the absence of unspecific staining on 22RV1, LNCaP and PC3 cells. (**b**) Upper panel: ICC staining performed using the anti-ARV7 antibody [EPR15656] (209491 Abcam) detected by a FITC-conjugated secondary antibody. A positive nuclear staining is observable on all three cell lines (in green), indicating unspecific signal of the antibody in PC3. Medium panel: ICC staining performed with the Rabbit IgG, monoclonal [EPR25A]-Isotype Control (172730 Abcam) detected with FITC-conjugated secondary antibody (97050, Abcam) showing negativity on 22RV1, LNCaP, and PC3 cells. Lower panel: ICC staining performed with FITC-conjugated secondary antibody (97050, Abcam) showing negativity on 22RV1, LNCAP, and PC3 cells.

In conclusion, none of the tested antibodies were deemed suitable for characterization of ARV7 protein on CTCs via the CellSearch® system. Additionally, the most intensively tested anti-ARV7 antibody [EPR15656] [25], described to give a specific nuclear and unspecific cytoplasmic staining did not show reliable results in our hands (Figure 1b), giving unspecific nuclear staining signals in ARV7− PC3 cells, even when neglecting the cytoplasmic staining and considering the described, relevant nuclear staining.

3.2. Development of a qPCR Based Assay to Detect ARV7 mRNA

As an alternative to protein-based detection we subsequently aimed at establishing a qPCR-based approach to detect *ARV7* on an mRNA level. We added an additional prostate cancer cell line to the analysis, to further confirm the robustness of our method. VCaP cells show similar *AR-FL* and *ARV7* expression profiles as 22RV1 cells (AR-FL+/AR-V7+) and were used as a second *ARV7*+ cell line during method establishment. An overview of the *AR-FL* and *ARV7* status for all four cell lines is listed in Figure 2a. Initially, we planned on using the *AR-FL* and *ARV7* primer sets already published [20] for our qPCR-based detection and then modifying the CTC pre-enrichment steps. However, when testing the primers using PCR according to the published protocol, it became clear that while the *AR-FL* primers showed specific bands at the correct expected size of 125 bp (Figure 2b), an additional, undescribed PCR fragment of around 250 bp was detected using the *ARV7* primers in 22Rv1 but not in LNCaP cells (Figure 2b). To ensure optimal primer quality for *ARV7* detection, additional *ARV7* primer sets described in literature [35,36] as well as an own design (UKE), were employed. To exclude that the unspecific PCR fragments detected were generated due to incorrect annealing temperature or incorrect cDNA synthesis, we tested all four primer sets in a gradient PCR on freshly generated 22Rv1 and LNCaP cDNA (Figure 2c). Again, an additional PCR product was detected for the Antonarakis [20] and Guo [35] primer sets across all annealing temperatures in 22Rv1 cells but not LNCaP cells (Figure 2c, lines 1,2). This could represent an additional AR splice variant, similar to *ARV7* [30]. Using the original protocol of 40 amplification cycles [20,23] this additional transcript could come up in clinical samples, especially those with high CTC counts, and result in an unaccounted bias. In contrast, the Liu [36] and UKE primer sets, resulted in specific PCR fragments at 112 bp and 73 bp, respectively (Figure 2c, lines 3,4). The fragment signal intensity appeared slightly higher for the UKE primers (Figure 2c, line 4) in comparison to the Liu primers [36] (Figure 2c, line 3), which could indicate a higher amount of generated PCR product. However, this cannot be conclusively deduced from qualitative PCR. Decreasing the PCR cycles from 39 to 30 (Figure 2d), reduced the unspecific PCR signals down to hardly visible levels for the Antonarakis [20] and Guo [35] primer sets (Figure 2d). However, as quantitative PCR represents a much more sensitive method than qualitative PCR, both primer sets were discarded for further experiments. Both the Liu [36] and UKE primer sets displayed cell line specific PCR results and PCR fragments at correct sizes, resulting in further evaluation of these two primer sets via qPCR.

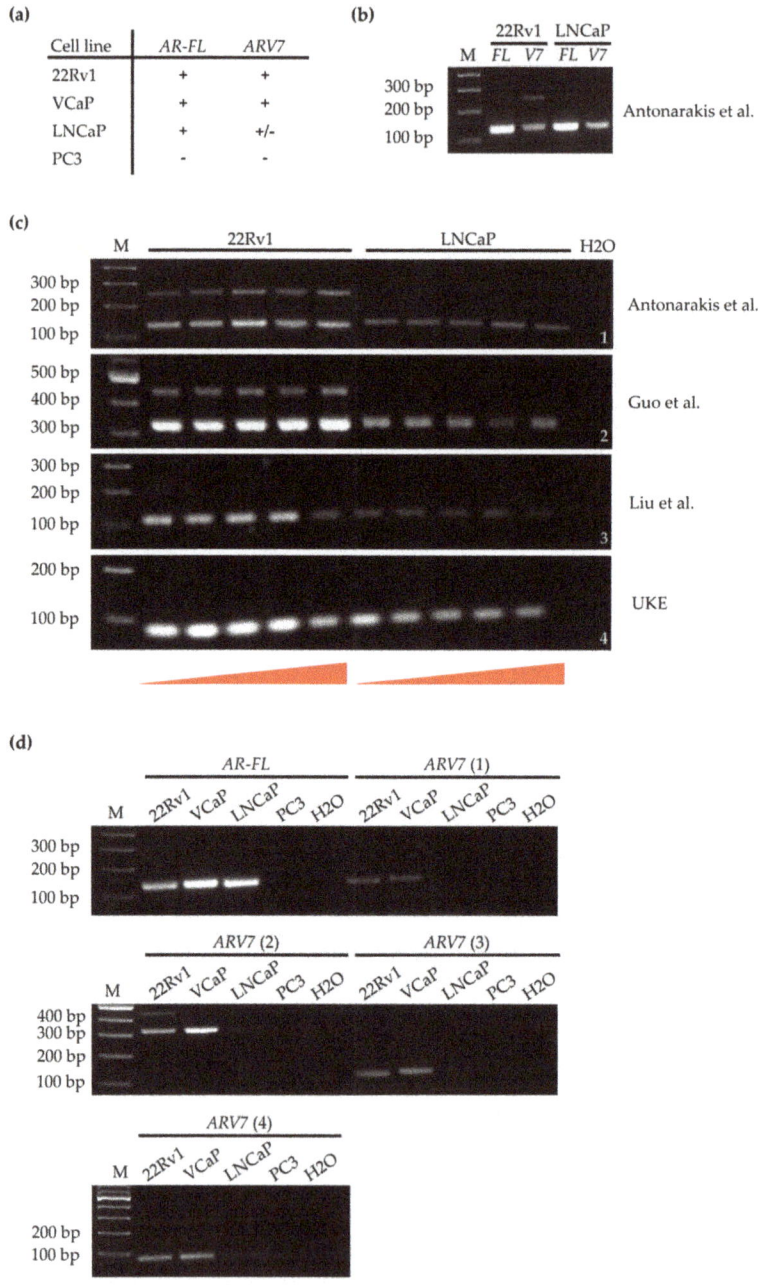

Figure 2. PCR-based detection of *AR-FL* and androgen receptor splice variant 7 (*ARV7*) in selected prostate cancer cell lines. Letter M indicating DNA ladder (marker) lanes. Ten nanograms of cDNA were analyzed for each PCR sample. (**a**) Schematic overview of *AR-FL* and *ARV7* positivity (+) and negativity (−) for established prostate cancer cell lines, as described in literature. (**b**) Agarose gels of a PCR detecting *AR-FL* and *ARV7* in cDNA isolated from 22RV1 and LNCaP cells. *ARV7* cDNA was detected using the primers described by Antonarakis et al. [20]. PCRs were performed for 39 cycles.

125 bp PCR products are expected for both *AR-FL* and *ARV7*. The *ARV7* PCR shows an additional, uncharacterized band at between 250–300 bp for 22RV1 cells, but not for LNCaP cells. (**c**) Agarose gel of a gradient PCR for *ARV7* on 22RV1 and LNCaP cDNA using different primer pairs. PCRs were performed for 39 cycles. Temperatures increasing from 58.5 °C to 65.5 °C, indicated by red triangles below gel images. Antonarakis (1) and Guo (2) primers both show secondary PCR bands on 22Rv1 cDNA (between 200–300bp and between 400–500 bp, respectively). Liu (3) and UKE (4) primers both give expected PCR bands for *ARV7* at 112 bp and 73 bp. Signal intensity appears higher, possibly indicating more generated PCR product, for UKE primers. (**d**) Agarose gels of PCRs detecting *AR-FL* and *ARV7* in cDNA of 22RV1, VCaP, LNCaP, and PC3 prostate cancer cell line cells. PCRs were performed for 30 cycles. AR-FL primer set, results in specific PCR signals in AR^+ and AR^- cell lines. *ARV7* (1) corresponds to Antonarakis et al., *ARV7* (2) corresponds to Guo et al., *ARV7* (3) corresponds to Liu et al., and *ARV7* (4) corresponds to our newly developed UKE primer sets.

Gene expression levels of *AR-FL* and *ARV7* (using the Liu and UKE primers) were assessed for 22Rv1, LNCaP, and VCaP cells in relation to their respective expression in PC3 cells (Figure 3a). As expected, both *AR-FL* and *ARV7* gene expression were dramatically increased in all three cell lines compared to PC3 cells. Additionally, the UKE primers showed most effective detection of *ARV7* (Figure 3a). All further experiments were therefore carried out using the newly designed UKE primers.

Apart from *AR-FL* and *ARV7*, *K19*, and *Actin* gene expression were also measured via qPCR. *Actin* functioning as a gene for normalization and a confirmation of successful cDNA synthesis, and *K19* as an established marker for CTC detection in blood [39,40] thus allowing confirmation of the presence of CTCs in future clinical samples. Figure 3b shows representative qPCR curves for all four cell lines (in different colors) for each gene. Due to the high sensitivity of qPCR analysis, *ARV7* expression can be detected at around 36 cycles for cDNA inputs generated from high PC3 cell counts (Figure 3b). This is an enormous difference to the approximately 22 cycles necessary for detection of $ARV7^+$ cell lines (Figure 3b). Despite the fact that such high CTC cell counts are extremely rarely to be expected in clinical samples, a cut-off of ≤35 cycles was established for gene expression to be counted as positive for the analyzed genes in all further analysis.

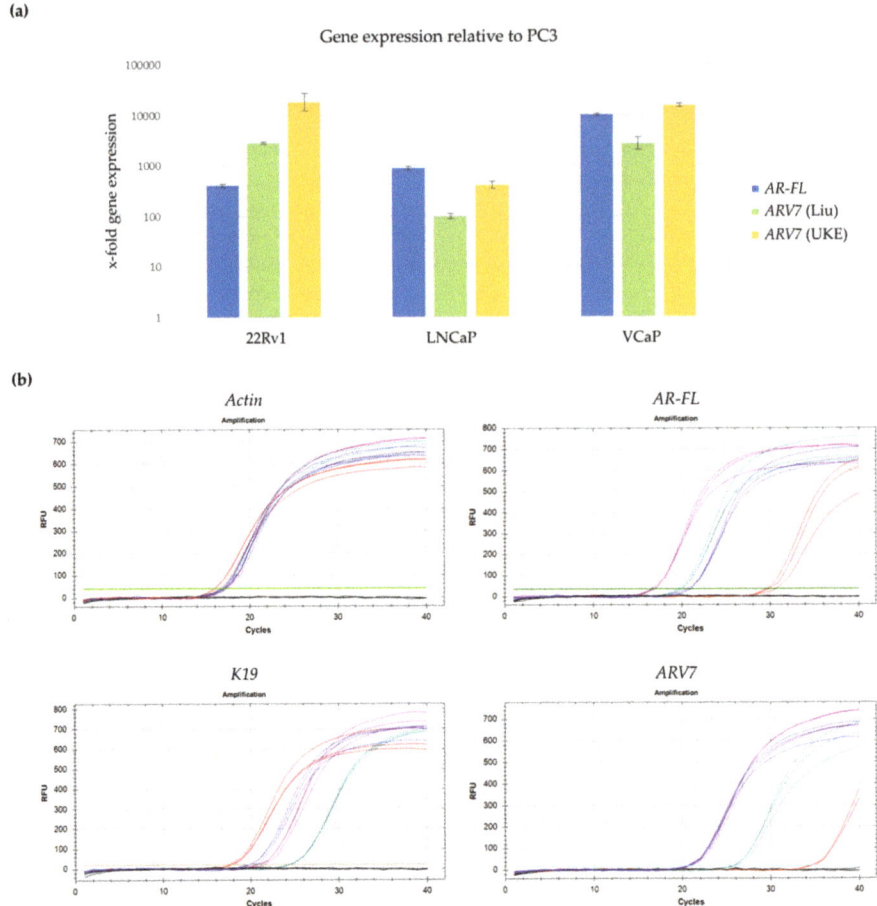

Figure 3. qPCR-based characterization of selected prostate cancer cell line cDNA. cDNA was generated from RNA isolated from 22RV1, VCaP, LNCaP, and PC3 cells and analyzed via qPCR. (**a**) Relative gene expression of *AR-FL* and *ARV7* using primers by Liu et al. and our newly developed primers (UKE). Gene expression was first normalized to actin and subsequently displayed relative to PC3 gene expression. Standard deviation is indicated as brackets. (**b**) Representative qPCR expression profiles for different target genes (*Actin, AR-FL, K19* and *ARV7*) across all four chosen cell lines: 22RV1 (blue), VCaP (purple), LNCaP (green), and PC3 (red). All samples were applied in triplicates.

3.3. Combining Profile-Kit-Based CTC Enrichment with ARV7 mRNA Detection

To allow for use of the CellSearch® system to isolate prostate cancer CTCs for *ARV7* detection on the one hand and enable parallel CTC quantification on the other, a two-armed approach was designed (Figure 4). 7.5 mL of whole blood was taken in parallel into standard EDTA tubes for RNA isolation and CellSave® blood preservation tubes for CTC enumeration, respectively. From EDTA blood, CTCs were enriched via the CellSearch® Profile Kit for subsequent RNA analysis. RNA was isolated and cDNA synthesized prior to analysis of *ARV7, K19,* and *Actin* via qPCR. In parallel CellSave® preserved blood was processed using the CellSearch® CXC Kit thus allowing for parallel AR-FL protein characterization in the fourth fluorescent channel of the device (Figure 4).

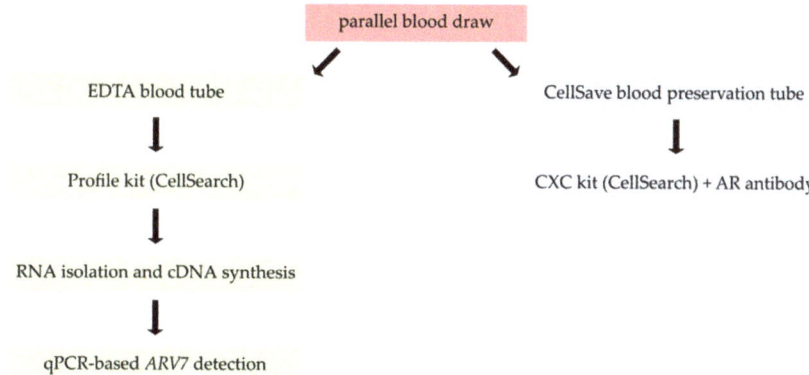

Figure 4. Schematic workflow of *ARV7* detection combined with parallel circulating tumor cell (CTC) enumeration.

"Mock" samples were generated to mimic clinical sample handling. Differing amounts of *ARV7*⁻ and *ARV7*⁺ cell line cells were manually spiked into healthy donor (HD) blood and directly processed by our workflow (Table 1). Following the qPCR run, generated products were applied to a gel electrophoresis allowing final confirmation of gene expression status (data not shown).

Table 1. Validation of protocol specificity and sensitivity. Titration experiments of spiked cell lines in blood from healthy donor (HD) samples. Indicated cell counts of *ARV7*⁺ (22Rv1) and *ARV7*⁻ (PC3) cells were manually spiked into HD blood and processed by our protocol. *ARV7* status is highlighted as "+" for positive and "−" for negative samples. Detection of gene expression was confirmed when at least 2/3 triplicates were positive in qPCR analysis. Unamplified qPCR samples are marked as N/D (not detected). HD samples were processed in parallel as a negative control for *ARV7* and *K19*. The bar (-) in the respective table column of detection indicates no further experiments were conducted.

Cell Line	ARV7 Status	Cell Amount	Target	Detection	Detection (n = 2)	Detection (n = 3)
HD	−	0	ARV7	N/D	N/D	N/D
			K19	N/D	N/D	N/D
			Actin	yes	yes	yes
PC3	−	50	ARV7	N/D	-	-
			K19	yes	-	-
			Actin	yes	-	-
		100	ARV7	N/D	-	-
			K19	yes	-	-
			Actin	yes	-	-
22RV1	+	50	ARV7	yes	-	-
			K19	yes	-	-
			Actin	yes	-	-
		20	ARV7	yes	-	-
			K19	yes	-	-
			Actin	yes	-	-
		10	ARV7	yes	yes	yes
			K19	yes	yes	yes
			Actin	yes	yes	yes
		5	ARV7	yes	yes	yes
			K19	yes	yes	yes
			Actin	yes	yes	yes

All HD samples measured (n = 3) were *ARV7* and *K19* negative (Table 1). PC3 samples were negative for *ARV7* and positive for *K19*, confirming the specificity of the established assay. *ARV7* and keratin 19 were still detectable down to 5 *ARV7*$^+$ 22RV1 cells using our protocol (n = 3), demonstrating high sensitivity (Table 1).

3.4. Assessment of Sample Storage Parameters

mRNA instability represents a common issue for RNA analysis. Sample processing time frames and optimal blood collection tubes therefore need to be carefully assessed to allow for reliable mRNA detection. As cells are not fixed in EDTA blood tubes, which is essential for subsequent RNA isolation, potential CTCs could deteriorate over time. This is especially crucial when calculating time frames for shipment of clinical samples. EDTA blood spiked with cell lines was left at room temperature (RT) for 24 h (Table 2) and 48 h (Table 3), respectively, to test processing windows. Following the qPCR run, generated products were applied to a gel electrophoresis allowing final confirmation of gene expression status (data not shown).

Table 2. Validation of protocol specificity and sensitivity after 24 h. Influence of sample storage on *ARV7* detection limits and assay robustness. *ARV7* status is highlighted as "+" for positive and "−" for negative samples. Detection of gene expression was confirmed when at least 2/3 triplicates were positive in qPCR analysis. N represents the number of repetitions performed per experimental setting. The ratio is defined as the frequency at which any specific gene was detected out of the N repetitions. N/D signifies no gene expression or gene expression above the set threshold of 35 cycles.

Cell Line	ARV7 Status	Cell Amount	Target	Detection	N	Ratio [detection/N]
HD	−	0	ARV7	N/D	3	3/3
			K19	N/D		
			Actin	yes		
22RV1	+	10	ARV7	yes	1	1/1
			K19	yes		
			Actin	yes		
		5	ARV7	yes	2	2/2
			K19	yes		
			Actin	yes		
		3	ARV7	yes	4	1/4
			K19	yes		3/4
			Actin	yes		4/4
		1	ARV7	yes	3	1/3
			K19	yes		2/3
			Actin	yes		3/3
VCaP	+	10	ARV7	yes	2	2/2
			K19	yes		
			Actin	yes		
		5	ARV7	yes	1	1/1
			K19	N/D		
			Actin	yes		

After 24 h of sample storage at RT, 5 *ARV7*$^+$ cells were still reliably detected using the assay (Table 2). This was confirmed on two *ARV7*$^+$ cell lines (22Rv1 and VCaP). Additionally, as low as 3 and down to 1 *ARV7*$^+$ cells were detectable (Table 2). With these low cell counts, detection frequency is more variable as cell enrichment from whole blood and extremely careful sample handling play crucial roles. Still, correct detection down to a single *ARV7*$^+$ cell is possible. After 48 h, detection of *ARV7* and *K19* transcripts is subject to even higher fluctuation and increased cell counts would be needed to robustly detect transcripts of interest from these samples (Table 3). The specificity of our assay was demonstrated as no signals for *ARV7* or *K19* were seen in blood samples from healthy individuals in EDTA blood tested for 24h (3/3) as well as 48 h (3/3) of sample storage (Tables 2 and 3).

Blood tube types vary and some may be more suitable for our assay than others. Therefore, we additionally tested the performance of AdnaCollect blood collection tubes (Qiagen, Hilden, Germany), designed for mRNA characterization by the AdnaTest Prostate Cancer (Qiagen, Hilden, Germany) with our assay. This tube has been used for PCR-based detection of RNA transcripts from whole blood and could therefore provide an alternative to EDTA, potentially prolonging the sample processing window. Again, different cell counts were spiked into HD blood, this time in AdnaCollect blood collection tubes, and processed after 48 h of storage with our protocol. In our hands, these tubes were able to detect *ARV7* in spiked samples, down to 5 ARV7$^+$ cells (Table 3). However, as *ARV7* and *K19* signals were seen in all three tested HD samples (Table 3) indicating low specificity, the use of this blood tube type was not further continued.

Table 3. Influence of sample tubes and sample storage times on ARV7 detection limits and assay specificity. ARV7 status is highlighted as "+" for positive and "−" for negative samples. Detection of gene expression was confirmed when at least 2/3 triplicates were positive in qPCR analysis. N represents the number of repetitions performed per experimental setting. The ratio is defined as the frequency at which any specific gene was detected out of the N repetitions. N/D signifies no gene expression or gene expression above the set threshold of 35 cycles.

Tube	Cell Line	ARV7 Status	Cell Amount	Target	Detection	N	Ratio [detection/N]
EDTA	HD	−	0	ARV7	N/D	3	3/3
				K19	N/D		
				Actin	yes		
	22RV1	+	10	ARV7	N/D	1	1/1
				K19	yes		
				Actin	yes		
			5	ARV7	yes	3	2/3
				K19	yes		2/3
				Actin	yes		3/3
Adnagen	HD	−	0	ARV7	yes	3	1/3
				K19	yes		2/3
				Actin	yes		3/3
	22RV1	+	10	ARV7	N/D	1	1/1
				K19	N/D		
				Actin	yes		
			5	ARV7	yes	3	3/3
				K19	yes		1/3
				Actin	yes		3/3

Our protocol ensures specific detection of tumor cell transcripts in 7.5 mL of blood down to a single cell level even after 24 h of sample storage (Table 2). Conclusively, a sample preparation window of 24 h was determined for the evaluation of clinical samples taken into EDTA blood to allow for sample shipment while ensuring robust detection of *ARV7* from CTCs.

3.5. Clinical Feasibility of the Complete ARV7 Detection Workflow

The clinical feasibility of our assay was demonstrated by analyzing blood samples of 26 metastatic prostate cancer (mPCa) patients. Detailed clinical patient data is listed in Table S1. qPCR based *ARV7* analysis was performed within 24 h of sample collection from 7.5 ml of EDTA blood for all 26 patients. Parallel blood draws to assess CTC counts via CellSearch® were collected and processed for 23/26 patients. AR-FL staining in the fourth fluorescent channel was available for 12/23 patient samples processed via CellSearch® (Table 4).

Table 4. Correlation of qPCR results, AR-FL detection and CTC enumeration via CellSearch for 26 mPCa patients analyzed. Detection of a gene was confirmed when at least 2/3 triplicates were positive in qPCR analysis. N/D signifies no gene expression or gene expression above the set threshold of 35 cycles. CTC enumeration via CellSearch® was not conducted for the first three patient samples, indicated by a bar in the respective table column (-). This also applies to 14 samples collected regarding AR occurrence. The number of CTCs with detectable AR-FL expression is indicated in brackets.

	CellSearch			qPCR		
Sample	CTC Count	AR (nucl.)	AR (cytopl.)	ARV7	K19	Actin
UKE-1	-	-	-	yes	yes	yes
UKE-2	-	-	-	N/D	N/D	yes
UKE-3	-	-	-	N/D	N/D	yes
UKE-4	0	-	-	N/D	N/D	yes
UKE-5	0	-	-	N/D	N/D	yes
UKE-6	0	-	-	N/D	N/D	yes
UKE-7	0	0	0	N/D	N/D	yes
UKE-8	1	0	yes (1)	N/D	N/D	yes
UKE-9	1	yes (1)	0	N/D	N/D	yes
UKE-10	1	0	0	N/D	yes	yes
UKE-11	1	0	yes (1)	N/D	yes	yes
UKE-12	1	-	-	yes	yes	yes
UKE-13	1	0	0	yes	yes	yes
UKE-14	2	yes (1)	yes (1)	N/D	yes	yes
UKE-15	6	0	0	N/D	N/D	yes
UKE-16	6	-	-	N/D	N/D	yes
UKE-17	8	yes (3)	yes (4)	yes	yes	yes
UKE-18	9	0	yes (9)	yes	yes	yes
UKE-19	11	0	yes (11)	yes	yes	yes
UKE-20	11	0	yes (11)	yes	yes	yes
UKE-21	14	-	-	N/D	yes	yes
UKE-22	16	-	-	yes	yes	yes
UKE-23	22	-	-	yes	N/D	yes
UKE-24	80	-	-	yes	yes	yes
UKE-25	156	-	-	yes	yes	yes
UKE-26	398	-	-	yes	yes	yes

Of the patient samples analyzed via CellSearch® 86.2% (19/23) were found to have ≥1 CTC in 7.5 mL of blood. In 52.2% (12/23) of patients ≥5 CTCs were detected in 7.5 ml of whole blood, reaching the clinically prognostic cut-off value for worse overall survival for metastatic mPCa patients [8]. The median of detected CTCs for our cohort is 6 (range: 0–398 CTCs) and the average is 32 CTCs/7.5 mL of blood. *ARV7* mRNA was detected in 46.2% (12/26) of mPCa patients, *K19* was detected is 57.7% (15/26) of samples and *Actin* was detected in all samples (26/26), indicating effective cDNA transcription. Four measured patients were negative for the androgen receptor splice variant and positive for *K19* (e.g., samples UKE-10 and UKE-11). Additionally, one patient was positive for *ARV7* expression and negative for *K19* (UKE-23). No *ARV7* or *K19* gene expression was found in samples classified as CTC negative by the CellSearch® system. Evaluation of the first 26 clinical samples resulted in 42.3% of $ARV7^+/K19^+$ of all cases (11/26) and 52.6% of $ARV7^+/K19^+$ cases (10/19) with ≥1 detectable CTC. Representative CellSearch® images of AR-FL staining are shown in Figure 5.

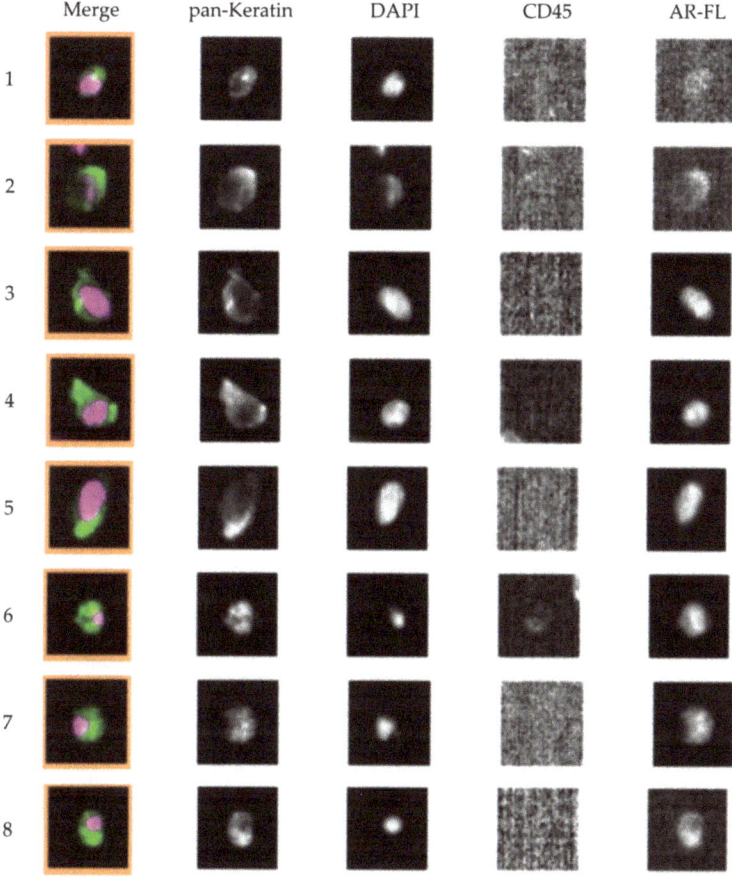

Figure 5. Representative CellSearch® images of CTCs and AR-FL staining from a single prostate cancer patient case. This patient had detectable AR-FL negative (**1**), weakly positive (**2**), nuclear AR-FL positive (**3–5**) and cytoplasmatically AR-FL positive (**6–8**) CTCs.

Only three of the 12 cases in which AR-FL protein staining was assessed in the CellSearch® (UKE-9, UKE-14, UKE-17) had detectable nuclear AR-FL protein levels (Table 4). Two of these three patients (UKE-14, UKE-17) had a mixed CTC population of nuclear and cytoplasmic AR-FL+ CTCs. Of these three patients, two were *ARV7* negative with our assay (UKE-9, UKE-14). Additional five patients showed cytoplasmic AR-FL protein expression, more than half of these patients were ARV7 positive (3/5).

For the majority of patient samples tested (88.4% or 23/26), the CTC count as measured by the CellSearch® system was in accordance to *K19* detection in parallel samples (Table 4). Detection of *ARV7* was possible in 2/6 patients with only a single CTC detected in the patient's blood (UKE-12, UKE-13) confirming the assays sensitivity (Table 4). *K19* was detected in 4/6 patients with only a single CTC indicating careful and effective sample handling (Table 4).

4. Discussion

The CellSearch® Profile technology allows a reliable, standardized, and automated enrichment of EpCAM-positive cancer cells. ARV7 expression in CTCs of prostate cancer patients has been linked to resistance toward AR-targeted therapy, in particular enzalutamide and abiraterone [20,25]. Our novel approach ensures specific detection of *ARV7* transcripts in CTCs isolated by the CellSearch® system down to the single cell level. The specificity of our assay was indicated as no signals for *ARV7* or *K19* were seen in 9 blood samples from healthy, male individuals (Tables 1–3). Our protocol ensures specific detection of tumor cell transcripts in 7.5 mL of blood even after 24 h of sample storage (Table 2). Robust *ARV7* and *K19* detection is feasible in as low as 5 *ARV7$^+$/K19$^+$* cells (Table 2). Transcript expression below 5 cells, even down to 1 *ARV7$^+$/K19$^+$* cell, was possible (Table 2). The clinical feasibility of our assay and its high sensitivity (down to a single CTC) was demonstrated in a cohort of 26 mPCa patients (Table 4).

Antonarakis et al. linked *ARV7* mRNA expression on CTCs of mCRPC patients receiving enzalutamide and/or abiraterone therapy to lower PSA response rates, as well as shorter progression free and overall survival [20]. Following this initial study, conducted with a combination of bead-based CTC enrichment and subsequent qPCR multiplexing, the group confirmed their finding in a larger cohort of 202 CRPC patients [23]. In their study, CTC$^-$ patients were found to have the best outcome (judged by best PSA-response, PSA progression-free survival, progression-free survival, and overall survival), followed by CTC$^+$/ARV7$^-$ and finally CTC$^+$/ARV7$^+$ patients [23]. Additionally, it was demonstrated that *ARV7* status can change in the course of hormone therapy [25,28,29] and that within one patient *ARV7* status on CTCs can be heterogeneous [41].

The CellSearch® system enables validated and automated enrichment of EpCAM-positive cancer cells [8,42–45]. Ideally, adding a specific and sensitive anti-ARV7 antibody to the fourth fluorescent channel of the CellSearch® device would therefore represent a valuable alternative to allow parallel CTC enumeration and the assessment of ARV7 status for each respective CTC. Unfortunately, detection of ARV7 protein using the CellSearch® technology was dramatically hampered by lacking specificity of most existing ARV7 antibodies (Figure 1, Figure S1). Recently, a novel commercially available antibody has been tested and validated for immunohistochemistry on primary tumor tissue, showing specific ARV7 staining results [32]. Whether this antibody might represent a promising novel candidate for immunocytochemical analysis and combination with CellSearch® needs to be investigated in future studies. However, so far most sources of CTC-related ARV7 information stems from RNA measurements.

The meaningful clinical impact of *ARV7* expression of CTCs [20,46] has led to the development of a multitude of different assays targeting ARV7 protein [25] or *ARV7* transcripts [30,31,47,48]. Primarily the developed methods are based on the analysis of pooled lysate of an enriched CTC fraction [26,30,48], only few perform whole blood gene expression analysis [47]. CTC are enriched by bead-based approaches [20,48], or the CellSearch® Profile kit and analyzed by subsequent qPCR or RNA-seq [26,30]. While these approaches effectively asses *ARV7* status, they give no additional information on the abundance of CTCs in a patient at the time point of blood draw. This could, however, prove to be valuable information allowing more precise interpretation of the qualitative ARV7$^+$ or ARV7$^-$ status of a patient. Without CTC count, an ARV7$^-$ status may refer to no available CTCs within the blood draw or to high amounts of ARV7$^-$ CTCs, respectively. The clinical information to be gained from both results is, however, very different, as no CTCs indicate good and ≥5 CTCs indicate poor outcome for the patient [8]. Multiplexing of additional genes such as prostate specific antigen (PSA) or prostate specific membrane antigen (PSMA), as well as epithelial genes is commonly used as a means of circumventing this issue and attempting to detect ARV7$^-$ CTCs [20,26,49]. While this is a feasible approach, it is limited by heterogeneous expression of these markers [31,41,49–51] and the required pre-amplification step can introduce bias.

Using our novel approach (Figure 4) information on both CTC count, AR-FL and *ARV7* status is collected. One could argue that the amount of CTCs present in the blood tube destined for *ARV7* assessment is also not directly assessed by our assay. However, studies have shown that CTC counts determined with the CellSearch® technology do not significantly fluctuate depending on circadian rhythm or serial blood draws [52,53], thereby indicating that stochastically, similar to equal CTC amounts would be expected in two sequential blood draws from the same patient at the same time (as is necessitated by our protocol). The importance of integrated CTC enumeration becomes apparent when looking at clinical cases such as UKE-23 (Table 4). While this patient had clearly detectable *ARV7* transcripts, he did not show *K19* positivity in our assay. Without the additional information of 22 CTCs being detected via CellSearch® analysis, interpretation of the qPCR results would have been impaired. This case also highlights the inert limitation of qPCR multiplexing, which lies in the before mentioned heterogeneity of gene and protein expression in CTCs [41,50,51]. In addition, CTC detection via the CellSearch® allows for morphological assessment of the CTCs in circulation and in our case, parallel characterization of AR-FL protein as well as its cellular location. Both represent important factors in resistance to androgen deprivation therapy [54]. The localization of the full length AR within the cell has been shown to be associated with disease progression on novel hormone therapies (e.g., enzalutamide and abiraterone) [55]. Therefore, it was critical for our assay to be able to distinguish both cytoplasmic and nuclear fractions of AR to support differentiation between "AR-on" and "AR-off" patients [54]. Apart from the AR-FL targeting antibody (by Janssen Diagnostics) used in this study, other well-established alternative antibodies have been published for AR-FL detection in the fourth channel of the CellSearch® [55].

To our knowledge, only two assays have been developed allowing parallel CTC enumeration and *ARV7* protein [24,25] or transcript detection [31] on the same cell so far. El-Heliebi et al. isolated CTCs via the CellSearch® Profile kit or the size-dependent Parsortix™ platform (ANGLE plc, Guildford, UK) [56] and subsequently characterized them for *ARV7*, *AR-FL*, and *PSA* expression via in situ padlock probe technology [31]. This approach allows for absolute transcript quantification while keeping cell morphology intact and thereby enabling tumor cell enumeration [31]. In regards to CellSearch® Profile kit pre-enrichment, a single patient with high CTC load was included in this study to demonstrate general feasibility of the approach [31]. Additional technical validation will therefore be required to ensure sufficient sensitivity and specificity of this method for future clinical application.

The ARV7 assay most advanced in regards to clinical validation is the EPICs approach [24,25]. Here, the nuclear cell fraction is placed on slides, stained via ICC and automatically screened and evaluated. The assay focusses on nuclear ARV7 protein expression using the same antibody clone EPR15656 (Abcam) that we tested in our present study. While the EPICS approach allows for parallel CTC enumeration and ARV7 protein assessment, it requires sample shipment to a centralized lab in the US, a costs intensive approach when conducting larger clinical studies or when shipping patient samples for routine testing. A nuclear ARV7 staining has been postulated to be relevant to predict therapy outcome of AR-targeted therapies as well as taxanes in a cohort of 161 mCRPC patients, leading to a favorable coverage recommendation and certification of the approach in the state of California (USA) [57,58]. However, in our hands, the EPR15656 antibody did not result in specific nuclear staining signals for tested cancer cell line cells on chamber slides or cytospins (Figure 1, Figure S1).

While the *ARV7* detection assay established in this study is highly specific and sensitive, some limitations require mentioning. The main limitation is the fact that our assay does not allow for simultaneous morphological and molecular *ARV7* characterization of each single CTC. However, this is somewhat compensated by the use of a clinically validated CTC enrichment method, adding weight to the clinical relevance of the CTCs analyzed. Additionally, *ARV7* and *K19* transcript detection cannot be guaranteed down to a single CTC level in all patient samples. Nevertheless, we can secure determination of *ARV7* status from ≥5 CTCs which is the prognostic cut-off value for patients with metastatic prostate cancer. Due to the high sensitivity and specificity of our *ARV7* detection assay and the parallel nature of the CellSearch® CTC-enumeration, *K19* detection is not a mandatory prerequisite for robust *ARV7* assessment and result interpretation. However, we believe *K19* adds further valuable information in positive cases and represents an additional confirmation of successful CTC analysis.

Taken together, the use of a FDA-cleared enrichment technology, high assay sensitivity and specificity, a shipment window of 24 h and comparably low necessity of elaborate additional laboratory equipment (standard qPCR cycler) corroborate the value of our established method. Inclusion into prospective clinical trials will be now necessary to demonstrate clinical validity and utility. Furthermore, additional age-matched healthy donors and other control cohorts (e.g., prostatitis patients) should be included into future studies to further corroborate assay specificity. Head-to-head comparison with other ARV7/CTC technologies is desirable to assess to which extent different assays are redundant or complementary.

5. Conclusions

The novel workflow developed in this study allows for a semi-automated enrichment of CTCs followed by a qPCR assay measuring the *ARV7* status of CTCs. This approach can now be integrated into future clinical trials assessing treatment responses to antiandrogen therapies in prostate cancer patients.

Supplementary Materials: The following are available online at http://www.mdpi.com/2073-4409/8/9/1067/s1. Figure S1: Assessment of ARV7 antibody performance on selected prostate cancer cell lines via Western Blot and immunocytochemical (ICC) staining; Table S1: Clinical data of 26 mPCa patients.

Author Contributions: C.H. collected and primarily analyzed data, developed the methodology, visualized the results and wrote the manuscript draft. T.M.G. conceptualized and supervised the project and contributed to the method establishment. S.R. and M.M. collected and analyzed data. M.M. additionally participated in writing of the manuscript. T.S., G.v.A., F.K. and S.P. contributed clinical as well as healthy donor samples and clinical data. They furthermore aided in the clinical interpretation of the results. C.A.-P. and K.P. participated in writing of the manuscript and supported data analysis. K.P. furthermore supervised and conceptualized the project and acquired funding. All authors contributed to the editing and writing of the manuscript. Conceptualization, T.M.G., C.A.-P., and K.P.; Funding acquisition, K.P.; Investigation, C.H., S.R., M.M., T.S., G.v.A., F.K. and S.P.; Methodology, C.H. and T.M.G.; Supervision, T.M.G.; Visualization, C.H.; Writing—original draft, C.H. and K.P.; Writing—review and editing, C.H., T.M.G., S.R., M.M., T.S., G.v.A., F.K., S.P., C.A.-P., and K.P.

Funding: Klaus Pantel and Catherine Alix-Panabières received funding from the European TRANSCAN PROLIPSY grant (Nr. 01KT1810).

Acknowledgments: The authors acknowledge Cornelia Coith, Antje Andreas and Oliver Mauermann from UKE Hamburg for technical support.

Conflicts of Interest: Klaus Pantel has ongoing patent applications related to circulating tumour cells. Klaus Pantel has received honoraria from Agena, Novartis, Roche, and Sanofi and research funding from European Federation of Pharmaceutical Industries and Associations (EFPIA) partners (Angle, Menarini and Servier) of the CANCER-ID programm (www.cancer-id.eu) of the European Union–EFPIA Innovative Medicines Initiative. CAP has received honoraria from Janssen and grant support from Menarini. The remaining authors declare no conflict of interest.

References

1. World Cancer Research Fund Inthernational, American Institute for Cancer Research. Prostate Cancer. 2018. Available online: https://www.wcrf.org/dietandcancer/prostate-cancer (accessed on 30 July 2019).
2. Ladjevardi, S.; Auer, G.; Castro, J.; Ericsson, C.; Zetterberg, A.; Häggman, M.; Wiksell, H.; Jorulf, H. Prostate Biopsy Sampling Causes Hematogenous Dissemination of Epithelial Cellular Material. *Dis. Markers* **2014**, *2014*, 1–6. [CrossRef] [PubMed]
3. Alix-Panabières, C.; Pantel, K. Clinical Applications of Circulating Tumor Cells and Circulating Tumor DNA as Liquid Biopsy. *Cancer Discov.* **2016**, *6*, 479–491. [CrossRef] [PubMed]
4. Pantel, K.; Speicher, M.R. The biology of circulating tumor cells. *Oncogene* **2016**, *35*, 1216–1224. [CrossRef] [PubMed]
5. Bardelli, A.; Pantel, K. Liquid Biopsies, What We Do Not Know (Yet). *Cancer Cell* **2017**, *31*, 172–179. [CrossRef] [PubMed]
6. Alix-Panabieres, C.; Pantel, K. Circulating tumor cells: Liquid biopsy of cancer. *Clin. Chem.* **2013**, *59*, 110–118. [CrossRef] [PubMed]

7. Hille, C.; Pantel, K. Circulating tumour cells in prostate cancer. *Nat. Rev. Urol.* **2018**, *15*, 265–266. [CrossRef] [PubMed]
8. De Bono, J.S.; Scher, H.I.; Montgomery, R.B.; Parker, C.; Miller, M.C.; Tissing, H.; Doyle, G.V.; Terstappen, L.W.W.M.; Pienta, K.J.; Raghavan, D. Circulating tumor cells predict survival benefit from treatment in metastatic castration-resistant prostate cancer. *Clin. Cancer Res.* **2008**, *14*, 6302–6309. [CrossRef] [PubMed]
9. Scher, H.I.; Jia, X.Y.; de Bono, J.S.; Fleisher, M.; Pienta, K.J.; Raghavan, D.; Heller, G. Circulating tumour cells as prognostic markers in progressive, castration-resistant prostate cancer: A reanalysis of IMMC38 trial data. *Lancet. Oncol.* **2009**, *10*, 233–239. [CrossRef]
10. Scher, H.I.; Heller, G.; Molina, A.; Attard, G.; Danila, D.C.; Jia, X.; Peng, W.; Sandhu, S.K.; Olmos, D.; Riisnaes, R.; et al. Circulating Tumor Cell Biomarker Panel As an Individual-Level Surrogate for Survival in Metastatic Castration-Resistant Prostate Cancer. *J. Clin. Oncol.* **2015**, *33*, 1348–1355. [CrossRef]
11. Heller, G.; Fizazi, K.; McCormack, R.; Molina, A.; MacLean, D.; Webb, I.J.; Saad, F.; de Bono, J.S.; Scher, H.I. The Added Value of Circulating Tumor Cell Enumeration to Standard Markers in Assessing Prognosis in a Metastatic Castration-Resistant Prostate Cancer Population. *Clin. Cancer Res.* **2017**, *23*, 1967–1973. [CrossRef]
12. Miller, M.C.; Doyle, G.V.; Terstappen, L.W. Significance of Circulating Tumor Cells Detected by the CellSearch System in Patients with Metastatic Breast Colorectal and Prostate Cancer. *J. Oncol.* **2010**, *2010*, 617421. [CrossRef] [PubMed]
13. Goodman, O.B.; Symanowski, J.T.; Loudyi, A.; Fink, L.M.; Ward, D.C.; Vogelzang, N.J. Circulating Tumor Cells as a Predictive Biomarker in Patients With Hormone-sensitive Prostate Cancer. *Clin. Genitourin. Cancer* **2011**, *9*, 31–38. [CrossRef] [PubMed]
14. Danila, D.C.; Fleisher, M.; Scher, H.I. Circulating tumor cells as biomarkers in prostate cancer. *Clin. Cancer Res.* **2011**, *17*, 3903–3912. [CrossRef] [PubMed]
15. Gorges, T.M.; Pantel, K. Circulating tumor cells as therapy-related biomarkers in cancer patients. *Cancer Immunol. Immunother.* **2013**, *62*, 931–939. [CrossRef] [PubMed]
16. Singhal, U.; Wang, Y.; Henderson, J.; Niknafs, Y.S.; Qiao, Y.; Gursky, A.; Zaslavsky, A.; Chung, J.-S.; Smith, D.C.; Karnes, R.J.; et al. Multigene Profiling of CTCs in mCRPC Identifies a Clinically Relevant Prognostic Signature. *Mol. Cancer Res.* **2018**, *16*, 643–654. [CrossRef]
17. Pantel, K.; Hille, C.; Scher, H.I. Circulating Tumor Cells in Prostate Cancer: From Discovery to Clinical Utility. *Clin. Chem.* **2019**, *65*, 87–99. [CrossRef] [PubMed]
18. Luo, J.; Attard, G.; Balk, S.P.; Bevan, C.; Burnstein, K.; Cato, L.; Cherkasov, A.; De Bono, J.S.; Dong, Y.; Gao, A.C.; et al. Role of Androgen Receptor Variants in Prostate Cancer: Report from the 2017 Mission Androgen Receptor Variants Meeting. *Eur. Urol.* **2018**, *73*, 715–723. [CrossRef]
19. Antonarakis, E.S.; Armstrong, A.J.; Dehm, S.M.; Luo, J. Androgen receptor variant-driven prostate cancer: Clinical implications and therapeutic targeting. *Prostate Cancer Prostatic Dis.* **2016**, *19*, 231–241. [CrossRef]
20. Antonarakis, E.S.; Lu, C.; Wang, H.; Luber, B.; Nakazawa, M.; Roeser, J.C.; Chen, Y.; Mohammad, T.A.; Chen, Y.; Fedor, H.L.; et al. AR-V7 and Resistance to Enzalutamide and Abiraterone in Prostate Cancer. *New Engl. J. Med.* **2014**, *371*, 1028–1038. [CrossRef]
21. Antonarakis, E.S.; Luo, J. Blood Based Detection of Androgen Receptor Splice Variants in Patients with Advanced Prostate Cancer. *J. Urol.* **2016**, *196*, 1606–1607. [CrossRef]
22. Lokhandwala, P.M.; Riel, S.L.; Haley, L.; Lu, C.; Chen, Y.; Silberstein, J.; Zhu, Y.; Zheng, G.; Lin, M.-T.; Gocke, C.D.; et al. Analytical Validation of Androgen Receptor Splice Variant 7 Detection in a Clinical Laboratory Improvement Amendments (CLIA) Laboratory Setting. *J. Mol. Diagn.* **2017**, *19*, 115–125. [CrossRef] [PubMed]
23. Antonarakis, E.S.; Lu, C.; Luber, B.; Wang, H.; Chen, Y.; Zhu, Y.; Silberstein, J.L.; Taylor, M.N.; Maughan, B.L.; Denmeade, S.R.; et al. Clinical Significance of Androgen Receptor Splice Variant-7 mRNA Detection in Circulating Tumor Cells of Men With Metastatic Castration-Resistant Prostate Cancer Treated With First- and Second-Line Abiraterone and Enzalutamide. *J. Clin. Oncol.* **2017**, *35*, 2149–2156. [CrossRef]
24. Scher, H.I.; Graf, R.P.; Schreiber, N.A.; Jayaram, A.; Winquist, E.; McLaughlin, B.; Lu, D.; Fleisher, M.; Orr, S.; Lowes, L.; et al. Assessment of the Validity of Nuclear-Localized Androgen Receptor Splice Variant 7 in Circulating Tumor Cells as a Predictive Biomarker for Castration-Resistant Prostate Cancer. *JAMA Oncol.* **2018**, *4*, 1179–1186. [CrossRef]

25. Scher, H.I.; Lu, D.; Schreiber, N.A.; Louw, J.; Graf, R.P.; Vargas, H.A.; Johnson, A.; Jendrisak, A.; Bambury, R.; Danila, D.; et al. Association of AR-V7 on Circulating Tumor Cells as a Treatment-Specific Biomarker With Outcomes and Survival in Castration-Resistant Prostate Cancer. *JAMA Oncol.* **2016**, *2*, 1441–1449. [CrossRef] [PubMed]
26. Onstenk, W.; Sieuwerts, A.M.; Kraan, J.; Van, M.; Nieuweboer, A.J.; Mathijssen, R.H.; Hamberg, P.; Meulenbeld, H.J.; De Laere, B.; Dirix, L.Y.; et al. Efficacy of Cabazitaxel in Castration-resistant Prostate Cancer Is Independent of the Presence of AR-V7 in Circulating Tumor Cells. *Eur. Urol.* **2015**, *68*, 939–945. [CrossRef] [PubMed]
27. Antonarakis, E.S.; Lu, C.; Luber, B.; Wang, H.; Chen, Y.; Nakazawa, M.; Nadal, R.; Paller, C.J.; Denmeade, S.R.; Carducci, M.A.; et al. Androgen Receptor Splice Variant 7 and Efficacy of Taxane Chemotherapy in Patients With Metastatic Castration-Resistant Prostate Cancer. *JAMA Oncol.* **2015**, *1*, 582–591. [CrossRef] [PubMed]
28. Bernemann, C.; Schnoeller, T.J.; Luedeke, M.; Steinestel, K.; Boegemann, M.; Schrader, A.J.; Steinestel, J. Expression of AR-V7 in Circulating Tumour Cells Does Not Preclude Response to Next Generation Androgen Deprivation Therapy in Patients with Castration Resistant Prostate Cancer. *Eur. Urol.* **2017**, *71*, 1–3. [CrossRef]
29. Nakazawa, M.; Lu, C.; Chen, Y.; Paller, C.J.; Carducci, M.A.; Eisenberger, M.A.; Luo, J.; Antonarakis, E.S. Serial blood-based analysis of AR-V7 in men with advanced prostate cancer. *Ann. Oncol.* **2015**, *26*, 1859–1865. [CrossRef]
30. De Laere, B.; Van Dam, P.-J.; Whitington, T.; Mayrhofer, M.; Diaz, E.H.; Eynden, G.V.D.; Vandebroek, J.; Del-Favero, J.; Van Laere, S.; Dirix, L.; et al. Comprehensive Profiling of the Androgen Receptor in Liquid Biopsies from Castration-resistant Prostate Cancer Reveals Novel Intra-AR Structural Variation and Splice Variant Expression Patterns. *Eur. Urol.* **2017**, *72*, 192–200. [CrossRef]
31. El-Heliebi, A.; Attard, G.; Balk, S.P.; Bevan, C.; Burnstein, K.; Cato, L.; Cherkasov, A.; De Bono, J.S.; Dong, Y.; Gao, A.C.; et al. In Situ Detection and Quantification of AR-V7, AR-FL, PSA, and KRAS Point Mutations in Circulating Tumor Cells. *Clin. Chem.* **2018**, *64*. [CrossRef]
32. Sharp, A.; Welti, J.C.; Lambros, M.B.; Dolling, D.; Rodrigues, D.N.; Pope, L.; Aversa, C.; Figueiredo, I.; Fraser, J.; Ahmad, Z.; et al. Clinical Utility of Circulating Tumour Cell Androgen Receptor Splice Variant-7 Status in Metastatic Castration-resistant Prostate Cancer. *Eur. Urol.* **2019**. [CrossRef]
33. Riethdorf, S.; O'Flaherty, L.; Hille, C.; Pantel, K. Clinical applications of the CellSearch platform in cancer patients. *Adv. Drug Deliv. Rev.* **2018**, *125*, 102–121. [CrossRef] [PubMed]
34. Riethdorf, S.; Fritsche, H.; Müller, V.; Rau, T.; Schindlbeck, C.; Rack, B.; Janni, W.; Coith, C.; Beck, K.; Jänicke, F.; et al. Detection of Circulating Tumor Cells in Peripheral Blood of Patients with Metastatic Breast Cancer: A Validation Study of the CellSearch System. *Clin. Cancer Res.* **2007**, *13*, 920–928. [CrossRef] [PubMed]
35. Guo, Z.; Yang, X.; Sun, F.; Jiang, R.; Linn, D.E.; Chen, H.; Chen, H.; Kong, X.; Melamed, J.; Tepper, C.G.; et al. A novel androgen receptor splice variant is up-regulated during prostate cancer progression and promotes androgen depletion-resistant growth. *Cancer Res.* **2009**, *69*, 2305–2313. [CrossRef] [PubMed]
36. Liu, L.L.; Xie, N.; Sun, S.; Plymate, S.; Mostaghel, E.; Dong, X. Mechanisms of the androgen receptor splicing in prostate cancer cells. *Oncogene* **2014**, *33*, 3140–3150. [CrossRef] [PubMed]
37. Untergasser, A.; Cutcutache, I.; Koressaar, T.; Ye, J.; Faircloth, B.C.; Remm, M.; Rozen, S.G. Primer3—New capabilities and interfaces. *Nucleic Acids Res.* **2012**, *40*, e115. [CrossRef]
38. Livak, K.J.; Schmittgen, T.D. Analysis of relative gene expression data using real-time quantitative PCR and the 2(-Delta Delta C(T)) Method. *Methods* **2001**, *25*, 402–408. [CrossRef]
39. Stathopoulou, A.; Angelopoulou, K.; Perraki, M.; Georgoulias, V.; Malamos, N.; Lianidou, E. Quantitative RT-PCR luminometric hybridization assay with an RNA internal standard for cytokeratin-19 mRNA in peripheral blood of patients with breast cancer. *Clin. Biochem.* **2001**, *34*, 651–659. [CrossRef]
40. Stathopoulou, A.; Gizi, A.; Perraki, M.; Apostolaki, S.; Malamos, N.; Mavroudis, D.; Georgoulias, V.; Lianidou, E.S. Real-time quantification of CK-19 mRNA-positive cells in peripheral blood of breast cancer patients using the lightcycler system. *Clin. Cancer Res.* **2003**, *9*, 5145–5151.
41. Gorges, T.M.; Kuske, A.; Röck, K.; Mauermann, O.; Müller, V.; Peine, S.; Verpoort, K.; Novosadova, V.; Kubista, M.; Riethdorf, S.; et al. Accession of Tumor Heterogeneity by Multiplex Transcriptome Profiling of Single Circulating Tumor Cells. *Clin. Chem.* **2016**, *62*, 1504–1515. [CrossRef]

42. Cristofanilli, M.; Budd, G.T.; Ellis, M.J.; Stopeck, A.; Matera, J.; Miller, M.C.; Reuben, J.M.; Doyle, G.V.; Allard, W.J.; Terstappen, L.W.; et al. Circulating Tumor Cells, Disease Progression, and Survival in Metastatic Breast Cancer. *New Engl. J. Med.* **2004**, *351*, 781–791. [CrossRef] [PubMed]
43. Cohen, S.J.; Punt, C.J.; Iannotti, N.; Saidman, B.H.; Sabbath, K.D.; Gabrail, N.Y.; Picus, J.; Morse, M.; Mitchell, E.; Miller, M.C.; et al. Relationship of Circulating Tumor Cells to Tumor Response, Progression-Free Survival, and Overall Survival in Patients With Metastatic Colorectal Cancer. *J. Clin. Oncol.* **2008**, *26*, 3213–3221. [CrossRef] [PubMed]
44. Ligthart, S.T.; Coumans, F.A.W.; Attard, G.; Cassidy, A.M.; De Bono, J.S.; Terstappen, L.W.M.M. Unbiased and Automated Identification of a Circulating Tumour Cell Definition That Associates with Overall Survival. *PLoS ONE* **2011**, *6*, e27419. [CrossRef] [PubMed]
45. Scholtens, T.M.; Schreuder, F.; Ligthart, S.T.; Swennenhuis, J.F.; Greve, J.; Terstappen, L.W. Automated identification of circulating tumor cells by image cytometry. *Cytometry A* **2012**, *81A*, 138–148. [CrossRef] [PubMed]
46. Markowski, M.C.; Silberstein, J.L.; Eshleman, J.R.; Eisenberger, M.A.; Luo, J.; Antonarakis, E.S. Clinical Utility of CLIA-Grade AR-V7 Testing in Patients With Metastatic Castration-Resistant Prostate Cancer. *JCO Precis. Oncol.* **2017**, 1–9.
47. Todenhöfer, T.; Azad, A.; Stewart, C.; Gao, J.; Eigl, B.J.; Gleave, M.E.; Joshua, A.M.; Black, P.C.; Chi, K.N. AR-V7 Transcripts in Whole Blood RNA of Patients with Metastatic Castration Resistant Prostate Cancer Correlate with Response to Abiraterone Acetate. *J. Urol.* **2017**, *197*, 135–142. [CrossRef] [PubMed]
48. Tommasi, S.; Pilato, B.; Carella, C.; Lasorella, A.; Danza, K.; Vallini, I.; De Summa, S.; Naglieri, E. Standardization of CTC AR-V7 PCR assay and evaluation of its role in castration resistant prostate cancer progression. *Prostate* **2018**, *79*, 54–61. [CrossRef] [PubMed]
49. Sieuwerts, A.M.; Mostert, B.; Van Der Vlugt-Daane, M.; Kraan, J.; Beaufort, C.M.; Van, M.; Prager, W.J.; De Laere, B.; Beije, N.; Hamberg, P.; et al. An In-Depth Evaluation of the Validity and Logistics Surrounding the Testing of AR-V7 mRNA Expression in Circulating Tumor Cells. *J. Mol. Diagn.* **2018**, *20*, 316–325. [CrossRef]
50. Gorges, T.M.; Riethdorf, S.; Von Ahsen, O.; Nastały, P.; Röck, K.; Boede, M.; Peine, S.; Kuske, A.; Schmid, E.; Kneip, C.; et al. Heterogeneous PSMA expression on circulating tumor cells—A potential basis for stratification and monitoring of PSMA-directed therapies in prostate cancer. *Oncotarget* **2016**, *7*, 34930–34941. [CrossRef] [PubMed]
51. Scher, H.I.; Graf, R.P.; Schreiber, N.A.; McLaughlin, B.; Jendrisak, A.; Wang, Y.; Lee, J.; Greene, S.; Krupa, R.; Lu, D.; et al. Phenotypic Heterogeneity of Circulating Tumor Cells Informs Clinical Decisions between AR Signaling Inhibitors and Taxanes in Metastatic Prostate Cancer. *Cancer Res.* **2017**, *77*, 5687–5698. [CrossRef]
52. Martín, M.; García-Sáenz, J.A.; Casas, M.L.M.D.L.; Vidaurreta, M.; Puente, J.; Veganzones, S.; Rodríguez-Lajusticia, L.; De La Orden, V.; Oliva, B.; De La Torre, J.-C.; et al. Circulating tumor cells in metastatic breast cancer: timing of blood extraction for analysis. *Anticancer. Res.* **2009**, *29*, 4185–4187. [PubMed]
53. Tibbe, A.G.J.; Miller, M.C.; Terstappen, L.W.M.M. Statistical considerations for enumeration of circulating tumor cells. *Cytom. Part. A* **2007**, *71*, 154–162. [CrossRef]
54. Miyamoto, D.T.; Lee, R.J.; Stott, S.L.; Ting, D.T.; Wittner, B.S.; Ulman, M.; Smas, M.E.; Lord, J.B.; Brannigan, B.W.; Trautwein, J.; et al. Androgen receptor signaling in circulating tumor cells as a marker of hormonally responsive prostate cancer. *Cancer Discov.* **2012**, *2*, 995–1003. [CrossRef] [PubMed]
55. Crespo, M.; Van Dalum, G.; Ferraldeschi, R.; Zafeiriou, Z.; Sideris, S.; Lorente, D.; Bianchini, D.; Rodrigues, D.N.; Riisnaes, R.; Miranda, S.; et al. Androgen receptor expression in circulating tumour cells from castration-resistant prostate cancer patients treated with novel endocrine agents. *Br. J. Cancer* **2015**, *112*, 1166–1174. [CrossRef] [PubMed]
56. Hvichia, G.; Parveen, Z.; Wagner, C.; Janning, M.; Quidde, J.; Stein, A.; Müller, V.; Loges, S.; Neves, R.; Stoecklein, N.; et al. A novel microfluidic platform for size and deformability based separation and the subsequent molecular characterization of viable circulating tumor cells. *Int. J. Cancer* **2016**, *138*, 2894–2904. [CrossRef]

57. Services, C.f.M.M. 2018. Available online: https://med.noridianmedicare.com/documents/10546/6990981/MolDX+Circulating+Tumor+Cell+Marker+Assays+LCD/8eaf89f0-9970-455c-b048-ebaeaf42bd7d (accessed on 30 July 2019).
58. Services C.f.M.M. 2018. Available online: https://www.cms.gov/medicare-coverage-database/details/lcd-details.aspx?LCDId=37914&ver=2&Cntrctr=All&UpdatePeriod=796&bc=AQAAEAAAAAAA& (accessed on 30 July 2019).

 © 2019 by the authors. Licensee MDPI, Basel, Switzerland. This article is an open access article distributed under the terms and conditions of the Creative Commons Attribution (CC BY) license (http://creativecommons.org/licenses/by/4.0/).

Article

Capture and Detection of Circulating Glioma Cells Using the Recombinant VAR2CSA Malaria Protein

Sara R. Bang-Christensen [1,2], Rasmus S. Pedersen [1], Marina A. Pereira [1], Thomas M. Clausen [1], Caroline Løppke [1], Nicolai T. Sand [1], Theresa D. Ahrens [1], Amalie M. Jørgensen [1], Yi Chieh Lim [3], Louise Goksøyr [1], Swati Choudhary [1], Tobias Gustavsson [1], Robert Dagil [1], Mads Daugaard [4], Adam F. Sander [1], Mathias H. Torp [5], Max Søgaard [6], Thor G. Theander [1], Olga Østrup [5], Ulrik Lassen [7], Petra Hamerlik [3], Ali Salanti [1,*] and Mette Ø. Agerbæk [1,*]

[1] Centre for Medical Parasitology at Department for Immunology and Microbiology, Faculty of Health and Medical Sciences, University of Copenhagen and Department of Infectious Disease, Copenhagen University Hospital, 2200 Copenhagen, Denmark
[2] VarCT Diagnostics, 2200 Copenhagen, Denmark
[3] Danish Cancer Society Research Center, 2100 Copenhagen, Denmark
[4] Department of Urologic Sciences, University of British Columbia, and Vancouver Prostate Centre, Vancouver, BC V6H 3Z6, Canada
[5] Centre for Genomic Medicine, Copenhagen University Hospital, 2100 Copenhagen, Denmark
[6] ExpreS²ion Biotechnologies, SCION-DTU Science Park, 2970 Hørsholm, Denmark
[7] Department of Oncology, Copenhagen University Hospital, 2100 Copenhagen, Denmark
* Correspondence: salanti@sund.ku.dk (A.S.); mettea@sund.ku.dk (M.Ø.A.)

Received: 8 July 2019; Accepted: 25 August 2019; Published: 28 August 2019

Abstract: Diffuse gliomas are the most common primary malignant brain tumor. Although extracranial metastases are rarely observed, recent studies have shown the presence of circulating tumor cells (CTCs) in the blood of glioma patients, confirming that a subset of tumor cells are capable of entering the circulation. The isolation and characterization of CTCs could provide a non-invasive method for repeated analysis of the mutational and phenotypic state of the tumor during the course of disease. However, the efficient detection of glioma CTCs has proven to be challenging due to the lack of consistently expressed tumor markers and high inter- and intra-tumor heterogeneity. Thus, for this field to progress, an omnipresent but specific marker of glioma CTCs is required. In this article, we demonstrate how the recombinant malaria VAR2CSA protein (rVAR2) can be used for the capture and detection of glioma cell lines that are spiked into blood through binding to a cancer-specific oncofetal chondroitin sulfate (ofCS). When using rVAR2 pull-down from glioma cells, we identified a panel of proteoglycans, known to be essential for glioma progression. Finally, the clinical feasibility of this work is supported by the rVAR2-based isolation and detection of CTCs from glioma patient blood samples, which highlights ofCS as a potential clinical target for CTC isolation.

Keywords: circulating tumor cells (CTCs); glioma; biomarker; rVAR2; malaria; enrichment and detection technologies

1. Introduction

Diffuse gliomas are the most common primary malignant brain tumors [1]. As the name implies, a general trait of these tumors is their diffuse invasion into the brain parenchyma, which impedes complete surgical resection and most likely explains the poor prognosis and frequent local recurrence [2]. A precise classification of diffuse gliomas is needed for the optimal diagnosis, stratification, and treatment of patients [3,4]. During the past decades, technologies for biopsy-based classification of diffuse gliomas have increased in their complexity [5–7]. However, repeated access to information regarding tumor progression remains challenging, due to the risk and inconvenience that are associated with performing patient brain biopsies. Several studies indicate that the cells constituting the infiltrative and invasive front of gliomas harbor tumor-initiating capacity and may be responsible for drug resistance and tumor recurrence [8–10]. Most likely, these migrating cells would also be the ones accessing the blood stream. Therefore, the isolation of circulating tumor cells (CTCs) from a liquid biopsy, such as blood or cerebrospinal fluid, could provide non-invasive, repeatable access to primary glioma cells for molecular analysis.

Tumors of the central nervous system were until recently not considered to be metastatic. However, organ recipients receiving organs from patients who succumbed to glioblastoma multiforme (GBM) have developed extracranial metastases, which strongly suggests that these organs harbored disseminated GBM cells [11]. In line with this, a few studies using different isolation and detection methods have detected CTCs in blood from glioma patients [12–15]. Taken together, these studies provide evidence that invasive glioma cells successfully intravasate to the blood circulation and may therefore potentially become an important and easily available source of information on the mutational and phenotypic state of the primary tumor. Molecular analysis of the circulating glioma cells could provide basis for the design and monitoring of personalized treatment strategies, as it has been the case with breast, prostate, and lung cancer [16–18]. However, the high degree of heterogeneity within gliomas constitutes a hindrance for the effective isolation and detection of such CTCs. The use of antibodies towards one or few protein surface markers will render the detection fragile to changes in the expression level of the selected marker. On the other hand, targeting several proteins by using an antibody cocktail increases the risk of false positives and high background levels due to healthy cells expressing one or more of the included markers. Hence, a single marker to distinguish a broad repertoire of glioma CTCs from healthy white blood cells (WBCs) is needed.

Notably, little attention has been given to cancer specific glycosylation patterns on CTCs and strategies for targeting these. Glycosaminoglycans (GAGs) are carbohydrate structures, which are added to proteins, called proteoglycans, as secondary modifications. Chondroitin sulfate (CS) is one type of GAG that is built up by repeated disaccharide units made up of N-acetyl-D-galactosamine and D-glucuronic acid units [19]. While the CS backbone structure is simple, an immense heterogeneity is achieved through additional modifications, such as alternate sulfation of component hydroxyl groups [20]. The long structures of repeated disaccharide units are implicated in the regulation of many oncogenic processes and CS up-regulation or modifications have been associated with cancer progression [21]. In the case of glioma, several chondroitin sulfate proteoglycans (CSPGs), including versican and NG2/CSPG4, have been shown to be up-regulated and involved in tumor cell growth, migration, and invasion, as well as in promoting angiogenesis [22–24].

We have previously shown that the recombinantly expressed VAR2CSA malarial protein (rVAR2) specifically binds a distinct CS structure, termed oncofetal chondroitin sulfate (ofCS), which is present in the placenta and on almost all cancer cells with limited expression in other normal tissues [25]. Although CS is present elsewhere in the vasculature, parasite infected erythrocytes that express VAR2CSA only bind in the placenta [26]. Thus, the protein has been evolutionary refined to specifically bind to ofCS and not to CS present in other organs.

We recently published a CTC isolation method demonstrating the use of rVAR2 protein on magnetic beads for the capture of CTCs from prostate, pancreatic, and hepatic cancer patient blood samples [27]. However, in terms of CTC detection after enrichment, this assay was still dependent on antibody staining using the epithelial marker cytokeratin (CK), thus limiting the applicability to cancers of epithelial origin. In this study, we investigate whether the rVAR2 protein can be applied in both the capture and detection step and thereby broaden the use of our CTC-isolation platform to include circulating glioma cells. We show that rVAR2 binds glioma cells of both adult and pediatric origin. We find that rVAR2 interacts with ofCS on several CSPGs that have shown to be up-regulated in GBM, including CD44, APLP2, CSPG4, PTPRZ1, versican, and syndecan 1. Furthermore, we confirm that rVAR2 binding is retained on a low-grade pediatric glioma cell line (Res259) after incubation with

Transforming Growth Factor beta (TGF-β). We validate that the rVAR2-based CTC capture enables capture of rare glioma cells spiked into blood, and show proof-of-concept of using rVAR2 for both the capture and downstream detection of such glioma cells. Importantly, we capture and detect glioma CTCs from glioma patient blood samples. Finally, CTCs from three patient samples are analyzed by whole exome sequencing (WES), which confirms the presence of glioma-associated mutations.

2. Materials and Methods

2.1. Production of Proteins

The recombinant DBL1-ID2a subunit or the shorter version, ID1-ID2a, of VAR2CSA (rVAR2) was expressed in SHuffle T7 Express Competent *E. coli* (NEB) and purified using affinity chromatography (HisTrap HP, GE Healthcare, Uppsala, Sweden), followed by cation exchange chromatography (HiTrap IMAC SP HP, GE Healthcare). Both constructs included a C-terminal 6x His-tag and V5-tag, as well as an N-terminal SpyTag. For the staining of CTCs, we produced the recombinant ID1-ID2a subunit of VAR2CSA in S2 insect, which encoded an N-terminal twin-strep affinity tag. Protein that was expressed in S2 cells was captured from the supernatant by Streptactin XT chromatography (Iba, GmbH, Germany) and polished by size exclusion (Superdex 200pg, GE).

Subsequently, purified monomeric proteins were identified by SDS-PAGE. All of the proteins were quality tested by decorin binding in ELISA and by ofCS binding on cancer cells using flow cytometry to ensure specificity.

The SpyCatcher domain was produced in *E. coli* BL21 as a soluble poly-HIS tagged protein, and purified using affinity chromatography (HisTrap, GE Healthcare), followed by anion exchange (HiTrap IMAC Q HP column, GE Healthcare). Purity was determined by SDS page and quality of protein was ensured by testing the capacity to form an isopeptide bond to the Spy-tagged rVAR2 protein. The SpyCatcher was biotinylated using NHS-biotin (Sigma-Aldrich, Steinheim, Germany). NHS-Biotin was dissolved in DMSO and added in 10 molar excess to the SpyCatcher. After a 1-h incubation at room temperature, the biotinylated SpyCatcher was purified using a zeba spin column with a 7 kDa cut off.

2.2. Cell Cultures

Janine Erler and Lara Perryman (Biotech Research & Innovation Centre, University of Copenhagen, Denmark) kindly provided the KNS-42, Res259, U87mg, and U118mg cell lines [28]. The U87mg cells were grown in EMEM, Res259 and KNS-42 were grown in DMEM/F12, and U118mg were grown in DMEM GlutaMAX. All culture media were supplemented with 10% fetal bovine serum, penicillin, streptomycin, and L-glutamine (except DMEM). The primary GBM cell, GBM02, was maintained as an in vivo model in NOG mice with ethical approval (2012-15-2934-00636). Tumor xenograft was dissociated using a papain dissociation kit (Worthington). Isolated ex-vivo GBM02 cells were authenticated by STR profiling and grown as neurospheres in Neurobasal media containing B-27 supplement (Gibco), GlutaMax (Gibco), 10 ng/mL EGF, and 10 ng/mL FGF, as described previously [29]. All of the cell lines were passaged at a regular basis and maintained at 5% CO_2 at 37 °C.

2.3. Flow Cytometry

The cells were grown to 70–80% confluency in appropriate growth media and then harvested in an EDTA detachment solution (Cellstripper®, Corning™). 100,000 cancer cells, WBCs from 100 µL RBC lysed blood, or a mixture of both (according to the description in the Results section) were added to each well in a 96 well plate. Cells were incubated with rVAR2 (400 nM–25 nM) for 30 min. at 4 °C. Subsequently, cells were washed twice and then incubated with FITC-labelled anti-V5 antibody (Invitrogen, 1:500) for 30 min. at 4 °C. Finally, the cells were washed twice and analyzed in a LSR-II (BD Biosciences) for staining intensity. Geometric mean fluorescent intensity (MFI) values were normalized to signals that were obtained when only adding the FITC-labelled anti-V5 antibody.

2.4. TGF-β Treatment of Res259 Cells

Res259 were seeded in a density of 2400–5200 cells/cm^2 in DMEM/F12 that was supplemented with 10% FBS in a T25 culture flask. Cells were allowed to attach for 24 h. After this, cells were treated with TGF-β (Cat. no. T7039, Sigma-Aldrich) at a concentration of 20 ng/mL or equal volumes of TGF-β suspension buffer as control (0.2 µm filtered distilled water) for 72 h to induce the mesenchymal transition. Transition was confirmed by changes in the expression of mesenchymal protein markers using western blot as well as changes in morphology.

For western blot analysis, the cells were lysed with EBC lysis buffer containing PhosSTOP (Sigma-Aldrich) and cOmplete EDTA-free Protease Inhibitor Cocktail (Roche) for 30 min. Protein extract was balanced using Bradford assay. An equal amount of protein lysates were loaded onto a NuPAGE 4–12% Bis-Tris gel (ThermoFisher Scientific), after which the samples were transferred to a nitrocellulose membrane (Biorad). Membranes were blocked in 5% skimmed milk powder in TBS-T. Anti-GAPDH (14C10) antibody (Cell Signaling, 1:1000), and anti-β-catenin (1:500), anti-N-cadherin (1:500), and anti-vimentin (1:1000) primary antibodies from the EMT Antibody Sampler Kit (Cell Signaling) were added to the membranes in TBST-T supplemented with 2% skimmed milk powder and incubated overnight at 4 °C. Following three washes in TBS-T, the membranes were incubated with HRP-linked goat anti-rabbit IgG (Cat. no. P0448, Dako, 1:2000) for 1 h at room temperature and the reactivity was detected using LumiGlo Reserve Chemiluminescent Substrate (KPL). Uncropped images of the membranes can be seen in Figure S6.

For fluorescent visualization of changes in morphology, the cells were grown on glass slides and fixed in 4% paraformaldehyde (PFA), washed three times in PBS, blocked with 1% BSA in PBS, and stained with Alexa Fluor® 594 Phalloidin (ThermoFisher, 1:40) for 20 min. at room temperature. Cells were subsequently stained with DAPI (Life Technologies) and mounted using FluorSave Reagent (Merck Millipore, Darmstadt, Germany). Staining was analyzed using Nikon TE2000-E C1 confocal microscope with 60× oil immersion objective lens (DIC).

For flow cytometry-based analysis of cells before and after induction with TGF-β, applying the same procedure as in "2.3. Flow Cytometry".

2.5. Immunoprecipitation and Proteomics

Membrane proteins were extracted by lysing the cells with EBC lysis buffer supplemented with a protease inhibitor cocktail (Roche). Biotinylated rVAR2 was added to the lysate and the mix was incubated overnight at 4 °C. The rVAR2 and bound protein was pulled down on streptavidin-coated dynabeads (MyOne C1, Invitrogen).

The pulled down lysate was dissolved in non-reducing LDS loading buffer (Invitrogen). The protein was reduced in 1 mM DTT and alkylated with 5.5 mM iodoacetamide. The samples were then run 1 cm into Bis-Tris gels and stained with coomasie blue. The protein was cut out, washed, and in-gel digested with trypsin. The resulting peptides were captured and washed using a C18 resin stage-tipping [30]. The peptides were sequenced using a Phusion Orbitrap Mass Spectrometer. Sample analysis and hit verification was performed using the MaxQuant software. All of the samples were verified against the control samples of cell lysates without rVAR2.

2.6. Proximity Ligation Assay (PLA)

The PLA protocol was run according to the manufacturer's instructions (Sigma-Aldrich). U87mg, U118mg, and KNS-42 cells were seeded on laminin-coated coverslips and fixed in 4% PFA. Unspecific binding of antibodies was minimized by incubating with a blocking solution with 1% BSA and 5% FBS in PBS for 1 h at room temperature. The samples were incubated with primary antibodies together with rVAR2 or rDBL4 over night at 4 °C in the following concentrations: rVAR2 (50 nM), rDBL4 (50 nM), anti-NRP1 (Cat. no. ab81321, 1:250), anti-NRP2 (Cat. no. sc-13117, 1:50), anti-PTPRZ1 (Cat. no. HPA015103, 1:61), anti-VCAN (Cat. no. HPA004726, 1:50), anti-CSPG4 (Cat. no. ab20156,

1:200), anti-DCN (Cat. no. PA5-27370, 1:100), anti-CD44 (Cat. no. BBA10, 1:200), anti-SDC1 (Cat. no. ab34164, 1:50), and anti-SDC4 antibody (Cat. no. HPA005716, 1:80). Between incubations, the cells were washed in Wash Buffer A (DUO82049, Sigma-Aldrich). An anti-V5 antibody (Invitrogen, 1:500) was used to detect rVAR2. The cells were then stained with Duolink® In Situ PLA® Probe Anti-Mouse MINUS (DUO92004) and Duolink® In Situ PLA® Probe Anti-Rabbit PLUS (DUO92002) diluted in Antibody Diluent (DUO82008). The cells were then treated with the ligation solution, followed by incubation with the amplification solution, which were provided with the kit Duolink® In Situ Detection Reagents Orange (DUO92007). The cells were washed with Wash Buffer B (DUO82048). The slides were mounted using Duolink® In Situ Mounting Medium with DAPI (DUO82040), and the results were then analyzed under a Nikon TE2000-E C1 confocal microscope with a 60× oil objective. A total of 75–100 cells were imaged per sample. The images were analyzed using the BlobFinder software (version 3.2.). Negative controls using the recombinant DBL4 domain of VAR2CSA are found in Figure S1.

2.7. Immunocytochemistry of Cancer Cells Mixed with White Blood Cells (WBCs)

Two mL blood from a healthy donor was drawn in a LBgard® vacutainer (BioMatrica). The blood sample was diluted 10 times in Red Blood Cell (RBC) lysis buffer resulting in a final concentration of 0.155M ammonium chloride, 0.01M potassium hydrogen carbonate and 0.1 mM EDTA, and incubated for 13 min. After centrifugation at 400× g for 8 min. the pelleted cells were resuspended in 2 mL Dulbecco's PBS (without Ca^{2+} and Mg^{2+}) supplemented with 2% Fetal Bovine Serum (FBS) (Gibco) and transferred to eppendorf tubes in aliquots of 0.5 mL. Res529, KNS-42 and U87mg were detached using 1 mL CellStripper (Corning™) and resuspended in their respective media. Approximately 2000 cells were added to each aliquot of WBCs. The samples were washed once, prior to incubation with a cocktail of 200 nM rVAR2, CF488-labelled anti-V5 (Cat.no. 20440, Biotium, 1:150), PE-labelled anti-CD45 [5B1] (Cat.no. 170-078-081, MACS Miltenyi Biotec, 1:40) and PE-labelled anti-CD66b [REA306] antibodies (Cat.no. 130-104-396, MACS Miltenyi BioTec, 1:40) in Dulbecco's PBS with 2% FBS for 30 min. at 4 °C. Finally, the cells were fixed with 4% PFA, stained with DAPI (Cat. no. D1306, Life Technologies) and mounted using Faramount Aquous Mounting Media (Dako). The slides were imaged using the 10× objective of Cytation 3 Cell Imaging Multi-Mode Reader (BioTek, Europe).

2.8. Preparation of rVAR2-Coated Beads

The Spy-tagged DBL1-ID2a or ID1-ID2a (rVAR2) was mixed with the biotinylated SpyCatcher in a 1.2:1 ratio and then incubated at room temperature for 1 h. After this step, the biotinylated rVAR2 protein was incubated with CELLection™ Biotin Binder Dynabeads® (4.5 μm) at room temperature for at least 30 min. resulting in rVAR2-coated beads (0.43 μg biotinylated protein per μL bead suspension). The remaining protein or antibody was removed by carefully washing the beads in Pierce™ Protein-Free (PBS) blocking buffer (Cat. no. 37572, ThermoFisher) three times, each time using a neodymium magnet (10 × 12 mm) for dragging beads into a pellet.

2.9. Spike-In Experiments

Prior to the spike-in experiments, the cancer cells were harvested with CellStripper (Corning™) or TrypLe (Cat. no. 12604013, Gibco) (only used for GBM02 cells) and resuspended in culture medium. For spike-in experiments measuring the efficiency of recovery, cancer cells were prestained using CellTracker™ Green CMFDA Dye (Cat. no. C7025, ThermoFisher), according to manufacturer's protocol for cells in suspension. Following staining, the cells were resuspended in complete growth media and incubated for 30 min. under normal growth conditions in order to recover.

Cell concentration was measured by manually counting the number of viable cells in a 1:1 mixture with Trypan Blue solution (Sigma-Aldrich). Subsequently, the cells were diluted to 10,000 cells/mL in Dulbecco's PBS and the desired number of cells were spiked into 3 mL blood. Triplicates of the spike-in volume (e.g., 10 μL for 100 cells) were placed on a glass slide and cells were manually counted under a light microscope (10× objective) in order to confirm the exact number of cells spiked into the blood. The average of the cell counts was used when calculating the percentage of recovery. Each of the spike-in experiments were repeated at least twice with 2–4 replicates per test.

When spiking in low cell numbers (5–10 cells), serial dilutions were made using cell culture media. The cell suspension was transferred to a 96-well plate and counted under a light microscope (10× objective) to ensure precise cell count before spike-in. Finally, cells were directly added from the well to the 3 mL blood sample.

After adding the cancer cells to the blood, the samples were immediately processed, as described in Section 2.11.

2.10. Patient Samples

Up to 9 mL blood samples from glioma patients were collected under ethical approval (journal no. H-3-2009-136). Informed written consent was obtained for all of the enrolled subjects. Blood was received in K2 EDTA-tubes and processed within 2 h of collection.

2.11. CTC Isolation from Blood

Three mL blood samples were lysed in 27 mL RBC lysis buffer reaching a final concentration of 0.155 M ammonium chloride, 0.01 M potassium hydrogen carbonate, and 0.1 mM EDTA for 13 min. After centrifugation at 400× g for 8 min., the cell pellet was gently washed in Dulbecco's PBS. The centrifugation step was repeated, and finally the cells were resuspended in 0.6 mL Pierce™ Protein-Free (PF) PBS blocking buffer (Cat. no. 37572, ThermoFisher) and then transferred to a low retention microcentrifuge tube (Fisherbrand). Under these conditions, the cells were incubated with ~1.8 million rVAR2-coated magnetic beads at 4 °C for 20 min. A neodymium cylinder magnet was used to drag cells bound to beads towards the side of the tube, enabling removal of supernatant. Cells were then fixed in 4% PFA for 5 min. and resuspended in Pierce™ Protein-Free (PBS) blocking buffer diluted 1:10 in Dulbecco's PBS.

2.12. CTC Staining and Enumeration

For spike-in experiments where the cells were prestained with CellTracker Green, cells were stained with DAPI (Cat. no. D1306, Life Technologies) diluted in Dulbecco's PBS with 0.5% BSA and 2 mM EDTA for 5 min. at room temperature. Following one wash, the bead-bound cells were added to a SensoPlate™ (24-well, glass bottom) (Cat. no. 662892, Greiner Bio-One). Excess liquid was removed by holding the bead-bound cells in place with a magnet underneath the plate, and the samples were mounted using Faramount Aqueous Mounting Media (DAKO). The entire well was scanned for DAPI and CellTracker signal using the Cytation™ 3 Cell Imaging Multi-Mode Reader and manually enumerated using the Gen5 software (BioTek).

For patient samples and spike-in experiments with non-prestained cells, the isolated cells were briefly blocked in Dulbecco's PBS containing 2% FBS, followed by incubation in a non-protein based blocking solution. The cells were then incubated with a mixture of 200 nM fluorophore-conjugated (Oregon Green® 488) rVAR2 and PE-labelled anti-CD45 [5B1] (Cat. no. 170-078-081, MACS Miltenyi Biotec, 1:10) and PE-labelled anti-CD66b [REA306] (Cat. no. 130-104-396, MACS Miltenyi BioTec, 1:20) antibodies for 30 min. at room temperature and then washed once in PBS with 2% FBS to remove excess staining reagents. Finally, the cells were DAPI-stained and mounted on a Sensoplate. Duplicates of 3 mL patient blood were imaged using the Cytation™ 3 Cell Imaging Multi-Mode Reader. Additional 1–2 mL blood was processed according to the above description, except that the exclusion marker that was used in this setup was APC-labelled anti-CD45 [HI30] antibody (Cat. no. 17-0459-42, ThermoFisher), and analyzed using the CellCelector™ (Automated Lab Solutions).

For spike-in experiments testing anti-CSPG4 antibody as a staining reagent after rVAR2 bead pull-down, cells were blocked in 10% Normal Donkey Serum (NDS), 0.5% BSA, and 2 mM EDTA in Dulbecco's PBS for 10 min. After this, the cells were incubated with anti-CSPG4 (Cat. no. ab20156,

Abcam, 1:100) antibody diluted in Dulbecco's PBS with 1% NDS, 0.5% BSA, and 2 mM EDTA for 30 min. at room temperature. The cells were washed once in Dulbecco's PBS containing 0.5% BSA and 2 mM EDTA and then incubated with anti-mouse IgG-FITC (Vector, Cat. no. FI-2000, 1:400) for 30 min. at room temperature. Finally, the cells were washed in Dulbecco's PBS containing 0.5% BSA and 2 mM EDTA, DAPI stained, and mounted on a Sensoplate.

2.13. Classification and Enumeration of rVAR2-Stained Cancer Cells or CTCs

The samples were scanned on a 10× objective using the Cytation™ 3 Cell Imaging Multi-Mode Reader and manually enumerated using the Gen5 software (BioTek). Putative CTCs were defined as DAPI+, CD45/CD66b−, and rVAR2+. The signal to noise ratios were adjusted according to the fluoresence of the CELLection Biotin Binder beads, such that a staining was only regarded as positive if the intensity was above the fluorescence from the beads. Furthermore, all the cells with a DAPI area below 4 μm were excluded from enumeration.

2.14. Single Cell Picking and Whole Genome Amplification

Cell samples were resuspended in 200 μL PBS and then loaded onto a CellCelector™ magnetic slide (Automated Lab Solutions) to align and preserve the localization of the magnetic beads and cells during scanning. Employing the CellCelector™, the samples were then screened for coinciding Origon Green® 488 and DAPI signals as well as absent APC fluorescent signals, thereby detecting potential CTCs. Single cells were picked by the CellCelector™ and then pooled into PCR tubes containing 5 μL lysis buffer and enzyme from the MALBAC® Single Cell WGA Kit (Yikon Genomics, Cat. no. EK100101210). Each tube was prepared with 10–20 cells. Whole genome amplification (WGA) was performed on picked CTCs or WBC controls, according to manufacturer's instructions. The quality of the WGA products was verified by agarose gel electrophoresis and concentrations were measured by Qubit™ dsDNA BR Assay Kit (Cat. no. Q32850, Thermofisher Scientific).

2.15. Whole Exome Sequencing

Whole exome sequencing on whole genome amplified DNA from isolated CTCs and patient-matched WBCs was performed as previously described [31]. The Exome sequencing data was aligned against the human reference genome (hg19/GRCh37) using bwa mem 0.7.15 and somatic SNVs and small indels were called using Mutect2 according to the GATK best practices for somatic short variant discovery using GATK 4.1.0.0. Variants outside a selected glioblastoma-related target region containing 95 candidate genes were excluded from the call set. Mutect2 was provided with data from picked, patient-specific WBCs as a matched reference sample to reduce the amount of germline variants in the call set, hence obtaining a list of somatic mutations only for the CTC samples. The most relevant mutations (described in somatic mutation databases or being frameshift/stop-gain) were further manually inspected by looking at aligned reads sequences. The Integrative Genomics Viewer (Broad Institute, UK) was used for the visualization of variants (Figure S4) [32].

3. Results

3.1. rVAR2 Binds to ofCS on Glioma Cells

We have previously shown that the rVAR2 protein interacts with ofCS present on cancer cell lines representing almost all known cancers [25,33,34]. We tested rVAR2 binding to a panel of cell lines, including low-grade (WHO grade II) diffuse glioma (Res259) as well as high-grade (WHO grade IV) GBM (U87mg, KNS-42, and U118mg), to test for the presence of ofCS in glioma. All of the cell lines were positive for rVAR2 binding by flow cytometry, indicating that the glioma lines expressed ofCS (Figure 1A and Figure S2).

Cell cultures poorly represent the phenotypic plasticity of cancer cells in vivo, where the tumor cells continuously respond to signals from the microenvironment. TGF-β, for instance, is known to enhance the migratory and invasive capability of glioma cells, most likely by pushing these cells towards a more mesenchymal phenotype [35,36]. Therefore, we evaluated whether TGF-β exposure of the low-grade Res259 glioma cell line affected the expression of ofCS, as measured by rVAR2 binding. It should be noted that, although glial cells originate from ectodermal tissue, these cells exhibit a more mesenchymal appearance, such as the expression of vimentin [28]. Thus, the transition is measured as an increased expression of mesenchymal markers, rather than a down-regulation of epithelial markers [36]. After incubation with TGF-β, the Res259 cells showed increased expression of the mesenchymal markers β-catenin and N-cadherin in accordance with a transition towards a more mesenchymal state (Figure 1B). This was accompanied by a clear change in morphology as cells tended to become more elongated, which confirms the occurrence of a transition [36,37] (Figure 1C). Importantly, when testing for rVAR2 binding to Res259 cells in flow cytometry before and after incubation with TGF-β, rVAR2 binding was not reduced (Figure 1D).

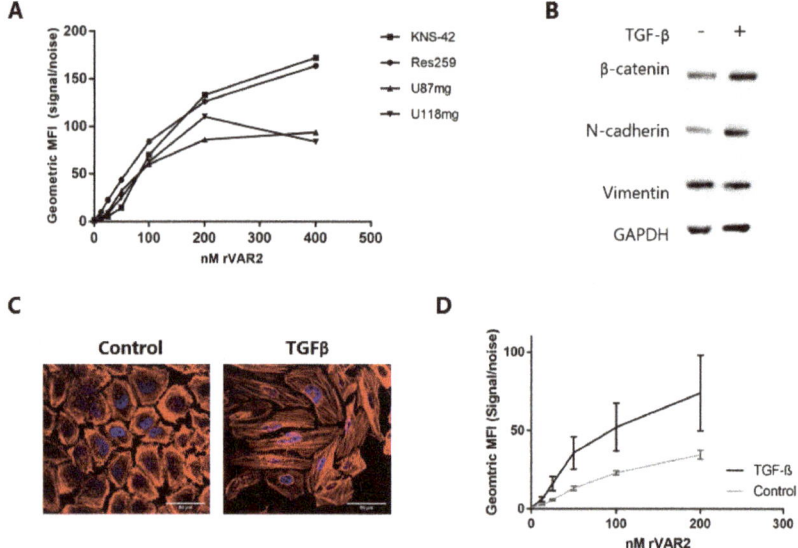

Figure 1. Recombinant malaria VAR2CSA protein (rVAR2) binds to glioma cells and the binding is unaffected by phenotypic changes. (**A**) rVAR2 binding to the glioma cell lines KNS-42, Res259, U87mg and U118mg was measured by flow cytometry using a FITC-conjugated anti-V5 antibody. Geometric Mean Fluorescence Intensity (MFI) was measured after incubation of cells with various rVAR2 concentrations. Results are displayed as signal/noise ratio. Figure represents data from one experiment, replicates are found in Figure S2. (**B**) Western blot of Res259 cell lysates after 72 h incubation with Transforming Growth Factor-beta (TGF-β) or buffer control. Membranes were incubated with anti-β-catenin, anti-N-cadherin, anti-Vimentin or anti-GAPDH antibodies and detected by anti-rabbit HRP antibody. (**C**) Representative images of fixed Res259 cells after 72 h incubation with TGF-β or buffer control. Cells were stained with phalloidin to stain F-actin (red) and DAPI to stain nuclei (blue). Scale bars, 50 μm. (**D**) rVAR2 binding to Res259 incubated with TGF-β or buffer control for 72 h measured by flow cytometry ($p < 0.001$, generalized least squares regression model). Geometric MFI was measured after incubation of cells with various rVAR2 concentrations and a FITC-conjugated anti-V5 antibody. Results are displayed as signal/noise ratio. Bars show standard deviation (n = 3).

3.2. rVAR2 Captures Glioma Cancer Cells Spiked Into Blood

To examine whether rVAR2 can be used for targeting circulating glioma cells, it is pivotal to ensure the specificity of rVAR2 binding to glioma cells in a background of normal white blood cells

(WBCs). Therefore, WBCs from 0.5 mL blood were mixed with 2000 U87mg, Res259, or KNS-42 cells and incubated with rVAR2 and a CF488-labeled anti-V5 antibody. rVAR2 binding showed a clear and specific membrane staining of all three cancer cell lines with minimal staining of the surrounding WBCs (Figure 2A). This was further tested by flow cytometry analysis showing specific binding to U87mg cells when mixed with WBCs (Figure 2B,C).

To examine whether rVAR2 binding to glioma cell lines could support magnetic capture and isolation of these cells from whole blood, 100 glioma cells were prestained using a CellTracker™ Green CMFDA Dye and spiked into a 3 mL blood sample from a healthy individual. This strategy allowed us to directly assess the recovery of spiked cells independent of downstream staining and detection biases. After lysis of the erythrocytes, cancer cells were isolated using rVAR2-coated magnetic beads (see Method section for details) [27]. By this procedure, we achieved an average recovery of 76%, 41%, 11%, and 64% for U87mg, Res259, KNS-42, and U118mg cells, respectively (Figure 2D). Furthermore, we spiked 3 mL blood samples with 100, 50, 10, or 5 U87mg cells to assess the sensitivity of the assay (Figure 2E). The average recovery ranged between 54–75% with no obvious association to the number of cells spiked into the blood ($p = 0.31$, one-way ANOVA). Although the recovery varied between the different cell lines, this data confirms that rVAR2 can be used as a capture molecule for the isolation of various glioma cell types.

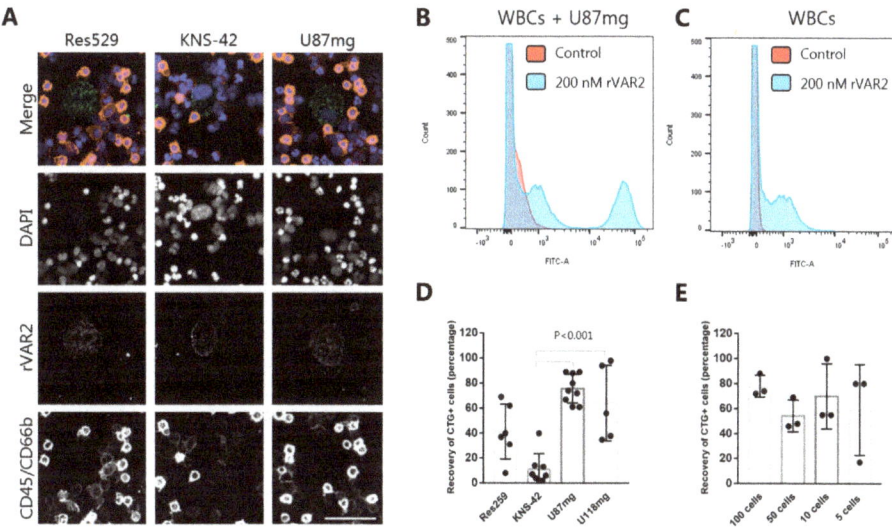

Figure 2. rVAR2 specifically binds to glioma cells and enables their retrieval from blood. (**A**) rVAR2 stains glioma cell lines in a background of white blood cells (WBCs). Res529 (left), KNS-42 (middle) and U87mg (right) cells were mixed with WBCs and stained with V5-tagged rVAR2 in combination with a CF488-conjugated anti-V5 antibody (green), PE-conjugated anti-CD45 and anti-CD66b antibodies (red), and DAPI (blue). Scale bar, 50 µm. (**B**) Flow cytometry analysis showing WBCs (100 µL RBC lysed blood) mixed with U87mg (50,000) cells and detected with either 200 nM rVAR2 and a FITC-conjugated anti-V5 antibody or with a FITC-conjugated anti-V5 antibody alone (control). (**C**) Same as in (**B**) but with no U87mg cells added. (**D**) Recovery of CellTracker Green-stained glioma cells from blood. 100 cells were spiked into 3 mL blood and recovered using rVAR2-coupled beads. Cells were stained with DAPI and scanned on the Cytation 3 Imager. Each dot represents the percentage of recovered cells from one sample. Bars represent mean recoveries and error bars show +/− standard deviation (n ≥ 2) (one-way ANOVA with Bonferroni correction). (**E**) Recovery of CellTracker Green-stained U87mg cells from blood. The indicated number of U87mg cells was spiked into 3 mL blood and captured using rVAR2-coupled beads. Enumeration of cells and data presentation were done as in (**D**).

3.3. rVAR2 Interacts with Several GBM-Associated Proteoglycans

We have previously described that single cancer cells simultaneously display the ofCS modification on several proteoglycans [33]. We analyzed rVAR2-based pull-down of lysates from KNS-42, U118mg, and U87mg cell lines to investigate the proteoglycan display on glioma cells. The mass spectrometry results showed the pull-down of multiple key cancer-related proteoglycans (Table 1). Among the hits were several chondroitin sulfate proteoglycans (CSPGs) that have been described for GBM, such as CSPG4, CD44, APLP2, and SDC1 [38–41]. To validate these findings, we studied the co-localization of ofCS and selected protein cores from the pull-down proteomic list by proximity ligation assay (PLA) (Figure 3A). Indeed, compared to other proteoglycans rVAR2 binding and CD44 showed a strong co-localization on each evaluated glioma cell line ($p < 0.001$, one-way ANOVA) (Figure 3B). Despite the high PLA signal, CD44 was not further examined as a potential CTC marker, as anti-CD44 antibodies are also found to target a subset of healthy WBCs [42]. Similarly, ofCS and CSPG4 were clearly co-localizing on U87mg and U188mg cells ($p < 0.001$, one-way ANOVA) (Figure 3B). Interestingly, anti-CSPG4 antibodies are already being used for the capture and detection of circulating melanoma cells [43–46]. Since CSPG4 is also an emerging target for GBM CAR-T immunotherapy, we examined whether CSPG4 is still accessible for antibody staining after rVAR2-based capture [47,48]. Indeed, the captured U87mg cells showed clear and specific CSPG4 staining in a background of WBCs (Figure 3C). Hence, the capture of glioma CTC might be useful for predicting response to anti-CSPG4 CAR-T therapy.

Table 1. rVAR2-based protein pull-down hits from cell lysates.

Protein Name	Gene	Peptides Count	Seq. Coverage (%)	Ratio to Neg
KNS-42				
Amyloid-like protein 2	APLP2	17	27	NA
CD44	CD44	9	37.4	41.17
Glypican 1	GPC1	10	23.7	NA
Glypican 4	GPC4	10	23	7.15
Integrin beta 1	ITGB1	11	15.2	7.50
Neuropilin 1	NRP1	12	21.1	NA
Neuropilin 2	NRP2	4	6	NA
Receptor-type tyrosine-protein phosphatase zeta	PTPRZ1	17	8.6	NA
Syndecan 1	SDC1	6	14.5	NA
Syndecan 2	SDC2	5	25.5	NA
Testican 1	SPOCK1	8	22.6	NA
Versican	VCAN	39	12.8	540.37
U118mg				
Amyloid-like protein 2	APLP2	4	5.2	NA
CD44	CD44	8	37.4	17.26
Decorin	DCN	8	29.8	41.89
Neuropilin 1	NRP1	9	15.9	16.74
Versican	VCAN	13	4	NA
U87mg				
Amyloid-like protein 2	APLP2	4	5.1	NA
Amyloid precursor protein	APP	3	4.3	NA
Carbonic anhydrase 9	CA9	1	3.3	NA
CD44	CD44	4	6.1	12.92
Chondroitin sulfate proteoglycan 4	CSPG4	7	3.1	NA
HLA class II histocompatibility antigen gamma chain	CD74	3	12.2	NA
Sushi repeat-containing protein SRPX	SRPX	7	15.7	NA
Syndecan-1	SDC1	3	15.2	9.71
Syndecan-4	SDC4	2	12.6	NA

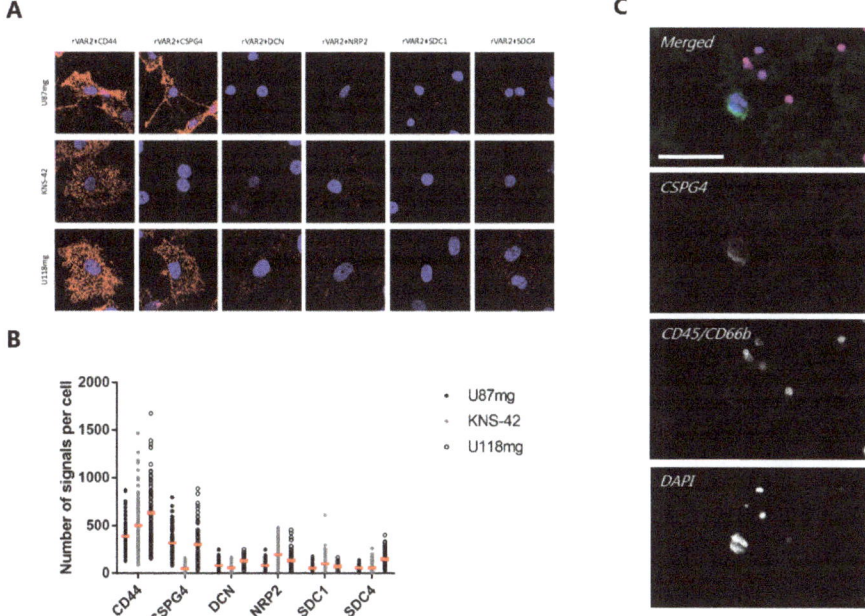

Figure 3. Evaluation of protein pull-down hits from mass spectrometry by proximity ligation assay (PLA). (**A**) Representative images of PLA assays on U87mg, KNS-42, and U118mg cells showing co-localization between rVAR2 and a panel of CSPGs as red dots. Cells were counterstained with DAPI (blue) and analyzed by confocal microscopy. All of the images are shown in same magnification using a 60× objective. (**B**) Quantification of the PLA co-localization signals between rVAR2 and each of the CSPGs analyzed. Data is shown as the number of signals per cell. Red bars represent the mean number of signals per cell. (**C**) Representative image showing specific CSPG4 staining (green) of an rVAR2-captured U87mg cell in a background of WBCs stained for CD45 and CD66b (both red). Cells were stained with DAPI (blue) and visualized on the Cytation 3 Imager with a 20× objective. Scale bar, 50 µm.

3.4. rVAR2 Detects Cancer Cells Spiked Into Blood Samples

All of the CTC capture protocols will, even in the best of circumstances, capture some normal WBCs along with the CTCs. A major challenge associated with the analysis of glioma CTCs is how to validate which captured cells are indeed cancer cells and not WBCs. With regard to carcinoma-derived CTCs the most widely used markers are EpCAM or CK. While gliomas are EpCAM negative, the results regarding CK-positivity are less consistent [12,49–51]. Thus, these markers are not optimal for the detection of glioma CTCs. Therefore, we established a platform where rVAR2-coupled beads were used for capturing, while a fluorophore-conjugated rVAR2 was used for microscopic detection of the captured cells. U87mg, Res259, and KNS-42 cells all showed rVAR2 staining after capture with rVAR2-coupled beads (Figure 4A and Figure S3). However, it was noticed that rVAR2-staining of U87mg after magnetic capture was somewhat reduced compared to the Res259 and KNS-42 cells. Next, we applied the same workflow to U87mg cells that were spiked into 3 mL healthy donor blood in order to mimic patient blood samples. The rVAR2 staining enabled detection of U87mg cells and their separation from CD45- and/or CD66b-positive WBCs (Figure 4B). The strategy was also effective with cells from a primary glioblastoma cell culture, GBM02 (Figure 4B).

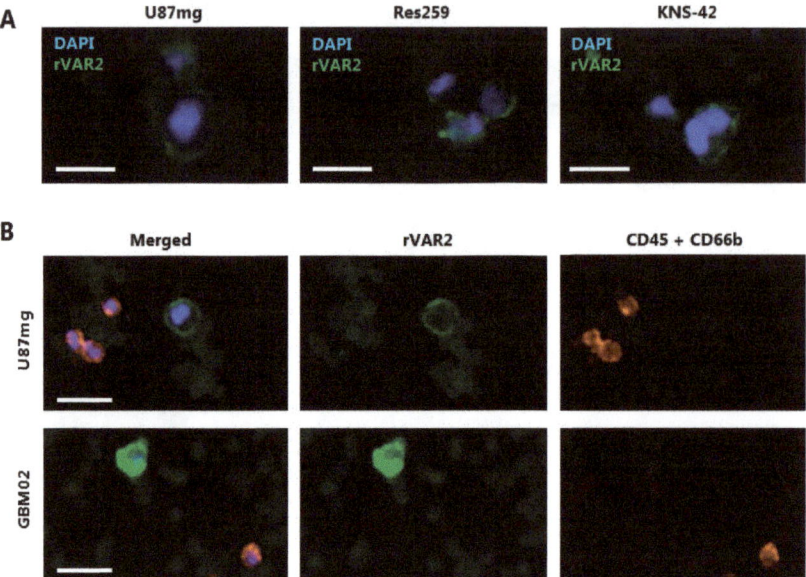

Figure 4. Using rVAR2 to stain glioma cells after capture with rVAR2-coupled beads. (**A**) Glioma cell lines (U87mg, Res259, and KNS-42) were incubated with rVAR2-coupled beads and stained with a fluorophore-conjugated rVAR2 (green) and DAPI (blue). Representative images were obtained using the Cytation 3 Imager with a 10× objective. Scale bars, 20 µm. (**B**) U87mg cells and GBM02 cells were spiked into 3 mL blood, retrieved using rVAR2-coupled beads, and stained using fluorescent rVAR2 (green), anti-CD45/CD66b antibodies (red) and DAPI (blue). Scale bars, 20 µm.

3.5. rVAR2 Captures and Detects CTCs in Glioma Patient Blood Samples

We tested for the presence of CTCs in blood samples from glioma patients using the combined rVAR2 capture and detection protocol. Duplicates of 3 mL blood samples from 10 glioma patients, suffering from oligodendroglioma (grade II), anaplastaic oligodendroglioma (grade III), or GBM (grade IV), were processed and visualized for enumeration. CTCs were manually enumerated as rVAR2+, CD45/CD66b−, and DAPI+ cells. The range of identified CTCs per 3 mL blood was 0.5–42 (Figure 5A). There was no obvious correlation between grade or type of diagnosis and CTC number. Interestingly, one patient who had progressed from an initial diagnosis of oligodendroglioma (grade II) to anaplastic oligodendroglioma (grade III) within a time span of 15 years had a relatively high number of CTCs (22 CTCs per 3 mL blood). In a patient with the reverse clinical history regressing from an initial diagnosis of anaplastic oligodendroglioma (grade III) to eight years later having oligodendroglioma (grade II) we detected an average of only 0.5 CTCs per 3 mL. Representative images of rVAR2+ cells from one of the GBM patients are shown (Figure 5B) and the full list of detected CTCs is found in Figure S5.

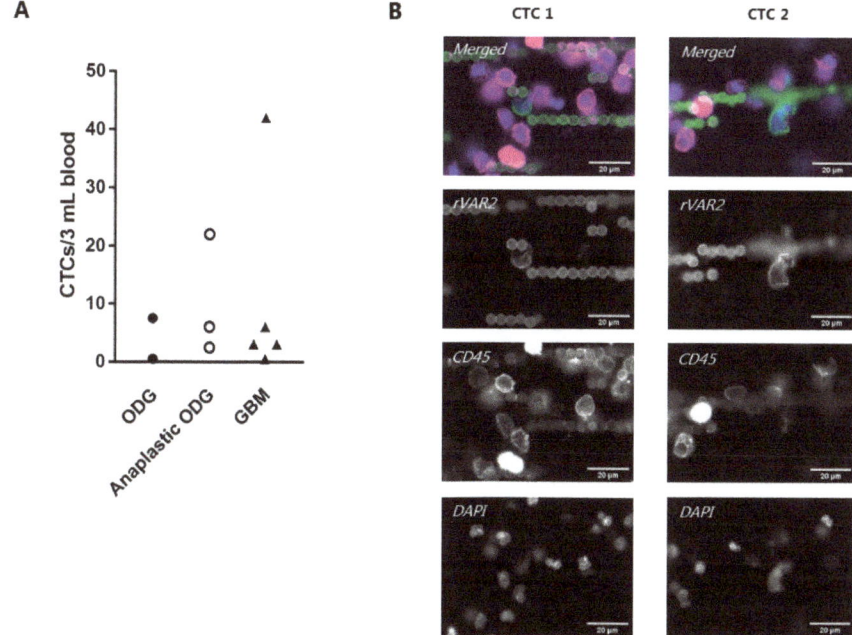

Figure 5. rVAR2 enables capture and detection of glioma circulating tumor cells (CTCs) from patient blood samples. (**A**) Average CTC count in blood samples from ten glioma patients. 3 mL blood samples were processed by using rVAR2-coupled beads followed by staining using a mixture of fluorophore-conjugated rVAR2 (green), anti-CD45/CD66b antibodies (red) and DAPI (blue). CTCs were defined as rVAR2+, CD45/CD66b−, DAPI+ cells. Each dot represents the average number of detected CTCs per 3 mL patient blood sample. The x-axis shows whether the patient was diagnosed with GBM, Anaplastic Oligodendroglioma (ODG), or Oligodendroglioma (ODG). (**B**) Representative images of identified CTCs from a patient diagnosed with GBM. The sample was stained with flurophore-conjugated rVAR2 (green), anti-CD45 antibody (magenta), and DAPI (blue). Images were obtained using the CellCelector™ (ALS) with a 40× objective. Scale bars, 20 µm.

3.6. Captured Glioma CTCs Show Cancer-Indicative Mutations

To confirm that the VAR2+, CD45− cells detected in the patient blood samples were indeed glioma-derived CTCs, we performed targeted whole exome sequencing (WES) searching for glioma relevant mutations. For three patient samples we single cell picked rVAR2+, CD45− cells, and patient-matched WBCs as germline controls using an ALS CellCelector™. For each patient, 2–4 CTCs were pooled into one sample and whole genome amplification (WGA) was performed (Figure 6). However, since WBCs were located close to some of the selected CTCs, the cell picking procedure resulted in samples containing CTCs together with some WBCs (Table 2). The WGA product was then used for WES. The WES results were filtered to only include glioma relevant mutations and each hit was visually confirmed by evaluating the IGV screen shots (Figure S4). Indeed, we identified genes with cancer-indicative mutations in all CTC samples: RB1, TP53/EPM2AIP1, and TP53/ALK for patient 1, 3, and 4, respectively (Table 2). Thus, the molecular profiling supports the tumor origin of the picked patient-derived CTCs.

Table 2. Mutations detected in patient CTCs by whole exome sequencing (WES).

	Patient Information					Confirmed Mutation by WES				
ID	Diagnosis	Molecular Features in Tumor Biopsy	Sample	CTCs	WBCs	Gene	Transcript ID	Transcript Variant	Allele Fraction (%)	Protein Variant	Translation Impact
1	Anaplastic oligodendro-glioma	IDH1 mutation LOH 1p/19q MGMT methylation	1	2	19	RB1	NM_000321.2	c.1644delA	16.88	p.K548fs*3	Frameshift
			2	3	34	RB1	NM_000321.2	c.1644delA	1.59	p.K548fs*3	Frameshift
3	GBM	IDH1 wild type	1	2	23	TP53	NM_000546.5	c.892G>T	21.39	p.E298*	Stop-gain
						EPM2AIP1	NM_014805.3	c.128G>T	54.17	p.R43L	Missense
4	Anaplastic oligodendro-glioma	IDH1 mutation 1p/19q deletion MGMT methylation	1	4	14	TP53	NM_000546.5	c.493C>T	31.43	p.Q165*	Stop-gain
						ALK	NM_004304.4	c.3824G>T	33.33	p.R1275L	Missense

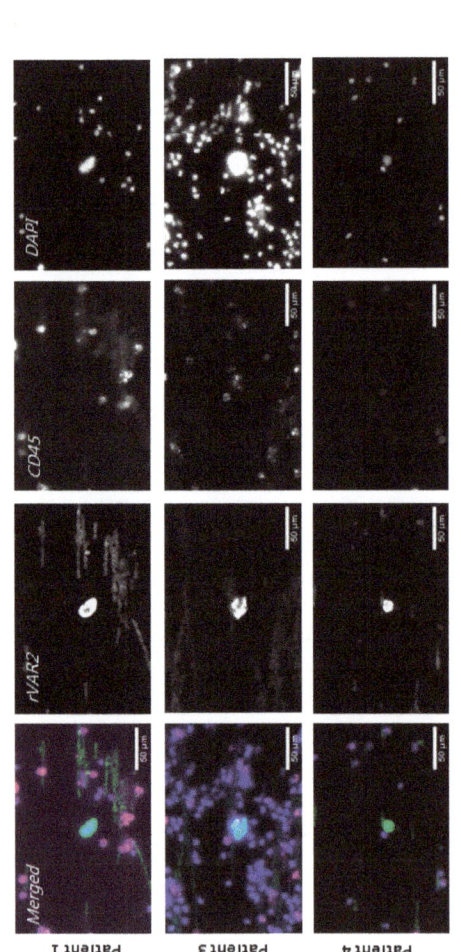

Figure 6. Identification of CTCs for whole exome sequencing (WES). Representative images of identified CTCs from three patients. Cells were stained with fluorophore-conjugated rVAR2 (green), anti-CD45 antibody (magenta) and DAPI (blue) and classified as CTCs if rVAR2+, CD45−, and DAPI+. Images were obtained using the CellCelector™ (ALS) with a 10× objective. Scale bars, 50 μm.

4. Discussion

The isolation and characterization of glioma CTCs have proven to be challenging, especially concerning the detection and validation of the tumor origin of the isolated cells. To date, only a few studies have shown the presence of circulating glioma cells utilizing either a single antibody marker or an antibody mixture for CTC detection [13,14]. Here, we present a novel strategy for glioma CTC capture and detection based on targeting the unique cancer-specific glycosaminoglycan structure ofCS. We show that rVAR2, which binds ofCS with high affinity, specifically targets a panel of glioma cell lines in a background of white blood cells (WBCs). Furthermore, we show that rVAR2 can be used for the capture of glioma cells that are spiked into blood by coupling the protein to magnetic beads. In addition, the staining of glioma cells with a fluorophore-conjugated rVAR2 after magnetic pull-down facilitates their detection and separation from WBCs. This workflow was applied to blood samples that were derived from ten glioma patients and established proof-of-concept for identification of glioma CTCs. In three of the patients, potential CTCs were picked and molecular analysis supported their tumor origin.

Flow cytometry analysis showed rVAR2 binding to all tested glioma cell lines. Interestingly, the cell lines showed varying maximum intensity at saturation indicating different levels of ofCS display. In all of the experiments, U87mg showed the lowest level of rVAR2-binding. This is interesting, since U87mg cells had the highest recovery when spiked into blood and isolated with rVAR2-coupled beads. This could indicate that efficiency of recovery, not only depends on the level of target expression but is also influenced by other factors, such as the capability of a given cell line to survive through the experimental workflow. When using spike-in of cancer cells in healthy donor blood, the experimental procedure, among others, includes detachment from the culture plate and exposure to the various components of a foreign immune system when spiked into blood. Patient-derived CTCs do indeed experience dramatic changes in physical conditions upon entering the circulation, such as shear stress forces and the loss of cell-cell or cell-matrix attachment. However, it could be debated how comparable this sequence of events is to the in vitro spike-in models, and thus how well cell line spike-in samples reflect the phenotypes of CTCs in patient-derived liquid biopsies.

Elevated levels of TGF-β in the tumor microenvironment and a mesenchymal phenotype of the glioma cells have independently been shown to be associated with a poor prognosis in glioma patients [36]. TGF-β is known to induce increased motility and invasive behavior, which underlines a potential link between cellular plasticity and intravasation of cancer cells into the bloodstream [52]. Notably, the expression of surface markers might be altered during such phenotypic changes, and this process should be taken into consideration when deciding on a capture and detection reagent for CTC capture. In line with other studies, we have previously shown that an EMT-like process can be induced in U87mg cells by incubating with TGF-β for 72 h [27]. Importantly, we also confirmed that rVAR2 binding to U87mg cells was maintained after the transition. However, several studies indicate that the EMT-like processes also play a role in the progression from low-grade to high-grade gliomas [53,54]. Here, we confirmed that ofCS display is retained when the Res259 low-grade glioma cell line is pushed towards a more mesenchymal morphology and protein expression pattern by incubation with TGF-β. This strengthens the potential of using ofCS as a target, not only for the capture, but also for the detection of glioma CTCs.

When considering previously published data showing that ofCS is presented by nearly all cancer cells of epithelial, mesenchymal, and hematopoietic origin, the use of rVAR2 staining reagent for CTC detection would be beneficial over traditional single-surface markers [25,27,33,34]. Here, we show that captured glioma CTCs can be identified by an rVAR2 stain. However, one should be cautious when using the same target for both capture and detection, as the general assumption is that the use of two independent markers would lead to a better exclusion of false positive hits. Furthermore, as CS is a common GAG that is displayed on all cell types, including WBCs, an extremely high degree of ofCS-specificity is needed to successfully capture and distinguish CTCs from WBCs. Interestingly, the naturally occurring VAR2CSA that is expressed by malaria-infected erythrocytes serves exactly this purpose, since binding to normal WBCs would result in parasite clearance [26]. However, the use of exclusion markers is highly important to exclude potential false positives. In the workflow presented here, we included CD45 and CD66b as exclusions markers to identify and reject a broad repertoire of WBCs. The optimized and combined rVAR2 capture and detection workflow enabled us to isolate and detect circulating glioma cells in glioma patients. In this very limited dataset, the number of CTCs detected did not correlate with type of diagnosis or WHO grade.

A potential clinical application of rVAR2-based CTC detection could be patient stratification based on the expression of therapeutically relevant CSPGs on CTC subsets. In this study, we sought to identify ofCS-modified proteoglycans in glioma by using rVAR2-based protein pull-down of lysates from KNS-42, U118mg, and U87mg cell lines. Indeed, the subsequent proteomics analysis showed the pull-down of multiple cancer-related proteoglycans with key roles in the pathogenesis of glioma. Unlike our previous study showing syndecan 1 to be the main VAR2CSA receptor in the placental syncytium [55], we found several interesting hits on the glioma cell lines, including syndicans, glypicans, neuropilins, decorin, versican, CSPG4, and PTPRZ1. The two last mentioned are currently being explored as potential anti-cancer targets in GBM [56–58]. In this study, we tested the use of anti-CSPG4 antibodies for staining and detection of cancer cells after rVAR2 capture, which could potentially be of future interest in the monitoring of anti-CSPG4 CAR-T therapies. Finally, CD44 was identified as a hit on all of the glioma cell lines. High CD44 expression is common in GBM and is used to identify GBM with particular poor survival chance [59,60]. Along this line, CD44 is expressed by GBM cancer stem cells, which promotes aggressive GBM growth [61]. Thus, adding a CD44-stain to rVAR2 captured CTCs could provide additional information regarding predicted outcome if sufficient exclusion markers are included. Another interesting application of the captured CTCs could be to culture and further characterize the CTCs in terms of responsiveness to relevant treatments. Liu et al. has shown proof-of-concept by culturing CTCs that are captured from a mouse GBM model [62]. To our knowledge, no one has to date been able to culture the sparse number of CTCs found in glioma patient blood samples.

We picked CTCs and performed WGA followed by WES against a panel of known glioma mutations to confirm that the detected rVAR2+, CD45−, and DAPI+ cells were actual CTCs derived from the brain tumors. Patient-matched WBCs were used as germline subtractions. Patient I, which was diagnosed with anaplastic oligodendroglioma, had CTCs with mutation in the *RB1* gene, which results in a frameshift with premature stop codon. Alterations in genes that are associated with the retinoblastoma pathway is a predictor of poor chance of survival in gliomas [63]. Interestingly, the somatic mutation pattern found in the tumor biopsy from this patient showed mutation of the *IDH1* gene, a common feature of lower grade gliomas [64], which was not detected in the CTC sample. However, CTCs could represent a minority of subclones in the primary tumor, which are not detectable by current standard NGS methods [65].

A *TP53* mutation was found in the CTCs from both patient 3 (GBM) and 4 (anaplastic oligodendroglioma). *TP53* encodes the p53 tumor suppressor protein, and this pathway is often deregulated in diffuse gliomas [66]. Another detected mutation in patient 3 was a missense mutation in the *EPM2AIP1* gene. The *EPM2AIP1* mutations have previously been described in different gastrointestinal cancers [67,68]. Interestingly, *EPM2AIP1* is part of a bidirectional promotor with *MLH1* and epimutations causing hypermethylation has been linked to hereditary colorectal cancers [69,70]. However, little is known regarding the functional role of *EPM2AIP1* silencing, as research has primarily focused on *MLH1*. In patient 4 the WES analysis also detected mutations in the *ALK* gene, which encodes a receptor tyrisone kinase. *ALK* is frequently mutated in neuroblastoma and indeed the detected NM_004304.4_p.R1275L variant is a described hot spot locus within the kinase domain. This hotspot mutation hinders the auto-inhibition of *ALK* and acts transformative. Consequently, neuroblastoma patients with *ALK* mutations show poorer overall survival [71]. Importantly, small molecules for targeted therapy of *ALK* have been developed and neuroblastoma cell lines harboring p.R1275 mutations show sensitivity towards *ALK* inhibitors, such as crizotinib [72,73]. Altogether, the specific detection of glioma-related mutation patterns in the CTC samples strongly indicates that the detected cells originate from a glioma site.

In summary, we present a method for enriching and staining CTCs from glioma patients. After a complete clinical validation the method could provide a powerful tool for non-invasive pheno- and genotyping of gliomas. Finally, the technology could potentially be used to monitor progression and recurrence in cancer patients.

Supplementary Materials: The following are available online at http://www.mdpi.com/2073-4409/8/9/998/s1, Figure S1: Proximity ligation assay controls, Figure S2: ofCS expression on glioma cell lines, Figure S3: rVAR2 staining of rVAR2-captured glioma cells, Figure S4: Screenshots from the Integrative Genomic Viewer, Figure S5: Images of detected patient CTCs, Figure S6: In vitro EMT western blot.

Author Contributions: Conceptualization, Conceptualization, S.R.B.-C. and M.Ø.A.; Formal analysis, S.R.B.-C., T.D.A., M.H.T., T.G.T., O.Ø. and M.Ø.A.; Funding acquisition, A.S. and M.Ø.A.; Investigation, S.R.B.-C., R.S.P., M.A.P., T.M.C., C.L., N.T.S., T.D.A., A.M.J., M.H.T., O.Ø. and M.Ø.A.; Methodology, S.R.B.-C., M.A.P., T.M.C. and M.Ø.A.; Project administration, M.Ø.A.; Resources, Y.C.L., L.G., S.C., T.G., R.D., M.D., A.F.S., M.S., U.L. and P.H.; Supervision, S.R.B.-C., T.M.C., A.S. and M.Ø.A.; Validation, S.R.B.-C., R.S.P., C.L. and N.T.S.; Visualization, S.R.B.-C.; Writing—original draft, S.R.B.-C. and M.Ø.A.; Writing—review & editing, S.R.B.-C., T.M.C., T.G.T. and A.S.

Funding: This research was funded by the European Research Council (Grant no 766544), Independent Research Fund Denmark (Grant no 8020-00446A), Innovation Fund Denmark (Grant no 7038-00212B), Læge Sofus Carl Emil Friis og Hustru Olga Doris Friis' Legat, Svend Andersen Fonden, and Toyota Fund Denmark. T.DA. is supported by the Bridge Translational Excellence Programme, funded by the Novo Nordisk Foundation (Grant no NNF18SA0034956).

Acknowledgments: The authors would like to thank Andreas Frederiksen and Sofie Amalie Schandorff for their great technical assistance. We would also like to thank Janine Erler and Lara Perryman (Biotech Research & Innovation Centre, University of Copenhagen, Denmark) for kindly providing cell lines for this study, and The Core Facility for Flow Cytometry (Faculty of Health and Medical Sciences, University of Copenhagen, Denmark) for the use of the LSR II flow cytometer.

Conflicts of Interest: The technology to diagnose cancer through rVAR2 is owned by VarCT Diagnostics through a license from VAR2Pharmaceuticals. AS, MØA, TGT, MD and TMC are cofounders of VAR2Pharmaceuticals.

References

1. Wesseling, P.; Kros, J.M.; Jeuken, J.W.J.D.H. The pathological diagnosis of diffuse gliomas: Towards a smart synthesis of microscopic and molecular information in a multidisciplinary context. *Diagn. Histopathol.* **2011**, *17*, 486–494. [CrossRef]
2. Claes, A.; Idema, A.J.; Wesseling, P. Diffuse glioma growth: A guerilla war. *Acta Neuropathol.* **2007**, *114*, 443–458. [CrossRef] [PubMed]
3. Olar, A.; Aldape, K.D. Using the molecular classification of glioblastoma to inform personalized treatment. *J. Pathol.* **2014**, *232*, 165–177. [CrossRef] [PubMed]
4. Tabatabai, G.; Stupp, R.; van den Bent, M.J.; Hegi, M.E.; Tonn, J.C.; Wick, W.; Weller, M. Molecular diagnostics of gliomas: The clinical perspective. *Acta Neuropathol.* **2010**, *120*, 585–592. [CrossRef] [PubMed]
5. Louis, D.N.; Perry, A.; Reifenberger, G.; von Deimling, A.; Figarella-Branger, D.; Cavenee, W.K.; Ohgaki, H.; Wiestler, O.D.; Kleihues, P.; Ellison, D.W. The 2016 World Health Organization Classification of Tumors of the Central Nervous System: A summary. *Acta Neuropathol.* **2016**, *131*, 803–820. [CrossRef]
6. Wang, Q.; Hu, B.; Hu, X.; Kim, H.; Squatrito, M.; Scarpace, L.; deCarvalho, A.C.; Lyu, S.; Li, P.; Li, Y.; et al. Tumor Evolution of Glioma-Intrinsic Gene Expression Subtypes Associates with Immunological Changes in the Microenvironment. *Cancer Cell* **2017**, *32*, 42–56. [CrossRef] [PubMed]
7. Ceccarelli, M.; Barthel, F.P.; Malta, T.M.; Sabedot, T.S.; Salama, S.R.; Murray, B.A.; Morozova, O.; Newton, Y.; Radenbaugh, A.; Pagnotta, S.M.; et al. Molecular Profiling Reveals Biologically Discrete Subsets and Pathways of Progression in Diffuse Glioma. *Cell* **2016**, *164*, 550–563. [CrossRef]
8. Piccirillo, S.G.; Dietz, S.; Madhu, B.; Griffiths, J.; Price, S.J.; Collins, V.P.; Watts, C. Fluorescence-guided surgical sampling of glioblastoma identifies phenotypically distinct tumour-initiating cell populations in the tumour mass and margin. *Br. J. Cancer* **2012**, *107*, 462–468. [CrossRef]

9. Mariani, L.; Beaudry, C.; McDonough, W.S.; Hoelzinger, D.B.; Kaczmarek, E.; Ponce, F.; Coons, S.W.; Giese, A.; Seiler, R.W.; Berens, M.E. Death-associated protein 3 (Dap-3) is overexpressed in invasive glioblastoma cells in vivo and in glioma cell lines with induced motility phenotype in vitro. *Clin. Cancer Res.* **2001**, *7*, 2480–2489.
10. Angelucci, C.; D'Alessio, A.; Lama, G.; Binda, E.; Mangiola, A.; Vescovi, A.L.; Proietti, G.; Masuelli, L.; Bei, R.; Fazi, B.; et al. Cancer stem cells from peritumoral tissue of glioblastoma multiforme: The possible missing link between tumor development and progression. *Oncotarget* **2018**, *9*, 28116–28130. [CrossRef]
11. Jimsheleishvili, S.; Alshareef, A.T.; Papadimitriou, K.; Bregy, A.; Shah, A.H.; Graham, R.M.; Ferraro, N.; Komotar, R.J. Extracranial glioblastoma in transplant recipients. *J. Cancer Res. Clin. Oncol.* **2014**, *140*, 801–807. [CrossRef] [PubMed]
12. Macarthur, K.M.; Kao, G.D.; Chandrasekaran, S.; Alonso-Basanta, M.; Chapman, C.; Lustig, R.A.; Wileyto, E.P.; Hahn, S.M.; Dorsey, J.F. Detection of brain tumor cells in the peripheral blood by a telomerase promoter-based assay. *Cancer Res.* **2014**, *74*, 2152–2159. [CrossRef] [PubMed]
13. Muller, C.; Holtschmidt, J.; Auer, M.; Heitzer, E.; Lamszus, K.; Schulte, A.; Matschke, J.; Langer-Freitag, S.; Gasch, C.; Stoupiec, M.; et al. Hematogenous dissemination of glioblastoma multiforme. *Sci. Transl. Med.* **2014**, *6*, 247ra101. [CrossRef] [PubMed]
14. Sullivan, J.P.; Nahed, B.V.; Madden, M.W.; Oliveira, S.M.; Springer, S.; Bhere, D.; Chi, A.S.; Wakimoto, H.; Rothenberg, S.M.; Sequist, L.V.; et al. Brain tumor cells in circulation are enriched for mesenchymal gene expression. *Cancer Discov.* **2014**, *4*, 1299–1309. [CrossRef] [PubMed]
15. Gao, F.; Cui, Y.; Jiang, H.; Sui, D.; Wang, Y.; Jiang, Z.; Zhao, J.; Lin, S. Circulating tumor cell is a common property of brain glioma and promotes the monitoring system. *Oncotarget* **2016**, *7*, 71330–71340. [CrossRef] [PubMed]
16. Appierto, V.; Di Cosimo, S.; Reduzzi, C.; Pala, V.; Cappelletti, V.; Daidone, M.G. How to study and overcome tumor heterogeneity with circulating biomarkers: The breast cancer case. *Semin. Cancer Biol.* **2017**, *44*, 106–116. [CrossRef] [PubMed]
17. Miyamoto, D.T.; Lee, R.J.; Stott, S.L.; Ting, D.T.; Wittner, B.S.; Ulman, M.; Smas, M.E.; Lord, J.B.; Brannigan, B.W.; Trautwein, J.; et al. Androgen receptor signaling in circulating tumor cells as a marker of hormonally responsive prostate cancer. *Cancer Discov.* **2012**, *2*, 995–1003. [CrossRef] [PubMed]
18. Pawlikowska, P.; Faugeroux, V.; Oulhen, M.; Aberlenc, A.; Tayoun, T.; Pailler, E.; Farace, F. Circulating tumor cells (CTCs) for the noninvasive monitoring and personalization of non-small cell lung cancer (NSCLC) therapies. *J. Thorac. Dis.* **2019**, *11* (Suppl. 1), S45–S56. [CrossRef]
19. Mikami, T.; Kitagawa, H. Biosynthesis and function of chondroitin sulfate. *Biochim. Biophys. Acta* **2013**, *1830*, 4719–4733. [CrossRef]
20. Gama, C.I.; Tully, S.E.; Sotogaku, N.; Clark, P.M.; Rawat, M.; Vaidehi, N.; Goddard, W.A., III; Nishi, A.; Hsieh-Wilson, L.C. Sulfation patterns of glycosaminoglycans encode molecular recognition and activity. *Nat. Chem. Biol.* **2006**, *2*, 467–473. [CrossRef]
21. Afratis, N.; Gialeli, C.; Nikitovic, D.; Tsegenidis, T.; Karousou, E.; Theocharis, A.D.; Pavao, M.S.; Tzanakakis, G.N.; Karamanos, N.K. Glycosaminoglycans: Key players in cancer cell biology and treatment. *FEBS J.* **2012**, *279*, 1177–1197. [CrossRef] [PubMed]
22. Onken, J.; Moeckel, S.; Leukel, P.; Leidgens, V.; Baumann, F.; Bogdahn, U.; Vollmann-Zwerenz, A.; Hau, P. Versican isoform V1 regulates proliferation and migration in high-grade gliomas. *J. Neurooncol.* **2014**, *120*, 73–83. [CrossRef] [PubMed]
23. Al-Mayhani, M.T.; Grenfell, R.; Narita, M.; Piccirillo, S.; Kenney-Herbert, E.; Fawcett, J.W.; Collins, V.P.; Ichimura, K.; Watts, C. NG2 expression in glioblastoma identifies an actively proliferating population with an aggressive molecular signature. *Neuro. Oncol.* **2011**, *13*, 830–845. [CrossRef] [PubMed]
24. Stallcup, W.B.; Huang, F.J. A role for the NG2 proteoglycan in glioma progression. *Cell Adh. Migr.* **2008**, *2*, 192–201. [CrossRef] [PubMed]
25. Salanti, A.; Clausen, T.M.; Agerbaek, M.O.; Al Nakouzi, N.; Dahlback, M.; Oo, H.Z.; Lee, S.; Gustavsson, T.; Rich, J.R.; Hedberg, B.J.; et al. Targeting Human Cancer by a Glycosaminoglycan Binding Malaria Protein. *Cancer Cell* **2015**, *28*, 500–514. [CrossRef] [PubMed]
26. Agerbaek, M.O.; Bang-Christensen, S.; Salanti, A. Fighting Cancer Using an Oncofetal Glycosaminoglycan-Binding Protein from Malaria Parasites. *Trends Parasitol.* **2019**, *35*, 178–181. [CrossRef] [PubMed]

27. Agerbaek, M.O.; Bang-Christensen, S.R.; Yang, M.H.; Clausen, T.M.; Pereira, M.A.; Sharma, S.; Ditlev, S.B.; Nielsen, M.A.; Choudhary, S.; Gustavsson, T.; et al. The VAR2CSA malaria protein efficiently retrieves circulating tumor cells in an EpCAM-independent manner. *Nat. Commun.* **2018**, *9*, 3279. [CrossRef] [PubMed]
28. Bax, D.A.; Little, S.E.; Gaspar, N.; Perryman, L.; Marshall, L.; Viana-Pereira, M.; Jones, T.A.; Williams, R.D.; Grigoriadis, A.; Vassal, G.; et al. Molecular and phenotypic characterisation of paediatric glioma cell lines as models for preclinical drug development. *PLoS ONE* **2009**, *4*, e5209. [CrossRef]
29. Rasmussen, R.D.; Gajjar, M.K.; Tuckova, L.; Jensen, K.E.; Maya-Mendoza, A.; Holst, C.B.; Mollgaard, K.; Rasmussen, J.S.; Brennum, J.; Bartek, J.; et al. BRCA1-regulated RRM2 expression protects glioblastoma cells from endogenous replication stress and promotes tumorigenicity. *Nat. Commun.* **2016**, *7*, 13398. [CrossRef]
30. Rappsilber, J.; Mann, M.; Ishihama, Y. Protocol for micro-purification, enrichment, pre-fractionation and storage of peptides for proteomics using StageTips. *Nat. Protoc.* **2007**, *2*, 1896–1906. [CrossRef]
31. Mogensen, M.B.; Rossing, M.; Ostrup, O.; Larsen, P.N.; Heiberg Engel, P.J.; Jorgensen, L.N.; Hogdall, E.V.; Eriksen, J.; Ibsen, P.; Jess, P.; et al. Genomic alterations accompanying tumour evolution in colorectal cancer: Tracking the differences between primary tumours and synchronous liver metastases by whole-exome sequencing. *BMC Cancer* **2018**, *18*, 752. [CrossRef] [PubMed]
32. Thorvaldsdottir, H.; Robinson, J.T.; Mesirov, J.P. Integrative Genomics Viewer (IGV): High-performance genomics data visualization and exploration. *Brief Bioinform.* **2013**, *14*, 178–192. [CrossRef] [PubMed]
33. Clausen, T.M.; Pereira, M.A.; Al Nakouzi, N.; Oo, H.Z.; Agerbaek, M.O.; Lee, S.; Orum-Madsen, M.S.; Kristensen, A.R.; El-Naggar, A.; Grandgenett, P.M.; et al. Oncofetal Chondroitin Sulfate Glycosaminoglycans Are Key Players in Integrin Signaling and Tumor Cell Motility. *Mol. Cancer Res.* **2016**, *14*, 1288–1299. [CrossRef] [PubMed]
34. Agerbaek, M.O.; Pereira, M.A.; Clausen, T.M.; Pehrson, C.; Oo, H.Z.; Spliid, C.; Rich, J.R.; Fung, V.; Nkrumah, F.; Neequaye, J.; et al. Burkitt lymphoma expresses oncofetal chondroitin sulfate without being a reservoir for placental malaria sequestration. *Int. J. Cancer* **2017**, *140*, 1597–1608. [CrossRef] [PubMed]
35. Platten, M.; Wick, W.; Weller, M. Malignant glioma biology: Role for TGF-beta in growth, motility, angiogenesis, and immune escape. *Microsc. Res. Tech.* **2001**, *52*, 401–410. [CrossRef]
36. Joseph, J.V.; Conroy, S.; Tomar, T.; Eggens-Meijer, E.; Bhat, K.; Copray, S.; Walenkamp, A.M.; Boddeke, E.; Balasubramanyian, V.; Wagemakers, M.; et al. TGF-beta is an inducer of ZEB1-dependent mesenchymal transdifferentiation in glioblastoma that is associated with tumor invasion. *Cell Death Dis.* **2014**, *5*, e1443. [CrossRef] [PubMed]
37. Shankar, J.; Messenberg, A.; Chan, J.; Underhill, T.M.; Foster, L.J.; Nabi, I.R. Pseudopodial actin dynamics control epithelial-mesenchymal transition in metastatic cancer cells. *Cancer Res.* **2010**, *70*, 3780–3790. [CrossRef] [PubMed]
38. Nevo, I.; Woolard, K.; Cam, M.; Li, A.; Webster, J.D.; Kotliarov, Y.; Kim, H.S.; Ahn, S.; Walling, J.; Kotliarova, S.; et al. Identification of molecular pathways facilitating glioma cell invasion in situ. *PLoS ONE* **2014**, *9*, e111783. [CrossRef] [PubMed]
39. Qiao, D.; Meyer, K.; Friedl, A. Glypican 1 stimulates S phase entry and DNA replication in human glioma cells and normal astrocytes. *Mol. Cell. Biol.* **2013**, *33*, 4408–4421. [CrossRef] [PubMed]
40. Xu, Y.; Yuan, J.; Zhang, Z.; Lin, L.; Xu, S. Syndecan-1 expression in human glioma is correlated with advanced tumor progression and poor prognosis. *Mol. Biol. Rep.* **2012**, *39*, 8979–8985. [CrossRef] [PubMed]
41. Schiffer, D.; Mellai, M.; Boldorini, R.; Bisogno, I.; Grifoni, S.; Corona, C.; Bertero, L.; Cassoni, P.; Casalone, C.; Annovazzi, L. The Significance of Chondroitin Sulfate Proteoglycan 4 (CSPG4) in Human Gliomas. *Int. J. Mol. Sci.* **2018**, *19*. [CrossRef] [PubMed]
42. Senbanjo, L.T.; Chellaiah, M.A. CD44: A Multifunctional Cell Surface Adhesion Receptor Is a Regulator of Progression and Metastasis of Cancer Cells. *Front. Cell Dev. Biol.* **2017**, *5*, 18. [CrossRef] [PubMed]
43. Faye, R.S.; Aamdal, S.; Hoifodt, H.K.; Jacobsen, E.; Holstad, L.; Skovlund, E.; Fodstad, O. Immunomagnetic detection and clinical significance of micrometastatic tumor cells in malignant melanoma patients. *Clin. Cancer Res.* **2004**, *10 Pt 1*, 4134–4139. [CrossRef] [PubMed]
44. Ulmer, A.; Schmidt-Kittler, O.; Fischer, J.; Ellwanger, U.; Rassner, G.; Riethmuller, G.; Fierlbeck, G.; Klein, C.A. Immunomagnetic enrichment, genomic characterization, and prognostic impact of circulating melanoma cells. *Clin. Cancer Res.* **2004**, *10*, 531–537. [CrossRef] [PubMed]

45. Rao, C.; Bui, T.; Connelly, M.; Doyle, G.; Karydis, I.; Middleton, M.R.; Clack, G.; Malone, M.; Coumans, F.A.; Terstappen, L.W. Circulating melanoma cells and survival in metastatic melanoma. *Int. J. Oncol.* **2011**, *38*, 755–760. [PubMed]
46. Gray, E.S.; Reid, A.L.; Bowyer, S.; Calapre, L.; Siew, K.; Pearce, R.; Cowell, L.; Frank, M.H.; Millward, M.; Ziman, M. Circulating Melanoma Cell Subpopulations: Their Heterogeneity and Differential Responses to Treatment. *J. Investig. Dermatol.* **2015**, *135*, 2040–2048. [CrossRef] [PubMed]
47. Pellegatta, S.; Savoldo, B.; Di Ianni, N.; Corbetta, C.; Chen, Y.; Patane, M.; Sun, C.; Pollo, B.; Ferrone, S.; DiMeco, F.; et al. Constitutive and TNFalpha-inducible expression of chondroitin sulfate proteoglycan 4 in glioblastoma and neurospheres: Implications for CAR-T cell therapy. *Sci. Transl. Med.* **2018**, *10*. [CrossRef]
48. Rodriguez, A.; Brown, C.; Badie, B. Chimeric antigen receptor T-cell therapy for glioblastoma. *Transl. Res.* **2017**, *187*, 93–102. [CrossRef]
49. Terada, T. Expression of cytokeratins in glioblastoma multiforme. *Pathol Oncol Res.* **2015**, *21*, 817–819. [CrossRef]
50. Goswami, C.; Chatterjee, U.; Sen, S.; Chatterjee, S.; Sarkar, S. Expression of cytokeratins in gliomas. *Indian J. Pathol. Microbiol.* **2007**, *50*, 478–481.
51. Fanburg-Smith, J.C.; Majidi, M.; Miettinen, M. Keratin expression in schwannoma; a study of 115 retroperitoneal and 22 peripheral schwannomas. *Mod. Pathol.* **2006**, *19*, 115–121. [CrossRef] [PubMed]
52. Pastushenko, I.; Brisebarre, A.; Sifrim, A.; Fioramonti, M.; Revenco, T.; Boumahdi, S.; Van Keymeulen, A.; Brown, D.; Moers, V.; Lemaire, S.; et al. Identification of the tumour transition states occurring during EMT. *Nature* **2018**, *556*, 463–468. [CrossRef] [PubMed]
53. Myung, J.; Cho, B.K.; Kim, Y.S.; Park, S.H. Snail and Cox-2 expressions are associated with WHO tumor grade and survival rate of patients with gliomas. *Neuropathology* **2010**, *30*, 224–231. [CrossRef] [PubMed]
54. Liu, Y.; Hu, H.; Wang, K.; Zhang, C.; Wang, Y.; Yao, K.; Yang, P.; Han, L.; Kang, C.; Zhang, W.; et al. Multidimensional analysis of gene expression reveals TGFB1I1-induced EMT contributes to malignant progression of astrocytomas. *Oncotarget* **2014**, *5*, 12593–12606. [CrossRef] [PubMed]
55. Ayres Pereira, M.; Mandel Clausen, T.; Pehrson, C.; Mao, Y.; Resende, M.; Daugaard, M.; Riis Kristensen, A.; Spliid, C.; Mathiesen, L.; Knudsen, L.E.; et al. Placental Sequestration of Plasmodium falciparum Malaria Parasites Is Mediated by the Interaction Between VAR2CSA and Chondroitin Sulfate A on Syndecan-1. *PLoS Pathog.* **2016**, *12*, e1005831. [CrossRef]
56. Foehr, E.D.; Lorente, G.; Kuo, J.; Ram, R.; Nikolich, K.; Urfer, R. Targeting of the receptor protein tyrosine phosphatase beta with a monoclonal antibody delays tumor growth in a glioblastoma model. *Cancer Res.* **2006**, *66*, 2271–2278. [CrossRef]
57. Higgins, S.C.; Bolteus, A.J.; Donovan, L.K.; Hasegawa, H.; Doey, L.; Al Sarraj, S.; King, A.; Ashkan, K.; Roncaroli, F.; Fillmore, H.L.; et al. Expression of the chondroitin sulphate proteoglycan, NG2, in paediatric brain tumors. *Anticancer Res.* **2014**, *34*, 6919–6924.
58. Wang, J.; Svendsen, A.; Kmiecik, J.; Immervoll, H.; Skaftnesmo, K.O.; Planaguma, J.; Reed, R.K.; Bjerkvig, R.; Miletic, H.; Enger, P.O.; et al. Targeting the NG2/CSPG4 proteoglycan retards tumour growth and angiogenesis in preclinical models of GBM and melanoma. *PLoS ONE* **2011**, *6*, e23062. [CrossRef]
59. Phillips, H.S.; Kharbanda, S.; Chen, R.; Forrest, W.F.; Soriano, R.H.; Wu, T.D.; Misra, A.; Nigro, J.M.; Colman, H.; Soroceanu, L.; et al. Molecular subclasses of high-grade glioma predict prognosis, delineate a pattern of disease progression, and resemble stages in neurogenesis. *Cancer Cell* **2006**, *9*, 157–173. [CrossRef]
60. Colman, H.; Zhang, L.; Sulman, E.P.; McDonald, J.M.; Shooshtari, N.L.; Rivera, A.; Popoff, S.; Nutt, C.L.; Louis, D.N.; Cairncross, J.G.; et al. A multigene predictor of outcome in glioblastoma. *Neuro Oncol.* **2010**, *12*, 49–57. [CrossRef]
61. Pietras, A.; Katz, A.M.; Ekstrom, E.J.; Wee, B.; Halliday, J.J.; Pitter, K.L.; Werbeck, J.L.; Amankulor, N.M.; Huse, J.T.; Holland, E.C. Osteopontin-CD44 signaling in the glioma perivascular niche enhances cancer stem cell phenotypes and promotes aggressive tumor growth. *Cell Stem Cell* **2014**, *14*, 357–369. [CrossRef] [PubMed]
62. Liu, T.; Xu, H.; Huang, M.; Ma, W.; Saxena, D.; Lustig, R.A.; Alonso-Basanta, M.; Zhang, Z.; O'Rourke, D.M.; Zhang, L.; et al. Circulating Glioma Cells Exhibit Stem Cell-like Properties. *Cancer Res.* **2018**, *78*, 6632–6642. [CrossRef] [PubMed]
63. Aoki, K.; Nakamura, H.; Suzuki, H.; Matsuo, K.; Kataoka, K.; Shimamura, T.; Motomura, K.; Ohka, F.; Shiina, S.; Yamamoto, T.; et al. Prognostic relevance of genetic alterations in diffuse lower-grade gliomas. *Neuro Oncol.* **2018**, *20*, 66–77. [CrossRef] [PubMed]

64. D'Alessio, A.; Proietti, G.; Sica, G.; Scicchitano, B.M. Pathological and Molecular Features of Glioblastoma and Its Peritumoral Tissue. *Cancers* **2019**, *11*. [CrossRef] [PubMed]
65. Heitzer, E.; Auer, M.; Gasch, C.; Pichler, M.; Ulz, P.; Hoffmann, E.M.; Lax, S.; Waldispuehl-Geigl, J.; Mauermann, O.; Lackner, C.; et al. Complex tumor genomes inferred from single circulating tumor cells by array-CGH and next-generation sequencing. *Cancer Res.* **2013**, *73*, 2965–2975. [CrossRef] [PubMed]
66. Zhang, Y.; Dube, C.; Gibert, M., Jr.; Cruickshanks, N.; Wang, B.; Coughlan, M.; Yang, Y.; Setiady, I.; Deveau, C.; Saoud, K.; et al. The p53 Pathway in Glioblastoma. *Cancers* **2018**, *10*. [CrossRef] [PubMed]
67. Muzny, D.M.; Bainbridge, M.N.; Chang, K.; Dinh, H.H.; Drummond, J.A.; Fowler, G.; Kovar, C.L.; Lewis, L.R.; Morgan, M.B.; Newsham, I.F.; et al. Comprehensive molecular characterization of human colon and rectal cancer. *Nature* **2012**, *487*, 330–337.
68. Dulak, A.M.; Stojanov, P.; Peng, S.Y.; Lawrence, M.S.; Fox, C.; Stewart, C.; Bandla, S.; Imamura, Y.; Schumacher, S.E.; Shefler, E.; et al. Exome and whole-genome sequencing of esophageal adenocarcinoma identifies recurrent driver events and mutational complexity. *Nat. Genet.* **2013**, *45*, 478–486. [CrossRef]
69. Hitchins, M.; Williams, R.; Cheong, K.; Halani, N.; Lin, V.A.; Packham, D.; Ku, S.; Buckle, A.; Hawkins, N.; Burn, J.; et al. MLH1 germline epimutations as a factor in hereditary nonpolyposis colorectal cancer. *Gastroenterology* **2005**, *129*, 1392–1399. [CrossRef]
70. Hitchins, M.P.; Rapkins, R.W.; Kwok, C.T.; Srivastava, S.; Wong, J.J.L.; Khachigian, L.M.; Polly, P.; Goldblatt, J.; Ward, R.L. Dominantly Inherited Constitutional Epigenetic Silencing of MLH1 in a Cancer-Affected Family Is Linked to a Single Nucleotide Variant within the 5' UTR. *Cancer Cell* **2011**, *20*, 200–213. [CrossRef]
71. Bresler, S.C.; Weiser, D.A.; Huwe, P.J.; Park, J.H.; Krytska, K.; Ryles, H.; Laudenslager, M.; Rappaport, E.F.; Wood, A.C.; McGrady, P.W.; et al. ALK mutations confer differential oncogenic activation and sensitivity to ALK inhibition therapy in neuroblastoma. *Cancer Cell* **2014**, *26*, 682–694. [CrossRef] [PubMed]
72. Chen, L.D.; Humphreys, A.; Turnbull, L.; Bellini, A.; Schleiermacher, G.; Salwen, H.; Cohn, S.L.; Bown, N.; Tweddle, D.A. Identification of different ALK mutations in a pair of neuroblastoma cell lines established at diagnosis and relapse. *Oncotarget* **2016**, *7*, 87301–87311. [CrossRef] [PubMed]
73. Bresler, S.C.; Wood, A.C.; Haglund, E.A.; Courtright, J.; Belcastro, L.T.; Plegaria, J.S.; Cole, K.; Toporovskaya, Y.; Zhao, H.Q.; Carpenter, E.L.; et al. Differential Inhibitor Sensitivity of Anaplastic Lymphoma Kinase Variants Found in Neuroblastoma. *Sci. Transl. Med.* **2011**, *3*, 108ra114. [CrossRef] [PubMed]

 © 2019 by the authors. Licensee MDPI, Basel, Switzerland. This article is an open access article distributed under the terms and conditions of the Creative Commons Attribution (CC BY) license (http://creativecommons.org/licenses/by/4.0/).

Article

Leukocyte-Derived Extracellular Vesicles in Blood with and without EpCAM Enrichment

Afroditi Nanou *, Leonie L. Zeune and Leon W.M.M. Terstappen *

Department of Medical Cell BioPhysics, Faculty of Science and Technology, University of Twente, 7522 NH Enschede, The Netherlands
* Correspondence: a.nanou@utwente.nl (A.N.); l.w.m.m.terstappen@utwente.nl (L.W.M.M.T.); Tel.: +31-53-489-2425 (L.W.M.M.T.)

Received: 29 June 2019; Accepted: 15 August 2019; Published: 20 August 2019

Abstract: Large tumor-derived Extracellular Vesicles (tdEVs) detected in blood of metastatic prostate, breast, colorectal, and non-small cell lung cancer patients after enrichment for Epithelial Cell Adhesion Molecule (EpCAM) expression and labeling with 4′,6-diamidino-2-phenylindole (DAPI), phycoerythrin-conjugated antibodies against Cytokeratins (CK-PE), and allophycocyanin-conjugated antibody against the cluster of differentiation 45 (CD45-APC), are negatively associated with the overall survival of patients. Here, we investigated whether, similarly to tdEVs, leukocyte-derived EVs (ldEVs) could also be detected in EpCAM-enriched blood. Presence of ldEVs and leukocytes in image data sets of EpCAM-enriched samples of 25 healthy individuals and 75 metastatic cancer patients was evaluated using the ACCEPT software. Large ldEVs could indeed be detected, but in contrast to the 20-fold higher frequency of tdEVs as compared to Circulating Tumor Cells (CTCs), ldEVs were present in a 5-fold lower frequency as compared to leukocytes. To evaluate whether these ldEVs pre-exist in the blood or are formed during the CellSearch procedure, the blood of healthy individuals without EpCAM enrichment was labelled with the nuclear dye Hoechst and fluorescently tagged monoclonal antibodies recognizing the leukocyte-specific CD45, platelet-specific CD61, and red blood cell-specific CD235a. Fluorescence microscopy imaging using a similar setup as the CellSearch was performed and demonstrated the presence of a similar population of ldEVs present at a 3-fold lower frequency as compared to leukocytes.

Keywords: leukocyte-derived extracellular vesicles; immunofluorescence imaging; EpCAM enrichment; CellSearch; EasyCount slides; ACCEPT

---

1. Introduction

During the last decades, Extracellular Vesicles (EVs) have emerged as promising disease biomarkers bearing similar membrane and cargo composition as their originating cells [1–3]. Importantly, for nucleic acid analysis, the membrane encapsulated nucleic acid cargo is protected from enzymatic degradation, and consequently, it can circulate for a longer time compared to cell-free DNA (cfDNA) [4,5]. In the case of cancer, the presence of nucleic acids (DNA, mRNA, and miRNA) within tumor-derived EVs (tdEVs) and proteins within or on tdEV membranes could provide information of the predisposition of the tumor to metastasize in specific organs and guide treatment monitoring of patients to block metastasis and cancer progression [5–9]. It has been demonstrated that EVs in biofluids of cancer patients are significantly elevated when compared to the respective numbers of healthy individuals [10,11]. However, to our knowledge, there is no data available in regards to the composition of the redundant EVs in the blood of cancer patients. The recent *in vivo* studies of Ricklefs et al. using brain tumors expressing the green fluorescent protein (GFP) in mice showed that less than 0.5% of the total circulating EVs were GFP+ [10]. That finding implies that more cell types secrete EVs in response to the present tumor contributing to the final EV pool detected in biofluids of cancer patients. Furthermore, the pre-analytical steps of sample processing determine the EV populations to be analyzed and could lead to biased conclusions. The majority of research groups is only interested in exosomes that constitute the smallest subclass of EVs as they consider them products of active cell secretion; therefore, they are using differential centrifugation steps to get rid of other EV subclasses, collect the exosome fraction as a pellet from the final ultracentrifugation step and label them with antibodies recognizing generic exosome-enriched biomarkers, mainly tetraspanins, such as the clusters of differentiation CD81, CD9, and CD63 to identify them [10,12]. Nevertheless, EV subclasses of larger size (microvesicles, oncosomes, and apoptotic bodies) have been reported to be bioactive with a wide spectrum of functions depending on their cells of origin [13]. Importantly, Vagner et al. reported the presence of DNA in large tdEVs reflecting the genetic aberrations of the tumor; a finding that highlights their promising potential in the liquid biopsy field [9]. Padda et al. also demonstrated that the majority of prostate-specific membrane antigen (PSMA) expressing EVs in plasma of prostate cancer patients derive directly from the plasma membrane and have a larger size [14]; hence, these clinically important populations are missed by solely the exosome analysis. Very few studies have investigated the isolation and downstream characterization of specific tdEVs from patient samples using immuno-affinity techniques [15,16]. Recently, we showed that large tdEVs, immunomagnetically isolated based on their EpCAM expression together with Circulating Tumor Cells (CTCs) by the CellSearch system from the blood of metastatic prostate, breast, colorectal, and non-small cell lung cancer patients have equivalent prognostic power to CTCs [16,17]. These observations were enabled through the availability of the open-source ACCEPT image analysis program, which allows for the exploration and enumeration in a single level of all different classes of objects detected in the fluorescence images in an automated, fast and reproducible manner, free of the subjectivity and bias of different operators [18,19]. However, it is not clear whether our previously reported large tdEVs are a result of the fragmentation of CTCs during the immunomagnetic EpCAM enrichment and washing steps that the CellSearch system is using or whether they pre-exist in the blood samples of cancer patients. Their rare frequency in combination with the abundance of blood cells and EVs of different origins prevent us from addressing that question by labeling of blood samples without any pre-enrichment steps and subsequent enumeration from fluorescence images. In this study, we identified in the digitally stored CellSearch images some CD45+, DAPI-, CK- objects of similar size to tdEVs that we baptized leukocyte-derived Extracellular Vesicles (ldEVs). We addressed the question of whether large ldEVs pre-exist in the blood of individuals without EpCAM enrichment or they are by-products of cell fragmentation by the CellSearch procedure. Towards that direction, we labeled blood samples of healthy individuals with the nuclear dye Hoechst and fluorophore-conjugated antibodies against the leukocyte-specific CD45, the platelet-specific CD61, and the red blood cell-specific CD235a without any pre-enrichment or pre-analytical steps. The samples were imaged using a fluorescence microscope with a 10×/0.45 numerical aperture (NA) objective to enable fair comparison of the image datasets acquired by the CellTracks Analyzer II of the CellSearch system [20].

2. Materials and Methods

2.1. Immunofluorescence Image Data Sets of EpCAM-Enriched Cells and Extracellular Vesicles of 25 Healthy Individuals and 75 Metastatic Cancer Patients

One-hundred digitally stored CellSearch image data sets corresponding to EpCAM-enriched blood samples of 25 healthy individuals, 25 metastatic prostate (CRPC), 25 colorectal (mCRC), and 25 non-small cell lung (NSCLC) cancer patients before the initiation of a new therapy, were used for this analysis. The EpCAM-enriched leukocytes and large leukocyte-derived EVs present in these fluorescence images were enumerated. The CRPC and mCRC patients had participated in the retrospective IMMC38 (*NCT00133900*) and CAIRO 2 (*NCT00208546*) clinical studies, respectively.

Briefly, the EpCAM+ Circulating Tumor Cells (CTCs) and tdEVs were positively selected by ferrofluids conjugated to an antibody recognizing the extracellular epitope of Epithelial Cell Adhesion Molecule (EpCAM, clone VU1D9) from 7.5 mL of the blood of cancer patients using the CellSearch system (Menarini Silicon Biosystems, Huntingdon Valley PA, USA), as previously described [21]. Following EpCAM immunomagnetic enrichment, the suspension was incubated with the nuclear dye DAPI and antibodies against the epithelial-specific cytokeratins 8, 18, and 19 (clone C11) conjugated to phycoerythrin (PE) and an antibody against the leukocyte-specific cluster of differentiation CD45 conjugated to allophycocyanin (APC). The suspension was transferred to a cartridge placed within a magnest that allowed for a homogeneous distribution of the ferrofluids and the EpCAM-enriched objects on a focal plane [22]. The cartridges were imaged using a semi-automated 10×/0.45 NA objective fluorescence microscope, the CellTracks Analyzer II, as previously described [21].

2.2. Blood Samples of 10 Healthy Individuals

Blood samples from 10 anonymous healthy individuals were collected in ethylenediaminetetraacetic acid (EDTA) tubes after written informed consent from the Experimental Centre for Technical Medicine (ECTM) donor service (University of Twente, Enschede, The Netherlands). The frequencies of white blood cells, red blood cells, and platelets were assessed using a hematology analyzer (Beckman Coulter, Brea, CA, USA). The samples were processed on the same day of the drawing.

2.3. Immunofluorescence Imaging of Cells and Extracellular Vesicles in Whole Blood Samples

10–20 µL of EDTA blood samples of 10 different healthy individuals were 10× diluted in 0.2 µm filtered 1% *w/v* bovine serum albumin (BSA) in phosphate- buffered saline (PBS) solution. Blood was incubated with the nuclear dye Hoechst 33342 (Invitrogen, cat. # H3570), the fluorescently tagged monoclonal antibodies CD45-PerCP (clone HI30 Invitrogen, cat. # MHCD4531) recognizing leukocytes, CD235a-Alexa Fluor® 647 (clone YTH89.1, bio-rad, MCA 506A647) antibodies recognizing erythrocytes and CD61-Alexa Fluor® 488 (clone Y2.51, bio-rad, cat. # MCA 2588A488) antibodies recognizing platelets. The final concentrations used were 4.0 µg/mL Hoechst, 0.5 µg/mL CD45-PerCP, 2.5 µg/mL CD235a-Alexa 647, and 0.6 µg/mL CD61-Alexa 488. The samples were incubated with the antibodies at 37 °C for 1–2 h and stored at 4 °C until further processing. Subsequently, the samples were further diluted to a final dilution of 500×. 10 µL of the diluted sample (corresponding to 0.02 µL of undiluted blood) were loaded in a well of EasyCount™ Slide-6™ (Immunicon Corp., Huntingdon Valley, PA, USA). Four–six technical replicates of samples were used to assess the reproducibility of the measurements. Image data sets of 55–65 frames/channel were acquired to cover the whole surface of each well using a semi-automated inverted fluorescence microscope (Eclipse Ti-E, Nikon Instruments, Amsterdam, The Netherlands) equipped with a 10×/0.45 numerical aperture (NA) objective, a camera (Orca flash 4.0 LT, C11440, Hamamatsu, Almere, The Netherlands) and fluorescence filter cubes (DAPI, FITC, PerCP, APC filter sets for the detection of Hoechst, CD61-Alexa 488, CD45-PerCP and CD235a-Alexa 647, respectively). The operator determined three corners of the surface to be scanned and adjusted the focus at four points distributed throughout each well. The exposure times used for the imaging were 20 ms for DAPI, 400 ms for PerCP, 500 ms for FITC, 1000 ms for APC, and 500 ms for brightfield. Few images were obtained using a 60×/0.70 NA objective and the same exposure times for comparison. However, only the images obtained with the 10× objective were used as an input for the enumeration of the different populations to allow a fair comparison with the images of the CellTracks Analyzer II. An example of a frame acquired with the 10× and 60× objectives is shown in Supplementary Figure S1.

2.4. Automated Enumeration of Objects in Immunofluorescence Images Using the Open-Source ACCEPT Software

All immunofluorescence image data sets, obtained with the 10×/0.45 NA objective microscopes, were processed with the open-source software for Automated CTC Classification, Enumeration and Phenotyping (ACCEPT) (http://github.com/LeonieZ/ACCEPT). The software detects all present objects, larger than four pixels, and extracts for each of them 10 morphological and fluorescence signal intensity measurements per fluorescence channel [18]. The user can design linear gates based on these features to define the classes of their interest and enumerate the objects falling within them [16,23]. The application of the same gates for all different samples allows the elimination of inter- and intra- operator variations leading subsequently, to a more objective consensus [24].

For the CellSearch generated images, gates for the enumeration of leukocyte-derived Extracellular Vesicles (ldEVs) and leukocytes were applied. The gates are summarized in Table 1. For the image data sets corresponding to the EasyCount Slides-6, gates for the enumeration of red blood cells, leukocytes, platelets and ldEVs were used and are summarized in Table 1.

Table 1. ACCEPT gates used for the automated enumeration of leukocytes, leukocyte-derived Extracellular Vesicles (ldEVs), platelets, and red blood cells in blood A with EpCAM enrichment and B without EpCAM enrichment.

		A. EpCAM Enrichment		B. No Enrichment	
Leukocytes	DAPI/Hoechst [a]	Mean Intensity	>30	Mean Intensity	>30
		Max Intensity	>50	Max Intensity	>50
		Size	>16 μm²	Size	>16 μm²
	CD45	Mean Intensity	>30	Mean Intensity	>30
		Max Intensity	>50	Max Intensity	>50
				Size	≤400 μm²
	CK	Standard Deviation	≤5	n/a [b]	
	CD61	n/a [b]		Standard Deviation	≤5
	CD235a	n/a [b]		Standard Deviation	≤5
	Extra channel	Standard Deviation	≤5	Standard Deviation	≤5
ldEVs	DAPI/Hoechst [a]	Standard Deviation	≤5	Standard Deviation	≤5
	CD45	Mean Intensity	>30	Mean Intensity	>30
		Max Intensity	>50	Max Intensity	>50
		Perimeter	>5 pixels	Perimeter	>5 pixels
		Size	≤150	Size	≤150 μm²
		Eccentricity	≤0.85	Eccentricity	≤0.85
	CK	Standard Deviation	≤5	n/a [b]	
	CD61	n/a [b]		Standard Deviation	≤5
	CD235a	n/a [b]		Standard Deviation	≤5
	Extra channel	Standard Deviation	≤5	Standard Deviation	≤5
Platelets	CD45			Standard Deviation	≤5
	CD61	n/a [b]		Mean Intensity	>30
				Max Intensity	>50
				Perimeter	>5 pixels
				Size	≤150 μm²
				Eccentricity	≤0.85
	CD235a			Standard Deviation	≤5
				Standard Deviation	≤5
	Extra Channel			Standard Deviation	≤5
Red blood cells	Hoechst	n/a [b]		Standard Deviation	≤5
	CD45			Standard Deviation	≤5
	CD61			Standard Deviation	≤5
	CD235a			Mean Intensity	>30
				Max Intensity	>50
				Perimeter	>5 pixels
	Extra Channel			Standard Deviation	≤5

[a] DAPI was used in EpCAM-enriched blood and Hoechst in the blood without pre-enrichment, [b] n/a: not applicable.

3. Results

3.1. Detection of ldEVs in EpCAM-Enriched Blood Samples of Healthy Individuals and Metastatic Cancer Patients

After careful examination of the immunofluorescence images of the CellSearch cartridges, CD45+, DAPI-, CK- objects, that resemble in size our previously reported CD45-, DAPI-, CK+ tdEVs [16,25], could be observed. We baptized these objects leukocyte-derived Extracellular Vesicles (ldEVs). Examples of single ldEV events in EpCAM-enriched samples by the CellSearch system are shown in Figure 1 next to some examples of leukocytes as a reference to their size and CD45 phenotype. The observation of the presence of these ldEVs in the CellSearch cartridges raised questions about their formation: are they fragments of leukocytes formed during the CellSearch procedure or do they pre-exist in the blood circulation?

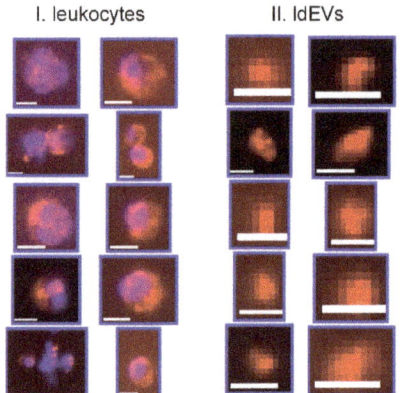

Figure 1. Thumbnails of I. leukocytes and II. Leukocyte-derived Extracellular Vesicles (ldEVs) detected in EpCAM-enriched blood samples. The red color represents CD45 and blue represents DAPI. Scale bars indicate 6.4 µm.

3.2. Detection of Cell and Extracellular Vesicle Classes in the Blood Of Healthy Individuals without EpCAM Enrichment

In order to address the aforementioned question, blood samples of healthy individuals were labeled with Hoechst, CD45-PerCP, CD61-Alexa 488, and CD235a-Alexa 647 and were imaged with a similar fluorescence microscope as the CellTracks Analyzer II. No pre-enrichment or washing steps were used in order to minimize the cell fragmentation or activation. The inclusion of the aforementioned antibodies allowed the detection of four different classes of objects in the whole blood of healthy individuals, namely leukocytes, platelets, red blood cells, and ldEVs, as shown in Figure 2. Leukocytes are defined as nucleated CD45+, CD61-, CD235a- cells of a size between 7 and 20 µm (Panel A); leukocyte-derived Extracellular Vesicles (ldEVs) as CD45+, CD61-, CD235- objects without a nucleus and of undefined size as shown in the respective brightfield image (Panel B); platelets as CD45-, CD61+, CD235a- objects without a nucleus of size between 2 and 5 µm (Panel C) and red blood cells as CD45-, CD61-, CD235a+ cells without a nucleus and a size range between 6 and 10 µm (Panel D). In Panel C, three platelets are shown, of which the middle one is clearly smaller and with a lower expression of CD61; examination at higher magnification would allow for better identification of the smaller size platelets, but no discrimination could be made between small platelets and larger platelet-derived EVs. The presence of ldEVs in blood samples without pre-enrichment (Panel B) confirmed their pre-existence in whole blood.

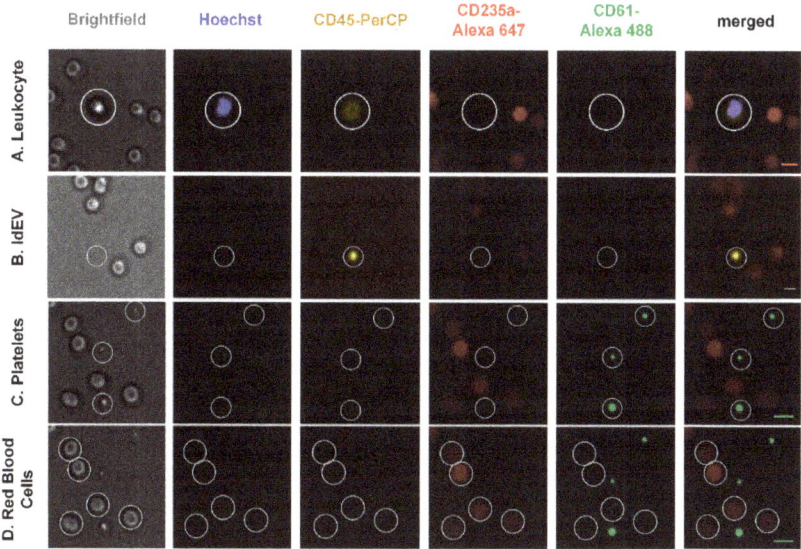

Figure 2. Bright field and immunofluorescence (IF) images of leukocytes, ldEVs, platelets and red blood cells in blood samples without EpCAM enrichment. Scale bars in the merged IF images indicate 10 μm.

3.3. ACCEPT Gates for the Automated Enumeration of Different Classes in the Blood with and without EpCAM Enrichment.

In order to acquire the absolute counts of the different classes from each data set of the healthy individuals in a fast and unbiased manner, we processed all data sets with the open-source ACCEPT software. Based on the aforementioned characteristics of the different classes, we developed linear gates to enumerate the objects falling within each class automatically. The gates are summarized in Table 1. Three examples of objects per class are shown in Figure 3 (Panel A). The objects that fall into each class are depicted in blue dots in the scatter plots (Panel B) showing the mean Hoechst intensity versus the mean CD45-PerCP intensity and the mean CD61-Alexa 488 intensity versus the mean CD235a-Alexa 647 intensity. Objects falling in the "leukocyte" gate are double-positive for CD45 and Hoechst and negative for CD61 and CD235a, (Panel B1); ldEVs are positive only for CD45 (Panel B2); platelets are only positive for CD61 (Panel B3) and red blood cells are only positive for CD235a (Panel B4).

In order to achieve a fair comparison between the leukocyte and ldEV counts detected in the blood with and without EpCAM enrichment, very similar ACCEPT gates were developed for the automated enumeration of the two classes applied in the different image data sets. The gates can be found in Table 1.

The size threshold of 20 × 20 μm^2 in the case of the "leukocyte" gate that was applied in blood samples with no enrichment was removed in the respective gate of EpCAM-enriched samples because within the CellSearch cartridges, there are many leukocytes present in close proximity to each other, as shown in Figure 1, that are segmented as one object by the ACCEPT software. Therefore, the inclusion of such a parameter in the EpCAM-enriched samples would lead to an even higher underestimation of this population compared to the underestimation already introduced by cell clusters counted as one object. On the other hand, the removal of that parameter in case of blood samples with no enrichment (Figure 3), where it is very rare to find two or more leukocytes in close proximity, leads to the inclusion of artefacts and an overestimation of leukocytes.

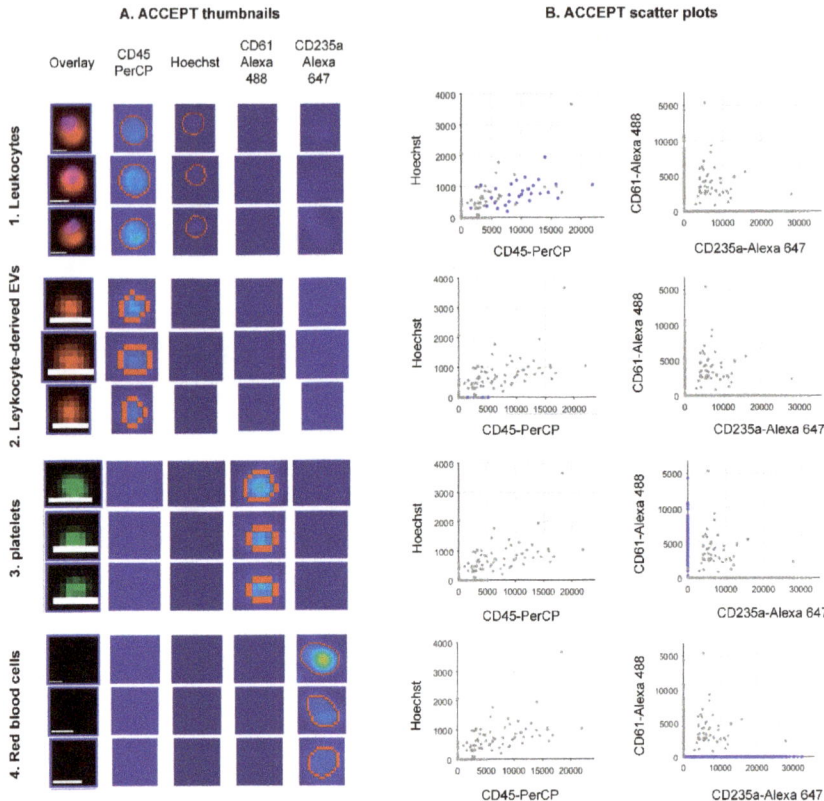

Figure 3. Examples of objects detected in the immunofluorescence image data sets of blood samples without EpCAM enrichment. The objects fall in the ACCEPT gates, the names of which are indicated vertically. Panel (**A**) shows examples of ACCEPT thumbnails. Scale bars indicate 6.4 µm. Panel (**B**) shows scatter plots of the mean intensity of the Hoechst versus the mean intensity of CD45-PerCP and the mean intensity of CD61-Alexa 488 versus the mean intensity of CD235a-Alexa 647. Blue dots represent single events falling in the respective gate.

For the acquisition of the fluorescence images of cells and EVs in blood without any pre-enrichment, the focus was set on four points distributed throughout each well to achieve optimal visualization of the cells, and the surface was afterwards automatically scanned. Since the objects were in suspension and not attached on a surface, most ldEVs and platelets, were out-of-focus with their blurring pattern influencing their perceived size, that seems much larger in the respective fluorescence images than it actually is. Hence, the size of ldEVs cannot be accurately derived using the immunofluorescence images (Figure 2). In case of EpCAM enriched samples, EVs are aligned on the same focal plane as the cells due to the design of the CellSearch magnets that result in a homogeneous cell distribution along the applied magnetic field [22]. Even in that case; however, the use of immunofluorescence images could lead to erroneous conclusions about the size of the EVs. That is even more profound when low magnification objectives are used in the fluorescence microscopes as in our case (10×/0.45 NA) limiting the determination of the size of EVs with confidence since each pixel of the acquired images corresponds to 0.64×0.64 µm². More examples of correlated bright field and immunofluorescence images of ldEVs in whole blood samples could be found in Supplementary Figure S2.

3.4. Absolute and Relative Frequencies of Leukocytes and ldEVs in 7.5 mL of EpCA- Enriched Blood of Healthy Individuals and Metastatic Cancer Patients.

The numbers of ldEVs and leukocytes in EpCAM-enriched 7.5 mL blood samples of 25 healthy individuals, 25 metastatic prostates, 25 colorectal, and 25 non-small cell lung cancer patients were determined and are presented in box plots (Figure 4). In addition, the number of ldEVs and leukocytes present in 0.02 µL of the blood of 10 healthy individuals with no enrichment was determined and extrapolated to 7.5 mL of blood for comparison (Figure 4). As it was expected, the leukocyte and ldEV frequencies are significantly depleted in the EpCAM-enriched blood samples of individuals, since EpCAM is an epithelial marker that is not expected to be expressed on the surface of leukocytes and ldEVs; therefore, leukocytes and ldEVs are not positively selected by the EpCAM ferrofluid. For each sample (with or without EpCAM enrichment), the relative frequencies of ldEVs over leukocytes were calculated. In the blood of healthy individuals with no enrichment, one ldEV was detected for every three leukocytes. In the EpCAM-enriched blood of both healthy individuals and metastatic cancer patients, the relative frequencies of ldEVs over leukocytes was found to be approximately half, with one ldEV being detected for every five (in case of samples from healthy individuals, prostate cancer, and non-small cell lung cancer patients) to six (in case of samples from colorectal cancer patients) leukocytes. The presence of ldEVs in higher relative frequencies in the the whole blood of individuals compared to EpCAM-enriched samples could be attributed to three reasons. Firstly, the blood samples are centrifuged at 800× g for 10 min, and the plasma fraction containing the majority of extracellular vesicles is not processed by the CellSearch system implying that half of the ldEVs detected in the blood samples without EpCAM enrichment end up in the plasma fraction. Secondly, the Fcγ receptors of leukocytes and ldEVs are expected to bind to the heavy chains rather than the antigen-binding sites of the antibodies against EpCAM that are conjugated to the ferrofluid. As ldEVs have fewer receptors due to their smaller surface, their carryover in the EpCAM-enriched sample should be lower than leukocytes. The third reason for that observation might be the lower CD45 signal of leukocytes and ldEVs in the images of the CellTracks Analyzer II compared to the imaging setup used in the case of the blood samples without EpCAM enrichment; the CellTracks Analyzer II uses a mercury arc lamp that results in a suboptimal excitation of the APC-conjugated antibody against CD45 in contrast to the other imaging setup that uses a light-emitting diode (LED) source. In combination with the smaller size of ldEVs, this could lead to ldEVs with a CD45 signal close to the background intensity not being considered as true events; thereby, underestimating the ldEV frequencies. In any case, the relative frequencies of ldEVs to leukocytes in blood samples with and without EpCAM enrichment are of a similar order of magnitude (1:3 and 1:5, respectively) supporting their pre-existence in the blood circulation of individuals and rejecting a possible hypothesis for their formation during the CellSearch procedure.

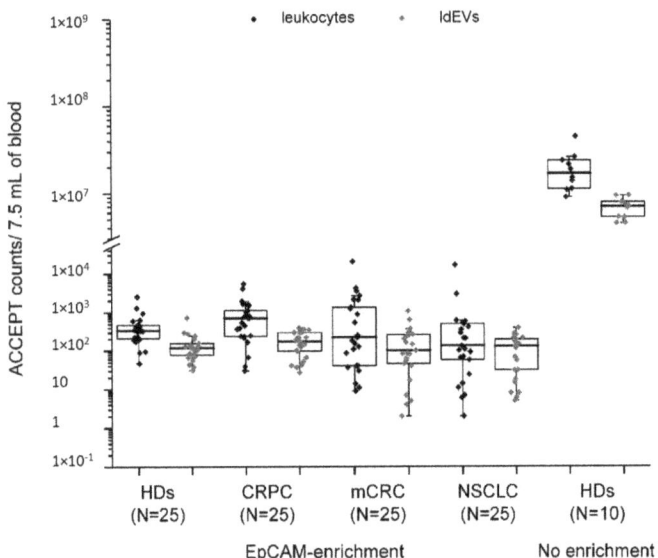

Figure 4. Absolute frequencies of leukocytes and large ldEVs in 7.5 mL of the blood of individuals with and without EpCAM enrichment. The interquartile range of the absolute frequencies of leukocytes (data in black dots) and ldEVs (data in grey dots) in whole blood of 10 healthy individuals and EpCAM enriched blood samples of 25 healthy individuals (HDs) and 75 EpCAM-enriched blood samples of metastatic prostate (CRPC), colorectal (mCRC) and non-small cell lung (NSCLC) cancer patients are shown in box plots. Whiskers indicate max and min values as estimated by Q3 + 1.5*IQR and Q1—1.5*IQR, respectively, where Q1: lower quartile, Q3: upper quartile and IQR: interquartile range. Each dot in the case of the blood of healthy individuals without EpCAM-enrichment corresponds to the mean values of 4–6 technical replicates.

3.5. The Reproducibility of Measurements by Fluorescence Imaging and the Correlation with the Frequencies of Blood Cells by Hematology Analyzer

The technical variability (N = 4–6 technical replicates) of measuring cell populations in 0.02 µL of blood of healthy individuals without any pre-enrichment was assessed by performing 4–6 replicates of 0.02 µL blood from 10 healthy individuals. An average standard error of 25%, 18%, and 23% was obtained for leukocytes, red blood cells, and platelets, respectively. The respective standard error for ldEVs from the technical replicates was found to be 50% because of the very low frequency of ldEVs in 0.02 µL of blood processed, that was found to be 18 ± 5 (mean value ± SD). We expect that processing larger blood volumes would lead to lower technical variations.

The averaged counts of the blood cell classes as estimated by the immunofluorescence imaging were extrapolated to 1 µL of blood and compared to the respective frequencies obtained by the hematology analyzer. The measurements of the fluorescence imaging were moderately correlated ($R^2 = 0.7$) with the counts from the hematology analyzer, as shown in Figure 5. However, all cell populations were underestimated by the fluorescence imaging approach when compared to the hematology analyzer. That can be justified by the low CD45 expression of neutrophils that comprise 60–70% of the whole leukocyte population, the overlap and aggregation of red blood cells (Supplementary Figure S1) that are considered as one when enumerated using the ACCEPT software and the detection limit of fluorescence imaging in case of smaller size platelets.

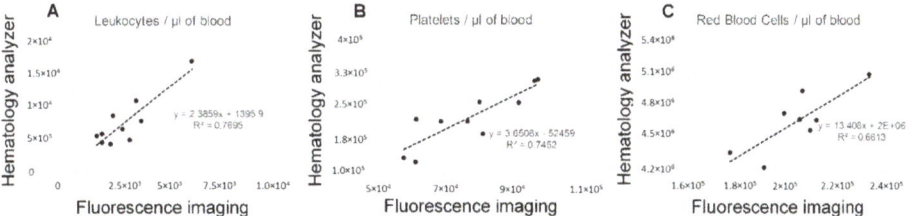

Figure 5. Scatter plots of leukocyte (Panel **A**), platelet (Panel **B**), and red blood cell frequencies (Panel **C**) in 1 µL of whole blood of 10 healthy individuals as estimated by fluorescence imaging and ACCEPT enumeration (x-axis) and by the hematology analyzer (y-axis). Correlation between the measurements of the two techniques was found as indicated by the R^2. The mean counts of each population of 4–6 technical replicates were used in the case of the fluorescence imaging approach.

4. Discussion

The Extracellular Vesicle field has focused so far on the biogenesis and functions of EVs with a size smaller than 1 µm secreted by various cells [26], including platelets [27], neutrophils [28], T and B lymphocytes [29], red blood cells [30], endothelial cells [31,32], and tumor cells [33]. However, the existing literature on the formation and frequencies of EVs larger than 1 µm in healthy individuals is very sparse, as they are considered to be apoptotic bodies, and as such not actively contributing in the intercellular communication. Nevertheless, recent findings in the cancer field shows the promising potential of large tdEVs as their load associated with clinical outcome in metastatic cancer patients [16,25] and their molecular cargo represents the mutational status of the tumor [9]. Our previous research on scanning electron microscopy imaging of CellSearch cartridges of castration-resistant prostate cancer patient samples after EpCAM enrichment [34] and the development of the open-source ACCEPT software for the automated enumeration of all fluorescently labeled objects from image data sets led to our first observations of the presence of DAPI-, CD45+, CK- objects [19]. We baptized these objects leukocyte-derived Extracellular Vesicles (ldEVs) and investigated their presence in digitally stored fluorescence images of CellSearch cartridges. The ldEVs had a similar size range to our previously reported DAPI-, CD45-, CK+ tumor-derived Extracellular Vesicles (tdEVs), that were detected after EpCAM enrichment in metastatic cancer patients [16,17]. Their detection raised questions regarding the pre-existence of these large ldEVs and tdEVs in the blood circulation of individuals or their formation as fragmentation by-products of leukocytes and CTCs, respectively during the CellSearch procedure.

Therefore, we decided to address the question of whether these EVs pre-exist in the blood circulation of individuals. Towards that direction, we enumerated ldEVs in blood samples of 10 healthy individuals without any pre-analytical or pre-enrichment steps and compared the frequencies of ldEVs and leukocytes in the whole blood to the frequency in EpCAM-enriched blood samples of 25 healthy individuals and 75 metastatic cancer patients. ldEVs and leukocytes were detected in a ratio of 1:3 in the blood of healthy individuals without any pre-enrichment and in 1:5 to 1:6 in the EpCAM-enriched blood of healthy individuals and metastatic cancer patients (Figure 4), supporting the pre-existence of these ldEVs in the blood circulation instead of their formation during the EpCAM enrichment. The lower relative frequencies of ldEVs to leukocytes in the EpCAM-enriched blood samples compared to the samples without EpCAM enrichment could be mainly explained by the blood fraction that is processed by the CellSearch system: blood samples are centrifuged at 800× g for 10 min, and the plasma fraction is discarded and not processed by the system. Using that centrifugation force, only EVs with a diameter above 1 µm will be in the blood fraction and will have the chance to come into contact with the EpCAM ferrofluid [35]. The measurements in the blood of healthy donors without EpCAM enrichment were reproducible, as confirmed by the standard deviations of the technical replicates, with a mean ± SD of 900 ± 254 ldEVs in 1 µL of the blood of healthy individuals.

These results do not deviate a lot from the previously reported ldEV frequencies (median value: 356, interquartile range: 268–529) of Simak et al. in the plasma of healthy donors; ldEVs were larger than 200 nm and were defined as CD45+, CD105-, CD235a- by flow cytometry [36]. The use of solely one specific but weakly expressed inclusion marker, namely CD45 to define them using either approaches results in the underestimation of the whole circulating ldEV population; a point also stressed out by Lacroix et al. [37]. Further investigation of ldEVs in terms of their size distribution and surface marker expression using established techniques in the EV field, such as nanoparticle tracking analysis, electron microscopy, flow cytometry, and surface plasmon resonance imaging, would lead to their better profiling [10,38,39]. Eventually, a similar test to the hematology analyzer having as an output the EV populations (of platelet, erythrocyte, leukocyte, endothelial, and epithelial origin) in the biofluids of individuals could serve as an important diagnostic tool in clinical practice, since EVs have been associated with numerous pathophysiological conditions, such as thrombogenicity, inflammation, angiogenesis, and cancer [26,40–44].

Our study has several limitations. Although the extrapolated counts of leukocytes, platelets, and red blood cells per µL of blood correlated to the respective values of the hematology analyzer (Figure 5), the detected frequencies of all the cell populations were consistently lower as compared to the respective ones measured by the hematology analyzer. The particularly lower detection of red blood cells could be explained by the overlap and aggregation of more than one red blood cell segmented as one object by the open-source ACCEPT software and the large range in the distribution of the fluorescence intensity of CD235-APC in the fluorescence images (Supplementary Figure S1, Panel A) through which part of the red blood cells fall outside the applied gate. The underestimation of the leukocytes could be explained by the lower expression of CD45 by the granulocytes that consist to 60–70% of white blood cells. The detection of the smaller platelet population and their secreted EVs is limited by our approach, because of the use of a 0.45 NA objective that results in a resolution of 0.64 µm/pixel. The abundance and high signal of the red blood cells prevented us from the detection and enumeration of low signal-to-background ratio red blood cell-derived EVs. Importantly, the Hoechst 33342 labeled nucleic acids in platelets could be detected by flow cytometry but not with our microscopy set-up. This observation implies that the zero DAPI signal detected with ACCEPT inside our previously reported tdEVs does not rule out the presence of nucleic acids within them. This is an important finding encouraging the further characterization of tdEVs that are immunomagnetically isolated based on their EpCAM expression [16,17,25]. That would come into agreement with the findings of Vagner et al. that the DNA cargo of large EVs reflect the genotype of prostate cancer patients [9]. The use of a membrane permeable dye, binding to both DNA and RNA, with a higher sensitivity, like SYTO13 [45] could also facilitate the detection of nucleic acids within the isolated EVs.

Interestingly, platelets have a similar size to ldEVs based on the immunofluorescence images of CD61 and CD45, respectively, with a minimum area of the detected objects being 9 µm^2 based on our observations from the ACCEPT scatter plots. That area corresponds to a circular object of an approximate radius of 1.7 µm. However, it was not possible to confirm the size of ldEVs from the respective bright field images, because opposite to platelets, the contrast between the background and ldEV intensity was inadequate to detect them (Figure 2 and Supplementary Figure S2) suggesting that their physical properties (absorption coefficient, scattering coefficient, scattering anisotropy, refractive index) differ from the ones of platelets.

It is worth mentioning that ldEVs were found in 5–6 lower frequencies compared to leukocytes in EpCAM enriched samples, whereas tdEVs in our previous studies were detected in 10–20 times higher frequencies compared to CTCs [16]. That observation could be explained after taking into consideration some technical and biological facts. From a technical perspective, our study was limited by the resolution of a 10×/0.45 NA objective fluorescence microscope, implying that only the larger EVs with a larger than 1 µm diameter or the ones with a high expression of inclusion markers could be detected. CK is the inclusion marker used for the detection of tdEVs, whereas the detection of ldEVs is accomplished by the inclusion of CD45. Since the CK expression is intracellular and proportional to the volume instead of the surface, as in the case of the CD45 expression of ldEVs, CK is easier to detect in smaller tdEVs than CD45 in similar size ldEVs. Consequently, the CD45 expression may be present in more particles in blood samples but not exceeding the detection limit to be considered positive. Further characterization of the size distribution and the surface marker expression profile could elaborate on the detection limits of our technique. From a biological perspective, tdEVs could be found in higher frequencies either because of the increased apoptosis and fragmentation of CTCs in the blood circulation or because of different EV secretion pattern between normal and cancerous cells. Regarding the first hypothesis, the lifespan of neutrophils is around 24 h [46], whereas the circulation lifetime of CTCs has been estimated to be 1 to 2.4 h [47]; that; however, does not imply that CTCs are fragmented and cleared by the blood. On the contrary, *in vivo* animal studies showed the trap of more than 80% of viable CTCs by the liver and lung, that serve as "filter" organs, and the survival of CTCs for the prolonged time in a dormant state [48,49]. The survival of CTCs in the bloodstream is further supported by studies on their mechanical phenotype. Atomic force microscopy studies on cell lines suggest that cells with increased metastatic potential are more deformable (as expressed by Young's modulus), compared to less metastatic or non-malignant cells [50,51]. These results were further confirmed in clinical samples from pleural effusions, where metastatic cells had lower stiffness compared to benign cells from the same effusions and leukocytes [52,53]. Interestingly, Sun et al. demonstrated that deformable cancer cells engulf neighboring ones with higher stiffness via entosis further encouraging the increased survivorship of CTCs [54]. Regarding the second hypothesis of different EV secretion pattern of CTCs and leukocytes, it is well known that cancer cells have reprogrammed metabolism and acquired traits that promote their survival and growth [55,56]. Recent findings of independent research groups converge into the survival of tumor cells regardless of their phenotypic characteristics of possible apoptosis, such as caspases activation, amoeboid phenotype and membrane blebbing [57,58]. Instead of undergoing apoptosis, cells with such traits have a more tumorigenic and invasive phenotype [59]. Hence, tumor cells may actively secrete EVs similar in size to apoptotic bodies, but without special receptors on their surface to be recognized and ingested by macrophages for their clearance as in case of healthy cells (e.g., white blood cells).

5. Conclusions

In conclusion, the relative frequencies of large (above 1 μm) leukocyte-derived Extracellular Vesicles (ldEVs) to leukocytes are similar in EpCAM-enriched blood samples of healthy individuals and cancer patients (1:6 to 1:5) as in the blood of healthy individuals without EpCAM enrichment (1:3), implying their pre-existence in the blood circulation rather than their formation from activated or apoptotic leukocytes using the CellSearch system. Furthermore, the Hoechst signal of platelets could not be detected using a similar fluorescence microscope as the CellTracks Analyzer II. These two findings have important implications for our previously reported tumor-derived Extracellular Vesicles (tdEVs), that were immunomagnetically co-isolated with CTCs based on their EpCAM expression from metastatic cancer patients [17]. Firstly, tdEVs are most likely not a result of CTC fragmentation during the CellSearch procedure and secondly, the presence of undetectable nucleic acids within ldEVs and tdEVs should not be excluded but instead further investigated. No conclusions could be drawn in regards to the smaller ldEV population, namely exosomes, since they are not expected to be detected with our imaging setup. Last but not least, our results do not allow for comparison of ldEVs between healthy individuals and cancer patients since the available image data sets of patients corresponded to only EpCAM enriched samples.

Supplementary Materials: The following are available online at http://www.mdpi.com/2073-4409/8/8/937/s1, Supplementary Figure S1: Examples of composite immunofluorescence images of blood samples without EpCAM enrichment, obtained with an inverted scanning fluorescence microscope using a 10×/0.45 NA (Panel A) and a 60×/0.7 NA objective (Panel B). Supplementary Figure S2: Examples of brightfield and immunofluorescence images of leukocyte- derived Extracellular Vesicles, enclosed within circles.

Author Contributions: Conceptualization—L.W.M.M.T. and A.N.; methodology—A.N., L.L.Z., and L.W.M.M.T.; Validation—A.N.; formal analysis, A.N.; Investigation—A.N.; Resources—A.N., L.L.Z., L.W.M.M.T.; Data curation—A.N.; Writing—original draft preparation—A.N.; Writing—review and editing—A.N., L.L.Z., L.W.M.M.T.; Visualization—A.N.; Supervision—L.W.M.M.T.; Project administration—L.W.M.M.T.; funding acquisition—L.W.M.M.T.

Funding: This research was funded by NWO Applied and Engineering Sciences, grant number 14190.

Acknowledgments: The authors acknowledge the Experimental Centre for Technical Medicine (ECTM) of the University of Twente for providing us blood samples of healthy individuals. We would also like to thank all the patients and healthy individuals, the blood samples of whom were used for the accomplishment of the present study. Special acknowledgments to C. Breukers and J. Weersink for their technical support with the inverted fluorescence microscope whenever needed.

Conflicts of Interest: The authors declare no conflict of interest. The funders had no role in the design of the study; in the collection, analyses, or interpretation of data; in the writing of the manuscript, or in the decision to publish the results.

References

1. Van Niel, G.D.; Angelo, G.; Raposo, G. Shedding light on the cell biology of extracellular vesicles. *Nat. Rev. Mol. Cell. Biol.* **2018**, *19*, 213–228. [CrossRef]
2. Fais, S.O.; Driscoll, L.; Borras, F.E.; Buzas, E.; Camussi, G.; Cappello, F.; Carvalho, J.; Cordeiro da Silva, A.; Del Portillo, H.; El Andaloussi, S. Evidence-based clinical use of nanoscale extracellular vesicles in nanomedicine. *ACS Nano* **2016**, *10*, 3886–3899. [CrossRef]
3. Yanez-Mo, M.; Siljander, P.R.; Andreu, Z.; Zavec, A.B.; Borras, F.E.; Buzas, E.I.; Buzas, K.; Casal, E.; Cappello, F.; Carvalho, J.; et al. Biological properties of extracellular vesicles and their physiological functions. *J. Extracell. Vesicles* **2015**, *4*, 27066. [CrossRef]
4. Jia, S.; Zhang, R.; Li, Z.; Li, J. Clinical and biological significance of circulating tumor cells, circulating tumor DNA, and exosomes as biomarkers in colorectal cancer. *Oncotarget* **2017**, *8*, 55632–55645. [CrossRef]
5. Krug, A.K.; Enderle, D.; Karlovich, C.; Priewasser, T.; Bentink, S.; Spiel, A.; Brinkmann, K.; Emenegger, J.; Grimm, D.G.; Castellanos-Rizaldos, E.; et al. Improved EGFR mutation detection using combined exosomal RNA and circulating tumor DNA in NSCLC patient plasma. *Ann. Oncol. Off. J. Eur. Soc. Med. Oncol.* **2018**, *29*, 2143. [CrossRef]
6. Costa-Silva, B.; Aiello, N.M.; Ocean, A.J.; Singh, S.; Zhang, H.; Thakur, B.K.; Becker, A.; Hoshino, A.; Mark, M.T.; Molina, H.; et al. Pancreatic cancer exosomes initiate pre-metastatic niche formation in the liver. *Nat. Cell Biol.* **2015**, *17*, 816–826. [CrossRef]
7. Hoshino, A.; Costa-Silva, B.; Shen, T.L.; Rodrigues, G.; Hashimoto, A.; Tesic Mark, M.; Molina, H.; Kohsaka, S.; Di Giannatale, A.; Ceder, S.; et al. Tumour exosome integrins determine organotropic metastasis. *Nature* **2015**, *527*, 329–335. [CrossRef]
8. Kosaka, N.; Yoshioka, Y.; Fujita, Y.; Ochiya, T. Versatile roles of extracellular vesicles in cancer. *J. Clin. Investig.* **2016**, *126*, 1163–1172. [CrossRef]
9. Vagner, T.; Spinelli, C.; Minciacchi, V.R.; Balaj, L.; Zandian, M.; Conley, A.; Zijlstra, A.; Freeman, M.R.; Demichelis, F.; De S Posadas, E.M. Large extracellular vesicles carry most of the tumour DNA circulating in prostate cancer patient plasma. *J. Extracell. Vesicles* **2018**, *7*, 1505403. [CrossRef]
10. Ricklefs, F.L.; Maire, C.L.; Reimer, R.; Duhrsen, L.; Kolbe, K.; Holz, M.; Schneider, E.; Rissiek, A.; Babayan, A.; Hille, C.; et al. Imaging flow cytometry facilitates multiparametric characterization of extracellular vesicles in malignant brain tumours. *J. Extracell. Vesicles.* **2019**, *8*, 1588555. [CrossRef]
11. Konig, L.; Kasimir-Bauer, S.; Bittner, A.K.; Hoffmann, O.; Wagner, B.; Santos Manvailer, L.F.; Kimmig, R.; Horn, P.A.; Rebmann, V. Elevated levels of extracellular vesicles are associated with therapy failure and disease progression in breast cancer patients undergoing neoadjuvant chemotherapy. *Oncoimmunology* **2017**, *7*, e1376153. [CrossRef]
12. Kanwar, S.S.; Dunlay, C.J.; Simeone, D.M.; Nagrath, S. Microfluidic device (ExoChip) for on-chip isolation, quantification and characterization of circulating exosomes. *Lab Chip* **2014**, *14*, 1891–1900. [CrossRef]
13. Slomka, A.; Urban, S.K.; Lukacs-Kornek, V.; Zekanowska, E.; Kornek, M. Large Extracellular Vesicles: Have We Found the Holy Grail of Inflammation? *Front. Immunol.* **2018**, *9*, 2723. [CrossRef]

14. Padda, R.S.; Deng, F.K.; Brett, S.I.; Biggs, C.N.; Durfee, P.N.; Brinker, C.J.; Williams, K.C.; Leong, H.S. Nanoscale flow cytometry to distinguish subpopulations of prostate extracellular vesicles in patient plasma. *Prostate* **2019**, *79*, 592–603. [CrossRef]
15. Reategui, E.; van der Vos, K.E.; Lai, C.P.; Zeinali, M.; Atai, N.A.; Aldikacti, B.; Floyd, F.P.A.H.K., Jr.; Thapar, V.; Hochberg, F.H.; Sequist, L.V.; et al. Engineered nanointerfaces for microfluidic isolation and molecular profiling of tumor-specific extracellular vesicles. *Nat. Commun.* **2018**, *9*, 175. [CrossRef]
16. Nanou, A.; Coumans, F.A.W.; van Dalum, G.; Zeune, L.L.; Dolling, D.; Onstenk, W.; Crespo, M.; Fontes, M.S.; Rescigno, P.; Fowler, G.; et al. Circulating tumor cells, tumor-derived extracellular vesicles and plasma cytokeratins in castration-resistant prostate cancer patients. *Oncotarget* **2018**, *9*, 19283–19293. [CrossRef]
17. Nanou, A.; Zeune, L.L.; de Wit, S.; Miller, C.M.; Punt, C.J.A.; Groen, H.J.M.; Hayes, D.F.; de Bono, J.S.L.W.M.M.T. *Tumor-Derived Extracellular Vesicles in Blood of Metastatic Breast, Colorectal, Prostate and Non-Small Cell Lung Cancer Patients Associate with Worse Survival*; American Association for Cancer Research: Philadelphia, PA, USA, 2019.
18. Zeune, L.; van Dalum, G.; Terstappen, L.W.M.M.; van Gils, S.A.; Brune, C. Multiscale Segmentation via Bregman Distances and Nonlinear Spectral Analysis. *SIAM J. Imaging Sci.* **2017**, *10*, 111–146. [CrossRef]
19. Zeune, L. *Automated CTC Classification, Enumeration and Pheno Typing: Where Marh Meets Biology. Medical Cell BioPhysics*; University of Twente: Overijssel, The Netherland, 2019.
20. Coumans, F.; Terstappen, L. Detection and Characterizati on of Circulating Tumor Cells by the CellSearch Approach. *Methods Mol. Biol.* **2015**, *1347*, 263–278.
21. Allard, W.J.; Matera, J.; Miller, M.C.; Repollet, M.; Connelly, M.C.; Rao, C.; Tibbe, A.G.; Uhr, J.W.; Terstappen, L.W. Tumor cells circulate in the peripheral blood of all major carcinomas but not in healthy subjects or patients with nonmalignant diseases. *Clin. Cancer Res.* **2004**, *10*, 6897–6904. [CrossRef]
22. Tibbe, A.G.; de Grooth, B.G.; Greve, J.; Dolan, G.J.; Rao, C.; Terstappen, L.W. Magnetic field design for selecting and aligning immunomagnetic labeled cells. *Cytometry* **2002**, *47*, 163–172. [CrossRef]
23. De Wit, S.; Zeune, L.L.; Hiltermann, T.J.N.; Groen, H.J.M.; Dalum, G.V.; Terstappen, L.W.M.M. Classification of Cells in CTC-Enriched Samples by Advanced Image Analysis. *Cancers* **2018**, *10*, 377. [CrossRef]
24. Zeune, L.L.; de Wit, S.; Berghuis, A.M.S.M.J.I.J.; Terstappen, L.; Brune, C. How to Agree on a CTC: Evaluating the Consensus in Circulating Tumor Cell Scoring. *Cytom. A* **2018**, *93*, 1202–1206. [CrossRef]
25. De Wit, S.; Rossi, E.; Weber, S.; Tamminga, M.; Manicone, M.; Swennenhuis, J.F. Groothuis-Oudshoorn, C.G.M.; Vidotto, R.; Facchinetti, A.; Zeune, L.L.; et al. Single tube liquid biopsy for advanced non-small cell lung cancer. *Int. J. Cancer* **2019**, *144*, 3127–3137. [CrossRef]
26. Van der Pol, E.; Boing, A.N.; Harrison, P.; Sturk, A.; Nieuwland, R. Classification, functions, and clinical relevance of extracellular vesicles. *Pharmacol. Rev.* **2012**, *64*, 676–705. [CrossRef]
27. Heijnen, H.F.; Schiel, A.E.; Fijnheer, R.; Geuze, H.J.; Sixma, J.J. Activated platelets release two types of membrane vesicles: Microvesicles by surface shedding and exosomes derived from exocytosis of multivesicular bodies and alpha-granules. *Blood* **1999**, *94*, 3791–3799.
28. Pitanga, T.N.; de Aragao Franca, L.; Rocha, V.C.; Meirelles, T.; Borges, V.M.; Goncalves, M.S.; Pontes-de-Carvalho, L.C.; Noronha-Dutra, A.A.; dos-Santos, W.L. Neutrophil-derived microparticles induce myeloperoxidase-mediated damage of vascular endothelial cells. *BMC Cell Biol.* **2014**, *15*, 21. [CrossRef]
29. Baka, Z.; Senolt, L.; Vencovsky, J.; Mann, H.; Simon, P.S.; Kittel, A.; Buzas, E.; Nagy, G. Increased serum concentration of immune cell derived microparticles in polymyositis/dermatomyositis. *Immunol. Lett.* **2010**, *128*, 124–130. [CrossRef]
30. Canellini, G.; Rubin, O.; Delobel, J.; Crettaz, D.; Lion, N.; Tissot, J.D. Red blood cell microparticles and blood group antigens: An analysis by flow cytometry. *Blood Transfus.* **2012**, *10*, s39–s45.
31. Wheway, J.; Latham, S.L.; Combes, V.; Grau, G.E.R. Endothelial microparticles interact with and support the proliferation of T cells. *J. Immunol.* **2014**, *193*, 3378–3387. [CrossRef]
32. Deregibus, M.C.; Cantaluppi, V.; Calogero, R.; Lo Iacono, M.; Tetta, C.; Biancone, L.; Bruno, S.; Bussolati, B.; Camussi, G. Endothelial progenitor cell derived microvesicles activate an angiogenic program in endothelial cells by a horizontal transfer of mRNA. *Blood* **2007**, *110*, 2440–2448. [CrossRef]
33. Ruhen, O.; Meehan, K. Tumor-Derived Extracellular Vesicles as a Novel Source of Protein Biomarkers for Cancer Diagnosis and Monitoring. *Proteomics* **2019**, *19*, e1800155. [CrossRef]
34. Nanou, A.; Crespo, M.; Flohr, P.; De Bono, J.S.; Terstappen, L. Scanning Electron Microscopy of Circulating Tumor Cells and Tumor-Derived Extracellular Vesicles. *Cancers* **2018**, *10*, 416. [CrossRef]

35. Rikkert, L.G.; van der Pol, E.; van Leeuwen, T.G.; Nieuwland, R.; Coumans, F.A.W. Centrifugation affects the purity of liquid biopsy-based tumor biomarkers. *Cytometry A* **2018**, *93*, 1207–1212. [CrossRef]
36. Simak, J.; Holada, K.; Risitano, A.M.; Zivny, J.H.; Young, N.S.; Vostal, J.G. Elevated circulating endothelial membrane microparticles in paroxysmal nocturnal haemoglobinuria. *Br. J. Haematol.* **2004**, *125*, 804–813. [CrossRef]
37. Lacroix, R.; Robert, S.; Poncelet, P.; Dignat-George, F. Overcoming limitations of microparticle measurement by flow cytometry. *Semin. Thromb. Hemost.* **2010**, *36*, 807–818. [CrossRef]
38. Van der Pol, E.; Coumans, F.A.; Grootemaat, A.E.; Gardiner, C.; Sargent, I.L.; Harrison, P.; Sturk, A.; van Leeuwen, T.G.; Nieuwland, R. Particle size distribution of exosomes and microvesicles determined by transmission electron microscopy, flow cytometry, nanoparticle tracking analysis and resistive pulse sensing. *J. Thromb. Haemost.* **2014**, *12*, 1182–1192. [CrossRef]
39. Gool, E.L.; Stojanovic, I.; Schasfoort, R.B.M.; Sturk, A.; van Leeuwen, T.G.; Nieuwland, R.; Terstappen, L.; Coumans, F.A.W. Surface Plasmon Resonance is an Analytically Sensitive Method for Antigen Profiling of Extracellular Vesicles. *Clin. Chem.* **2017**, *63*, 1633–1641. [CrossRef]
40. Halim, A.T.A.; Ariffin, N.A.F.M.; Azlan, M. Review: The Multiple Roles of Monocytic Microparticles. *Inflammation* **2016**, *39*, 1277–1284. [CrossRef]
41. Tissot, J.D.; Canellini, G.; Rubin, O.; Angelillo-Scherrer, A.; Delobel, J.; Prudent, M.; Lion, N. Blood microvesicles: From proteomics to physiology. *Transl. Proteom.* **2013**, *1*, 38–52. [CrossRef]
42. Burnier, L.; Fontana, P.; Kwak, B.R.; Angelillo-Scherrer, A. Cell-derived microparticles in haemostasis and vascular medicine. *Thromb. Haemost.* **2009**, *101*, 439–451. [CrossRef]
43. Julich-Haertel, H.; Urban, S.K.; Krawczyk, M.; Willms, A.; Jankowski, K.; Patkowski, W.; Kruk, B.; Krasnodebski, M.; Ligocka, J.; Schwab, R.; et al. Cancer-associated circulating large extracellular vesicles in cholangiocarcinoma and hepatocellular carcinoma. *J. Hepatol.* **2017**, *67*, 282–292. [CrossRef] [PubMed]
44. Ullal, A.J.; Pisetsky, D.S.; Reich, C.F., III. Use of SYTO 13, a fluorescent dye binding nucleic acids, for the detection of microparticles in in vitro systems. *Cytom. Part A J. Int. Soc. Anal. Cytol.* **2010**, *7*, 294–301. [CrossRef] [PubMed]
45. McCracken, J.M.; Allen, L.A. Regulation of human neutrophil apoptosis and lifespan in health and disease. *J. Cell Death* **2014**, *7*, 15–23. [CrossRef] [PubMed]
46. Meng, S.; Tripathy, D.; Frenkel, E.P.; Shete, S.; Naftalis, E.Z.; Huth, J.F.; Beitsch, P.D.; Leitch, M.; Hoover, S.; Euhus, D.; et al. Circulating tumor cells in patients with breast cancer dormancy. *Clin. Cancer Res.* **2004**, *10*, 8152–8162. [CrossRef] [PubMed]
47. Luzzi, K.J.; MacDonald, I.C.; Schmidt, E.E.; Kerkvliet, N.; Morris, V.L.; Chambers, A.F.; Groom, A.C. Multistep nature of metastatic inefficiency: Dormancy of solitary cells after successful extravasation and limited survival of early micrometastases. *Am. J. Pathol.* **1998**, *153*, 865–873. [CrossRef]
48. Cameron, M.D.; Schmidt, E.E.; Kerkvliet, N.; Nadkarni, K.V.; Morris, V.L.; Groom, A.C.; Chambers, A.F.; MacDonald, I.C. Temporal progression of metastasis in lung: Cell survival, dormancy, and location dependence of metastatic inefficiency. *Cancer Res.* **2000**, *60*, 2541–2546. [PubMed]
49. Li, Q.S.; Lee, G.Y.H.; Ong, C.N.; Lim, C.T. AFM indentation study of breast cancer cells. *Biochem. Biophys. Res. Commun.* **2008**, *374*, 609–613. [CrossRef]
50. Zhang, W.; Kai, K.; Choi, D.S.; Iwamoto, T.; Nguyen, Y.H.; Wong, H.; Landis, M.D.; Ueno, N.T.; Chang, J.; Qin, L. Microfluidics separation reveals the stem-cell-like deformability of tumor-initiating cells. *Proc. Natl. Acad. Sci. USA* **2012**, *109*, 18707–18712. [CrossRef]
51. Cross, S.E.; Jin, Y.S.; Rao, J.; Gimzewski, J.K. Nanomechanical analysis of cells from cancer patients. *Nat. Nanotechnol.* **2007**, *2*, 780–783. [CrossRef]
52. Gossett, D.R.; Tse, H.T.K.; Lee, S.A.; Ying, Y.; Lindgren, A.G.; Yang, O.O.; Rao, J.; Clark, A.T.; Di Carlo, D. Hydrodynamic stretching of single cells for large population mechanical phenotyping. *Proc. Natl. Acad. Sci. USA* **2012**, *109*, 7630–7635. [CrossRef]
53. Sun, Q.; Luo, T.; Ren, Y.; Florey, O.; Shirasawa, S.; Sasazuki, T.; Robinson, D.N.; Overholtzer, M. Competition between human cells by entosis. *Cell Res.* **2014**, *24*, 1299–1310. [CrossRef] [PubMed]
54. Hanahan, D.; Weinberg, R.A. Hallmarks of cancer: The next generation. *Cell* **2011**, *144*, 646–674. [CrossRef] [PubMed]
55. Hanahan, D.; Weinberg, R.A. The hallmarks of cancer. *Cell* **2000**, *100*, 57–70. [CrossRef]

56. Jinesh, G.G.; Choi, W.; Shah, J.B.; Lee, E.K.; Willis, D.L.; Kamat, A.M. Blebbishields, the emergency program for cancer stem cells: Sphere formation and tumorigenesis after apoptosis. *Cell Death Differ.* **2013**, *20*, 382–395. [CrossRef] [PubMed]
57. Di Vizio, D.; Morello, M.; Dudley, A.C.; Schow, P.W.; Adam, R.M.; Morley, S.; Mulholland, D.; Rotinen, M.; Hager, M.H.; Insabato, L.; et al. Large oncosomes in human prostate cancer tissues and in the circulation of mice with metastatic disease. *Am. J. Pathol.* **2012**, *181*, 1573–1584. [CrossRef] [PubMed]
58. Di Vizio, D.; Kim, J.; Hager, M.H.; Morello, M.; Yang, W.; Lafargue, C.J.; True, L.D.; Rubin, M.A.; Adam, R.M.; Beroukhim, R.; et al. Oncosome formation in prostate cancer: Association with a region of frequent chromosomal deletion in metastatic disease. *Cancer Res.* **2009**, *69*, 5601–5609. [CrossRef] [PubMed]
59. Reis-Sobreiro, M.; Chen, J.F.; Novitskaya, T.; You, S.; Morley, S.; Steadman, K.; Gill, N.K.; Eskaros, A.; Rotinen, M.; Chu, C.Y.; et al. Emerin Deregulation Links Nuclear Shape Instability to Metastatic Potential. *Cancer Res.* **2018**, *78*, 6086–6097. [CrossRef] [PubMed]

© 2019 by the authors. Licensee MDPI, Basel, Switzerland. This article is an open access article distributed under the terms and conditions of the Creative Commons Attribution (CC BY) license (http://creativecommons.org/licenses/by/4.0/).

Article

Molecular Characterization of Circulating Tumor Cells Enriched by A Microfluidic Platform in Patients with Small-Cell Lung Cancer

Eva Obermayr [1,*], Christiane Agreiter [1], Eva Schuster [1], Hannah Fabikan [2], Christoph Weinlinger [2], Katarina Baluchova [3], Gerhard Hamilton [4], Maximilian Hochmair [2] and Robert Zeillinger [1]

1. Molecular Oncology Group, Department of Obstetrics and Gynecology, Comprehensive Cancer Center, Medical University of Vienna, Waehringer Guertel 18-20, 1090 Vienna, Austria
2. Department of Respiratory and Critical Care Medicine, Sozialmedizinisches Zentrum Baumgartner Höhe, Sanatoriumstrasse 2, 1140 Vienna, Austria
3. Division of Oncology, Biomedical Center Martin, Jessenius Faculty of Medicine in Martin, Comenius University in Bratislava, Malá Hora 4C, 036 01 Martin, Slovakia
4. Department of Surgery, Medical University of Vienna, Waehringer Guertel 18-20, 1090 Vienna, Austria
* Correspondence: eva.obermayr@meduniwien.ac.at

Received: 25 June 2019; Accepted: 31 July 2019; Published: 13 August 2019

Abstract: At initial diagnosis, most patients with small-cell lung cancer (SCLC) present with metastatic disease with a high number of tumor cells (CTCs) circulating in the blood. We analyzed RNA transcripts specific for neuroendocrine and for epithelial cell lineages, and Notch pathway delta-like 3 ligand (*DLL3*), the actionable target of rovalpituzumab tesirine (Rova-T) in CTC samples. Peripheral blood samples from 48 SCLC patients were processed using the microfluidic Parsortix™ technology to enrich the CTCs. Blood samples from 26 healthy donors processed in the same way served as negative controls. The isolated cells were analyzed for the presence of above-mentioned transcripts using quantitative PCR. In total, 16/51 (31.4%) samples were CTC-positive as determined by the expression of epithelial cell adhesion molecule 1 (*EpCAM*), cytokeratin 19 (*CK19*), chromogranin A (*CHGA*), and/or synaptophysis (*SYP*). The epithelial cell lineage-specific *EpCAM* and/or *CK19* gene expression was observed in 11 (21.6%) samples, and positivity was not associated with impaired survival. The neuroendocrine cell lineage-specific *CHGA* and/or *SYP* were positive in 13 (25.5%) samples, and positivity was associated with poor overall survival. *DLL3* transcripts were observed in four (7.8%) SCLC blood samples and *DLL3*-positivity was similarly associated with poor overall survival (OS). CTCs in SCLC patients can be assessed using epithelial and neuroendocrine cell lineage markers at the molecular level. Thus, the implementation of liquid biopsy may improve the management of lung cancer patients, in terms of a faster diagnosis, patient stratification, and on-treatment therapy monitoring.

Keywords: small-cell lung carcinoma; circulating tumor cells; microfluidics; gene expression analysis; synaptophysin; chromogranin A; rovalpituzumab tesirine

---

1. Introduction

Lung cancer is the most common cancer worldwide. In 2018, a total of 2.1 million new cases were estimated, accounting for 11.6% of all new cancer diagnoses [1,2]. In general, two major types of lung cancer exist: non-small-cell lung cancer (NSCLC), which accounts for about 85% of all lung cancer cases, and small-cell lung cancer (SCLC), which is diagnosed in approximately 15% of all lung cancers. For patients with early-stage NSCLC, a surgical resection offers the best opportunity for cure, while in advanced cases a systemic therapy is the standard of care. SCLC, however, is usually diagnosed rather late when the cancer has already disseminated. In this case a multimodal therapy which includes chemotherapy and radiotherapy is considered the gold standard [3]. Due to these different therapeutic approaches, it is of utmost importance to have a reliable diagnostic platform to differentiate between SCLC and NSCLC.

SCLC belongs to the group of neuroendocrine tumors of the lung. It is diagnosed using hematoxylin and eosin stained sections of the biopsied tissue. However, the histopathological diagnosis of SCLC based on its distinctive morphology may be difficult due to limited material supply from biopsied tissue or aspirated cytological specimens [4]. In some cases the diagnosis of SCLC may be further confirmed by immunohistochemistry using the neuroendocrine markers chromogranin (CHGA), synaptophysin (SYP), and neural cell adhesion molecule 1 (NCAM1) [5,6]. In recent years, the Notch pathway delta-like 3 ligand (DLL3) has gradually gained more interest since it is frequently and selectively expressed on tumorous tissue in SCLC patients and hence it has been associated with neuroendocrine tumorigenesis. Most importantly it is the therapeutic target of the antibody-drug conjugate rovalpituzumab tesirine (Rova-T) [7].

In contrast to conventional tissue biopsies or cytological preparations, liquid biopsies that contain circulating tumor cells (CTCs) and/or circulating tumor DNA, represent a novel approach that illuminates the whole molecular profile of a tumor at the time of sampling [8,9]. Liquid biopsies are taken by a simple blood draw and, thus, are less stressful for the patient, more conventionally used and less expensive than tissue biopsies. For this reason, liquid biopsies can be taken several times to monitor the temporal heterogeneity of the tumor. Especially in lung cancer, liquid biopsies may outperform tissue biopsies with respect to the tumor's accessibility at resection. In addition, small tissue samples are often already exhausted after routine histological staining and hence no longer available for advanced analysis. Furthermore, longitudinal sampling for monitoring of any development of therapy resistance is almost impossible with tissue biopsies [10].

The presence and clinical significance of CTCs has already been shown in many types of malignancies, among them e.g., breast, colorectal, prostate, and lung cancer. In contrast to most other cancer types, SCLC is characterized by a large number of CTCs in the circulation [11]. Several studies have shown the prognostic value of CTC counts in SCLC, most of them using the US Food and Drug Administration (FDA) cleared CellSearch test [11–17]. In addition to the number of CTCs found, their molecular characterization may be a part of a more comprehensive approach providing further information on e.g., downregulation of epithelial markers or presence of druggable targets. Recently, we have demonstrated that processing blood samples using the microfluidic Parsortix™ technology considerably improved the molecular analysis of the enriched CTCs [18].

Considering the abundance of CTCs and the ease of obtaining/performing liquid biopsies extends the possibilities for differential diagnosis and patient stratification. For these reasons we believe that the molecular characterization of CTCs in SCLC may be of uppermost importance for this type of lung cancer. In the present study we applied a recently developed workflow which combines a microfluidic enrichment of CTCs and a qPCR-based analysis for evaluating the gene expression levels of markers of the epithelial (epithelial cell adhesion molecule 1, *EpCAM* and cytokeratin 19, *CK19*) and neuroendocrine (*CHGA*, *SYP*, *NCAM1* and enolase 2 *ENO2*) cell lineage origin, in addition to the druggable target *DLL3*.

2. Materials and Methods

Blood samples were taken from patients with SCLC at the Department of Respiratory and Critical Care Medicine at Sozialmedizinisches Zentrum Baumgartner Höhe, Vienna, Austria. Control blood samples came from healthy donors without a history of cancer. All donors signed an informed consent. The study was approved by the Ethics Committee of the Medical University of Vienna, Austria (EK366/2003 and EK2266/2018).

The blood was collected in Vacuette EDTA tubes (Greiner Bio-One) and processed on the same day in accordance with a recently published protocol employing the label-free microfluidic Parsortix™ technology (Angle plc., UK) [18]. The key component of the device is a microscope slide sized disposable separation cassette, which contains a series of steps with a precisely defined height. Rare cells (e.g., CTCs) are captured within the separation cassette based on their less deformable nature and usually larger size compared to blood cells. Before separation, the blood was diluted with an equal volume of phosphate buffered saline (PBS) and directly processed using a Parsortix™ technology. In this study a separation cassette with a critical step size of 6.5 µm was used, and the separation was performed at 99 mbar pressure. After the separation was completed the captured cells were recovered using a back-flush cycle and immediately lysed by adding 350 µl RLT lysis buffer (Qiagen). The lysates were stored at −80 °C until RNA extraction.

Total RNA was extracted from the cell lysates using the RNeasy Micro Kit (Qiagen) without DNase treatment. The total amount of RNA was converted into cDNA using the SuperScript VILO Mastermix (Invitrogen). qPCR was done in duplicates in a 10 µL total reaction volume on the ViiA7 Real-Time PCR System using the TaqMan® Universal Mastermix II and exon spanning TaqMan® assays specific for *EpCAM*, *NCAM1*, *CHGA*, *SYP*, *DLL3*, *ENO2*, and *CDKN1B* (Life Technologies) with thermal cycling parameters (50 °C for 2 min; 95 °C for 10 min followed by 40 cycles at 95 °C for 15 s and 60 °C for 1 min). A qPCR specific for *CK19* was performed at 65 °C annealing/extension with forward and reverse primers that correspond to published primer sequences and with a FAM™ labeled hydrolysis probe (5′-TgTCCTgCAgATCgACAACgCCC-3′) [19]. Raw data were analyzed using the ViiA7 Software v1.1 with automatic threshold setting and baseline correction. If the fluorescent signal did not reach the threshold in both duplicate reactions, the sample was regarded as negative.

The SCLC CTC lines used for the spiking experiments were derived from patients' blood samples [20]. They were trypsinized at about 70% confluence and stained with CellTrace Violet (Invitrogen) according to the manufacturer's protocol. Subsequently, 100 stained cells were added manually to a 10 mL control blood sample, which was then processed using the Parsortix™ technology as described above. The tumor cells captured in the separation cassette were counted using a fluorescence microscope (Olympus BX50).

The Pearson's chi-square and Fisher's exact test were used to assess the relationship between clinicopathological characteristics of the patients and the presence or absence of the respective gene transcripts. Overall survival (OS) was defined as the period of time in months between blood draw and either death or the last date the patient was seen alive. Kaplan–Meier survival analyses and log-rank testing were used to compare survival outcomes [21]. Cox proportional-hazards regression was used to determine univariate and multiple hazards ratios (HR) for OS [7]. The included covariates were the stage of disease at blood draw (primary vs. progressive disease) and the presence vs. absence of the respective transcripts. The model was built using a forward stepwise method by entering all variables at a p value of less than 0.05 and removing them at a p value of greater than 0.10. The statistical analysis was performed with SPSS version 19.0 (SPSS Inc., Chicago, IL). The level of significance was set at $p < 0.05$. Graphs were done using GraphPad Prism version 6.01.

3. Results

3.1. Patients and Samples

The characteristics of 48 patients with a histopathologically confirmed diagnosis of SCLC are shown in Table 1. The SCLC patients were 51 to 78 years old (mean/median age at 64.6/63.5 years), and all patients but one were former or current smokers, with a median of 60 pack years (range 20 to 150). Thirty-four patients died within the observation period, with a median overall survival of 7 months (range 0 to 14 months). The 14 patients who were still alive at study completion were surveyed over a median period of 14 months (range 0 to 19 months). All blood samples were taken before treatment, either at primary diagnosis ($n = 27$), or when progression or recurrence of the disease was observed (n

= 24). In total, 51 blood samples were available, as blood samples from three patients with progressive disease were taken at two serial time points. The volume of blood was 18 mL in 58.8% of the samples, 17 mL to 14 mL in 33.3%, and 10 mL to 8 mL in 7.8% of the samples. In the control group, 18 mL of blood was taken from 26 healthy donors.

Table 1. Characteristics of 48 small-cell lung cancer (SCLC) patients included in the study.

Characteristics	n (%)
Age	
Mean (median)	63.5 y (64.6 y)
Range	51.0–78.0 y
Gender	
Male	30 (62.5%)
Female	18 (37.5%)
Tobacco abuse	
Current smokers	13 (27.1%)
Former smokers	26 (54.2%)
Never smokers	1 (2.1%)
Unknown	8 (16.7%)
UICC 8th edition TNM stage at diagnosis[1]	
III	4 (11.4%)
IV	31 (88.6%)
Unknown	13 (27.1%)
Outcome at study completion	
Dead	34 (70.8%)
Alive	14 (29.2%)
Blood draw for CTCs	
At primary diagnosis	27 (56.3%)
At progression/recurrence	21 (43.8%)

[1] UICC, International Union for Cancer Control.

3.2. Spiking Experiments

The efficiency of the microfluidic Parsortix™ system for capturing cultivated SCLC cells derived from four CTC lines [20] in a separation cassette with a critical gap size of 6.5 µm is shown in Figure 1. The overall mean capture efficiency of all four cell lines was 27.8% (SD 16.4%). The gene expression levels of the epithelial (*EpCAM* and *CK19*) and neuroendocrine (*CHGA* and *SYP*) cell lineage origins were assessed in the same four CTC lines using qPCR, showing a wide-ranging pattern of gene expression.

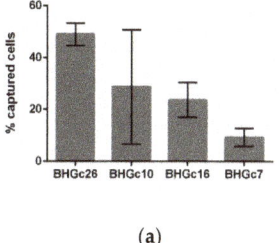

(a)

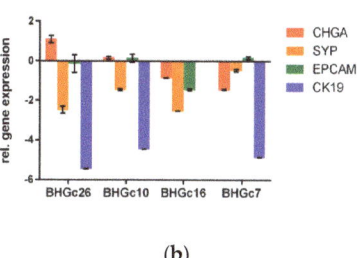

(b)

Figure 1. Characteristics of the microfluidic enrichment procedure and of the tumor cell lines used for the spiking experiments are illustrated. (**a**) Four SCLC tumor cell lines (BHGc26, BHGc10, BHGc16, and BHGc7) were fluorescently labeled and spiked into blood (100 cells per 10 mL) in triplicates. The graph depicts the mean percentage and the standard deviation of tumor cells captured in the Parsortix™ microfluidic cassette. (**b**) The gene expression levels of the epithelial and neuroendocrine cell lineage specific markers of the same cell lines are shown relative to the expression level of *cyclin dependent kinase inhibitor 1B* as reference gene. The graphs depict the means and the standard deviation from duplicate qPCRs amplifications.

3.3. Epithelial and Neuroendocrine Markers in Controls and SCLC Blood Samples

EpCAM, CK19, and CHGA transcripts were not detected in any of the control blood samples (Figure 2a–c). In contrast, SYP levels above the detection limit of qPCR were observed in 1/26 (3.8%), and ENO2 and NCAM1 transcripts in 24/26 (92.3%) and 19/26 (73.1%) controls, respectively (Figure 2d–f). Due to the high number of ENO2- and NCAM1-positive healthy donor samples, these markers were considered as less appropriate for CTC detection and thus excluded from further analyses.

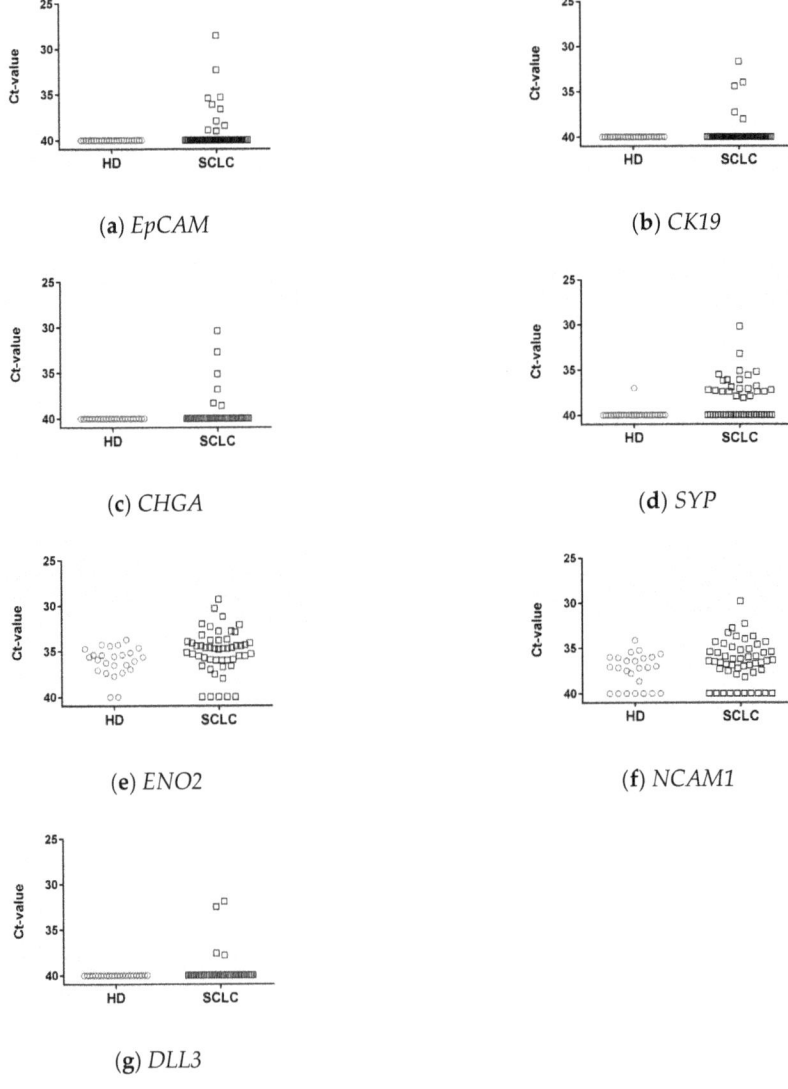

Figure 2. Mean cycle threshold (Ct-) values of the respective transcripts in blood samples taken from 26 healthy donors (HD) and 48 patients with small-cell lung cancer (SCLC). (**a**) EpCAM, epithelial cell adhesion molecule; (**b**) CK19, cytokeratin 19; (**c**) CHGA, chromogranin A; (**d**) SYP, synaptophysin; (**e**) ENO2, enolase 2; (**f**) NCAM1, neural cell adhesion molecule 1; (**g**) DLL3, Notch pathway delta-like 3 ligand.

In contrast, *EpCAM*-, *CK19*-, or *CHGA*-positivity above the detection limit of qPCR was observed in 10 (19.6%), 4 (7.8%), and 6 (11.8%) of the 51 samples obtained from SCLC patients. Due to the observed *SYP* gene expression in a single control blood sample, the threshold for *SYP*-positivity in the patients' samples was set at Ct = 37.0. Thus, *SYP* transcript levels below that threshold were observed in 40 (78.4%), and above that threshold in 11 (21.6%) of the 51 SCLC samples. These 11 samples were assigned as *SYP*-positive. In none of the gene transcripts did (*EpCAM*, *CK19*, *CHGA*, and *SYP*)-positivity differ significantly between the blood samples taken at diagnosis and disease progression.

In total, 16/51 (31.4%) samples were CTC-positive due to the expression of at least one of *EpCAM*, *CK19*, *CHGA*, and *SYP* markers (Figure 3). The expression of epithelial markers (*EpCAM* and/or *CK19*) was observed in 11 (21.6%), and of neuroendocrine markers (*CHGA* and/or *SYP*) in 13 (25.5%) samples. Among the 16 CTC-positive blood samples, three (18.8%) and five (31.3%) were characterized by the presence of just epithelial or neuroendocrine markers, respectively, and eight samples (50.0%) by both types.

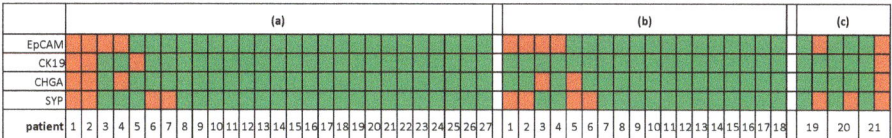

Figure 3. Heat map for *EpCAM*, *CK19*, *CHGA*, and *SYP* in the 51 microfluidic enriched blood samples of patients with small-cell lung cancer. (**a**) Twenty-seven samples were taken at diagnosis, (**b**) 18 samples were taken at progression/recurrence, and (**c**) displays serial blood draws taken from three patients during disease progression. Red and green squares indicate positive and negative gene expression per tested sample, respectively.

3.4. Alterations of Transcript Levels during Disease Progression

From three patients with progressive disease two serial blood samples were taken at the start of the consecutive lines of treatment. In two cases the second blood was taken two months after the first blood draw (patients 19 and 20 in Figure 3c), and in one case after three months (patient 21 in Figure 3c). At the first blood draw all patients were negative in all markers tested; however, at the second blood draw all patients were PCR-positive for at least *SYP* (see Figure 3c). All patients died within 1.5 months of the second blood draw.

3.5. Epithelial and Neuroendocrine Specific Markers and Patient Outcome

The blood samples were stratified on the basis of the epithelial cell lineage-specific gene transcripts *EpCAM* and *CK19* into the epi-positive (*n* = 11) and the epi-negative group (*n* = 40), and on the basis of neuroendocrine-specific transcripts *SYP* and *CHGA* into the nec-positive (*n* = 13) and nec-negative (*n* = 38) group. The presence of *EpCAM* and/or *CK19* transcripts in the epi-positive group at primary diagnosis may be associated with a shorter OS of the patients (Figure 4a); future studies with larger sample sizes may prove whether or not this difference is statistically significant. Similarly, the presence of *EpCAM* and/or *CK19* transcripts at disease progression was not related to OS (Figure 4b). In contrast, nec-positive patients had a significantly shorter OS than nec-negative patients, both at primary diagnosis and at disease progression. That association of *SYP* and/or *CHGA* transcripts with OS was observed with the presence of these neuroendocrine markers both at primary diagnosis (median OS 4 vs. 11 months, log-rank $p = 0.007$; Figure 4c), as well as at progression (median OS 1 vs. 5 months, log-rank $p = 0.014$; Figure 4d). Irrespective of whether the sample was taken at primary diagnosis or at disease progression, nec-positive patients had a high-risk of an early death (HR 3.475, 95% CI 1.685–7.164; $p = 0.001$).

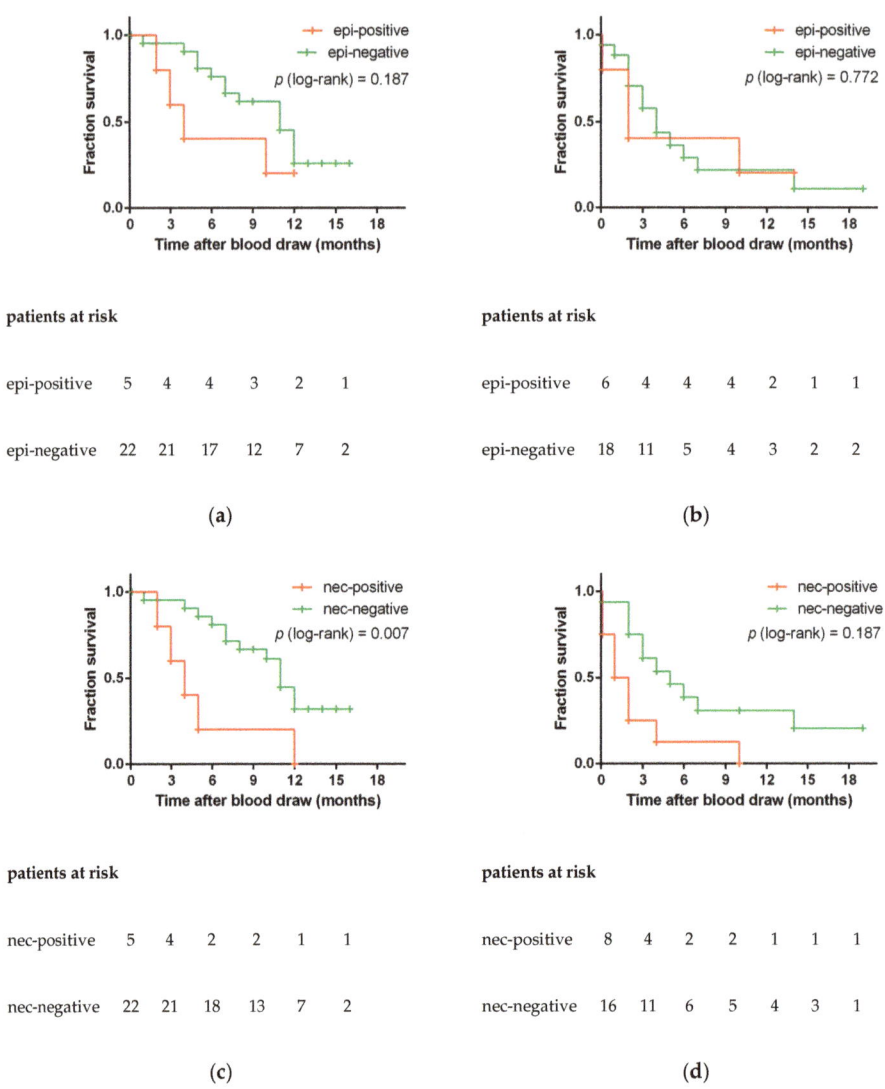

Figure 4. Overall survival of small-cell lung cancer patients according to presence (red) or absence (green) of epithelial (*EpCAM, CK19*) and neuroendocrine markers (*SYP, CHGA*). The figures (**a**) and (**c**) display samples taken at primary diagnosis, whereas the figures (**b**) and (**d**) display samples taken at progression. Log-rank testing was used to compare survival outcomes. epi, epithelial; nec, neuroendocrine.

3.6. DLL3 in Controls and SCLC Blood Samples

DLL3 transcripts were observed in 4/51 (7.8%) of the SCLC blood samples and in none of the 26 control blood samples (Figure 2g). Three *DLL3*-positive blood samples were taken at primary diagnosis, and one was taken from patient 19 at the second blood draw (see Figure 3c). Due to the small number of *DLL3*-positive patients, we did not stratify the patients into two groups depending on the time-point of blood draw. All four *DLL3*-positive patients had a significantly shorter OS than *DLL3*-negative patients (median OS 2 vs. 7 months, log-rank p = 0.003; Figure 5). The risk of dying earlier was 3.793 (95% CI 2.803–115.6) higher in the *DLL3*-positive group compared to the *DLL3*-negative group.

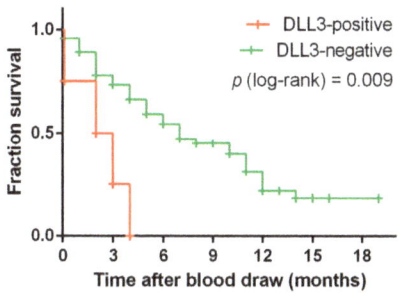

Figure 5. Overall survival according to presence (red) or absence (green) of *DLL3*. Log-rank testing was used to compare survival outcomes.

4. Discussion

We have applied a recently established workflow for molecular detection of CTCs [18] in blood samples taken from patients with SCLC, which is a highly aggressive neuroendocrine tumor of the lung. The enrichment of the CTCs was achieved with the microfluidic Parsortix™ technology, and the molecular analysis of the harvested cells was performed using markers that are specific to epithelial (*EpCAM* and *CK19*) and to neuroendocrine cell lineages (*SYP*, *CHGA*, *ENO2*, *NCAM1*), and to *DLL3*, an actionable target of antibody-drug conjugate rovalpituzumab tesirine (Rova-T). To the best of our knowledge, this is the first study that investigates neuroendocrine markers and *DLL3* in CTCs of SCLC patients at a molecular level.

We detected *EpCAM* and/or *CK19* transcripts in 21.6%, and neuroendocrine *CHGA* and/or *SYP* transcripts in 25.5% of the 51 SCLC blood samples. Interestingly, five of the 16 (31.3%) qPCR-positive samples were identified by the presence of neuroendocrine-specific transcripts alone. Similarly, three (18.8%) CTC-positive samples expressed the epithelial markers alone.

The percentage of CTC-positive samples due to the expression of epithelial markers of 21.6% in our cohort is smaller than reported by others in SCLC [22]. Applying the FDA approved CellSearch-based approach for the detection and enumeration of CTCs, positive findings were observed in 50% to 86% of the patients by other investigators [12,23,24]. The reason for the low detection rate of CTCs in our study may be the low overall sensitivity of our approach reflecting the need to split the sample into aliquots to analyze the expression of multiple genes individually. Improved sensitivity could be achieved by multi-plexing the gene expression analysis to avoid splitting the sample. A further improvement of the overall approach may also be achieved by employing a gene-specific pre-amplification of the respective targets prior to qPCR. In a recent study we have demonstrated that targeted pre-amplification in Parsortix™-enriched blood samples is feasible [18].

Another clear limitation of our study is the possibly low efficiency of the enrichment procedure to isolate CTCs from SCLC blood samples. The spiking experiments showed only moderate capture rates of SCLC CTC lines (mean capture efficiency 27.8%, SD 16.4%). These capture rates are lower than reported for the breast cancer cell line MCF-7 using the same type of separation cassette (average 63% [25]). In line with our observations, the capture rates varied depending on the type of cell line used from 30% to 87% in renal carcinoma [26]. The four cell lines used in the present study had been established from patients' CTCs [20]. Their diverse gene expression pattern (see Figure 1b) may reflect the initial heterogeneity of their provenance and might contribute to varying capture efficiencies. In the present study we did not check the number of tumor cells after harvesting; however, results from a recent study imply that the recovery rate may vary depending on the type of cell line from 62%–84% [27]. The number of harvested tumor cells can be increased by intensifying the back-flush cycle; however a higher recovery will only be achieved at the cost of a lower purity of the tumor cells.

Using larger volumes of blood may be a further attempt to increase the sensitivity of the assay in future studies. In our study all four samples with a volume of 10 mL blood or less were negative for all gene markers investigated. We did not exclude these few blood samples from the survival analyses shown in Figures 4 and 5, as that would not alter the significance of the analyses.

There is a single study applying the Parsortix™ technology for the enrichment of blood samples from SCLC patients. In that study, Chudziak et al. found CTCs in all 12 patients, as assessed by immune-fluorescent cytokeratin-specific staining of the enriched cells [16]. In contrast to our approach that group used blood collection tubes containing a preservative which is known to increase the rigidity of the cell, and thereby increasing the number of cells captured in the microfluidic cassette. This fact, along with the more advanced stage of disease in their study population may be the reason for the divergent CTC-positivity rate obtained in that study as compared to ours.

Another weakness of our study may be the limited sample size. Because of that and the low overall sensitivity we had very few positive samples. Thus we were not able to investigate association of the patients' prognosis and the respective gene expression levels, and the interplay of epithelial and neuroendocrine markers in a more detailed way.

High numbers of CTCs at diagnosis, as assessed by CellSearch, were associated with a poor prognosis (reviewed by [10]), yet the investigators reported divergent CTC numbers as a threshold for defining a group of patients with poor prognosis. [14,23]. However, we did not observe any significant impact of the expression of the epithelial markers on the OS. Nonetheless, patients who were epi-positive at primary diagnosis died earlier than epi-negative patients. A statistical significance may be reached by increasing the number of patients in future studies.

Studies investigating CTCs in other neuroendocrine tumors, such as those originating from the prostate, thyroid gland, or the intestine, mainly applied epithelial cell lineage-specific markers and protein-based technologies for the enrichment and analysis of CTCs [28]. However, CTCs may be missed when epithelial markers, such as *EpCAM*, are downregulated in the tumor tissue, as was shown in neuroendocrine tumors of the lung [29]. In addition, tumor cells can undergo epithelial-to-mesenchymal transition and lose their epithelial phenotype [24]. In the present study we detected *CHGA* and/or *SYP* transcripts in 13 samples; this absolute number corresponds to 25.5% of all 51 SCLC blood samples, and to 81.3% of all 16 CTC-positive samples. That the percentage is still not 100% may be because of low numbers of CTCs in some samples. Furthermore, Guinee et al. demonstrated the absence of neuroendocrine markers in just about 20% of the specimen by immunohistochemical staining [6]. The fact that one third of the qPCR-positive samples was identified by the presence of these neuroendocrine transcripts already indicates that epithelial markers alone may not be sufficient to detect CTCs in neuroendocrine tumors such as SCLC. One observation in this respect is of particular interest. The presence of selected neuroendocrine markers was associated with a worse outcome and not the presence of used epithelial markers. This even applies irrespective of the time the markers were detected—be it prior to treatment at initial diagnosis or when the disease has already progressed.

To the best of our knowledge there is just a single study investigating the clinical relevance of neuroendocrine markers in CTCs: Recently, Pal and colleagues quantified the percentage of

SYP-positive CTCs in blood samples taken from castration-resistant prostate cancer patients using the open fluorescent channel of the CellSearch platform [30]. They observed an increasing number of *SYP*-positive CTCs with the onset of resistance to androgen-receptor targeting drugs, which are assumed to stimulate the transition to the neuroendocrine phenotype [31].

5. Conclusions

Besides the neuroendocrine markers *SYP* and *CHGA*, our study also clearly shows that *DLL3* can be detected in CTC-enriched blood samples. Traditional patient stratification for personalized treatment options, such as Rova-T, is based the analysis of tissue samples that were taken long before the disease progression occurred. In contrast, liquid biopsy samples can be taken right before the start of treatment, and may thus provide a snapshot analysis of promising targets for personalized treatments, such as *DLL3*. Apart from treatment stratification, liquid biopsies can be taken at several consecutive points in time to assess the response to treatment. Despite the promising results of our study, the findings need to be validated in larger studies of SCLC patients. In conclusion, the molecular analysis of CTCs may add relevant information to traditional tissue biopsies or cytological specimens in small-cell lung cancer patients, especially in treatment selection and patient monitoring.

Author Contributions: Conceptualization, E.O. and R.Z.; data curation, K.B.; funding acquisition, R.Z.; investigation, E.O.; methodology, E.O., C.A., E.S., and C.W.; project administration, R.Z.; Resources, H.F. and M.H.; supervision, G.H., M.H., and R.Z.; writing—original draft, E.O.; writing—review and editing, K.B. and R.Z.

Funding: K.B. was supported by the Slovak Research and Development Agency Grant (No. APVV-15- 0181).

Acknowledgments: This study received support from ANGLE plc in the form of an in-kind contribution of Parsortix^{TM} devices and microfluidic separation cassettes. The authors thank Gabriele Klaming for language editing, and the team at ANGLE plc for their technical support. Last but not least, the authors would like to thank all patients and voluntary donors for providing blood samples.

Conflicts of Interest: C.A., E.S., H.F., C.W., K.B., G.H., and M.H. declare no conflict of interest. E.O. and R.Z. have a patent pending.

References

1. Wong, M.C.S.; Lao, X.Q.; Ho, K.F.; Goggins, W.B.; Tse, S.L.A. Incidence and mortality of lung cancer: Global trends and association with socioeconomic status. *Sci. Rep.* **2017**, *7*, 14300. [CrossRef] [PubMed]
2. Ferlay, J.; Colombet, M.; Soerjomataram, I.; Dyba, T.; Randi, G.; Bettio, M.; Gavin, A.; Visser, O.; Bray, F. Cancer incidence and mortality patterns in Europe: Estimates for 40 countries and 25 major cancers in 2018. *Eur. J. Cancer* **2018**. [CrossRef] [PubMed]
3. Fruh, M.; De Ruysscher, D.; Popat, S.; Crino, L.; Peters, S.; Felip, E. Small-cell lung cancer (SCLC): ESMO Clinical Practice Guidelines for diagnosis, treatment and follow-up. *Ann. Oncol.* **2013**, *24* (Suppl. 6), vi99–vi105. [CrossRef]
4. Travis, W.D. Update on small cell carcinoma and its differentiation from squamous cell carcinoma and other non-small cell carcinomas. *Mod. Pathol.* **2012**, *25* (Suppl. 1), S18–S30. [CrossRef]
5. Nicholson, S.A.; Beasley, M.B.; Brambilla, E.; Hasleton, P.S.; Colby, T.V.; Sheppard, M.N.; Falk, R.; Travis, W.D. Small cell lung carcinoma (SCLC): A clinicopathologic study of 100 cases with surgical specimens. *Am. J. Surg. Pathol.* **2002**, *26*, 1184–1197. [CrossRef] [PubMed]
6. Guinee, D.G., Jr.; Fishback, N.F.; Koss, M.N.; Abbondanzo, S.L.; Travis, W.D. The spectrum of immunohistochemical staining of small-cell lung carcinoma in specimens from transbronchial and open-lung biopsies. *Am. J. Clin. Pathol.* **1994**, *102*, 406–414. [CrossRef] [PubMed]
7. Saunders, L.R.; Bankovich, A.J.; Anderson, W.C.; Aujay, M.A.; Bheddah, S.; Black, K.; Desai, R.; Escarpe, P.A.; Hampl, J.; Laysang, A.; et al. A DLL3-targeted antibody-drug conjugate eradicates high-grade pulmonary neuroendocrine tumor-initiating cells in vivo. *Sci. Transl. Med.* **2015**, *7*, 302ra136. [CrossRef] [PubMed]
8. Kuhn, P.; Bethel, K. A fluid biopsy as investigating technology for the fluid phase of solid tumors. *Phys. Biol.* **2012**, *9*, 010301. [CrossRef] [PubMed]

9. Alberter, B.; Klein, C.A.; Polzer, B. Single-cell analysis of CTCs with diagnostic precision: Opportunities and challenges for personalized medicine. *Expert Rev. Mol. Diagn.* **2016**, *16*, 25–38. [CrossRef] [PubMed]
10. Kapeleris, J.; Kulasinghe, A.; Warkiani, M.E.; Vela, I.; Kenny, L.; O'Byrne, K.; Punyadeera, C. The Prognostic Role of Circulating Tumor Cells (CTCs) in Lung Cancer. *Front. Oncol.* **2018**, *8*, 311. [CrossRef]
11. Hou, J.M.; Krebs, M.G.; Lancashire, L.; Sloane, R.; Backen, A.; Swain, R.K.; Priest, L.J.; Greystoke, A.; Zhou, C.; Morris, K.; et al. Clinical significance and molecular characteristics of circulating tumor cells and circulating tumor microemboli in patients with small-cell lung cancer. *J. Clin. Oncol.* **2012**, *30*, 525–532. [CrossRef] [PubMed]
12. Normanno, N.; Rossi, A.; Morabito, A.; Signoriello, S.; Bevilacqua, S.; Di Maio, M.; Costanzo, R.; De Luca, A.; Montanino, A.; Gridelli, C.; et al. Prognostic value of circulating tumor cells' reduction in patients with extensive small-cell lung cancer. *Lung Cancer* **2014**, *85*, 314–319. [CrossRef] [PubMed]
13. Hiltermann, T.J.; Pore, M.M.; van den Berg, A.; Timens, W.; Boezen, H.M.; Liesker, J.J.; Schouwink, J.H.; Wijnands, W.J.; Kerner, G.S.; Kruyt, F.A.; et al. Circulating tumor cells in small-cell lung cancer: A predictive and prognostic factor. *Ann. Oncol* **2012**, *23*, 2937–2942. [CrossRef] [PubMed]
14. Naito, T.; Tanaka, F.; Ono, A.; Yoneda, K.; Takahashi, T.; Murakami, H.; Nakamura, Y.; Tsuya, A.; Kenmotsu, H.; Shukuya, T.; et al. Prognostic impact of circulating tumor cells in patients with small cell lung cancer. *J. Thorac. Oncol.* **2012**, *7*, 512–519. [CrossRef] [PubMed]
15. Cheng, Y.; Liu, X.Q.; Fan, Y.; Liu, Y.P.; Liu, Y.; Ma, L.X.; Liu, X.H.; Li, H.; Bao, H.Z.; Liu, J.J.; et al. Circulating tumor cell counts/change for outcome prediction in patients with extensive-stage small-cell lung cancer. *Future Oncol.* **2016**, *12*, 789–799. [CrossRef] [PubMed]
16. Chudziak, J.; Burt, D.J.; Mohan, S.; Rothwell, D.G.; Mesquita, B.; Antonello, J.; Dalby, S.; Ayub, M.; Priest, L.; Carter, L.; et al. Clinical evaluation of a novel microfluidic device for epitope-independent enrichment of circulating tumour cells in patients with small cell lung cancer. *Analyst* **2016**, *141*, 669–678. [CrossRef] [PubMed]
17. Huang, C.H.; Wick, J.A.; Sittampalam, G.S.; Nirmalanandhan, V.S.; Ganti, A.K.; Neupane, P.C.; Williamson, S.K.; Godwin, A.K.; Schmitt, S.; Smart, N.J.; et al. A multicenter pilot study examining the role of circulating tumor cells as a blood-based tumor marker in patients with extensive small-cell lung cancer. *Front. Oncol.* **2014**, *4*, 271. [CrossRef] [PubMed]
18. Obermayr, E.; Maritschnegg, E.; Agreiter, C.; Pecha, N.; Speiser, P.; Helmy-Bader, S.; Danzinger, S.; Krainer, M.; Singer, C.; Zeillinger, R. Efficient leukocyte depletion by a novel microfluidic platform enables the molecular detection and characterization of circulating tumor cells. *Oncotarget* **2018**, *9*, 812–823. [CrossRef] [PubMed]
19. Stathopoulou, A.; Ntoulia, M.; Perraki, M.; Apostolaki, S.; Mavroudis, D.; Malamos, N.; Georgoulias, V.; Lianidou, E.S. A highly specific real-time RT-PCR method for the quantitative determination of CK-19 mRNA positive cells in peripheral blood of patients with operable breast cancer. *Int. J. Cancer* **2006**, *119*, 1654–1659. [CrossRef]
20. Klameth, L.; Rath, B.; Hochmaier, M.; Moser, D.; Redl, M.; Mungenast, F.; Gelles, K.; Ulsperger, E.; Zeillinger, R.; Hamilton, G. Small cell lung cancer: Model of circulating tumor cell tumorospheres in chemoresistance. *Sci. Rep.* **2017**, *7*, 5337. [CrossRef]
21. Kaplan, E.L.; Meier, P. Nonparametric estimation from incomplete observations. *J. Am. Stat. Assoc.* **1958**, *53*, 457–481. [CrossRef]
22. Foy, V.; Fernandez-Gutierrez, F.; Faivre-Finn, C.; Dive, C.; Blackhall, F. The clinical utility of circulating tumour cells in patients with small cell lung cancer. *Transl. Lung Cancer Res.* **2017**, *6*, 409–417. [CrossRef] [PubMed]
23. Hou, J.M.; Greystoke, A.; Lancashire, L.; Cummings, J.; Ward, T.; Board, R.; Amir, E.; Hughes, S.; Krebs, M.; Hughes, A.; et al. Evaluation of circulating tumor cells and serological cell death biomarkers in small cell lung cancer patients undergoing chemotherapy. *Am. J. Pathol.* **2009**, *175*, 808–816. [CrossRef] [PubMed]
24. Messaritakis, I.; Politaki, E.; Kotsakis, A.; Dermitzaki, E.K.; Koinis, F.; Lagoudaki, E.; Koutsopoulos, A.; Kallergi, G.; Souglakos, J.; Georgoulias, V. Phenotypic characterization of circulating tumor cells in the peripheral blood of patients with small cell lung cancer. *PLoS ONE* **2017**, *12*, e0181211. [CrossRef] [PubMed]
25. Lampignano, R.; Yang, L.; Neumann, M.H.D.; Franken, A.; Fehm, T.; Niederacher, D.; Neubauer, H. A Novel Workflow to Enrich and Isolate Patient-Matched EpCAM(high) and EpCAM(low/negative) CTCs Enables the Comparative Characterization of the PIK3CA Status in Metastatic Breast Cancer. *Int. J. Mol. Sci.* **2017**, *18*, 1885. [CrossRef] [PubMed]

26. Maertens, Y.; Humberg, V.; Erlmeier, F.; Steffens, S.; Steinestel, J.; Bogemann, M.; Schrader, A.J.; Bernemann, C. Comparison of isolation platforms for detection of circulating renal cell carcinoma cells. *Oncotarget* **2017**, *8*, 87710–87717. [CrossRef] [PubMed]
27. Miller, M.C.; Robinson, P.S.; Wagner, C.; O'Shannessy, D.J. The Parsortix Cell Separation System-A versatile liquid biopsy platform. *Cytom. A* **2018**. [CrossRef]
28. Rizzo, F.M.; Meyer, T. Liquid Biopsies for Neuroendocrine Tumors: Circulating Tumor Cells, DNA, and MicroRNAs. *Endocrinol. Metab. Clin. N. Am.* **2018**, *47*, 471–483. [CrossRef]
29. Khan, M.S.; Tsigani, T.; Rashid, M.; Rabouhans, J.S.; Yu, D.; Luong, T.V.; Caplin, M.; Meyer, T. Circulating tumor cells and EpCAM expression in neuroendocrine tumors. *Clin. Cancer Res.* **2011**, *17*, 337–345. [CrossRef]
30. Pal, S.K.; He, M.; Chen, L.; Yang, L.; Pillai, R.; Twardowski, P.; Hsu, J.; Kortylewski, M.; Jones, J.O. Synaptophysin expression on circulating tumor cells in patients with castration resistant prostate cancer undergoing treatment with abiraterone acetate or enzalutamide. *Urol. Oncol.* **2018**, *36*, 162.e1–162.e6. [CrossRef]
31. Small, E.J.; Huang, J.; Youngren, J.; Sokolov, A.; Aggarwal, R.R.; Thomas, G.; True, L.D.; Zhang, L.; Foye, A.; Alumkal, J.J.; et al. Characterization of neuroendocrine prostate cancer (NEPC) in patients with metastatic castration resistant prostate cancer (mCRPC) resistant to abiraterone (Abi) or enzalutamide (Enz): Preliminary results from the SU2C/PCF/AACR West Coast Prostate Cancer Dream Team (WCDT). *J. Clin. Oncol.* **2015**, *33*, 5003. [CrossRef]

© 2019 by the authors. Licensee MDPI, Basel, Switzerland. This article is an open access article distributed under the terms and conditions of the Creative Commons Attribution (CC BY) license (http://creativecommons.org/licenses/by/4.0/).

Article

S100-EPISPOT: A New Tool to Detect Viable Circulating Melanoma Cells

Laure Cayrefourcq [1], Aurélie De Roeck [2], Caroline Garcia [2], Pierre-Emmanuel Stoebner [2], Fanny Fichel [2], Françoise Garima [1], Françoise Perriard [3], Jean-Pierre Daures [3], Laurent Meunier [2,†] and Catherine Alix-Panabières [1,*,†]

1. Laboratory of Rare Human Circulating Cells (LCCRH), University Medical Centre of Montpellier, UPRES EA2415, 34093 Montpellier, France
2. Department of Dermatology, Nîmes University Hospital, University of Montpellier, 30029 Nîmes, France
3. UPRES EA2415, University Institute of Clinical Research (IURC), Montpellier University, 34093 Montpellier, France
* Correspondence: c-panabieres@chu-montpellier.fr; Tel.: +33-411759931
† These authors have contributed equally to the work.

Received: 19 June 2019; Accepted: 18 July 2019; Published: 20 July 2019

Abstract: Metastatic melanoma is one of the most aggressive and drug-resistant cancers with very poor overall survival. Circulating melanoma cells (CMCs) were first described in 1991. However, there is no general consensus on the clinical utility of CMC detection, largely due to conflicting results linked to the use of heterogeneous patient populations and different detection methods. Here, we developed a new EPithelial ImmunoSPOT (EPISPOT) assay to detect viable CMCs based on their secretion of the S100 protein (S100-EPISPOT). Then, we compared the results obtained with the S100-EPISPOT assay and the CellSearch® CMC kit using blood samples from a homogeneous population of patients with metastatic melanoma. We found that S100-EPISPOT sensitivity was significantly higher than that of CellSearch®. Specifically, the percentage of patients with ≥2 CMCs was significantly higher using S100-EPISPOT than CellSearch® (48% and 21%, respectively; $p = 0.0114$). Concerning CMC prognostic value, only the CellSearch® results showed a significant association with overall survival ($p = 0.006$). However, due to the higher sensitivity of the new S100-EPISPOT assay, it would be interesting to determine whether this functional test could be used in patients with non-metastatic melanoma for the early detection of tumor relapse and for monitoring the treatment response.

Keywords: circulating tumor cells; melanoma; liquid biopsy; EPISPOT; CellSearch®

1. Introduction

Melanoma is the most malignant skin cancer, and its incidence rate is increasing worldwide. Early stage and localized melanoma can be cured by surgical resection. Conversely, metastatic melanoma is one of the most aggressive and drug-resistant cancers with very poor overall survival (OS) (six to nine months). Melanoma management has recently undergone revolutionary changes with the discovery of predictive tumor biomarkers (BRAF mutations and immune checkpoint inhibitors such as programmed cell death protein 1 (PD-1), its ligand (PD-L1), and cytotoxic T-lymphocyte antigen 4 (CTLA-4)) and the development of the associated treatments. These new treatments, alone or in combination, have dramatically improved the outcome of patients with metastatic melanoma. For example, the anti-PD-1 drug pembrolizumab has demonstrated benefits to progression-free survival (34–38%) and objective responses (21–25%) at six months compared with chemotherapy (16% and 4%; $p < 0.0001$) [1]. However, despite the good response rates, immunotherapy results in systemic toxicity, and it is not effective in all patients.

Circulating tumor cells (CTCs) are cancer cells that are shed from the primary and metastatic tumor(s). They can be detected in peripheral blood samples using different technologies, but their identification and characterization require extremely sensitive and specific analytical methods [2–6]. Their analysis is considered as a real-time liquid biopsy for patients with cancer [7–10]. In 2011, the U.S. Food and Drug Administration (FDA) cleared the CellSearch® system (Menarini Silicon Biosystems) for CTC analysis to monitor patients with metastatic breast, colorectal and prostate cancer [11–13]. The CellSearch® epithelial cell-based assay has clearly demonstrated its clinical significance and is now used as the gold standard in clinical studies evaluating different cancer types. Even though a very limited number of studies have evaluated melanoma CTCs using the CellSearch® Circulating Melanoma Cell Kit, they all provided similar results, reflecting the robustness and reproducibility of this assay. The detection of circulating melanoma cells (CMCs) was described for the first time in 1991. Since then, the many studies on CMCs from patients with melanoma at different stages and using different detection approaches have reported conflicting results [14]. Indeed, metastatic melanoma is a highly heterogeneous tumor and CMCs may display different phenotypes and functional states.

Moreover, CMC analysis with the CellSearch® detection kit does not allow discriminating between dead and viable CMCs, the only CMCs involved in metastatic development [15]. The functional EPithelial ImmunoSPOT (EPISPOT) assay was described in 2005 and allows the identification of viable CTCs in peripheral blood samples of patients with cancer (e.g., breast, prostate, and colon cancer) [16–20] by detecting proteins secreted/released/shed by single viable epithelial cancer cells [21].

The aim of this study was to compare CMC detection using the CellSearch® system and a new EPISPOT assay (S100-EPISPOT assay) designed to identify viable CMCs that secrete S100, a protein expressed and secreted by melanoma cells [22], in blood samples from patients with metastatic melanoma.

2. Materials and Methods

2.1. Patient Cohort

A prospective controlled observational comparative study (Circulating Tumor Cells and Melanoma: Comparing the EPISPOT and CellSearch Techniques; NCT01558349) was conducted at the Nîmes University Hospital, Nîmes, France, between June 2013 and June 2017. The main objective was to determine if we can observe more positive patients with the EPISPOT assay than the CellSearch® system. All patients with melanoma signed a written informed consent before enrolment in the CELLCIRC study and treatment initiation. The study was carried out in accordance with the World Medical Association Declaration of Helsinki. The experimental protocol was approved by the French bioethical committee "Sud Méditerranée III" (Approval reference No. 2012.06.10). Blood samples from healthy volunteers ($n = 38$) and patients with metastatic malignant melanoma ($n = 50$; before any treatment) were collected in the morning and processed within 24 h.

2.2. Melanoma Cell Lines

The melanoma cancer cell lines WM-266-4 (ATCC® CRL-1676™) and MV3 (kindly provided by Klaus Pantel, University of Tumor Biology, Hamburg, Germany) were used for optimizing the S100-EPISPOT assay. WM-266-4 cells were maintained in αMEM medium (22571, Gibco, Grand Island, USA) supplemented with 10% fetal calf serum (FCS), and MV3 cells in RPMI 1640 medium (L0501, Dominique Dutscher, Brumath, France), supplemented with 5mM L-glutamine (25030, Gibco, Grand Island, USA) and 10% FCS.

2.3. Flow Cytometry Experiments

Intracellular expression of the S100 protein in WM-266-4 and MV3 cells was determined by flow cytometry using a Cyan cytometer (Beckman-Coulter, Villepinte, France) and a fixation/permeabilization kit (Beckman Coulter, Brea, USA). The two anti-S100 antibodies (clones 8B10 and 6G1) used in the EPISPOT assay were tested to confirm S100 expression in these melanoma cell lines.

2.4. Immunofluorescence Assay

Melanoma cell lines were immunostained with the two anti-S100 antibodies (8B10 and 6G1), as described for the flow cytometry experiments. Then, cells were seeded on glass slides using a Cytospin 4 centrifuge (Shandon, Runcorn, England) and mounted with ProLong Gold Antifade reagent with 4',6-diamidino-2-phenylindole (DAPI) (Invitrogen). S100 expression was analyzed with a fluorescent microscope (Axio Imager M1, Carl Zeiss Vision, Halbermoos, Germany).

2.5. S100-EPISPOT Assay

As CMCs are rare in peripheral blood, a pre-enrichment step was performed before the EPISPOT assay. To separate erythrocytes and leukocytes from CMCs, the RosettSep^{TM} reagent (20 µL/mL) was added to 13–15 mL of blood collected in EDTA tubes, and enrichment was performed following the manufacturer's instructions (RosetteSep^{TM} CTC Enrichment Cocktail containing Anti-CD36, STEMCELL Technologies). During the S100-EPISPOT assay, enriched CMCs were cultured on a membrane coated with the anti-S100 8B10 antibody (10 ng/µL; Abcam) for 2 days. Then, secreted S100 captured by the 8B10 antibody was detected by incubation with another antibody against S100 (6G1: 3 ng/µL; Abcam) conjugated to AlexaFluor 488. Single fluorescent S100 immunospots were counted under a fluorescent microscope equipped with a camera and computer-assisted analysis (KS ELISPOT, Carl Zeiss Vision). The detailed procedure of the EPISPOT assay has been described by Soler et al. [21]. Results were corrected as "number of cells per 7.5 mL of blood" to be comparable with those obtained with the CellSearch system.

2.6. Cell Search® System

Patient blood samples were collected in special CellSave® tubes and the analysis was performed using the Circulating Melanoma Cell Kit (9594V, Menarini, Bologna, Italy), according to the manufacturer's instructions, as previously described [23]. Reagents consisted of ferrofluids coated with anti-CD146 antibodies to enrich melanoma cells and endothelial cells, a phycoerythrin-conjugated antibody that binds to high molecular weight melanoma-associated antigen (HMW-MAA) to identify melanoma cells, a mixture of two allophycocyanin-conjugated antibodies against CD45 to identify leukocytes and against CD34 to identify endothelial cells, and the nuclear dye DAPI to identify nucleated cells. The criteria to identify an object as a melanoma cell included: round to oval morphology, visible (DAPI-positive) nucleus, positive staining for HMW-MAA, and negative staining for CD45 and CD34. Results were expressed as number of cells per 7.5 mL of blood.

2.7. Statistical Analysis

The sensitivity and specificity of each technology (S100-EPISPOT and CellSearch®) were assessed by comparing the true positive/negative and false positive/negative results for healthy donors and patients with melanoma using the McNemar's test and the McNemar's test with continuity correction.

Patients' characteristics were described using medians and ranges (quantitative variables). The Fisher's test was used to evaluate the association between categorical clinical characteristics and CMC detection. The association between lactate dehydrogenase (LDH) value and CMC number was assessed using Spearman and Kendall rank correlation coefficients.

The median OS was analyzed using the Kaplan–Meier method. The end point was death by any cause. Survival curves were compared with the non-parametric log rank test ($p \leq 0.05$ considered as significant). The univariate Cox proportional hazard regression model was used to estimate the hazard ratio and 95% confidence intervals (CI) for the CellSearch® assay, and the risk of death in function of the CMC number (<2 vs. ≥2) after verifying the hazard proportional test.

3. Results

3.1. S100 Expression in Melanoma Cell Lines

To develop a new EPISPOT assay for CMC detection, two different melanoma cell lines (WM-266-4 and MV3), derived from metastatic sites, were used. Analysis of S100 protein expression by flow cytometry (FC) showed that both cell lines expressed S100, but at different levels (one log difference between cell lines) (Figure 1a). These results were confirmed by immunofluorescence (IF) experiments (Figure 1b). Both cell lines expressed S100, but the signal intensity was higher in WM-266-4 than in MV3 cells (same exposure time). Similar results were obtained with the two anti-S100 antibodies (8B10 and 6G1) used in the S100-EPISPOT assay. These findings indicated that both melanoma cell lines could be used to optimize the S100-EPISPOT assay.

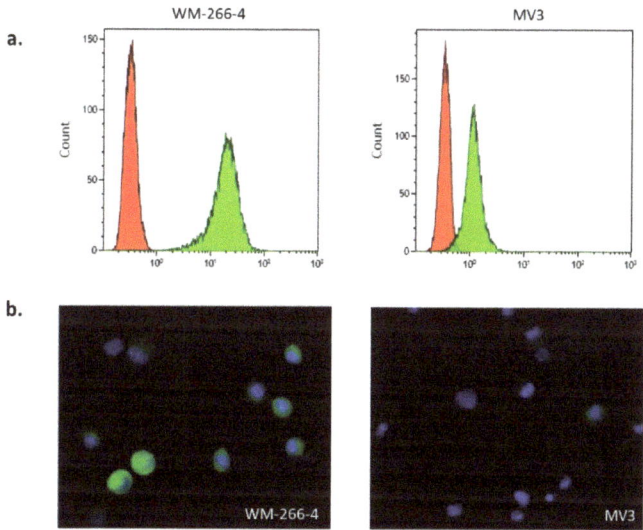

Figure 1. S100 protein expression in WM-266-4 and MV3 cells by (**a**) flow cytometry and (**b**) immunofluorescence analysis using the anti-S100 antibody 6G1 conjugated to AlexaFluor 488.

3.2. S100-EPISPOT Assay

Then, the two melanoma cell lines were used to evaluate the feasibility and detection threshold of the S100-EPISPOT assay. No immunospots (indicative of S100 secretion) could be detected when using MV3 cells, possibly due to their very weak S100 expression, as observed by FC and IF. Conversely, on average, 20–47% of WM-266-4 cells secreted S100 (Figure 2). This detection rate is considered normal for cell lines analyzed with the EPISPOT assay because of variations in cell cycle and protein productivity. In parallel, enriched peripheral blood mononuclear cells (PBMCs) from healthy donors were tested ($n = 3$) to confirm that S100 was not secreted by PBMCs (data not show) and validate this assay for CMC detection after blood enrichment.

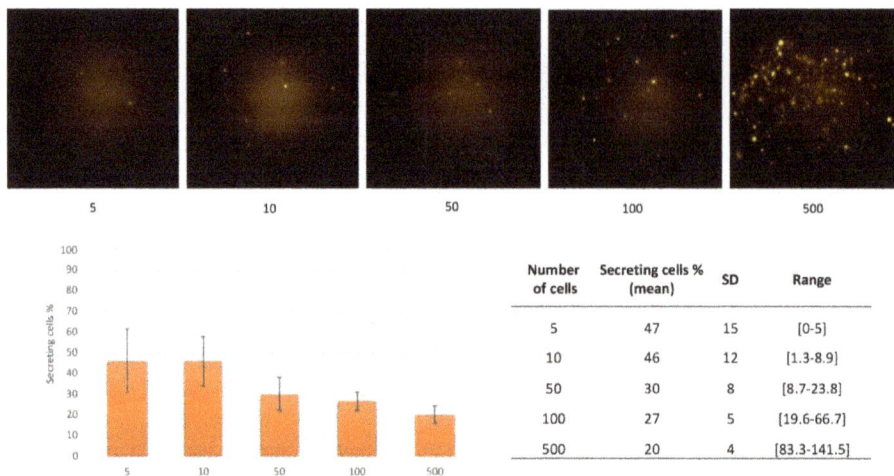

Figure 2. Melanoma cell detection using the S100-EPISPOT assay. Representative images of immunospots using WM-266-4 cells (upper panels) and percentage (n = 3) of S100-secreting cells recovered from serial dilutions of WM-266-4 cells (lower panels). SD = standard deviation.

Surprisingly, the detection rate was higher with smaller numbers of tested cells (Figure 2). These data demonstrate that the new S100-EPISPOT assay could be used to detect rare CMCs in peripheral blood. Previous studies also reported a better recovery rate for small number of cancer cells spiked in samples of healthy donors (1–20 cancer cells/10 mL) [24,25].

3.3. Specificity of the S100-EPISPOT Assay

Among the 38 healthy donors enrolled, six and four donors could not be evaluated by EPISPOT and CellSearch, respectively, for reasons listed in Figure 3a. The specificity of the S100-EPISPOT and CellSearch® assays were compared using blood samples from 38 healthy donors and two detection thresholds (≥1 and ≥2 CMCs) (Table 1a). The CellSearch® system gave similar results with both thresholds: 94% and 97%. The S100-EPISPOT assay specificity was lower for the ≥1 CMC threshold (78%), but reached 97%, as for CellSearch®, for the ≥2 CMC threshold. These results are consistent with previous reports and allowed for choosing a cut-off of 2 CMCs for the positive CMC detection [23,26].

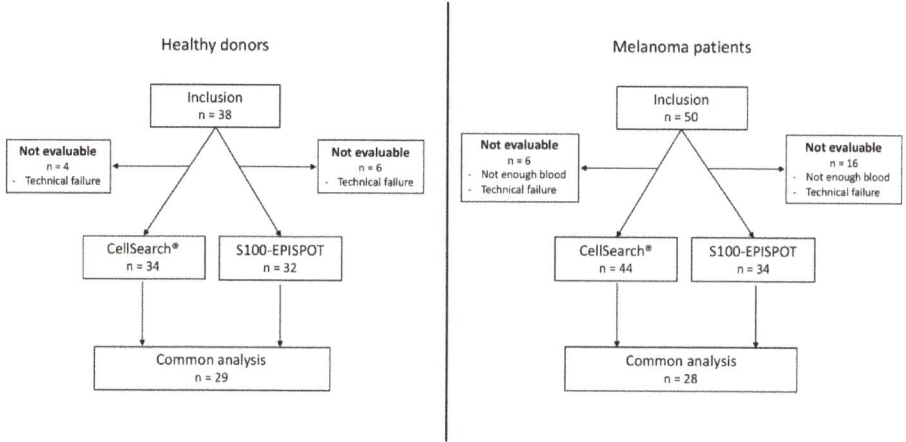

Figure 3. Study flowchart showing the number of included healthy donors and patients and the number of patients and healthy donors who underwent circulating melanoma cell (CMC) analysis using the S100- EPISPOT assay and CellSearch® system.

Table 1. (a) Specificity of the S100-EPISPOT and CellSearch CMC assays in healthy donors using two different CMC cut-offs. (b) Sensitivity and specificity of the two assays with the cut-off ≥2 CMCs (healthy donors: $n = 28$; patients with metastatic melanoma: $n = 29$).

a.				
	S100-EPISPOT (n = 32)		CellSearch CMC (n = 34)	
<1	25	(78%)	32	(94%)
≥1	7	(22%)	2	(6%)
<2	31	(97%)	33	(97%)
≥2	1	(3%)	1	

b.				
	S100 EPISPOT	CellSearch CMC	p value (Mac Nemar)	p value (corrected Mac Nemar)
Sensitivity (n = 29)	48%	21%	0.0114	0.0269
Specificity (n = 28)	100%	96%	0.3173	1

3.4. Circulating Melanoma Cell Detection in Patients with Metastatic Melanoma

Among the 50 patients enrolled, 16 and 6 could not be evaluated by EPISPOT and CellSearch, respectively, for reasons listed in Figure 3b. The patients included in this study (26 men and 24 women with a mean age of 64, range, 29–89) had metastatic melanoma, with a median survival of 6.72 months (95% CI = 4.24–12.16). CMCs were detected using the CellSearch® system and the S100-EPISPOT assay in blood samples from 44 and 34 patients, respectively. Using the cut-off of ≥2 CMCs per 7.5 mL of blood, 10/44 (23%) patients were positive for CMCs with the CellSearch® system, and 15/34 (44%) with the S100-EPISPOT assay (Table 2). The S100-EPISPOT assay gave a higher number of positive patients than the CellSearch® system, but with a smaller range: EPISPOT (0–450) and CellSearch® (0–4937). Analysis of the specificity and sensitivity using the results obtained for samples from healthy donors ($n = 28$) and patients with metastatic melanoma ($n = 29$) tested with both assays, showed that sensitivity was 48% for S100-EPISPOT and 21% for CellSearch® ($p = 0.0114$, corrected $p = 0.0269$), and specificity was 100% for S100-EPISPOT and 96% for CellSearch® (Table 1b). The correlation between both technologies was assessed using the Spearman ($r_s = 0.49$, $p = 0.0070$) and Kendall ($\tau = 0.44$, $p = 0.0055$) tests.

Table 2. Description of CMC detection for the S100-EPISPOT and CellSearch® assays in (**a**) all patients included in the study and (**b**) the subpopulation of patients tested with both assays.

a.

Nb Patients	Assay	Total	Failed	Mean	Standard Deviation	Median	Min	Max	Lower quartile	Upper quartile
50	S100-EPISPOT	34	16	21.59	80.35	1.00	0.00	450.00	0.00	3.00
	CellSearch®	44	6	142.57	752.01	0.00	0.00	4937.00	0.00	1.00

b.

Nb Patients	Assay	Total	Failed	Mean	Standard Deviation	Median	Min	Max	Lower quartile	Upper quartile
29	S100-EPISPOT	29	0	25.24	86.68	1.00	0.00	450.00	0.00	4.00
	CellSearch®	29	0	206.10	924.77	0.00	0.00	4937.00	0.00	1.00

3.5. Clinical Relevance of CMC Detection

No significant association between the patients' clinical characteristics and CMC detection with EPISPOT and CellSearch® was observed (Table 3), except for high LDH level (2-fold higher than normal) and CMC detection using CellSearch® ($p = 0.0315$). The correlation between LDH level and CTC detection by CellSearch® was significant using the Spearman ($r_s = 0.39$, $p = 0.0138$) and Kendall ($\tau = 0.30$, $p = 0.0132$) tests.

Table 3. Clinical characteristic of patients with metastatic melanoma in function of CTC detection with the S100-EPISPOT and CellSearch® assays.

	EPISPOT			CellSearch®		
	< 2 n = 19	≥ 2 n = 15	p value (Fisher)	< 2 n = 34	≥ 2 n = 10	p value (Fisher)
Sex						
Men	8 (42.11%)	9 (60.00%)	0.4905	18 (52.94%)	4 (40.00%)	0.7205
Women	11 (57.89%)	6 (40.00%)		16 (47.06%)	6 (60.00%)	
BRAF mutation						
No	8 (47.06%)	6 (54.55%)	1.0000	15 (48.39%)	2 (28.57%)	0.4267
Yes	9 (52.94%)	5 (45.45%)		16 (51.61%)	5 (71.43%)	
Ulceration						
Absence	4 (36.36%)	6 (85.71%)	0.0656	11 (52.38%)	4 (66.67%)	0.6618
Presence	7 (63.64%)	1 (14.29%)		10 (47.62%)	2 (33.33%)	
Metastatic sites						
Nb ≤ 2	10 (52.63%)	7 (46.67%)	1.0000	22 (64.71%)	3 (30.00%)	0.0738
Nb > 2	9 (47.37%)	8 (53.33%)		12 (35.29%)	7 (70.00%)	
LDH value						
Normal	6 (42.86%)	5 (35.71%)	0.2909	10 (34.48%)	1 (10.00%)	**0.0315**
> Normal	7 (50.00%)	4 (28.57%)		16 (55.17%)	4 (40.00%)	
2x > Normal	1 (7.14%)	5 (35.71%)		3 (10.34%)	5 (50.00%)	

Table 3. Cont.

		EPISPOT			CellSearch®		
		< 2 n = 19	≥ 2 n = 15	p value (Fisher)	< 2 n = 34	≥ 2 n = 10	p value (Fisher)
Primary tumor localization							
Other		1 (2.94%)	6 (17.65%)	0.0106	6 (13.95%)	3 (6.98%)	0.2017
Face		1 (2.94%)	0 (0.00%)		0 (0.00%)	1 (2.33%)	
Lower limb		8 (23.53%)	2 (5.88%)		12 (27.91%)	1 (2.33%)	
Upper limb		6 (17.65%)	1 (2.94%)		7 (16.28%)	1 (2.33%)	
Torso		3 (8.82%)	6 (17.65%)		9 (20.93%)	3 (6.98%)	

In univariate Kaplan–Meier analyses, OS was associated only with CMC detection by CellSearch® (CS) ($p = 0.0006$) (Figure 4a). Hazard ratio for death was 3.57 (95% CI 1.64–7.77; Cox regression analysis with CMC <2 vs. ≥2). No association was detected between OS and the classical clinical variables (age, sex, Breslow depth, Clark stage, LDH, BRAF mutation, and number of metastatic sites) (Table S1). Moreover, no significant association was observed for the S100-EPISPOT assay (EPI) alone (Figure S1). Analysis of the potential synergistic effect of combining both assays indicated that the "double negative" (EPI and CS < 2 CMCs) group was not associated with OS ($p = 0.3419$) compared with the "double and/or simple positive" (EPI and/or CS ≥2) group. Conversely, analysis of the "double positive" (EPI and CS ≥ 2 CMCs), "simple positive" (EPI or CS ≥2 CMCs) and "double negative" (EPI and CS < 2 CMCs) groups highlighted the association of the "double positive" group with OS ($p = 0.0081$) (Figure 4b).

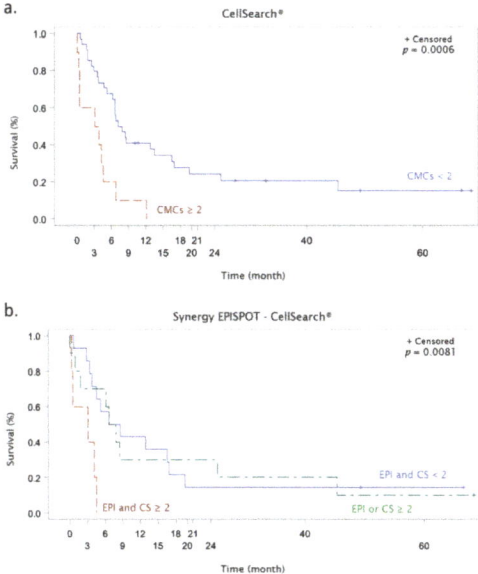

Figure 4. Kaplan–Meier survival curves of patients with metastatic melanoma according to the detection cut-off of 2 CMCs using the (**a**) CellSearch® system, and (**b**) when combining the S100-EPISPOT (EPI) and CellSearch® (CS) assay results.

4. Discussion

The cancer metastatic cascade is a complex process characterized by several events, including cell migration, local invasion, intravasation of tumor cells into the circulation and extravasation at distant sites to form detectable metastases [27]. The mechanisms involved in this multistep process are largely unknown, but recent studies suggested that some CTCs could be at the origin of distant metastases [28,29]. For this reason, CTC detection has become a great challenge for the personalized treatment of patients with cancer.

In the literature, the number of techniques used for CTC detection in patients with melanoma is almost equal to the number of studies published on this topic. These methods are based on the physical or phenotypical properties of melanoma cells [3,30,31]. Up to now, none of them has been validated because of the lack of reproducible results. Nestin, CD133 [32], receptor activator of NF-k B (RANK) [33], ATP-binding cassette sub-family B member 5 (ABCB5) [34], CD20 [35], and CD271 [36] have been identified as potential candidates for the identification of melanoma-initiating cells.

Compared with CTC detection techniques based only on the expression of surface markers, assays that identify viable/functional cells could be more interesting and more relevant for monitoring the response to treatment. Functional CTC analysis also offers the possibility to determine the biological properties of metastatic cells, including the identification of metastasis-initiating cells [37]. Currently, the EPISPOT assay is the only functional assay to detect viable, prognostically relevant CTCs at the single-cell level after enrichment by leukocyte depletion. This system has been used to test blood samples from hundreds of patients with different tumor types [17–19]. As this technique could be considered to be a protein secretion-profiling assay, the analyzed proteins should be specific and significantly produced and released by tumor cells. Among the proteins expressed and released by melanoma cells, the S100 family is the most studied [22,38]. S100B expression is increased in melanoma cells compared with melanocytes and can be used for the diagnosis of metastatic malignant melanoma by immunohistochemistry. Moreover, serum S100B level increases in patients with melanoma, independently of the cancer stage. Its expression is clearly correlated with the presence of metastases, tumor burden, prognosis, and survival [39,40]. Recent data demonstrated that isolated CMCs obtained from patients with metastatic melanoma uniformly express S100 [41].

In this study, we first described the new S100-EPISPOT assay to detect viable CMCs, and then compared its results with those obtained with the CellSearch® system (Circulating Melanoma Cell Kit). In two melanoma cell lines, we found that S100 expression and secretion were heterogeneous. Consequently, only the cell line that strongly expresses S100 (WM-266-4 cells) secreted enough protein to be detected by the S100-EPISPOT assay at the single-cell level. The subpopulation of cancer cells that weakly express S100 are missed using this assay. Moreover, WM-266-4 cell detection by the S100-EPISPOT system was more efficient when only a few cells were present in the sample. These experimental observations were confirmed by data obtained from blood samples from a homogeneous group of patients with metastatic melanoma, indicating that the S100-EPISPOT assay sensitivity is significantly higher than that of the CellSearch®system. Specifically, the percentage of patients with ≥2 CMCs was 48% with the S100-EPISPOT assay and 21% with the CellSearch® system (corrected $p = 0.0269$).

CMC detection (both methods) was not associated with Breslow depth and BRAF mutation status. Only LDH value was significantly correlated with CMC detection (CellSearch® method) in Spearman ($r_s = 0.39$, $p = 0.0138$) and Kendall ($k = 0.30$, $p = 0.0132$) tests.

Analysis of the prognostic value of CMC detection (both assays) indicated that only CMC ≥ 2 with CellSearch® was significantly associated with the OS ($p = 0.006$), as previously reported for the CellSearch® CMC detection system. By using this CellSearch® system, Rao et al. [23] detected at least two CMCs in 23% (18/79) of blood samples from 44 patients with metastatic melanoma. They also found that OS was shorter in patients with ≥2 CMCs per 7.5 mL of whole blood compared with the <2 CMC group. Similarly, Khoja et al. [26] showed that 26% of patients with metastatic skin melanoma had more than two CMCs per 7.5 mL of blood, and that the median OS was significantly shorter in this group.

In our patient cohort, no other clinical features, such as number of metastatic sites, Breslow depth, BRAF mutational status and even LDH level (routinely measured in the clinical practice) provided prognostic information. The S100-EPISPOT assay results alone did not predict the patients' clinical outcome, although this test was more sensitive than the CellSearch® system. Conversely, by combining the data obtained by both methods, OS was significantly associated ($p = 0.0005$) with "double positive results" (CellSearch® and S100-EPISPOT) compared with "simple positive result" (CellSearch® or EPISPOT) and "double negative result" (CellSearch® and S100-EPISPOT). This suggests that OS is shorter in patients in whom CMCs are detected with both methods.

However, in our patients with advanced disease and short survival probability (0.38, 95% CI 0.25–0.51 after 12 months and 0.16, 95% CI 0.07–0.27 after 24 months), measuring OS is not of clinical interest. The crucial aim of liquid biopsy is to obtain reliable and "real-time" information before, during, and after treatment to monitor the patient response. This is especially important for melanoma because currently no melanoma-specific blood-based biomarker is available for routine use in clinical practice. Another circulating biomarker of interest in advanced melanoma is ctDNA. The majority of studies have been applied to patients treated with BRAF inhibitors, via monitoring of the singular BRAFV600 mutation, predicting response to therapy and prognosis in metastatic melanoma [42]. Nevertheless, clinical trials that look at patient outcome as a result of ctDNA-guided clinical decisions are required before ctDNA can be successfully established as a melanoma-specific biomarker in clinical practice.

Finally, the main interest of our study is the finding that the new S100-EPISPOT assay has a good sensitivity (48%) for CMC detection and an acceptable specificity. It would be interesting to determine whether the S100-EPISPOT functional assay could be used for the early detection of tumor relapse or for monitoring therapy response in patients with non-metastatic melanoma.

Moreover, a completely new optimized EPISPOT assay, named EPIDROP (EPISPOT in a DROP) [43], is currently under development. This innovative micro-droplet technology allows not only the detection of viable S100-secreting CMCs at the single-cell level, but also the immunostaining of all CMCs before their encapsulation, for instance for the identification of tumor cells that could be targeted by immunotherapy (e.g., PD-L1-positive). This technology represents a new combination of standard CMC detection by IF, like the CellSearch® system, with a functional assay to identify the subset of functional and potentially metastasis-competent CMCs.

Supplementary Materials: The following are available online at http://www.mdpi.com/2073-4409/8/7/755/s1. Table S1: Univariate analysis of overall survival in patients with metastatic melanoma using the Kaplan–Meier method, Figure S1: Kaplan–Meier survival curves of patients with metastatic melanoma according to the detection cut-off of two CMCs using the S100-EPISPOT (EPI).

Author Contributions: Conceptualization, L.C., C.A.-P. and L.M.; Methodology, J-P.D.; Validation, L.C., C.A.-P. and L.M.; Formal analysis F.G., L.C. and F.P.; Investigation, L.C and L.M.; Resources, A.D.R., C.G., P-E.S., F.F. and L.M.; Writing—original draft preparation, L.C.; Writing—review and editing, F.G., C.A.-P., L.M. and F.P.; Supervision, C.A.-P. and L.M.; Funding acquisition, C.A.-P. and L.M.

Funding: This research was funded by Nîmes University Hospital, University of Montpellier, AOI Local 2011.

Acknowledgments: We thank Nathalie Bedos, Sabrina Nicolas and Lyamin Bendjeddou, research nurses at the Nîmes University Hospital, involved in this study. We thank Elisabetta Andermarcher for assistance with her comments and proof-reading that greatly improved the manuscript.

Conflicts of Interest: The authors declare no conflict of interest.

References

1. Dummer, R.; Daud, A.; Puzanov, I.; Hamid, O.; Schadendorf, D.; Robert, C.; Schachter, J.; Pavlick, A.; Gonzalez, R.; Hodi, F.S.; et al. A randomized controlled comparison of pembrolizumab and chemotherapy in patients with ipilimumab-refractory melanoma. *J. Transl. Med.* **2015**, *13*. [CrossRef]
2. Alix-Panabieres, C.; Pantel, K. Technologies for detection of circulating tumor cells: Facts and vision. *Lab Chip* **2014**, *14*, 57–62. [CrossRef] [PubMed]

3. Nezos, A.; Msaouel, P.; Pissimissis, N.; Lembessis, P.; Sourla, A.; Armakolas, A.; Gogas, H.; Stratigos, A.J.; Katsambas, A.D.; Koutsilieris, M. Methods of detection of circulating melanoma cells: A comparative overview. *Cancer Treat. Rev.* **2011**, *37*, 284–290. [CrossRef] [PubMed]
4. van der Toom, E.E.; Verdone, J.E.; Gorin, M.A.; Pienta, K.J. Technical challenges in the isolation and analysis of circulating tumor cells. *Oncotarget* **2016**. [CrossRef] [PubMed]
5. Joosse, S.A.; Gorges, T.M.; Pantel, K. Biology, detection, and clinical implications of circulating tumor cells. *EMBO Mol. Med.* **2015**, *7*, 1–11. [CrossRef] [PubMed]
6. Alix-Panabieres, C.; Schwarzenbach, H.; Pantel, K. Circulating tumor cells and circulating tumor DNA. *Annu. Rev. Med.* **2012**, *63*, 199–215. [CrossRef] [PubMed]
7. Pantel, K.; Alix-Panabieres, C. Liquid biopsy: Potential and challenges. *Mol. Oncol.* **2016**, *10*, 371–373. [CrossRef] [PubMed]
8. Alix-Panabieres, C.; Pantel, K. Clinical Applications of Circulating Tumor Cells and Circulating Tumor DNA as Liquid Biopsy. *Cancer Discov.* **2016**. [CrossRef]
9. Khoo, B.L.; Grenci, G.; Jing, T.; Lim, Y.B.; Lee, S.C.; Thiery, J.P.; Han, J.; Lim, C.T. Liquid biopsy and therapeutic response: Circulating tumor cell cultures for evaluation of anticancer treatment. *Sci. Adv.* **2016**, *2*, e1600274. [CrossRef]
10. Alix-Panabieres, C.; Pantel, K. Circulating tumor cells: Liquid biopsy of cancer. *Clin. Chem.* **2013**, *59*, 110–118. [CrossRef]
11. de Bono, J.S.; Scher, H.I.; Montgomery, R.B.; Parker, C.; Miller, M.C.; Tissing, H.; Doyle, G.V.; Terstappen, L.W.; Pienta, K.J.; Raghavan, D. Circulating tumor cells predict survival benefit from treatment in metastatic castration-resistant prostate cancer. *Clin. Cancer Res.* **2008**, *14*, 6302–6309. [CrossRef]
12. Cohen, S.J.; Punt, C.J.; Iannotti, N.; Saidman, B.H.; Sabbath, K.D.; Gabrail, N.Y.; Picus, J.; Morse, M.; Mitchell, E.; Miller, M.C.; et al. Relationship of circulating tumor cells to tumor response, progression-free survival, and overall survival in patients with metastatic colorectal cancer. *J. Clin. Oncol.* **2008**, *26*, 3213–3221. [CrossRef]
13. Cristofanilli, M.; Budd, G.T.; Ellis, M.J.; Stopeck, A.; Matera, J.; Miller, M.C.; Reuben, J.M.; Doyle, G.V.; Allard, W.J.; Terstappen, L.W.M.M.; et al. Circulating Tumor Cells, Disease Progression, and Survival in Metastatic Breast Cancer. *N. Engl. J. Med.* **2004**, *351*, 781–791. [CrossRef]
14. Khoja, L.; Lorigan, P.; Dive, C.; Keilholz, U.; Fusi, A. Circulating tumour cells as tumour biomarkers in melanoma: Detection methods and clinical relevance. *Ann. Oncol. Off. J. Eur. Soc. Med Oncol. ESMO* **2015**, *26*, 33–39. [CrossRef]
15. Pantel, K.; Alix-Panabieres, C. Functional Studies on Viable Circulating Tumor Cells. *Clin. Chem.* **2016**, *62*, 328–334. [CrossRef]
16. Alix-Panabieres, C. EPISPOT assay: Detection of viable DTCs/CTCs in solid tumor patients. *Recent Results Cancer Res.* **2012**, *195*, 69–76. [CrossRef]
17. Deneve, E.; Riethdorf, S.; Ramos, J.; Nocca, D.; Coffy, A.; Daures, J.P.; Maudelonde, T.; Fabre, J.M.; Pantel, K.; Alix-Panabieres, C. Capture of viable circulating tumor cells in the liver of colorectal cancer patients. *Clin. Chem.* **2013**, *59*, 1384–1392. [CrossRef]
18. Ramirez, J.M.; Fehm, T.; Orsini, M.; Cayrefourcq, L.; Maudelonde, T.; Pantel, K.; Alix-Panabieres, C. Prognostic relevance of viable circulating tumor cells detected by EPISPOT in metastatic breast cancer patients. *Clin. Chem.* **2014**, *60*, 214–221. [CrossRef]
19. Alix-Panabieres, C.; Pantel, K. Liquid biopsy in cancer patients: Advances in capturing viable CTCs for functional studies using the EPISPOT assay. *Expert Rev. Mol. Diagn.* **2015**, *15*, 1411–1417. [CrossRef]
20. Kuske, A.; Gorges, T.M.; Tennstedt, P.; Tiebel, A.K.; Pompe, R.; Preisser, F.; Prues, S.; Mazel, M.; Markou, A.; Lianidou, E.; et al. Improved detection of circulating tumor cells in non-metastatic high-risk prostate cancer patients. *Sci. Rep.* **2016**, *6*, 39736. [CrossRef]
21. Soler, A.; Cayrefourcq, L.; Mazel, M.; Alix-Panabieres, C. EpCAM-Independent Enrichment and Detection of Viable Circulating Tumor Cells Using the EPISPOT Assay. *Methods Mol. Biol.* **2017**, *1634*, 263–276. [CrossRef]
22. Alegre, E.; Sammamed, M.; Fernández-Landázuri, S.; Zubiri, L.; González, Á. Chapter Two-Circulating Biomarkers in Malignant Melanoma. In *Advances in Clinical Chemistry*; Makowski, G.S., Ed.; Elsevier: Amsterdam, The Netherlands, 2015; Volume 69, pp. 47–89.

23. Rao, C.; Bui, T.; Connelly, M.; Doyle, G.; Karydis, I.; Middleton, M.R.; Clack, G.; Malone, M.; Coumans, F.A.; Terstappen, L.W. Circulating melanoma cells and survival in metastatic melanoma. *Int. J. Oncol.* **2011**, *38*, 755–760. [CrossRef]
24. Alix-Panabieres, C.; Brouillet, J.P.; Fabbro, M.; Yssel, H.; Rousset, T.; Maudelonde, T.; Choquet-Kastylevsky, G.; Vendrell, J.P. Characterization and enumeration of cells secreting tumor markers in the peripheral blood of breast cancer patients. *J. Immunol. Methods* **2005**, *299*, 177–188. [CrossRef]
25. Alix-Panabières, C.; Rebillard, X.; Brouillet, J.-P.; Barbotte, E.; Iborra, F.; Segui, B.; Maudelonde, T.; Jolivet-Reynaud, C.; Vendrell, J.-P. Detection of Circulating Prostate-Specific Antigen–Secreting Cells in Prostate Cancer Patients. *Clin. Chem.* **2005**, *51*, 1538–1541. [CrossRef]
26. Khoja, L.; Lorigan, P.; Zhou, C.; Lancashire, M.; Booth, J.; Cummings, J.; Califano, R.; Clack, G.; Hughes, A.; Dive, C. Biomarker utility of circulating tumor cells in metastatic cutaneous melanoma. *J. Investig. Dermatol.* **2013**, *133*, 1582–1590. [CrossRef]
27. Pantel, K.; Speicher, M.R. The biology of circulating tumor cells. *Oncogene* **2015**. [CrossRef]
28. Baccelli, I.; Schneeweiss, A.; Riethdorf, S.; Stenzinger, A.; Schillert, A.; Vogel, V.; Klein, C.; Saini, M.; Bauerle, T.; Wallwiener, M.; et al. Identification of a population of blood circulating tumor cells from breast cancer patients that initiates metastasis in a xenograft assay. *Nat. Biotechnol.* **2013**, *31*, 539–544. [CrossRef]
29. Hodgkinson, C.L.; Morrow, C.J.; Li, Y.; Metcalf, R.L.; Rothwell, D.G.; Trapani, F.; Polanski, R.; Burt, D.J.; Simpson, K.L.; Morris, K.; et al. Tumorigenicity and genetic profiling of circulating tumor cells in small-cell lung cancer. *Nat. Med.* **2014**, *20*, 897–903. [CrossRef]
30. Rodic, S.; Mihalcioiu, C.; Saleh, R.R. Detection methods of circulating tumor cells in cutaneous melanoma: A systematic review. *Crit. Rev. Oncol. Hematol.* **2014**, *91*, 74–92. [CrossRef]
31. Xu, M.J.; Dorsey, J.F.; Amaravadi, R.; Karakousis, G.; Simone, C.B., 2nd; Xu, X.; Xu, W.; Carpenter, E.L.; Schuchter, L.; Kao, G.D. Circulating Tumor Cells, DNA, and mRNA: Potential for Clinical Utility in Patients With Melanoma. *Oncology* **2016**, *21*, 84–94. [CrossRef]
32. Fusi, A.; Reichelt, U.; Busse, A.; Ochsenreither, S.; Rietz, A.; Maisel, M.; Keilholz, U. Expression of the stem cell markers nestin and CD133 on circulating melanoma cells. *J. Investig. Dermatol.* **2011**, *131*, 487–494. [CrossRef]
33. Kupas, V.; Weishaupt, C.; Siepmann, D.; Kaserer, M.L.; Eickelmann, M.; Metze, D.; Luger, T.A.; Beissert, S.; Loser, K. RANK is expressed in metastatic melanoma and highly upregulated on melanoma-initiating cells. *J. Investig. Dermatol.* **2011**, *131*, 944–955. [CrossRef]
34. Schatton, T.; Murphy, G.F.; Frank, N.Y.; Yamaura, K.; Waaga-Gasser, A.M.; Gasser, M.; Zhan, Q.; Jordan, S.; Duncan, L.M.; Weishaupt, C.; et al. Identification of cells initiating human melanomas. *Nature* **2008**, *451*, 345–349. [CrossRef]
35. Fang, D.; Nguyen, T.K.; Leishear, K.; Finko, R.; Kulp, A.N.; Hotz, S.; Van Belle, P.A.; Xu, X.; Elder, D.E.; Herlyn, M. A tumorigenic subpopulation with stem cell properties in melanomas. *Cancer Res.* **2005**, *65*, 9328–9337. [CrossRef]
36. Boiko, A.D.; Razorenova, O.V.; van de Rijn, M.; Swetter, S.M.; Johnson, D.L.; Ly, D.P.; Butler, P.D.; Yang, G.P.; Joshua, B.; Kaplan, M.J.; et al. Human melanoma-initiating cells express neural crest nerve growth factor receptor CD271. *Nature* **2010**, *466*, 133–137. [CrossRef]
37. Alix-Panabieres, C.; Bartkowiak, K.; Pantel, K. Functional studies on circulating and disseminated tumor cells in carcinoma patients. *Mol. Oncol.* **2016**, *10*, 443–449. [CrossRef]
38. Palmer, S.R.; Erickson, L.A.; Ichetovkin, I.; Knauer, D.J.; Markovic, S.N. Circulating serologic and molecular biomarkers in malignant melanoma. *Mayo Clin. Proc.* **2011**, *86*, 981–990. [CrossRef]
39. Zarogoulidis, P.; Tsakiridis, K.; Karapantzou, C.; Lampaki, S.; Kioumis, I.; Pitsiou, G.; Papaiwannou, A.; Hohenforst-Schmidt, W.; Huang, H.; Kesisis, G.; et al. Use of proteins as biomarkers and their role in carcinogenesis. *J. Cancer* **2015**, *6*, 9–18. [CrossRef]
40. Kruijff, S.; Hoekstra, H.J. The current status of S-100B as a biomarker in melanoma. *Eur. J. Surg. Oncol. EJSO* **2012**, *38*, 281–285. [CrossRef]
41. Long, E.; Ilie, M.; Bence, C.; Butori, C.; Selva, E.; Lalvee, S.; Bonnetaud, C.; Poissonnet, G.; Lacour, J.P.; Bahadoran, P.; et al. High expression of TRF2, SOX10, and CD10 in circulating tumor microemboli detected in metastatic melanoma patients. A potential impact for the assessment of disease aggressiveness. *Cancer Med.* **2016**. [CrossRef]

42. Calapre, L.; Warburton, L.; Millward, M.; Ziman, M.; Gray, E.S. Circulating tumour DNA (ctDNA) as a liquid biopsy for melanoma. *Cancer Lett.* **2017**, *404*, 62–69. [CrossRef]
43. Pantel, K.; Alix-Panabieres, C. Liquid biopsy and minimal residual disease—Latest advances and implications for cure. *Nat. Rev. Clin. Oncol.* **2019**. [CrossRef]

 © 2019 by the authors. Licensee MDPI, Basel, Switzerland. This article is an open access article distributed under the terms and conditions of the Creative Commons Attribution (CC BY) license (http://creativecommons.org/licenses/by/4.0/).

Article

The Detection and Morphological Analysis of Circulating Tumor and Host Cells in Breast Cancer Xenograft Models

Loredana Cleris, Maria Grazia Daidone, Emanuela Fina *,† and Vera Cappelletti *,†

Biomarkers Unit, Department of Applied Research and Technological Development, Fondazione IRCCS Istituto Nazionale dei Tumori, 20133 Milan, Italy
* Correspondence: emanuela1.fina@gmail.com (E.F.); vera.cappelletti@istitutotumori.mi.it (V.C.)
† These authors share senior authorship.

Received: 4 June 2019; Accepted: 1 July 2019; Published: 5 July 2019

Abstract: Hematogenous dissemination may occur early in breast cancer (BC). Experimental models could clarify mechanisms, but in their development, the heterogeneity of this neoplasia must be considered. Here, we describe circulating tumor cells (CTCs) and the metastatic behavior of several BC cell lines in xenografts. MDA-MB-231, BT-474, MDA-MB-453 and MDA-MB-468 cells were injected at the orthotopic level in immunocompromised mice. CTCs were isolated using a size-based method and identified by cytomorphological criteria. Metastases were detected by COX IV immunohistochemistry. CTCs were detected in 90% of animals in each model. In MDA-MB-231, CTCs were observed after 5 weeks from the injection and step wisely increased at later time points. In animals injected with less aggressive cell lines, the load of single CTCs (mean ± SD CTCs/mL: 1.8 ± 1.3 in BT-474, 122.2 ± 278.5 in MDA-MB-453, 3.4 ± 2.5 in MDA-MB-468) and the frequency of CTC clusters (overall 38%) were lower compared to MDA-MB-231 (946.9 ± 2882.1; 73%). All models had lung metastases, MDA-MB-453 and MDA-MB-468 had ovarian foci too, whereas lymph nodal involvement was observed in MDA-MB-231 and MDA-MB-468 only. Interestingly, CTCs showed morphological heterogeneity and were rarely associated to host cells. Orthotopic xenograft of BC cell lines offers valid models of hematogenous dissemination and a possible experimental setting to study CTC-blood microenvironment interactions.

Keywords: circulating tumor cells; metastasis; xenograft models; breast cancer

1. Introduction

Metastasis is definitely a hallmark of cancer [1] and represents the main cause of cancer-related deaths [2] due to ineffective therapies. Unraveling the molecular mechanisms of tumor progression would help to anticipate disease outcome and to point the way for selecting personalized treatments. In breast cancer (BC), in particular, the timing of cancer cell dissemination has been largely discussed [3] and has proven to represent an early step in tumor progression [4,5]. In accordance with this, circulating tumor cells (CTCs) can be detected in patients without clinical evidence of secondary lesions [6–8] and, in several studies, the presence of dormant cells has been also reported even in the bone marrow of patients with ductal carcinoma in situ [9–11]. In addition to this grim scenario, BC is, in fact, a group of heterogeneous tumors [12–15], with cancer cells cross-talking with normal cells from the microenvironment [16,17]. More recently, based on copy number and gene expression data from over 2000 tumors, BCs were re-classified into ten clusters associated with distinct clinical outcomes [18,19], with implications for patient management. As the development of drug resistance is often interpreted as an inevitable consequence of tumor heterogeneity [20,21], efforts to address such interrelated themes are urgently needed, especially in non-operable and advanced-stage clinical settings.

At present, the biological events and molecular mechanisms that orchestrate the metastatic process are still not fully understood due to their complexity [22–24]. Functional assays to elucidate the biological meaning of a gene in tumor dissemination or the effect of a compound on metastasis outgrowth have to be necessarily set in organisms. In this field, scientists have largely based their studies on metastasis modeling on laboratory animals, including drosophila, zebrafish, mice, rats and, more rarely, rabbits, companion pets and monkeys with spontaneous onset of cancer [25]. Xenotransplantation of BC cell lines in mice with a compromised immune system is commonly used as a model for metastasis studies. In particular, direct injection into the systemic circulation of the MDA-MB-231 cell line and its derivatives generated several models of metastasis [26–30], either in basal conditions or after selection of organ-specific metastatic variants upon several rounds of transplantation [31,32], providing valuable knowledge on the genetic determinants of metastasis in BC. However, although forced hematogenous dissemination does enable to finely dissect the late steps of the metastatic cascade [31], this strategy is not adequate to recapitulate the initial events of the process as in spontaneous metastasis models, where cells are implanted at the orthotopic level. Moreover, the research mainly focused on a single model type might fail in addressing the heterogeneity issue in BC, thus limiting possible applications to the clinical context [33].

Since the molecular classification of BC has been established, researchers have paid attention to the similarities between cell lines and clinical samples. Studies have shown that the luminal, basal, HER2 and claudin-low clusters identified in BC are mirrored in BC cell lines [34–36]. However, the claudin-low and basal subtypes are over-represented among the BC cell lines used for xenograft models [37]. Indeed, spontaneous metastasis is a rare event when using cell lines belonging to less aggressive subtypes, and only a few models with variable frequencies of metastasis have been described in recent years for MCF7, BT-474 and MDA-MB-453 [38,39]. Not dissimilar from BC cell lines, which however ensure a high tumor take in mice, is the behavior of xenotransplanted BC specimens (PDXs, patient-derived xenografts), whose both development and metastatic organotropism in mice are variable and dependent on the aggressiveness of the tumor of origin. Indeed, despite PDXs representing important preclinical tools since proven to retain over serial passages histopathology, behavior and genomic features of the tumor of origin [40–45], in BC the tumor take efficiency of the luminal subtype in mice is low [46,47], thus generating a bias towards aggressive triple-negative BCs models.

On the basis of these considerations, we have reconsidered the use of xenograft models from BC cell lines for basic metastasis research studies. To this aim, we have (i) transplanted BC cell lines belonging to different molecular subtypes in the mammary fat pad of immunocompromised mice, (ii) set up a method to detect CTCs and small foci of metastatic cells in such xenograft models, and (iii) described the morphological features of BC cell line derived CTCs and host-derived circulating cells.

2. Materials and Methods

2.1. Cell Lines

Cell lines were purchased from the American Type Culture Collection organization and verified for identity via short tandem repeat (STR) profile analysis using the StemElite™ ID System kit (Promega, Madison, WI, USA), which yielded a 100% match on 9 STR loci, and on amelogenin for gender identification, in all cases.

BT-474, MDA-MB-453 and MDA-MB-468 BC cell lines were cultured in Dulbecco's Modified Eagles' Medium (DMEM)/F-12 medium (Lonza, Switzerland) supplemented with 10% South America Fetal Bovine Serum (FBS, Lonza). The MDA-MB-231 BC cell line was cultured in DMEM/F-12 medium supplemented with 5% FBS. Cells were grown at 37 °C in a 95% humidified 5% pCO_2 atmosphere.

All experiments were performed using cells from the second to the eighth in vitro passage from thawing, and showing at least 95% viability by 0.4% Trypan Blue solution exclusion test. Cell culture supernatants were regularly tested for Mycoplasma contamination using the MycoAlert™ Mycoplasma Detection Kit (Lonza) before each injection in mice.

2.2. Animal Models

Animal experiments were performed according to the Italian law D.L. 116/92, and the following additions, which enforced the 2010/63/EU Directive. The study protocols were approved by the Ethical Committee for Animal Experimentation at Fondazione IRCCS Istituto Nazionale dei Tumori (INT), in Milan, (INT_08/2012, and INT_01/2017, which was also approved by Italian Ministry of Health with approval number 452/2017-PR, following the receipt of the D.L. 26/2014). All efforts were deployed to minimize animal suffering [48], following the most recently published version of recommended ARRIVE guidelines [49]. Female NOD.CB17-$Prkdc^{scid}$/J (NOD scid) and NOD.Cg-$Prkdc^{scid}$ $Il2rg^{tm1Wjl}$/SzJ (NSG) mice were purchased from Charles River (Wilmington, MA, USA) and The Jackson Laboratory (Sacramento, CA, USA), respectively, and bred by the qualified personnel at INT Animal House Facility in individually ventilated cages, 3 to 5 animals per cage. Animals were anesthetized by intraperitoneal injection of a ketamine (100 mg/kg) and xylazine (5 mg/kg) cocktail before orthotopic injection of cancer cells and before animal sacrifice. Sacrifice procedure was cervical dislocation, performed at a priori set experimental time points or immediately upon signs of moderate suffering (e.g., decrease in activity, hunched appearance, ruffled hair coat, respiratory distress).

The tumor implant was performed under sterile conditions on healthy and normal-weight 7- to 16-week-old anesthetized mice using a 30G needle syringe. Eighty to ninety μL of Dulbecco's Phosphate Buffered Saline (DPBS, Lonza) cell suspensions mixed with 50% ECM Gel from Engelbreth-Holm-Swarm murine sarcoma (Sigma-Aldrich, St. Louis, MO, USA) Matrigel matrix (final concentration 4 mg/mL) were injected in the mammary fat pad (m.f.p.) of the axillary and/or the inguinal mammary gland, according to the scheme reported in Table 1:

Table 1. Scheme of breast cancer cell line xenotransplantation in immunocompromised mice.

Cell Line	Mouse Model	Number of Cells per Injection	Injection Sites (m.f.p.)
BT-474	NSG	5×10^6	4th left
MDA-MB-453	NSG	10^7	4th left
MDA-MB-468	NSG	5×10^6	2nd right and 4th left
MDA-MB-231	NOD scid	5×10^6	2nd right and 4th left

BT-474 cell injection was performed after 24–48 h from subcutaneous implantation of a 0.72 mg 90-day release 17-β-estradiol pellet (Innovative Research of America, Sarasota, FL, USA), performed on the neck lateral side using a trocar. The overall tumor take rate was 100%.

For the time-course experiments with MDA-MB-231 cells, 4 groups of 6 animals each were injected with cells according to the standard scheme. Animals were randomized before sacrifice at the defined time points (day 35, 50, 65 and 80) according to the tumor growth rate and the cage where they had been bred. Tumor take was obtained in 23/24 mice.

Tumor growth was monitored every week using a caliper and the tumor mass (g) was estimated by the $(D \times d^2)/2$ formula, where D and d represent the longest and the shortest diameter, respectively, of the nodule. The tumor load was lower than 10% of the body mass (range: 0.4–9.5%), except for two animals (10.2% and 15.7%) in which tumors had increased rapidly during the latest week.

An intravenous injection was performed using suspensions of 10^6 or 2×10^6 cells in 400 μL of DPBS.

Splenic leukocytes from BALB/c Nude mice were kindly provided by Dr. Claudia Chiodoni from the Molecular Immunology Unit at INT.

2.3. Collection of Tissues and Organs

Blood samples were drawn from anesthetized mice by cardiac puncture, using a 1 mL 26G needle EDTA conditioned syringe (1.8 mg/mL final concentration), stored at 4 °C and processed for CTC isolation within 30 min. Mice were immediately sacrificed and primary tumor nodules and organs (lung, axillary, inguinal subclavian or peritoneal lymph-nodes, ovaries, liver, kidneys, brain, and spleen) were collected and fixed in a 10% neutral buffered formalin solution (Bio-Optica, Milan, Italy) for 18–24 h; samples were then washed with distilled water and stored in 70% ethanol until paraffin embedding.

2.4. Circulating Tumor Cell Isolation and Detection

CTCs were isolated using the ScreenCell®Cyto kit (ScreenCell, Sarcelles, France), according to the manufacturer instructions. Briefly, blood was diluted in DPBS to reach 3 mL and subsequently mixed with 4 mL of the ScreenCell®FC2 proprietary buffer for red blood cell osmotic lysis and cell fixation. When the flux rate decreased due to a microcoagulation phenomenon or the presence of numerous CTCs, the residual blood was filtered on further devices. After filtration, the isolation supports (IS) were stained with Hematoxylin Solution S (Merck, Darmstadt, Germany) for 1 min and a Shandon Eosin Y Aqueous Solution (Thermo Fisher Scientific Inc., Waltham, MA, USA) for 30 s, or with a pure May-Grünwald solution for 2.5 min, followed by a 2.5-min incubation step with a May-Grünwald solution diluted 1:2 with pH 7-adjusted distilled water, and a 10-min incubation step with a Giemsa solution (Merck) 1:10 diluted with pH 7-adjusted distilled water. All samples were analyzed by a referral pathologist at ScreenCell. The cytomorphological analysis and CTC count were performed on the basis of the criteria of malignancy reported by Hofman et al. [50]. Major criteria for CTC identification were a high nucleus-to-cell ratio (i.e., cytoplasm area/whole cell area, ≥0.5) and large nuclear size (≥20 µm diameter), whereas minor criteria included irregular nuclear contours and nuclear hyperchromatism. CTC clusters were defined as groups of two or more CTCs, sometimes mixed with platelets and various leukocytes (i.e., circulating tumor microemboli, CTM), showing criteria of malignancy like those described for single CTCs. The nucleus-to-cell ratios in CTC aggregates are similar to those in single CTCs in [51]. Platelets appear as small, round eosinophilic or grayish particles, and can be found isolated or grouped in plaques, sometimes mixed with deposits of fibrin. Like CTCs, lymphocytes have a high nucleus-to-cell ratio, but they are smaller (7–8 µm diameter). Circulating atypical giant cells were defined as large cells (20–300 µm diameter), with generally voluminous and filamentary cytoplasm, various morphology (e.g., amorphous, round, elongated) and nucleus to cell ratio lower than that of CTCs [52,53]. Samples were defined as CTC-positive (+ve) when at least one single CTC and/or CTC cluster and/or CTM were observed in at least one stained IS.

2.5. Immunofluorescence and Immunohistochemical Staining

Immunofluorescence was performed on unstained ISs upon storage at −20 °C. ISs were incubated in an oven at 37 °C for 1 h, rehydrated in Tris Buffered Saline (TBS) 1× pH 7.4 (Bio-Optica) and blocked for 30 min with 5% bovine serum albumin (BSA, Sigma-Aldrich, St. Louis, MO, USA) in TBS 1X. Tumor cells were stained overnight at 4 °C using a rabbit monoclonal Alexa Fluor®488 conjugated antibody against human cytochrome c oxidase subunit IV (COX IV, clone 3E11, isotype IgG; Cell Signaling Technology, Danvers, MA, USA), diluted 1:100 in 5% BSA in TBS 1×. Nuclei were stained with a 5 µg/mL 4′,6-Diamidino-2-phenylindole (DAPI) dilactate solution (Sigma-Aldrich, St. Louis, MO, USA). ISs were mounted on glass slides and covered with a round coverslip using the Fluoroshield Mounting Medium (Abcam, Cambridge, UK). Images were acquired by Nikon Eclipse TE2000-S fluorescence (Nikon, Tokyo, Japan) microscope.

Four-micron thick formalin-fixed paraffin-embedded (FFPE) sections from tumor nodules and organs were deparaffinized by standard protocols and stained using a rabbit monoclonal antibody against human COX IV (clone 3E11, isotype IgG, Cell Signaling Technology, Danvers, MA, USA).

Antigen retrieval was performed at 95 °C for 30 min in a Sodium Citrate Buffer (10 mM Sodium Citrate, 0.05% Tween 20, pH 6.0). Endogenous biotin blocking was performed for liver sections only using the Dako Cytomation Biotin Blocking System (Dako, Troy, MI, USA). Samples were incubated with a 1:1000 diluted (Antibody Diluent, Dako) primary antibody at 4 °C overnight. Antibody visualization was obtained using the EnVision®+ System-HRP Labelled Polymer (Dako). Nuclei were counterstained with a Mayer's Hematoxylin Solution (Bio-Optica). Sections were observed and images acquired by a Nikon Eclipse E600 microscope.

For COX IV specificity verification, 4 consecutive sections from different organs of 3 non-tumor-bearing NOD scid mice were analyzed. For the MDA-MB-231 time-course experiment, 4 sections per lymph-node and 10 sections per lung sample were analyzed. Macroscopic inspection of organs at sacrifice and microscopic analysis by IHC on a series of non-adjacent FFPE sections (series of 4 consecutive stained and 8 consecutive unstained sections), for a total of 24 or 48, according to positivity, were performed for the preliminary assessment of metastasis formation in all kind of models (Experiment 1). For the quantitation of metastasis-positive (+ve) sections (Experiment 2), systematic IHC analysis was focused on a series of 24 or 48 non-adjacent FFPE sections (a series of 8 consecutive stained and 8 consecutive unstained sections) from lung, lymph-nodes and ovary samples. For the artificial metastasis experiment by tail-vein injection, 4 FFPE consecutive sections per lung sample were analyzed.

2.6. Statistical Analysis

Statistical analyses and graph constructions were performed using Graph Pad Prism v5. Differences in tumor mass between axillary and inguinal nodules were assessed using the point by point multiple Student's *t*-test, assuming that all time points were samples from populations with the same standard deviation, and the false discovery rate was set at 1% and determined using the two-stage linear step-up procedure of Benjamini, Krieger and Yekutieli [54].

3. Results

3.1. Technical Protocol for CTC Isolation and Species-Specificity-Based Detection of Tumor Cells in Xenograft Models

For CTC isolation, blood samples were drawn from anesthetized mice by cardiac puncture, which was proven to ensure the highest CTC yield compared to other approaches, according to Eliane et al. [55], and processed with the size-based CTC isolation device provided by ScreenCell®(Figure 1, Panel A; details are reported in Materials and Methods).

CTCs were identified on the basis of the cytomorphological criteria of malignancy already described for cancer patients [50]: in xenograft models, CTCs showed (i) a larger nucleus (generally 13 to 15 µm in diameter) compared to leukocytes, whose nuclei instead appeared slightly larger (about 7–8 µm in diameter) than membrane pores (6.5 ± 0.33 µm), (ii) a high nucleus-to-cell ratio (>0.5 for cell lines, rather than 0.75, cut-off used for clinical samples), (iii) a dense basophilic and irregularly outlined nucleus and (iv) a pale-bluish ring of cytoplasm, which generally appears as a thin rim encircling the nucleus.

Such a blood sampling approach coupled with filtration showed high efficiency in terms of sensitivity, as described in the following paragraphs, and adaptability to murine blood sample processing and cytological analysis for CTC detection, since the sample quality in terms of cellularity was adequate and the cell morphology was well preserved in about 80% (58/71) of samples.

Furthermore, a technical protocol was developed to effectively isolate and unambiguously identify tumor cells in tissues from xenograft mouse models, taking advantage of the human-murine species-specificity.

Given the weak metastatic ability expected in some models, an antibody-based staining protocol was set up in order to facilitate both the quantitation of rare CTCs and to enable screening for metastases in FFPE tissue sections from samples with microscopic and scattered metastatic foci. Immunofluorescence (for CTCs) and immunohistochemistry (IHC) analyses (for tissue sections) were performed using a commercially available antibody specific for the human mitochondrial marker cytochrome *c* oxidase subunit IV (COX IV). The non-cross-reactivity of the antibody with the murine counterpart has been preliminarily verified by immunofluorescence on peripheral blood mononuclear cells from a BALB/c nude mouse used as the control (Figure S1), and by IHC on FFPE sections of several organs from non-tumor-bearing NOD scid mice (Figure 1, Panel C), thus proving to be a reliable method to detect tumor cells in mouse xenografts.

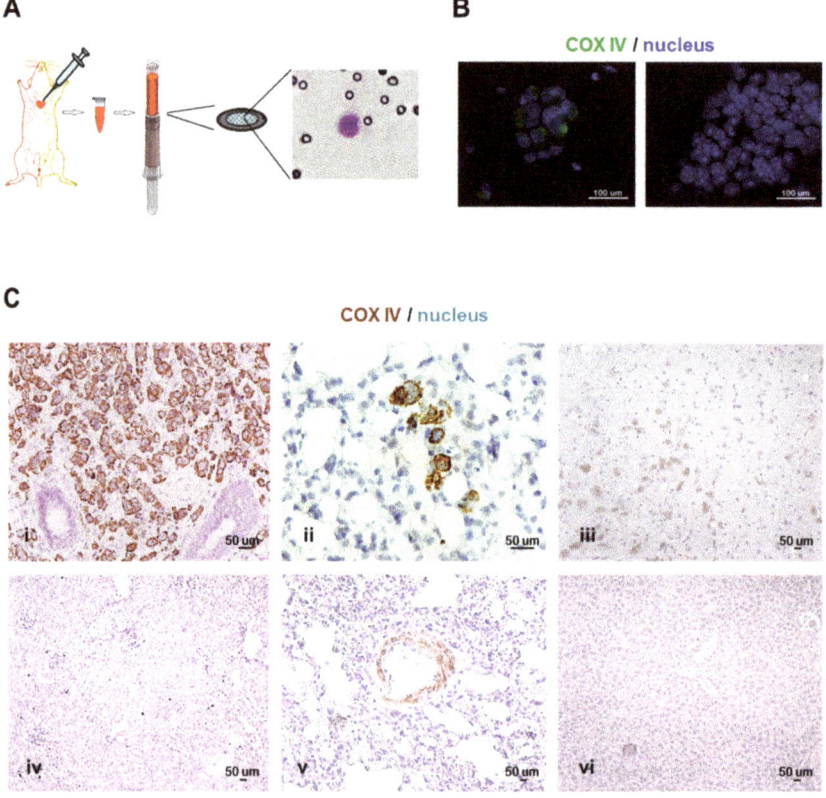

Figure 1. The methodology for the detection of circulating tumor cells (CTCs) and metastases in xenograft models. (**A**) The scheme illustrates a CTC isolation and detection technical approach for application in xenograft mouse models. Briefly, blood was drawn by cardiac puncture and cells were isolated by filtration on a porous membrane and identified by the morphological criteria (or immunostaining). Images represent (**B**) a cluster of CTCs (left) and a cluster of leukocytes (right) isolated from MDA-MB-231 xenografts, acquired by the 4′,6-Diamidino-2-phenylindole (DAPI) and the Fluorescein isothiocyanate (FITC) filters (60× oil immersion objective) and showing COX IV positive and negative staining, respectively; (**C**) COX IV immunohistochemistry stained formalin-fixed paraffin-embedded sections of (**i**) primary tumor nodule (20× objective) and (**ii**) lung metastases (40× objective) from MDA-MB-231 xenograft, and of (**iii**) brain, (**iv**) kidney, (**v**) lung, and (**vi**) liver (10× objective) collected from a non-tumor-bearing mouse.

CTCs were detectable by immunofluorescence and distinguished from leukocytes by the nucleus size and the typical staining pattern, as depicted in Figure 1, Panel B: CTCs organized in clusters have intense cytoplasmic-specific staining and are larger compared to the cluster of leukocytes, which instead have smaller nuclei and show negative staining for COX IV.

3.2. Cancer Cell Dissemination Can Be Monitored from the Early to Late Stages of Tumor Progression in the MDA-MB-231 Xenograft Model

The dynamics of dissemination in the MDA-MB-231 model was investigated in a time-course experiment, where the CTC load and the frequency of lymph-nodal and pulmonary metastases were measured at different time points after tumor cell injection. Overall, the load of single CTCs (mean ± SD: 0.40 ± 0.89; 0.33 ± 0.58; 79.33 ± 181.7; 1,993 ± 4,269; Figure 2, Panel B) and CTC clusters (mean ± SD: 2.33 ± 4.04; 1.75 ± 1.50; 62.00 ± 137.20; 1,229 ± 2,653; Figure 2, Panel C) showed a stepwise increase during progression, which mirrored the primary tumor growth (Figure 2, Panel A). Following a similar trend, the frequency of metastasis +ve cases, assessed in lymph-nodes (axillary, inguinal, subclavian or peritoneal) and lungs, increased during time, although, differently than lungs, metastases at lymph-nodes were detectable since the earliest phases from tumor injection (Figure 2, Panel D). At day 35 CTCs were found in 1 out of 5 assessable cases (2 CTCs, Figure 2, Panel B), consistently with the detection of few metastatic cells at lung in the same animal (Figure 2, Panel D). At day 50 lung metastases were found in 1 out of 3 CTC +ve cases only. On the contrary, 5 out of 6 cases at day 65 and 5 out of 5 cases at day 80 had both CTCs and pulmonary metastases.

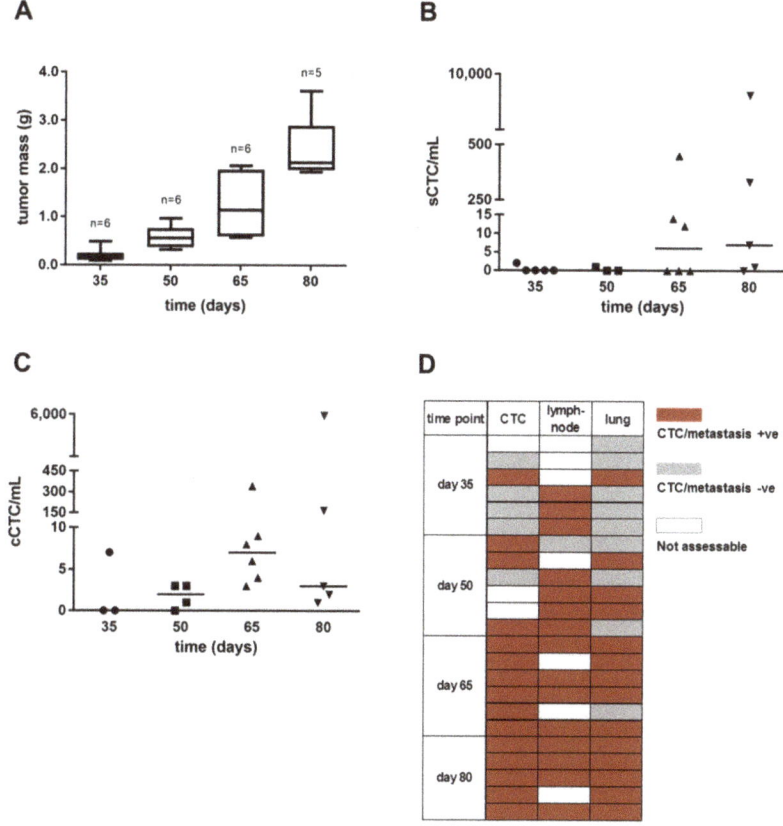

Figure 2. The detection of CTCs and metastases at the early and late stages of tumor progression in the MDA-MB-231 xenograft model. Box and whiskers plots and dot plots represent the distribution of (**A**) the total tumor mass, (**B**) single CTC (sCTC) and (**C**) CTC cluster or tumor microemboli (cCTC) numbers per milliliter of blood (horizontal line representing the median value), at different experimental time points. (**D**) the scheme represents the frequency of CTC-positive (+ve) and of lymph-nodal or pulmonary metastasis-positive (+ve) animals per group, at each experimental time point.

MDA-MB-231 cells were also injected in the tail vein of five animals and their presence in blood was monitored during time. Blood samples collected from two animals, injected with 10^6 or 2×10^6 cells, 1 h after injection contained 1 and 7 sCTCs per milliliter, respectively, thus indicating that the vast majority of cells had reached peripheral districts in short time from forced blood dissemination. The remaining three animals, two injected with 10^6 and one injected with 2×10^6 cells, were sacrificed after 78 days and were all CTC +ve and lung metastasis +ve. cCTCs were detected in all cases and ranged from 1 to 31 per milliliter, while sCTCs (about 280) were found in one animal only, injected with 10^6 cells. Lymph-nodes, ovaries and spleen were all negative for metastases by macroscopic examination and IHC analysis.

3.3. Breast Cancer Cell Lines with Different Subtypes Disseminate in Blood and Show Distinct Organotropism in Xenograft Models

CTC models were obtained by the orthotopic injection of BT-474, MDA-MB-453, MDA-MB-468 and MDA-MB-231 BC cell lines, performed in two independent experiments. The tumor take rate was 100% in all models and the growth rate of nodules was faster in MDA-MB-231 xenografts, where the total mass reached 500 mg after about 50 days from the cell injection, compared to the other models, which reached comparable masses over longer times (Figure S2), thus mirroring the expected level of aggressiveness according to the molecular subtype of each cell line. A significant difference (adjusted p-value <0.01) between the tumor mass of axillary and inguinal nodules was also observed in MDA-MB-468 (axillary versus inguinal mean ± SD tumor mass (g): 0.89 ± 0.22 versus 0.62 ± 0.27 and 1.00 ± 0.31 versus 0.69 ± 0.30, after 92 and 98 days from tumor implant, respectively) and MDA-MB-231 xenografts (axillary versus inguinal mean ± SD tumor mass (g): 1.02 ± 0.38 versus 0.69 ± 0.27, 1.12 ± 0.40 versus 0.77 ± 0.21, and 1.53 ± 0.64 versus 0.83 ± 0.27, after 75, 78, and 83 days from the tumor implant, respectively), with a general trend towards a faster growth rate in the axillary compared to the inguinal mammary fat pad injection site (Figure S2). Interestingly, MDA-MB-231 and the less aggressive BC cell lines were both able to disseminate in blood as sCTCs, (Figure 3, Panel A) found in about 90% of cases, and as cCTCs (both tumor cell clusters and microemboli, Figure 3, Panel B), detected at variable frequency according to the specific xenograft model. Overall, in both experiments, the sCTC and cCTC load per milliliter of blood was higher in the MDA-MB-231 (median(range): 2(0–9625) and 2(0–5973), respectively) compared to the other CTC models (Table S1), in keeping with the aggressiveness and high proliferation rate of these cells.

Among the weakly metastagenic models, MDA-MB-453 showed the highest numbers of sCTC/mL (median(range): 4.5(0–800)), while overall sCTC numbers for BT-474 and MDA-MB-468 ranged from 0 to 8. Moreover, cCTC positivity was approximately 2-fold lower in the less aggressive models (overall 10/26 cases) compared to the highly metastatic MDA-MB-231 xenografts (8/11 cases). Representative images of sCTCs and cCTCs from each model are reported in Figure 3, Panel C.

The metastatic potential of BC cell lines was also assessed in a preliminary exploratory experiment by macroscopic inspection and IHC analysis. Overall, organs presenting with metastasis were lung, lymph-nodes and ovaries, and those without metastasis +ve sections were the liver, brain and spleen. In a second experiment, systematic IHC analysis (Table 2) confirmed that ovarian metastases were detectable in 2 out of 7 MDA-MB-468 and the majority of MDA-MB-453 xenografts, but not in the BT-474, and likely MDA-MB-231 models, since they did not show ovary enlargement at macroscopic inspection. Lymph-nodal involvement was already macroscopically assessable in the MDA-MB-468 and MDA-MB-231 models in 100% of cases, as also confirmed by the IHC analysis, whereas lymph-nodes in BT-474 and MDA-MB-453 models were hardly detectable and collectible, suggesting the absence of massive dissemination via the lymphatic system. Instead, lungs were the metastatic site showing the highest tropism and frequency in all xenograft models. Consistently with CTC numbers, metastatic foci in weakly aggressive models consisted of single scattered cells or small foci of 3–30 cells each compared to the larger clusters, and macroscopic nodules in a few cases, observed in MDA-MB-231 xenografts (Figure S3).

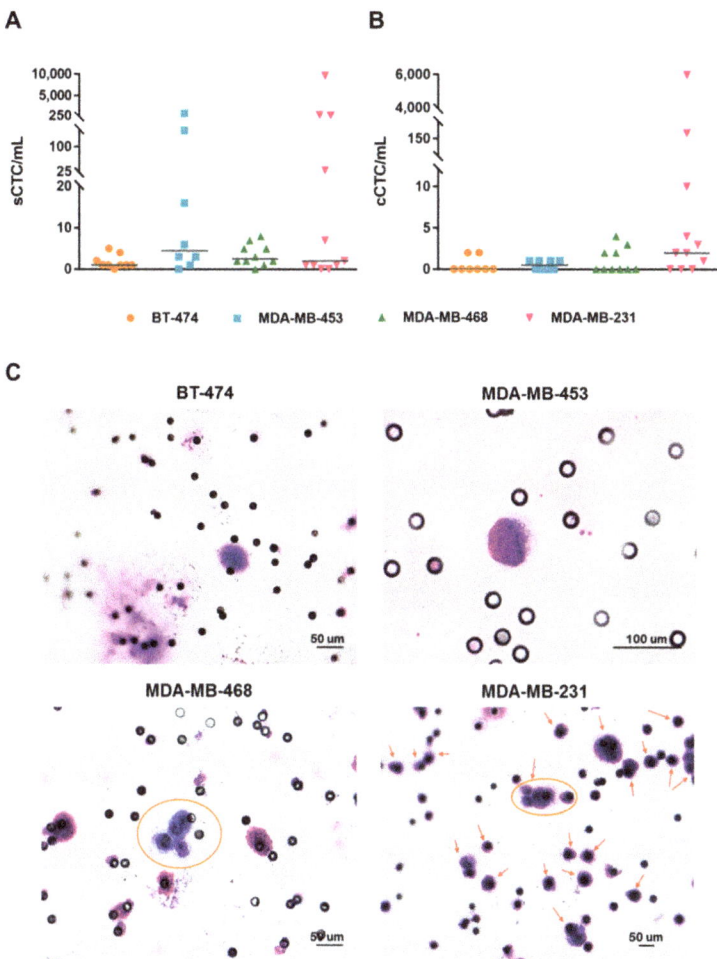

Figure 3. The single CTC (sCTC) and CTC cluster (cCTC) numbers in breast cancer xenograft models. Dot plots represent the distribution (horizontal line corresponding to the median value) of (**A**) sCTCs and (**B**) cCTCs per milliliter of blood. Images (**C**) represent May-Grünwald-Giemsa stained sCTC (40× objective) from BT-474, sCTC (60× oil immersion objective) from MDA-MB-453, cCTC consisting of four cells from MDA-MB-468 (circle, 40×), numerous sCTCs and a cCTC consisting of 4-to-5 cells (arrows, 20× objective) from MDA-MB-231 xenografts.

Table 2. The metastasis sites and frequencies in breast cancer xenograft models.

CTC-Model	N	Experiment 1								N	Experiment 2							
		Lymph-Node		Lungs		Ovary 1		Ovary 2			Lymph-Node		Lungs		Ovary 1		Ovary 2	
		+ve Cases	Positivity Frequency (%) *	+ve Cases	Positivity Frequency (%) *	+ve Cases	Positivity Frequency (%) *	+ve Cases	Positivity Frequency (%)		+ve Cases	Positivity Frequency (%) *	+ve Cases	Positivity frequency (%) *	+ve Cases	Positivity Frequency (%) *	+ve Cases	Positivity Frequency (%) *
BT-474	3	- §	-	3	25–100 †	0	-	0	-	7	- §	-	7	100	0	-	0	-
MDA-MB-453	3	- §	-	2	80–100 †	0	-	0	-	7	- §	-	7	75–100 †	6	75–100 †	5 ‡	43–100 †
MDA-MB-468	3	- §	-	3	100	1	21	0	-	7	7	100	7	100	2	23–100 †	1	90
MDA-MB-231	5	5	100	5	100	- ¥	-	- ¥	-	6	6	100	6	100	- ¥	-	- ¥	-

* range of positivity frequencies (number of metastasis-positive (+ve) out of 10 to 24 sections in Experiment 1, and out of 24 to 48 sections in Experiment 2). ‡ 1 out of 2 ovaries not assessable in one case. † positivity range. § not detectable or not involved at the macroscopic level. ¥ not assessed.

3.4. Circulating Tumor Cells in Breast Cancer Xenograft Models are Pleomorphic and Circulate with Cells of the Host

Cytological blood samples from different CTC models were analyzed and compared in order to highlight intra-sample and inter-model differences on the basis of morphological criteria (details reported in Materials and Methods). The identified cell subpopulations are hereafter described. As already appreciable in the MDA-MB-231 model (Figure 3, Panel C), single CTCs show a certain degree of morphological heterogeneity (i.e., pleomorphism), each cell with a more or less irregularly outlined nucleus of various sizes and shapes, in addition to the heterogeneity in the whole cell size and morphology (Figure 4, Panel A). The difference in sizes is particularly evident in the two BT-474- and the two MDA-MB-453-derived CTCs depicted in Figure 4, one of them smaller than the other. While in the BT-474 CTC model, the cells display an irregular nucleus, the MDA-MB-453-derived CTCs have a clearly round-shaped nucleus and, besides the larger size, the bigger cell also displays a higher nucleus-to-cell ratio (>0.90) compared to the other, as also indicated by the thinner cytoplasmic rim. sCTCs with low (<0.75) or high (>0.90) nucleus-to-cell ratios were also observed in MDA-MB-468, a few of them also presenting with a multilobulated nucleus. CTCs in the MDA-MB-231 model may present as either round-shaped or polygonal physically interacting cells and may have a widely variable whole size. Interestingly, CTC clusters intermingled with or surrounded by platelets were rarely detected in all models, sporadically also in direct contact with leukocytes, as observed in MDA-MB-453 and MDA-MB-231 (Figure 4, Panel B). Few numbers of cytological figures appearing like atypical giant cells with several shapes (morphological details reported in Materials and Methods), were detected in 13%, 30% and 17% of CTC positive cases from MDA-MB-453, MDA-MB-468 and MDA-MB-231 models, respectively. Despite being present in a minority of CTC +ve cases, atypical giant cells were never found in samples called CTC-negative. Images depicting all cell types described in cytological blood samples from xenograft models is reported in Figure S4. Additionally, a complete list of data describing the presence of circulating cells and metastases in all the analyzed animals for each model is reported in Table S2.

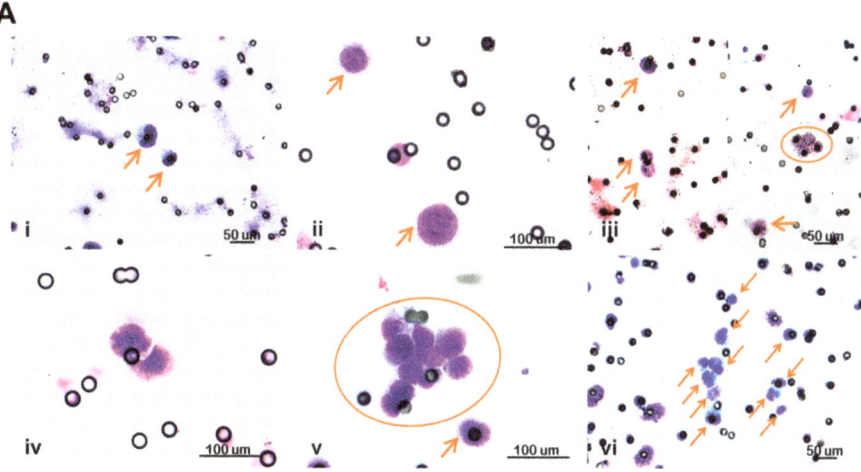

Figure 4. *Cont.*

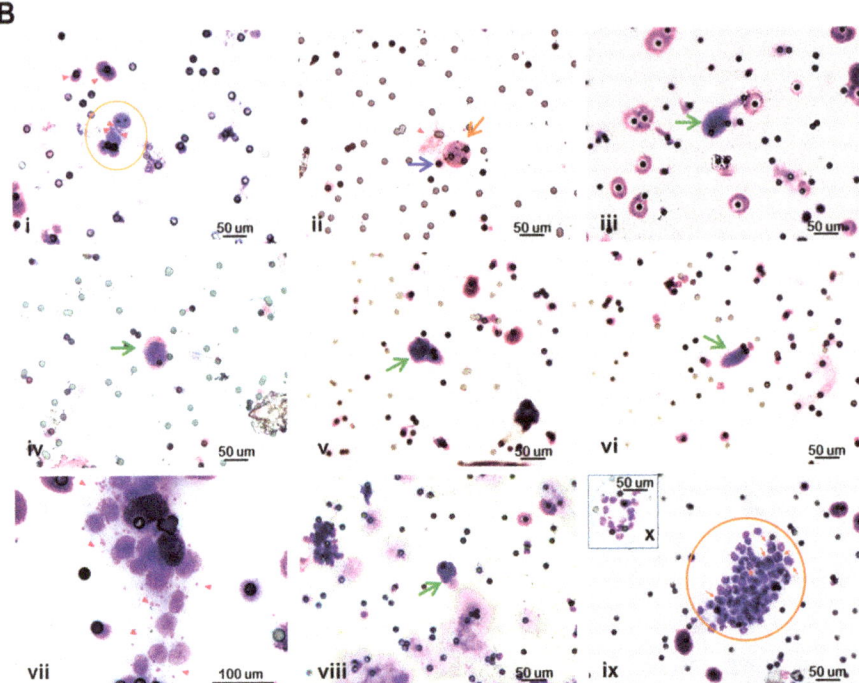

Figure 4. The morphological heterogeneity of CTCs and circulating host cells in BC xenograft models. (**A**) Images are representative of CTC pleomorphism: sCTCs (arrows) with different sizes from (**i**) BT-474 (40× objective) and from (**ii**) MDA-MB-453 (60× oil immersion); (**iii**) sCTCs (arrows) with low (top images) and high (bottom images) nucleus-to-cell ratios, and cCTC (circle) from MDA-MB-468 (40×); (**iv**) the cluster of two CTCs with multilobulated nucleus from MDA-MB-468 (60× oil immersion); (**v**) cCTC from MDA-MB-231 (60× oil immersion); (**vi**) sCTCs and CTCs in clusters (arrows) with different sizes from MDA-MB-231 (40×). (**B**) Images are representative of CTCs and circulating host cells: (**i**) a cluster of four CTCs (circle) and platelets (arrowheads) from BT-474 (40×); (**ii**) sCTC (orange arrow) in cluster with platelets (arrowhead) and one leukocyte (blue arrow) from MDA-MB-453 (40×); (**iii**) atypical giant cells cell from MDA-MB-453 (green arrow, 40×); (**iv,v,vi**) three atypical giant cells from MDA-MB-468 (green arrow, 40×); (**vii**) cCTC and platelets (arrowheads) from MDA-MB-231 (60× oil immersion); (**viii**) an atypical giant cell from MDA-MB-231 (green arrow, 40×); (**ix**) a cluster (circle) of CTCs (arrows showing clearly distinguishable tumor cells) combined with leukocytes (40×).

4. Discussion

Our study demonstrates the reliability of a technical protocol for CTC and metastasis detection in BC xenograft models based on the classical morphological features of malignancy and relying on the advantage of the species-specificity barrier for the identification of cells of human origin in both cytological blood samples and tissues compared to the host counterpart. The described methodology can be, in principle, applied to every kind of xenograft model for CTC and metastasis biology studies, thanks to its sensitivity and specificity in detecting rare circulating cells and also small metastatic foci in weakly metastatic models. Assessment of metastasis formation using a species-specific antibody was proven a reliable method to identify small and rare metastases, especially in weakly metastatic models.

As such, a preliminary validation test with the metastatic MDA-MB-231 BC cell line showed that hematogenous dissemination and metastases may occur even at the earliest stages upon tumor nodule appearances at the orthotopic site, and can be monitored in a time-course experiment. Here, studies were also performed to model BC metastases using BT-474, MDA-MB-453 and MDA-MB-468 cells, providing a comparative analysis, for the first time, of the hematogenous dissemination potential among BC cell lines belonging to different molecular subtypes, in addition to MDA-MB-231, upon injection at the orthotopic level rather than forced metastasis formation assays. Consistently with the growth rate of the primary tumors, xenografts in the murine model, generated using the HER2 positive BT-474 and MDA-MB-453 and the basal A MDA-MB-468 cell lines, according to Neve et al. [34], determined a lower CTC load compared to the numbers of CTCs generated by MDA-MB-231 xenografts. Interestingly, BC cell line xenografts can generate a pleomorphic CTC population, consisting of markedly heterogeneous cells in terms of the whole cellular and/or nuclear size and morphology, as also nucleus to cell ratio, and which also includes clusters of CTCs, released in blood at a reduced frequency compared to the sCTC subset. Intra-clonal size heterogeneity has been already reported in MDA-MB-231-derived clonal subpopulations in vitro [56], suggesting that upon injection in mice cells with different metastatic abilities and morphological features underwent clonal selection and that such clones might be more easily identifiable in the CTC population in xenografts. However, variable sizes and nuclear-cytoplasmic ratios have been also described in CTCs from prostate cancer patients compared to cultured prostate cancer cell lines [57], again indicating that a clonal selection process takes place at the primary or secondary sites during tumor progression. Therefore, CTC morphological heterogeneity might be interpreted as a hallmark of tumor cells which is likely to be more easily assessable among clones of the blood disseminating population.

We have also demonstrated that not only MDA-MB-231 but also other models may represent experimental tools suitable for CTC characterization and metastasis biology studies. Indeed, each cell line was shown to follow preferential dissemination routes, i.e., through blood or lymphatic vessels, as also distinct colonization patterns at distant sites. Lastly, despite the species-barrier, it was surprising to find, for the first time, circulating cells from the murine host in physical contact with tumor cells of human origin, as also atypical cytological figures.

MDA-MB-231 cells were already proven to induce lung metastases when injected in the tail vein of nude mice [58], and the success of transplantation experiments in the m.f.p. ranked them among the most aggressive BC cell lines [59,60]. Since the first reports, studies employing such a cell line have started to proliferate, and even nowadays they represent a large fraction of the literature on BC metastasis biology. On the contrary, ER-positive cell lines such as MCF7, T47D, and BT-474 are able to form tumors only in the presence of an exogenous source of estrogen. However, despite the metastatic origin of these cell lines, they have a limited ability to invade and metastasize [37], unless subjected to a selection of hormone-resistant variants or genetically modified [61]. More recently, severely immunocompromised mice, such as NSG and Rag2$^{-/-}$ γc$^{-/-}$ models, which exhibit T cell, B cell and natural killer cell immunodeficiency, were also explored to generate new metastatic models. This time, MCF7 were able to give rise to metastases at the lymph-node, lung, spleen and, sporadically, even at the renal level when injected in the m.f.p. of NSG mice [38]. BT-474 cells were instead less metastatic in these mice, generating macro-metastases in only a few cases (axillary lymph node in 17% of mice and the spleen in 8%. of mice). The latter result is in contrast with our model where, although no macrometastases were found, small foci were detected in all animals and in all FFPE sections we have analyzed. In another study, bioimaging analysis enabled the detection of multi-organ metastases in Rag2$^{-/-}$ γc$^{-/-}$ mice injected at the orthotopic level with the MDA-MB-453 and BT-474 cell lines [39].

Attempts to explore hematogenous dissemination in BC experimental models were made only recently. In a technical paper published in 2008 [55], different approaches for blood collection were tested to isolate CTCs from tumor-bearing mice, finally demonstrating that the cardiac puncture represents the most suitable approach to reach high yields without interference from contaminating normal murine epithelial cells. The authors also validated a method to enumerate CTCs by applying a modified version of an in vitro diagnostic system for quantifying CTCs in patients, obtaining numbers of CTCs ranging from ~100 to 1000 per milliliter of blood. In line with our results, the reported CTC concentration in the blood of MDA-MB-231 xenograft models was highly variable among different animals. Concerning CTC variability, despite the wide range of cells detected in this model, we have observed a correlation between CTC load and tumor burden, whereas the literature data from experiments with GFP-expressing MDA-MB-231 cells suggested that the primary tumor size is not a strong indicator of CTC load [62]. Hence, such a variability in CTC load in experimental models could also be the result of fluctuations in CTC release, as also suggested by results obtained in a melanoma CTC-model [62]. Here, the authors performed real-time continuous monitoring of CTCs and could estimate a release of 0 to 54 CTCs every 5 min, alternated to CTC-free phases. Differently from data obtained in the MDA-MB-231 CTC model [62], this peak in CTC level mirrored the increase in tumor growth at the same time point, suggesting again that the number of CTCs may indeed correlate with tumor size. In our MDA-MB-231 model, probably due to its pronounced aggressiveness, we had no possibility to define an optimal time point to isolate CTCs before they were able to colonize distant organs since in the time-course experiment CTC release and increase during time mirrored the onset, frequency and extent of pulmonary metastases in matched FFPE sections. On the contrary, dissemination via the lymphatic system was observed at the earliest time points (after 35 days from injection), suggesting that different molecular mechanisms are required to enter the lymphatic system compared to blood vessels. In agreement with this observation, Juratli and colleagues [62] also reported that metastatic foci at lymph nodes can be observed just two weeks after orthotopic inoculation. More recently, two seminal works demonstrated that in experimental models, tumor cells invading lymph nodes are able to enter local blood vessels and exit from them to invade distant organs [63,64]. However, commonalities in clinical tumors need to be demonstrated. The results of our dynamic studies with the MDA-MB-231 model also show that CTCs can be actively released from pulmonary metastases. At the same time, once in the bloodstream, MDA-MB-231 cells were able to rapidly reach peripheral districts and colonize the lung as CTC numbers rapidly dropped out upon intravenous injection and their presence was generally associated to the presence of lymph-node and lung metastases, even at early time points. Different results were observed in a MDA-MB-468 CTC-model already described by Bonnomet and colleagues [65], where CTCs were detectable as early as 8 days after injection and increased 36 days later, after which their levels remained quite constant. Indeed, Bonnomet and colleagues also found lung metastases at later time point only, despite CTC recovery was possible even a few days after cell injection in a time-course experiment [65].

An interesting result emerging from our study is the detection of clusters of CTCs in all the examined BC models. Overall, to our knowledge, the presence of cCTCs has been reported for the triple-negative LM2 MDA-MB-231 [66] and MDA-MB-435 [67] BC cell lines only. We were able to monitor the presence of CTCs using size-based isolation support also in xenograft models obtained from BC cells with different molecular subtypes. Consistently with the demonstration that cCTCs are endowed with higher efficiency in initiating metastasis [66,68], the frequency of cCTCs was 2-fold lower in models with weak metastatic potential compared to MDA-MB-231. Moreover, CTCs generated upon xenograft of BC cell lines were heterogeneous in morphology, even within the same cluster, thus suggesting further heterogeneity at the molecular level, probably as a result of a selection process of cells which are committed to disseminate in blood according to their functionality and ability to cooperate and survive in a foreign microenvironment. Unexpectedly, some clusters of tumor cells also presented with physically interacting platelets and/or leukocytes. CTCs in contact with blood or stromal cells were already described both in experimental models and clinical samples and some cell types, such as platelets, neutrophils and monocytes/macrophages, were also demonstrated to promote epithelial-to-mesenchymal transition and the pre-metastatic niche formation [69–72], even at early BC stages [73], or to assist CTCs during their transit in blood and organ colonization, increasing their metastatic ability [74]. However, to our knowledge, interactions between circulating tumor and host cells in xenografts have not been reported at present. Platelet depletion in a nude mouse model transplanted subcutaneously with SKOV3 ovarian cancer cells provided evidence that platelets are involved in cancer cell growth [75], hence suggesting that even cells from the murine compartment can be actively recruited to cooperate with human tumor cells in xenograft models. With reference to our data, the growth rate of matched axillary and inguinal m.f.p. nodules generally showed a different trend in both MDA-MB-468 and MDA-MB-231 models, with a higher tumor mass at the axillary compared to the inguinal site, even statistically significant during the latest time point measurements. Such data are consistent with the hypothesis that, despite the species specificity barrier, tumor cells can interact and possibly cross-talk with the murine microenvironment in xenograft models. Finally, atypical giant cells, presenting without features of malignancy and, therefore, expected to originate from the host, were also observed in our models. Similarly to clinical samples [52], such a cell type has been generally associated with the presence of CTCs and never found in CTC-negative samples. The origin of circulating atypical giant cells in xenografts has not been investigated here. Our hypothesis is that cells displaying not all the classical features of malignancy and presenting with unusual morphological patterns, which resemble those described for cancer-associated macrophage-like cells by Adams and colleagues [52], might derive from the interaction between tumor and host cells. Fusion hybrids, i.e., hybrid cells derived from fusion events between tumor cells and macrophages, were already described in murine experimental models [76]. If such cells originate in response to the attack from the immune system towards tumor cells, or as a strategy to increase the tumor cell viability and metastatic potential, is yet to be established. Literature data reported that murine peritoneal macrophages can phagocytize apoptotic BC tumor cells from cell lines in vitro, and acquire stem-like features in the following steps [77]. With reference to the xenograft milieu, another study reported the host macrophage invasion and the presence of multinuclear giant cells or foreign body giant cells at the implant site upon the injection of human mesenchymal stem cells with biopolymers in NOD scid mice, thus indicating that severely immunocompromised mice are able to retain a certain level of innate immune responsiveness [78]. On the other hand, cell fusion has been associated with the acquisition of increased metastatic capacity or enhanced drug resistance [79], and the presence of circulating hybrid cells was shown to correlate with the disease stage and patient survival [76]. Overall, we are aware that beside the identification based on morphological criteria although performed by a pathologist experienced in CTC detection in clinical samples, only a molecular characterization of such atypical populations of cells, including leukocytes interacting with CTCs, e.g., with species-specific antibodies and gene expression analyses, would definitely elucidate their nature and confirm the validity of xenograft models for new research lines.

5. Conclusions

In the end, CTCs and metastases can be in vivo modeled from BC cell lines with different subtypes and disseminating potential. In xenografts, several subpopulations of cells circulating in the blood can be identified by applying the classical morphological criteria, thus offering experimental models alternative to MDA-MB-231, as unusual and intriguing tools to investigate tumor-host interactions in the blood microenvironment.

Supplementary Materials: The following are available online at http://www.mdpi.com/2073-4409/8/7/683/s1, Figure S1: Species-specificity of anti-human COX IV antibody; Figure S2: Primary tumor growth in breast cancer xenograft models; Figure S3: Lung metastases in breast cancer xenograft models; Figure S4: Cell types observed in cytological blood samples from breast cancer xenograft models; Table S1: Circulating tumor cell load in breast cancer xenograft models; Table S2: Circulating cells and metastases in xenograft models from breast cancer cell lines.

Author Contributions: Conceptualization, E.F.; Methodology and Investigation, E.F. and L.C.; Formal analysis, E.F.; Project administration and Supervision, V.C. and M.G.D.; Funding acquisition, M.G.D. and E.F.; Visualization and Writing—original draft, E.F.; Writing—review & editing, M.G.D., V.C. and L.C.

Funding: This research was supported by Associazione Italiana per la Ricerca sul Cancro (AIRC Investigator Grant: 16900; Principal Investigator: M.G. Daidone) and by Fondazione Italiana per la Ricerca sul Cancro (FIRC, Three-year Fellowship Elda e Attilio Pandolfi, id. 16411; Principal Investigator: E. Fina).

Acknowledgments: The Authors thank Janine Wechsler, consultant pathologist at ScreenCell, for morphological analysis of cytological samples, and Gloria Morandi, Lucia Gioiosa and Lorena Ventura, technicians at INT IHC Facility, for performing IHC staining. A special thanks to INT Committee for animal welfare (OPBA, Organismo Per il Benessere Animale) for its support in writing the protocol for in vivo experiments.

Conflicts of Interest: The Authors declare no conflict of interest.

References

1. Hanahan, D.; Weinberg, R.A. Hallmarks of cancer: The next generation. *Cell* **2011**, *144*, 646–674. [CrossRef] [PubMed]
2. WHO. Cancer. Available online: https://www.who.int/en/news-room/fact-sheets/detail/cancer (accessed on 28 May 2019).
3. Klein, C.A. Parallel progression of primary tumours and metastases. *Nat. Rev. Cancer* **2009**, *9*, 302–312. [CrossRef] [PubMed]
4. Hüsemann, Y.; Geigl, J.B.; Schubert, F.; Musiani, P.; Meyer, M.; Burghart, E.; Forni, G.; Eils, R.; Fehm, T.; Riethmüller, G.; et al. Systemic spread is an early step in breast cancer. *Cancer Cell* **2008**, *13*, 58–68. [CrossRef] [PubMed]
5. Hosseini, H.; Obradović, M.M.S.; Hoffmann, M.; Harper, K.L.; Sosa, M.S.; Werner-Klein, M.; Nanduri, L.K.; Werno, C.; Ehrl, C.; Maneck, M.; et al. Early dissemination seeds metastasis in breast cancer. *Nature* **2016**, *540*, 552–558. [CrossRef] [PubMed]
6. Bidard, F.C.; Mathiot, C.; Delaloge, S.; Brain, E.; Giachetti, S.; de Cremoux, P.; Marty, M.; Pierga, J.Y. Single circulating tumor cell detection and overall survival in nonmetastatic breast cancer. *Ann. Oncol.* **2010**, *21*, 729–733. [CrossRef] [PubMed]
7. Lucci, A.; Hall, C.S.; Lodhi, A.K.; Bhattacharyya, A.; Anderson, A.E.; Xiao, L.; Bedrosian, I.; Kuerer, H.M.; Krishnamurthy, S. Circulating tumour cells in non-metastatic breast cancer: A prospective study. *Lancet Oncol.* **2012**, *13*, 688–695. [CrossRef]
8. Fina, E.; Reduzzi, C.; Motta, R.; Di Cosimo, S.; Bianchi, G.; Martinetti, A.; Wechsler, J.; Cappelletti, V.; Daidone, M.G. Did circulating tumor cells tell us all they could? The missed circulating tumor cell message in breast cancer. *Int. J. Biol. Markers* **2015**, *30*, 429–433. [CrossRef]
9. Sänger, N.; Effenberger, K.E.; Riethdorf, S.; Van Haasteren, V.; Gauwerky, J.; Wiegratz, I.; Strebhardt, K.; Kaufmann, M.; Pantel, K. Disseminated tumor cells in the bone marrow of patients with ductal carcinoma in situ. *Int. J. Cancer* **2011**, *129*, 2522–2526. [CrossRef]
10. Franken, B.; de Groot, M.R.; Mastboom, W.J.; Vermes, I.; van der Palen, J.; Tibbe, A.G.; Terstappen, L.W. Circulating tumor cells, disease recurrence and survival in newly diagnosed breast cancer. *Breast Cancer Res.* **2012**, *14*, R133. [CrossRef]
11. Banys, M.; Hahn, M.; Gruber, I.; Krawczyk, N.; Wallwiener, M.; Hartkopf, A.; Taran, F.A.; Röhm, C.; Kurth, R.; Becker, S.; et al. Detection and clinical relevance of hematogenous tumor cell dissemination in patients with ductal carcinoma in situ. *Breast Cancer Res. Treat.* **2014**, *144*, 531–538. [CrossRef]
12. Bloom, H.J.; Richardson, W.W. Histological grading and prognosis in breast cancer; a study of 1409 cases of which 359 have been followed for 15 years. *Br. J. Cancer* **1957**, *11*, 359–377. [CrossRef] [PubMed]
13. Sotiriou, C.; Pusztai, L. Gene-expression signatures in breast cancer. *N. Engl. J. Med.* **2009**, *360*, 790–800. [CrossRef] [PubMed]
14. Bertos, N.R.; Park, M. Breast cancer—One term, many entities? *J. Clin. Investig.* **2011**, *121*, 3789–3796. [CrossRef] [PubMed]
15. Russnes, H.G.; Navin, N.; Hicks, J.; Borresen-Dale, A.L. Insight into the heterogeneity of breast cancer through next-generation sequencing. *J. Clin. Investig.* **2011**, *121*, 3810–3818. [CrossRef] [PubMed]
16. Korkaya, H.; Liu, S.; Wicha, M.S. Breast cancer stem cells, cytokine networks, and the tumor microenvironment. *J. Clin. Investig.* **2011**, *121*, 3804–3809. [CrossRef] [PubMed]
17. Place, A.E.; Jin Huh, S.; Polyak, K. The microenvironment in breast cancer progression: Biology and implications for treatment. *Breast Cancer Res.* **2011**, *13*, 227. [CrossRef]
18. Curtis, C.; Shah, S.P.; Chin, S.F.; Turashvili, G.; Rueda, O.M.; Dunning, M.J.; Speed, D.; Lynch, A.G.; Samarajiwa, S.; Yuan, Y.; et al. The genomic and transcriptomic architecture of 2,000 breast tumours reveals novel subgroups. *Nature* **2012**, *486*, 346–352. [CrossRef]
19. Dawson, S.J.; Rueda, O.M.; Aparicio, S.; Caldas, C. A new genome-driven integrated classification of breast cancer and its implications. *EMBO J.* **2013**, *32*, 617–628. [CrossRef]
20. Dexter, D.L.; Leith, J.T. Tumor heterogeneity and drug resistance. *J. Clin. Oncol.* **1986**, *4*, 244–257. [CrossRef]
21. Dagogo-Jack, I.; Shaw, A.T. Tumour heterogeneity and resistance to cancer therapies. *Nat. Rev. Clin. Oncol.* **2018**, *15*, 81–94. [CrossRef]

22. Fidler, I.J. The pathogenesis of cancer metastasis: The 'seed and soil' hypothesis revisited. *Nat. Rev. Cancer* **2003**, *3*, 453–458. [CrossRef] [PubMed]
23. Gupta, G.P.; Massagué, J. Cancer metastasis: Building a framework. *Cell* **2006**, *127*, 679–695. [CrossRef] [PubMed]
24. Talmadge, J.E.; Fidler, I.J. AACR centennial series: The biology of cancer metastasis: Historical perspective. *Cancer Res.* **2010**, *70*, 5649–5669. [CrossRef] [PubMed]
25. Saxena, M.; Christofori, G. *Mol. Oncol.* **2013**, *7*, 283–296. [CrossRef] [PubMed]
26. Khanna, C.; Hunter, K. Modeling metastasis in vivo. *Carcinogenesis* **2005**, *26*, 513–523. [CrossRef] [PubMed]
27. Yin, J.J.; Selander, K.; Chirgwin, J.M.; Dallas, M.; Grubbs, B.G.; Wieser, R.; Massagué, J.; Mundy, G.R.; Guise, T.A. TGF-beta signaling blockade inhibits PTHrP secretion by breast cancer cells and bone metastases development. *J. Clin. Investig.* **1999**, *103*, 197–206. [CrossRef] [PubMed]
28. Kang, Y.; Siegel, P.M.; Shu, W.; Drobnjak, M.; Kakonen, S.M.; Cordón-Cardo, C.; Guise, T.A.; Massagué, J. A multigenic program mediating breast cancer metastasis to bone. *Cancer Cell* **2003**, *3*, 537–549. [CrossRef]
29. Harms, J.F.; Welch, D.R. MDA-MB-435 human breast carcinoma metastasis to bone. *Clin. Exp. Metastasis* **2003**, *20*, 327–334. [CrossRef]
30. Minn, A.J.; Gupta, G.P.; Siegel, P.M.; Bos, P.D.; Shu, W.; Giri, D.D.; Viale, A.; Olshen, A.B.; Gerald, W.L.; Massagué, J. Genes that mediate breast cancer metastasis to lung. *Nature* **2005**, *436*, 518–524. [CrossRef]
31. Nguyen, D.X.; Bos, P.D.; Massagué, J. Metastasis: From dissemination to organ-specific colonization. *Nat. Rev. Cancer* **2009**, *9*, 274–284. [CrossRef]
32. Bos, P.D.; Nguyen, D.X.; Massagué, J. Modeling metastasis in the mouse. *Curr. Opin. Pharmacol.* **2010**, *10*, 571–577. [CrossRef] [PubMed]
33. Vargo-Gogola, T.; Rosen, J.M. Modelling breast cancer: One size does not fit all. *Nat. Rev. Cancer* **2007**, *7*, 659–672. [CrossRef] [PubMed]
34. Neve, R.M.; Chin, K.; Fridlyand, J.; Yeh, J.; Baehner, F.L.; Fevr, T.; Clark, L.; Bayani, N.; Coppe, J.P.; Tong, F.; et al. A collection of breast cancer cell lines for the study of functionally distinct cancer subtypes. *Cancer Cell* **2006**, *10*, 515–527. [CrossRef] [PubMed]
35. Mackay, A.; Tamber, N.; Fenwick, K.; Iravani, M.; Grigoriadis, A.; Dexter, T.; Lord, C.J.; Reis-Filho, J.S.; Ashworth, A. A high-resolution integrated analysis of genetic and expression profiles of breast cancer cell lines. *Breast Cancer Res. Treat.* **2009**, *118*, 481–498. [CrossRef] [PubMed]
36. Prat, A.; Parker, J.S.; Karginova, O.; Fan, C.; Livasy, C.; Herschkowitz, J.I.; He, X.; Perou, C.M. Phenotypic and molecular characterization of the claudin-low intrinsic subtype of breast cancer. *Breast Cancer Res.* **2010**, *12*, R68. [CrossRef] [PubMed]
37. Lacroix, M.; Leclercq, G. Relevance of breast cancer cell lines as models for breast tumours: An update. *Breast Cancer Res. Treat.* **2004**, *83*, 249–289. [CrossRef] [PubMed]
38. Iorns, E.; Drews-Elger, K.; Ward, T.M.; Dean, S.; Clarke, J.; Berry, D.; El Ashry, D.; Lippman, M. A new mouse model for the study of human breast cancer metastasis. *PLoS ONE* **2012**, *7*, e47995. [CrossRef] [PubMed]
39. Nanni, P.; Nicoletti, G.; Palladini, A.; Croci, S.; Murgo, A.; Ianzano, M.L.; Grosso, V.; Stivani, V.; Antognoli, A.; Lamolinara, A.; et al. Multiorgan metastasis of human HER-2$^+$ breast cancer in Rag2$^{-/-}$;Il2rg$^{-/-}$ mice and treatment with PI3K inhibitor. *PLoS ONE* **2012**, *7*, e39626. [CrossRef] [PubMed]
40. Marangoni, E.; Vincent-Salomon, A.; Auger, N.; Degeorges, A.; Assayag, F.; de Cremoux, P.; de Plater, L.; Guyader, C.; De Pinieux, G.; Judde, J.G.; et al. A new model of patient tumor-derived breast cancer xenografts for preclinical assays. *Clin. Cancer Res.* **2007**, *13*, 3989–3998. [CrossRef]
41. Bergamaschi, A.; Hjortland, G.O.; Triulzi, T.; Sørlie, T.; Johnsen, H.; Ree, A.H.; Russnes, H.G.; Tronnes, S.; Maelandsmo, G.M.; Fodstad, O.; et al. Molecular profiling and characterization of luminal-like and basal-like in vivo breast cancer xenograft models. *Mol. Oncol.* **2009**, *3*, 469–482. [CrossRef]
42. DeRose, Y.S.; Wang, G.; Lin, Y.C.; Bernard, P.S.; Buys, S.S.; Ebbert, M.T.; Factor, R.; Matsen, C.; Milash, B.A.; Nelson, E.; et al. Tumor grafts derived from women with breast cancer authentically reflect tumor pathology, growth, metastasis and disease outcomes. *Nat. Med.* **2011**, *17*, 1514–1520. [CrossRef] [PubMed]

43. Kabos, P.; Finlay-Schultz, J.; Li, C.; Kline, E.; Finlayson, C.; Wisell, J.; Manuel, C.A.; Edgerton, S.M.; Harrell, J.C.; Elias, A.; et al. Patient-derived luminal breast cancer xenografts retain hormone receptor heterogeneity and help define unique estrogen-dependent gene signatures. *Breast Cancer Res. Treat.* **2012**, *135*, 415–432. [CrossRef] [PubMed]
44. Cassidy, J.W.; Caldas, C.; Bruna, A. Maintaining Tumor Heterogeneity in Patient-Derived Tumor Xenografts. *Cancer Res.* **2015**, *75*, 2963–2968. [CrossRef] [PubMed]
45. Bruna, A.; Rueda, O.M.; Greenwood, W.; Batra, A.S.; Callari, M.; Batra, R.N.; Pogrebniak, K.; Sandoval, J.; Cassidy, J.W.; Tufegdzic-Vidakovic, A.; et al. A Biobank of Breast Cancer Explants with Preserved Intra-tumor Heterogeneity to Screen Anticancer Compounds. *Cell* **2016**, *167*, 260–274. [CrossRef] [PubMed]
46. Zhang, X.; Claerhout, S.; Prat, A.; Dobrolecki, L.E.; Petrovic, I.; Lai, Q.; Landis, M.D.; Wiechmann, L.; Schiff, R.; Giuliano, M.; et al. A renewable tissue resource of phenotypically stable, biologically and ethnically diverse, patient-derived human breast cancer xenograft models. *Cancer Res.* **2013**, *73*, 4885–4897. [CrossRef] [PubMed]
47. Siolas, D.; Hannon, G.J. Patient-derived tumor xenografts: Transforming clinical samples into mouse models. *Cancer Res.* **2013**, *73*, 5315–5319. [CrossRef] [PubMed]
48. Workman, P.; Aboagye, E.O.; Balkwill, F.; Balmain, A.; Bruder, G.; Chaplin, D.J.; Double, J.A.; Everitt, J.; Farningham, D.A.; Glennie, M.J.; et al. Guidelines for the welfare and use of animals in cancer research. *Br. J. Cancer* **2010**, *102*, 1555–1577. [CrossRef]
49. ARRIVE Guidelines. Available online: https://www.nc3rs.org.uk/arrive-guidelines (accessed on 28 May 2019).
50. Hofman, V.J.; Ilie, M.I.; Bonnetaud, C.; Selva, E.; Long, E.; Molina, T.; Vignaud, J.M.; Fléjou, J.F.; Lantuejoul, S.; Piaton, E.; et al. Cytopathologic detection of circulating tumor cells using the isolation by size of epithelial tumor cell method: Promises and pitfalls. *Am. J. Clin. Pathol.* **2011**, *135*, 146–156. [CrossRef]
51. Cho, E.H.; Wendel, M.; Luttgen, M.; Yoshioka, C.; Marrinucci, D.; Lazar, D.; Schram, E.; Nieva, J.; Bazhenova, L.; Morgan, A.; et al. Characterization of circulating tumor cell aggregates identified in patients with epithelial tumors. *Phys. Biol.* **2012**, *9*, 016001. [CrossRef]
52. Adams, D.L.; Martin, S.S.; Alpaugh, R.K.; Charpentier, M.; Tsai, S.; Bergan, R.C.; Ogden, I.M.; Catalona, W.; Chumsri, S.; Tang, C.M.; et al. Circulating giant macrophages as a potential biomarker of solid tumors. *Proc. Natl. Acad. Sci. USA* **2014**, *111*, 3514–3519. [CrossRef]
53. Wechsler, J. *Atlas de Cytologie. Cellules Tumorales Circulantes Des Cancers Solides*; Sauramps Medical: Montpellier, France, 2015; ISBN 979-103030-009-3.
54. Benjamini, Y.; Krieger, A.M.; Yekutieli, D. Adaptive linear step-up procedures that control the false discovery rate. *Biometrika* **2006**, *93*, 491–507. [CrossRef]
55. Eliane, J.P.; Repollet, M.; Luker, K.E.; Brown, M.; Rae, J.M.; Dontu, G.; Schott, A.F.; Wicha, M.; Doyle, G.V.; Hayes, D.F.; et al. Monitoring serial changes in circulating human breast cancer cells in murine xenograft models. *Cancer Res.* **2008**, *68*, 5529–5532. [CrossRef] [PubMed]
56. Nguyen, A.; Yoshida, M.; Goodarzi, H.; Tavazoie, S.F. Highly variable cancer subpopulations that exhibit enhanced transcriptome variability and metastatic fitness. *Nat. Commun.* **2016**, *7*, 11246. [CrossRef] [PubMed]
57. Park, S.; Ang, R.R.; Duffy, S.P.; Bazov, J.; Chi, K.N.; Black, P.C.; Ma, H. Morphological differences between circulating tumor cells from prostate cancer patients and cultured prostate cancer cells. *PLoS ONE* **2014**, *9*, e85264. [CrossRef] [PubMed]
58. Fraker, L.D.; Halter, S.A.; Forbes, J.T. Growth inhibition by retinol of a human breast carcinoma cell line in vitro and in athymic mice. *Cancer Res.* **1984**, *44*, 5757–5763. [PubMed]
59. Price, J.E.; Polyzos, A.; Zhang, R.D.; Daniels, L.M. Tumorigenicity and metastasis of human breast carcinoma cell lines in nude mice. *Cancer Res.* **1990**, *50*, 717–721.
60. Zhang, R.D.; Fidler, I.J.; Price, J.E. Relative malignant potential of human breast carcinoma cell lines established from pleural effusions and a brain metastasis. *Invasion Metastasis* **1991**, *11*, 204–215.
61. Clarke, R. Animal models of breast cancer: Their diversity and role in biomedical research. *Breast Cancer Res. Treat.* **1996**, *39*, 1–6. [CrossRef]
62. Juratli, M.A.; Sarimollaoglu, M.; Nedosekin, D.A.; Melerzanov, A.V.; Zharov, V.P.; Galanzha, E.I. Dynamic Fluctuation of Circulating Tumor Cells during Cancer Progression. *Cancers* **2014**, *6*, 128–142. [CrossRef]

63. Brown, M.; Assen, F.P.; Leithner, A.; Abe, J.; Schachner, H.; Asfour, G.; Bago-Horvath, Z.; Stein, J.V.; Uhrin, P.; Sixt, M.; et al. Lymph node blood vessels provide exit routes for metastatic tumor cell dissemination in mice. *Science* **2018**, *359*, 1408–1411. [CrossRef]
64. Pereira, E.R.; Kedrin, D.; Seano, G.; Gautier, O.; Meijer, E.F.J.; Jones, D.; Chin, S.M.; Kitahara, S.; Bouta, E.M.; Chang, J.; et al. Lymph node metastases can invade local blood vessels, exit the node, and colonize distant organs in mice. *Science* **2018**, *359*, 1403–1407. [CrossRef] [PubMed]
65. Bonnomet, A.; Syne, L.; Brysse, A.; Feyereisen, E.; Thompson, E.W.; Noël, A.; Foidart, J.M.; Birembaut, P.; Polette, M.; Gilles, C. A dynamic in vivo model of epithelial-to-mesenchymal transitions in circulating tumor cells and metastases of breast cancer. *Oncogene* **2012**, *31*, 3741–3753. [CrossRef] [PubMed]
66. Aceto, N.; Bardia, A.; Miyamoto, D.T.; Donaldson, M.C.; Wittner, B.S.; Spencer, J.A.; Yu, M.; Pely, A.; Engstrom, A.; Zhu, H.; et al. Circulating tumor cell clusters are oligoclonal precursors of breast cancer metastasis. *Cell* **2014**, *158*, 1110–1122. [CrossRef] [PubMed]
67. Glinsky, V.V.; Glinsky, G.V.; Glinskii, O.V.; Huxley, V.H.; Turk, J.R.; Mossine, V.V.; Deutscher, S.L.; Pienta, K.J.; Quinn, T.P. Intravascular metastatic cancer cell homotypic aggregation at the sites of primary attachment to the endothelium. *Cancer Res.* **2003**, *63*, 3805–3811. [CrossRef] [PubMed]
68. Cheung, K.J.; Padmanaban, V.; Silvestri, V.; Schipper, K.; Cohen, J.D.; Fairchild, A.N.; Gorin, M.A.; Verdone, J.E.; Pienta, K.J.; Bader, J.S.; et al. Polyclonal breast cancer metastases arise from collective dissemination of keratin 14-expressing tumor cell clusters. *Proc. Natl. Acad. Sci. USA* **2016**, *113*, E854–E863. [CrossRef] [PubMed]
69. Labelle, M.; Begum, S.; Hynes, R.O. Direct signaling between platelets and cancer cells induces an epithelial-mesenchymal-like transition and promotes metastasis. *Cancer Cell* **2011**, *20*, 576–590. [CrossRef] [PubMed]
70. Coffelt, S.B.; Kersten, K.; Doornebal, C.W.; Weiden, J.; Vrijland, K.; Hau, C.S.; Verstegen, N.J.M.; Ciampricotti, M.; Hawinkels, L.J.A.C.; Jonkers, J.; et al. IL-17-producing γδ T cells and neutrophils conspire to promote breast cancer metastasis. *Nature* **2015**, *522*, 345–348. [CrossRef] [PubMed]
71. Wculek, S.K.; Malanchi, I. Neutrophils support lung colonization of metastasis-initiating breast cancer cells. *Nature* **2015**, *528*, 413–417. [CrossRef]
72. Hanna, R.N.; Cekic, C.; Sag, D.; Tacke, R.; Thomas, G.D.; Nowyhed, H.; Herrley, E.; Rasquinha, N.; McArdle, S.; Wu, R.; et al. Patrolling monocytes control tumor metastasis to the lung. *Science* **2015**, *350*, 985–990. [CrossRef]
73. Linde, N.; Casanova-Acebes, M.; Sosa, M.S.; Mortha, A.; Rahman, A.; Farias, E.; Harper, K.; Tardio, E.; Reyes Torres, I.; Jones, J.; et al. Macrophages orchestrate breast cancer early dissemination and metastasis. *Nat. Commun.* **2018**, *9*, 21. [CrossRef]
74. Szczerba, B.M.; Castro-Giner, F.; Vetter, M.; Krol, I.; Gkountela, S.; Landin, J.; Scheidmann, M.C.; Donato, C.; Scherrer, R.; Singer, J.; et al. Neutrophils escort circulating tumour cells to enable cell cycle progression. *Nature* **2019**, *566*, 553–557. [CrossRef] [PubMed]
75. Yuan, L.; Liu, X. Platelets are associated with xenograft tumor growth and the clinical malignancy of ovarian cancer through an angiogenesis-dependent mechanism. *Mol. Med. Rep.* **2015**, *11*, 2449–2458. [CrossRef] [PubMed]
76. Gast, C.E.; Silk, A.D.; Zarour, L.; Riegler, L.; Burkhart, J.G.; Gustafson, K.T.; Parappilly, M.S.; Roh-Johnson, M.; Goodman, J.R.; Olson, B.; et al. Cell fusion potentiates tumor heterogeneity and reveals circulating hybrid cells that correlate with stage and survival. *Sci. Adv.* **2018**, *4*, eaat7828. [CrossRef] [PubMed]
77. Zhang, Y.; Zhou, N.; Yu, X.; Zhang, X.; Li, S.; Lei, Z.; Hu, R.; Li, H.; Mao, Y.; Wang, X.; et al. Tumacrophage: Macrophages transformed into tumor stem-like cells by virulent genetic material from tumor cells. *Oncotarget* **2017**, *8*, 82326–82343. [CrossRef] [PubMed]
78. Xia, Z.; Ye, H.; Choong, C.; Ferguson, D.J.; Platt, N.; Cui, Z.; Triffitt, J.T. Macrophagic response to human mesenchymal stem cell and poly(epsilon-caprolactone) implantation in nonobese diabetic/severe combined immunodeficient mice. *J. Biomed. Mater. Res. A* **2004**, *71*, 538–548. [CrossRef] [PubMed]
79. Weiler, J.; Dittmar, T. Cell Fusion in Human Cancer: The Dark Matter Hypothesis. *Cells* **2019**, *8*, 132. [CrossRef] [PubMed]

© 2019 by the authors. Licensee MDPI, Basel, Switzerland. This article is an open access article distributed under the terms and conditions of the Creative Commons Attribution (CC BY) license (http://creativecommons.org/licenses/by/4.0/).

Article

Detection of AR-V7 in Liquid Biopsies of Castrate Resistant Prostate Cancer Patients: A Comparison of AR-V7 Analysis in Circulating Tumor Cells, Circulating Tumor RNA and Exosomes

Mohammed Nimir [1,2], Yafeng Ma [1,2], Sarah A. Jeffreys [1,3], Thomas Opperman [1,2], Francis Young [1,2], Tanzila Khan [1,3], Pei Ding [1,3,4], Wei Chua [4], Bavanthi Balakrishnar [4], Adam Cooper [1,3,4], Paul De Souza [1,3,4,5] and Therese M. Becker [1,2,3,*]

1. Centre for Circulating Tumour Cell Diagnostics and Research, Ingham Institute for Applied Medical Research, 1 Campbell St, Liverpool, NSW 2170, Australia
2. South Western Clinical School, University of New South Wales, Goulburn St, Liverpool, NSW 2170, Australia
3. School of Medicine, Western Sydney University, Campbelltown, NSW 2560, Australia
4. Liverpool Hospital, Elizabeth St & Goulburn St, Liverpool, NSW 2170, Australia
5. School of Medicine, University of Wollongong, Wollongong, NSW 2522, Australia
* Correspondence: t.becker@unsw.edu.au; Tel.: +61-287389033

Received: 31 May 2019; Accepted: 4 July 2019; Published: 8 July 2019

Abstract: Detection of androgen receptor (AR) variant 7 (AR-V7) is emerging as a clinically important biomarker in castrate resistant prostate cancer (CRPC). Detection is possible from tumor tissue, which is often inaccessible in the advanced disease setting. With recent progress in detecting AR-V7 in circulating tumor cells (CTCs), circulating tumor RNA (ctRNA) and exosomes from prostate cancer patients, liquid biopsies have emerged as an alternative to tumor biopsy. Therefore, it is important to clarify whether these approaches differ in sensitivity in order to achieve the best possible biomarker characterization for the patient. In this study, blood samples from 44 prostate cancer patients were processed for CTCs and ctRNA with subsequent AR-V7 testing, while exosomal RNA was isolated from 16 samples and tested. Detection of AR and AR-V7 was performed using a highly sensitive droplet digital PCR-based assay. AR and AR-V7 RNA were detectable in CTCs, ctRNA and exosome samples. AR-V7 detection from CTCs showed higher sensitivity and has proven specificity compared to detection from ctRNA and exosomes. Considering that CTCs are almost always present in the advanced prostate cancer setting, CTC samples should be considered the liquid biopsy of choice for the detection of this clinically important biomarker.

Keywords: prostate cancer; CTC; AR; AR-V7; ctRNA; exosome

1. Introduction

Advanced prostate cancer (PC) tends to be initially hormone sensitive and is treated with androgen deprivation therapy (ADT). However, resistance to first line therapy usually develops in approximately 20–40% of patients, referred to as castrate resistant prostate cancer (CRPC) [1]. Most commonly, PC cells become resistant through molecular changes of the androgen receptor (AR), such as mutations, gene amplification, and, more recently reported, the expression of transcript AR variants [2]. In particular, the expression of AR variant 7 (AR-V7), the most abundant and clinically relevant of all variants, has been implicated as a cause of CRPC [3,4]. The translated AR-V7 protein is truncated and lacks the ligand binding domain as well as sequences important for stability and maybe cellular localization [5,6]. Importantly, intracellular AR-V7 predominantly localizes to the nucleus and displays ligand independent transcriptional activity, which is thought to be fundamental in its ability to promote ADT resistance [3,6,7].

Due to its role in ADT resistance several therapies, such as Galeterone and EPI-506 have been developed to effectively target and reduce AR-V7 levels as shown in cell line studies, with clinical trials underway [3]. Consequently, expression of AR-V7 together with that of full-length AR (AR-FL) have emerged as clinically relevant molecular biomarkers for CRPC [8,9]. AR-V7 is detectable in tissue at the RNA and protein levels, and can be evaluated using RNA hybridization techniques and immunohistology [10,11]. AR-V7 expression is rare in hormone-sensitive PC but correlates with CRPC [12]. However, in the advanced PC setting, tissue biopsies are generally unavailable for biomarker testing and diagnostic decision making. Liquid biopsies have in recent years emerged as an alternative tumor source for biomarker testing [13]. To date, AR-V7 has been detectable in circulating tumor cells (CTCs) isolated from blood samples, whole blood mRNA, from ctRNA, tumor arisen cellular vesicles, so-called exosomes found in plasma or serum and even urine [14–19]. Significantly, AR-V7 detection in CTCs has been associated with non-response to novel anti-androgens such as abiraterone and enzalutamide [14,20]. In contrast, the response rates to taxane-based chemotherapy showed no significant difference between AR-V7-positive versus-negative CRPC patients [14]. As such, this has positioned AR-V7 as a potential predictive biomarker that can discriminate between the use of an anti-androgen therapy and taxane chemotherapy.

With AR-V7 emerging as a molecular biomarker of clinical importance, it is essential to determine the best strategy to detect this biomarker for predictive and prognostic purposes. Moreover, with blood biopsies emerging as a potential source to determine AR-V7 status, it is imperative to define which blood-based approach is the most sensitive and reliable. Herein we analyzed AR-V7 and AR-FL detectability in 44 PC patients using parallel blood samples for CTC isolation and ctRNA isolation. For 16 patients, parallel evaluation of AR-V7 and AR-FL status from exosomes was also possible (see Figure 1). Our data suggest that AR-V7 and AR-FL, while detectable from ctRNA and exosomal RNA, are most sensitively detected from CTC samples.

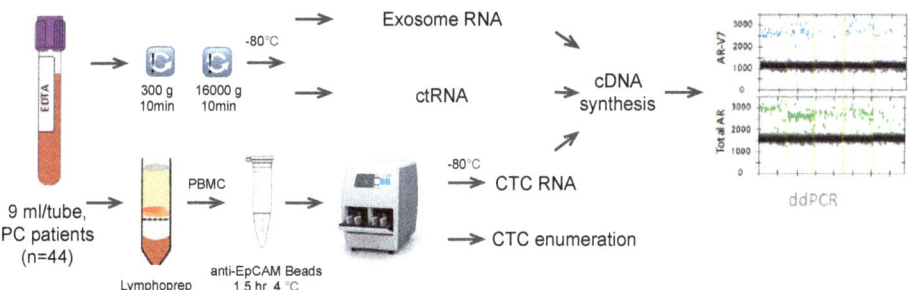

Figure 1. Work flow of the study. PC: prostate cancer, PBMC: peripheral blood mononuclear cells, ctRNA: circulating tumor ribonucleic acid, CTC: circulating tumor cell, cDNA: complementary deoxyribonucleic acid.

2. Materials and Methods

2.1. Patients

The study was undertaken in accordance with the Declaration of Helsinki with human ethics approval, HREC/13/LPOOL/158, from the South Western Sydney Local Health District Ethics Committee (approval Sep 2013, extension July 2018). A total of 44 patients with prostate cancer were recruited at Liverpool Public and St George Private Hospitals in Australia. Patient information is summarized in Table 1. Per patient, 3 × 9 mL blood draws into 9 mL EDTA vacutubes (BD, Franklin Lakes, NJ, USA) were collected, 2 for CTC isolation and 1 for plasma preparation.

Table 1. Patient Characteristics.

Patient	$n = 44$
Mean age (years)	77 (55–94)
Gleason score (%)	
≥7	34 (77%)
<7	4 (9%)
N.A.	6 (14%)
Metastatic status (%)	41 (93%)
Bone metastases	35 (79%)
Lymph metastases	22 (50%)
Visceral metastases	13 (29%)
Definitive therapy (%)	
Radical Prostatectomy	12 (27%)
Radiotherapy	12 (27%)
Both	2 (4%)
None	22 (50%)
Past systemic therapy (%)	
None	2 (4%)
ADT	42 (95%)
Chemotherapy	17 (38%)
Novel antiandrogens	
Enzalutamide	14 (31%)
Abiraterone	9 (20%)
Both	4 (9%)

N.A.: data not available, ADT: androgen deprivation therapy.

2.2. CTC Isolation

Peripheral blood mononuclear cells (PBMCs) including CTCs were isolated in parallel from two blood samples using Lymphoprep™ with SepMate™ tube-based gradient centrifugation (STEMCELL Technologies, Vancouver, BC, Canada) as per the manufacturer's protocol. PBMCs were recovered and washed once with 5 mL phosphate-buffered saline (PBS) (Lonza, Basel, Switzerland), resuspended in 800 µL binding buffer with 40 µL FC buffer and 40 µL anti-EpCAM antibody coupled beads (IsoFlux CTC Enrichment kit, Fluxion Bioscience, San Francisco, CA, USA) and incubated for 90 min at 4 °C under gentle rotation. CTCs were then enriched using the standard IsoFlux CTC isolation protocol (Fluxion Bioscience, CA, USA). One CTC isolate underwent CTC enumeration and the other was immediately frozen at −80 °C until RNA extraction and gene expression analysis.

2.3. CTC Enumeration

IsoFlux enriched CTCs were enumerated with an IsoFlux CTC enumeration kit containing all the staining antibodies (Fluxion Bioscience, Alameda, CA, USA). Briefly, after 10 min fixation with fixation buffer, cells were blocked for 10 min with 10% normal donkey serum (NDS) and probed with rabbit anti-human CD45 antibody (1:100) followed by detection with tetramethylrhodamine (TRITC) conjugated donkey-anti rabbit IgG (1:200). Cells were then washed twice with binding buffer and permeabilized using 0.2% Trition 100 before probing with fluorescein (FITC) conjugated anti-human Pan-cytokeratin antibody (1:10). After two further washes, cells were transferred to SensoPlate™ 24-well glass bottom plates (Greiner Bio-one, Kremsmünster, Austria) with Hoechst dye included in the mounting media. Fluorescent microscopy (ALS CellCelector™, ALS, Jena, Germany) was used for CTC enumeration.

2.4. Plasma Processing

All plasma was processed within 4 h of blood draw. Whole blood was spun at 280× g for 10 min and the supernatant spun again using a microfuge at 16,000× g for 10 min at 4 °C. The supernatant was transferred and frozen at −80 °C until RNA extraction.

2.5. RNA Extraction and cDNA Synthesis

RNA from CTC samples was extracted with Total RNA Purification Micro Kit (Norgen Biotek, Thorold, ON, Canada) and eluted in 30 µL elution buffer. RNA from plasma was extracted with QIAamp circulating nucleic acid kit (QIAGEN, Hilden, Germany) and eluted in 65 µL elution buffer. Samples processed with this kit are referred to as ctRNA even for healthy controls that would only have normal cell free RNA. RNA from exosomes was extracted with Qiagen exoRNeasy serum/plasma Midi kit (QIAGEN, Hilden, Germany) and eluted in 30 µL. A total of 15 µL eluted RNA was reverse transcribed using SensiFAST cDNA synthesis kit (Bioline, Alexandria, Australia).

2.6. Droplet Digital PCR (ddPCR)

A total of 7 µL of cDNA of either CTC, ctRNA or exosomal samples was used for detection of AR-FL and AR-V7 transcripts by ddPCR as previously described [15]. Glyceraldehyde 3-phosphate dehydrogenase (GAPDH) transcript abundance was similarly evaluated (see Table 2 for primers, probes) and the optimized annealing temperature of 55 °C using the Bio-Rad QX200 ddPCR instrument (Bio-Rad, Hercules, CA, USA). Positive (cDNA from 22RV1 prostate cancer cells) and negative controls (DNase and RNase free H_2O) were included in all ddPCR experiments.

Table 2. GAPDH ddPCR primers and probe.

Name	Sequence
GAPDH-fwrd	CGGGAAGCTTGTCATCAATGG
GAPDH-rev	CTCCACGACGTACTCAGCG
GAPDH-probe	FAM-5′-TCTTCCAGGAGCGAGATCCCT-3-BQ1

2.7. Statistics

Data were analyzed with GraphPad Prism8 (GraphPad Software, San Diego, CA, USA). As a normal distribution was not assumed, non-parametric statistics was utilized. A Fisher exact test was used to test the significance between two categorical variables; $p < 0.05$ represents statistical significance. Unpaired comparison of CTC numbers between hormone sensitive prostate cancer (HSPC) and CRPC and AR copy numbers between healthy controls and patient samples was performed with the non-parametric Mann–Whitney test.

3. Results

3.1. Patients

Twelve HSPC and 32 CRPC patients at various stages of treatment were recruited, all patients had been originally diagnosed with subtype adenocarcinoma PC. Patient baseline characteristics are listed in Table 1. CTCs were enumerated and the expression levels of AR-FL and AR-V7 transcripts were detected in enriched CTC samples using our previously established ddPCR assay [15]. Extracted ctRNA and exosomal RNA were also tested by ddPCR for AR-FL and AR-V7 transcripts. To verify the quality of extracted RNA the reference gene GAPDH was tested for and detected in all samples (median GAPDH: 13212 copies/ml plasma, 2920-120606).

3.2. Prevalence of CTCs

CTCs were detected in 97.7% (43/44) of patients. CTC counts varied from 0 to 184 per 9 mL blood with no significant difference in CTC counts between HSCP and CRPC patients (Figure 2).

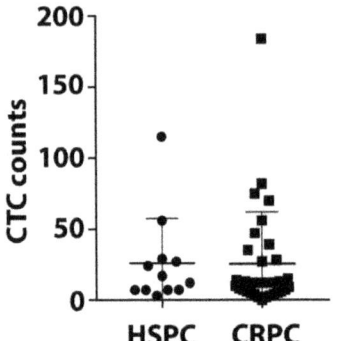

Figure 2. Circulating tumor cells (CTC) counts. CTCs were isolated using the IsoFlux CTC platform and enumerated. The range of CTC counts/9 mL blood for 12 HSPC and 32 CRPC patients is depicted. There is no significant difference of CTC numbers between HSPC and CRPC patients (p value = 0.59).

3.3. AR-V7 and AR-FL in CTC RNA and Plasma ctRNA

As expected, AR-FL and AR-V7 was only found in CTC processed blood samples with detectable CTCs. AR-FL was detected in 22 out of 44 patient CTC samples (50%). AR-FL copy number detected for CTC samples ranged from 0 to 13,714 copies/mL blood. A 62.5% share (20/32) of CRPC samples had AR-FL positive CTC samples, and only 2 HSPC patients, 16.7%, had low copy numbers of AR-FL (Figure 3, Supplementary Table S1). For AR-V7, 47.7% (21/44) of patient CTC samples tested positive with 0–146 copies/ml blood. A share of 53.1% (17/32) of CRPC patients had detectable AR-V7 (range 0–146 copies/mL blood), while AR-V7 was also identified in 4 HSPC patient CTC samples, at generally lower levels (range 0–4 copies/mL blood).

Interestingly, AR-FL detection from plasma-derived ctRNA was higher, with 70.5% (31/44) patients testing positive, although detection frequency was not significantly different from CTC samples (Fisher exact, p = 0.13). The detected AR-FL copy numbers found in ctRNA ranged from 0 to 180 per ml plasma and were overall less abundant compared to CTC samples. Detection of AR-FL in ctRNA samples was similarly common in HSPC patients at 66.7% (8/12) compared to CRPC patients at 71.9% (23/32), although copy numbers tended to be higher in CRPC samples ranging between 0 and 15.2 vs. 0–180, for HSPC and CRPC patients, respectively.

In contrast, AR-V7 detection in ctRNA was lower in comparison to CTC samples with only 15.9% (7/44) of patient ctRNA samples testing positive. The difference in AR-V7 detectability in CTCs vs. ctRNA was significant (p = 0.003). AR-V7 tends to be more frequently found in CRPC patient ctRNA 18.8% (6/32) vs. 8.3% (1/12) in HSPC patient ctRNA, but detectability and copy numbers are generally low for both CRPC and HSPC patients (range 0–8 per mL plasma). This data indicates that both AR-FL and AR-V7 are more readily detectable in CTC samples.

In the 44-patient cohort, AR-FL detection in CTCs correlated with CRPC (p = 0.02) while AR-V7 detection did not reach a significant correlation with CRPC (p = 0.32). In contrast, no correlation was found for AR-FL in ctRNA (p = 0.73) or AR-V7 (p = 0.65) (Supplementary Table S1).

The concordance of AR-FL positivity between CTC and ctRNA was 41% (18/44) while concordance of AR-V7 positivity was only 9.1% (4/44), indicating lack of sensitivity and possibly specificity for AR-V7 detection in ctRNA.

3.4. AR-V7 and AR-FL in Exosomal RNA, ctRNA and CTCs

From 16 of the analyzed patient samples, enough plasma was available to additionally extract exosomal RNA and analyze for AR-FL and AR-V7 transcripts.

Overall, 12.5% (2/16) of patients were positive for AR-V7 when analyzed using exosomal RNA, compared to 6.3% (1/16) being considered AR-V7 positive according to ctRNA analysis in that patient sub-cohort. AR-V7 was also detectable in all CTC samples from the 16 patient cohort that were found to have detectable AR-V7 either by ctRNA or exosomal. Importantly however, overall more (43.8%, 7/16) patients tested positive for AR-V7 by CTC RNA analysis. In contrast, AR-FL was detected in 68.8% (11/16) ctRNA samples, in 50% (8/16) CTC samples and 37.5% (6/16) of exosomal RNA samples (Table 3).

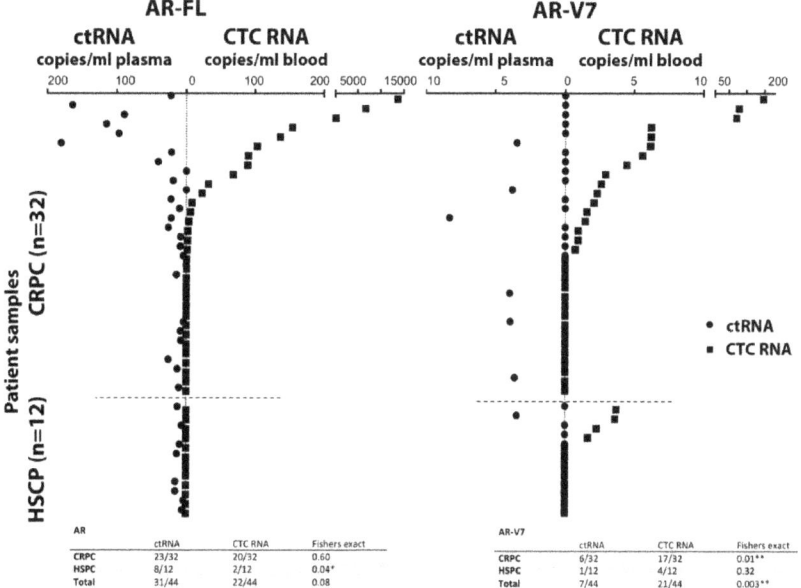

Figure 3. Detection of AR-FL (left) and AR-V7 (right) transcripts. Comparison of AR-FL and AR-V7 detection in CTC RNA and ctRNA from the same blood draw illustrated by mirrored scatter blot for the 44 patient samples (sorted in relation to detection in CTCs); summarized data are tabled below. *, ** indicate significance p-value ≤ 0.5 and ≤ 0.01, respectively.

Table 3. AR-FL/AR-V7 status in 16 patients tested from CTC, ctRNA and exosomal samples.

Patient	Status	AR-FL CTCs	AR-V7 CTCs	AR-FL ctRNA	AR-V7 ctRNA	AR-FL Exosomal	AR-V7 Exosomal
1	CRPC	−	−	−	−	+	−
2	CRPC	+	−	+	−	−	−
3	CRPC	+	−	+	−	−	−
4	CRPC	+	+	+	−	−	−
5	CRPC	−	+	+	−	+	+
6	CRPC	−	+	−	−	+	−
7	CRPC	+	+	+	−	+	+
8	CRPC	+	+	−	−	−	−
9	CRPC	+	−	+	−	−	−
10	HSPC	+	−	−	−	+	−
11	HSPC	+	−	+	−	−	−
12	HSPC	−	−	+	−	−	−
13	HSPC	−	+	+	−	−	−
14	HSPC	−	+	−	+	−	−
15	HSPC	−	−	+	−	+	−
16	HSPC	−	−	+	−	−	−

AR-V7 positive patients by CTC analysis are highlighted in grey.

3.5. Healthy Control Analysis and Implications for Assay Specificity

We previously have demonstrated that AR-FL and AR-V7 detection from CTC samples is not only highly sensitive but also highly specific. When establishing the ddPCR assay we found AR-V7 undetectable even in large numbers of healthy donor PBMCs and AR-FL detection was very low (the equivalent of 0.001 or 0.002 copies per cell were detected in only 2 of 6 healthy donor PBMC samples; considering the average residual lymphocyte number in a CTC sample that would be equivalent to 0.44 or 0.88 copies per ml blood) [15]. To evaluate sensitivity and specificity of our AR assays when screening ctRNA and exosomal RNA samples, we obtained blood from five healthy individuals, age and sex matched to our patient cohort, and extracted ctRNA and exosomal RNA for AR-FL and AR-V7 analysis. AR-FL was readily detectable in ctRNA from all healthy subjects tested with an average copy number of 14.1 per ml plasma (range: 9.6–21.5). Of note, AR-FL copy numbers in only 34.1% (15/44) of our patient ctRNA samples were above that healthy control copy number average. AR-FL was also detected in exosomal RNA from three of five healthy subjects, and crucially, the copy number range was comparable to that detected in the exosomal patient samples. AR-V7 was not as prevalent but still present with 11.1 copies/mL plasma for one healthy individual when analyzing exosomal RNA while we detected very similar copy numbers, 13.9 and 9.7 copies per ml plasma, in the two exosomal RNA AR-V7 positive patients. AR-V7 was not detected in healthy donor ctRNA samples in this small healthy control cohort. Overall, these data strongly suggest the presence of variable and sometimes high levels of AR mRNA in healthy male plasma and exosomes, which clearly impacts reliable detection of PC-derived AR-FL/AR-V7 from ctRNA and exosomes and indicates low specificity for tumor-derived AR when testing ctRNA and exosomal RNA (Figure 4).

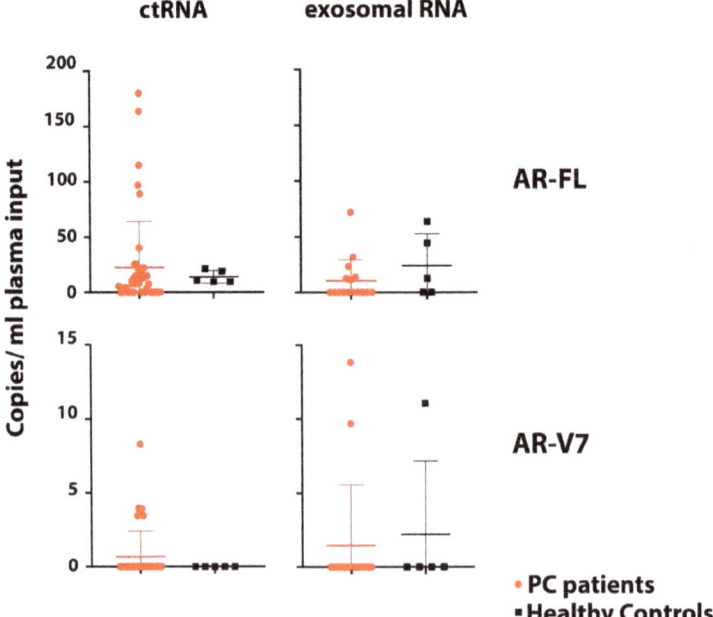

Figure 4. Comparison of AR-FL and AR-V7. AR-FL and AR-V7 detection in ctRNA (44 PC patients, 5 healthy controls) and exosomal RNA samples (16 PC patients and 5 healthy controls) indicates high background of AR transcript in plasma impacting specificity of tumor-derived AR-FL and AR-V7 detection. There is no significant difference of AR-FL and AR-V7 copy numbers detected in ctRNA and exosomal RNA between patients and healthy controls (p values: AR-FL ctRNA = 0.4; AR-FL exosomal = 0.32; AR-V7 ctRNA = 0.58; AR-V7 exosomal = 0.85).

4. Discussion

Advancements in molecular technology have aided the accurate detection and quantification of novel blood-based biomarkers paving the way for their use in clinical environments. AR-V7 is emerging as a biomarker with the potential as a predictive tool in treatment selection. Patients that have detectable AR-V7 are considered to be resistant to abiraterone and enzalutamide but respond to taxane chemotherapy, and potentially eligible for clinical trials of new generation ADT drugs. Advanced stage PC tissue samples are generally unavailable for testing and liquid biopsies are explored as a surrogate. Thus far, studies on AR-V7 have utilized various methods of detection, resulting in different degrees of correlation between data and disease parameters, highlighting that a clear "gold standard" still needs to be found [21]. While previously developing a sensitive and specific AR-FL/AR-V7 detection method for enriched CTC samples [15], here we compared this method for detection of AR-FL and AR-V7 from CTC samples and ctRNA samples of 44 PC patients and for a subset of patient samples we also were able to compare detection from exosomal RNA. Exosomes in particular have become an attractive source of tumor information. They are small double lipid membrane vesicles of endocytic origin, that contain proteins, nucleic acids and lipids released by cells. Since this includes cancer cells, exosomes can be extracted from liquid biopsies such as plasma, serum and urine to be tested for biomarker information. Importantly, both ctRNA and exosomal RNA require simpler processing protocols than CTCs with commercially available kits and thus AR-FL/AR-V7 testing using these tumor information sources would potentially be easier translated into a clinical setting.

However, our data indicate CTC-based detection is superior in sensitivity and specificity for AR-V7 with a detection rate of 48% as compared to 16% with ctRNA. Nevertheless, both assays demonstrated some association of AR-V7 detection with patients classified as castrate resistant, although our cohort was too small to find significant correlation. Previous studies which utilize CTC enrichment-based techniques reported broad detection rates ranging from 27% to 75% [15,20,22,23]. These studies demonstrated a link between baseline CTC derived AR-V7 status and disease burden, which was found to increase with subsequent lines of therapy. Our data follows this trend, with the majority of our AR-V7 patients having had at least two lines of previous treatment (Supplementary Table S1). Our study, with a relatively small study patient cohort, was mainly aimed at comparing the effectiveness of choosing simple blood-based tumor sources rather than CTC samples to screen for AR-V7. We achieved this aim and our data indicates that CTC sample testing for PC derived AR-V7 and AR-FL is more sensitive and specific and thus ultimately more reliable.

Interestingly, the few patients detected positive for AR-V7 by either ctRNA or exosomal RNA testing were also found to be AR-V7 positive by CTC testing. Some CTC AR-V7 positive patients were however not detected as positive by ctRNA or exosomal RNA testing, suggesting higher sensitivity of the AR-V7 testing in CTC samples. We also detected AR-V7 in one healthy donor exosomal RNA sample. The implications of that for the individual remain unclear and further follow up is not possible because the individual consented as a healthy volunteer. Since the AR-V7 levels were not negligible and ddPCR data showed convincing detection, we have to assume poorer specificity of AR-V7 detection from exosomes at this stage, in comparison to that shown in our previous study for CTC samples [15]. However, larger healthy control studies would be necessary to confirm this.

Not all CTC samples tested positive for AR-FL when either parallel ctRNA or exosomal RNA testing detected it. However, this finding has to be interpreted together with the fact that we found quite high levels of AR-FL transcripts in ctRNA and exosomal RNA samples from age- and sex-matched healthy control individuals. This may not be surprising as exosomes and cell free nucleic acids are thought to be released during normal tissue homeostasis, and it is quite conceivable that AR-FL transcript is released into the blood stream in that way from organs like the testis and prostate in male subjects. By analyzing AR-FL from CTC samples we seem to largely avoid such transcripts from non-tumor sources as AR expression in normal blood cells is known to be minimal or null [24].

There may be other issues underpinning how, in comparison to ctRNA-based detection, CTC-derived AR-V7 provides for a more reliable assay. Given that high CTC counts on their own have proven to be of prognostic value in PC and detection of CTCs indicate higher disease burden, CTC detection can be interpreted hand-in-hand with AR detection to exclude potential false positives, and in our hands our ddPCR assay never detected AR-FL or AR-V7 in the absence of CTCs in a parallel enumerated sample [25]. ctRNA, on the other hand, due to the high levels of nucleases in the blood, has to be considered highly unstable and is also far more susceptible to variations in pre-analytical handling. Meanwhile, CTCs in blood appear to demonstrate superior stability and we have demonstrated previously that AR-V7 copies can be detected from CTC samples from patient blood drawn in simple EDTA tubes and stored up to 48 h at room temperature, suggesting CTCs either protect the AR-V7 transcript and/or continue to express it in a drawn blood sample [26].

Since exosomes in blood are similarly believed to protect the intravesicular content including mRNAs it was important to also test AR-V7 and AR-FL detection from exosomes. We detected AR-V7 less frequently than in parallel CTC samples in only two patients. While both patients with detectable AR-V7 from exosomes are classified as castrate resistant, we also detected similar AR-V7 copy numbers in one healthy subject, challenging reliability of the finding in PC patients.

While our small patient cohort was enough to determine the best liquid biopsy entity for AR-V7 and the study was not intended to answer questions of AR-V7 biology, there are a few issues worth highlighting. Firstly, our sensitive assay detected AR-V7 in some HSPC patient CTC samples. This is not entirely unexpected and has been reported previously [27,28]. It will be interesting to see whether AR-V7 detection in HSPC patients may be an early predictor of developing CRPC, as it tends to be associated with longer time on treatment in our study (AR-V7 positive HSPC patients: between 24 month and 60 month on ADT versus AR-V7 negative HSPC patients: being maximally treated for 12 month with ADT (see Supplementary Table S1)). Secondly, AR-V7 was found in some patient CTC samples despite undetectability of AR-FL. This has been reported by others [29] and it would be interesting to investigate the impacts of AR-V7 potentially totally replacing AR-FL in these patients in future studies. Finally, there seems to be a trend which matches previous reports [30] that patients on second line ADT have more commonly detectable AR-V7 in CTCs. In our study, 10 of 14 (69%) CRPC patients receiving enzalutamide and 5 of 9 (55.5%) receiving abiraterone (some patients were treated consecutively with both) were AR-V7 positive by CTC testing.

Although the small patient cohort allowed us to answer the main question of which liquid biopsy entity is the better for detection of AR-V7 and AR-FL, a limitation of the study is that correlation with disease parameters was not as informative as a bigger cohort would have been. Larger cohorts of healthy control comparisons would be able to better define background AR-FL and AR-V7 in plasma.

5. Conclusions

This study compared AR-V7 and AR-FL detection in liquid biopsy (blood)-derived CTC samples, ctRNA samples and exosomes. Our data show that testing of these clinically highly relevant biomarkers for PC patients is most reliable performed from CTC samples in regards to sensitivity and specificity. It also should be noted that a recent report shows that at the protein level correlation with PC disease parameters is linked to the nuclear localization of AR-V7 protein in CTCs [30]. We usually analyze two parallel blood samples, one for CTC enumeration, and we have recently amended this protocol to incorporate AR-V7 immunocytostaining and cellular localization screening. The second blood sample is screened for AR-V7 and AR-FL transcripts by ddPCR. Both tests go hand-in-hand to confirm AR-V7 presence and highlight the importance of analyzing CTCs rather than other circulating tumor entities. Our future studies will clarify whether both CTC tests cooperatively show correlation with resistance to ADT.

Supplementary Materials: Supplementary materials can be found at http://www.mdpi.com/2073-4409/8/7/688/s1.

Author Contributions: Conceptualization: T.M.B., P.D.S., Y.M.; methodology: Y.M., M.N., S.A.J., T.O., F.Y., T.K.; formal analysis: T.M.B., P.D.S., Y.M., M.N., P.D.; resources: T.M.B., P.D.S., P.D., W.C., B.B., A.C., writing—original draft preparation: T.M.B., Y.M., M.N., T.K.; writing—review and editing: T.M.B., P.D.S., Y.M., M.N., S.A.J., T.O., F.Y., T.K., W.C., B.B., P.D., A.C., supervision: T.M.B., P.D.S.; project administration: T.M.B., P.D.S.

Funding: This research was funded by the Cancer Institute NSW grant number 13/TRC/1-01. SJ is recipient of an Ingham Institute PhD scholarship donated by the Liverpool Catholic Club, Australia.

Acknowledgments: Human ethics approval, HREC/13/LPOOL/158, was obtained and managed by the CONCERT Biobank.

Conflicts of Interest: The authors declare no conflict of interest.

References

1. Knudsen, K.E.; Scher, H.I. Starving the addiction: New opportunities for durable suppression of AR signaling in prostate cancer. *Clin. Cancer Res.* **2009**, *15*, 4792–4798. [CrossRef] [PubMed]
2. Wadosky, K.M.; Koochekpour, S. Molecular mechanisms underlying resistance to androgen deprivation therapy in prostate cancer. *Oncotarget* **2016**, *7*, 64447–64470. [CrossRef] [PubMed]
3. Antonarakis, E.S.; Armstrong, A.J.; Dehm, S.M.; Luo, J. Androgen receptor variant-driven prostate cancer: clinical implications and therapeutic targeting. *Prostate Cancer Prostatic Dis.* **2016**, *19*, 231–241. [CrossRef] [PubMed]
4. Li, Y.; Chan, S.C.; Brand, L.J.; Hwang, T.H.; Silverstein, K.A.; Dehm, S.M. Androgen receptor splice variants mediate enzalutamide resistance in castration-resistant prostate cancer cell lines. *Cancer Res.* **2013**, *73*, 483–489. [CrossRef] [PubMed]
5. Cao, B.; Qi, Y.; Zhang, G.; Xu, D.; Zhan, Y.; Álvarez, X.; Guo, Z.; Fu, X.; Plymate, S.R.; Sartor, O.; et al. Androgen receptor splice variants activating the full-length receptor in mediating resistance to androgen-directed therapy. *Oncotarget* **2014**, *5*, 1646–1656. [CrossRef]
6. Chan, S.C.; Li, Y.; Dehm, S.M. Androgen Receptor Splice Variants Activate Androgen Receptor Target Genes and Support Aberrant Prostate Cancer Cell Growth Independent of Canonical Androgen Receptor Nuclear Localization Signal*. *J. Boil. Chem.* **2012**, *287*, 19736–19749. [CrossRef] [PubMed]
7. Xu, D.; Zhan, Y.; Qi, Y.; Cao, B.; Bai, S.; Xu, W.; Gambhir, S.S.; Lee, P.; Sartor, O.; Flemington, E.K.; et al. Androgen receptor splice variants dimerize to transactivate target genes. *Cancer Res.* **2015**, *75*, 3663–3671. [CrossRef]
8. Henzler, C.; Li, Y.; Yang, R.; McBride, T.; Ho, Y.; Sprenger, C.; Liu, G.; Coleman, I.; Lakely, B.; Li, R.; et al. Truncation and constitutive activation of the androgen receptor by diverse genomic rearrangements in prostate cancer. *Nat. Commun.* **2016**, *7*, 13668. [CrossRef]
9. Conteduca, V.; Wetterskog, D.; Sharabiani, M.T.A.; Grande, E.; Fernandez-Perez, M.P.; Jayaram, A.; Salvi, S.; Castellano, D.; Romanel, A.; Lolli, C.; et al. Androgen receptor gene status in plasma DNA associates with worse outcome on enzalutamide or abiraterone for castration-resistant prostate cancer: a multi-institution correlative biomarker study. *Ann. Oncol.* **2017**, *28*, 1508–1516. [CrossRef]
10. Guedes, L.B.; Morais, C.L.; Almutairi, F.; Haffner, M.C.; Zheng, Q.; Isaacs, J.T.; Antonarakis, E.S.; Lu, C.; Tsai, H.; Luo, J.; et al. Analytic Validation of RNA In Situ Hybridization (RISH) for AR and AR-V7 Expression in Human Prostate Cancer. *Clin. Cancer Res.* **2016**, *22*, 4651–4663. [CrossRef]
11. Welti, J.; Rodrigues, D.N.; Sharp, A.; Sun, S.; Lorente, D.; Riisnaes, R.; Figueiredo, I.; Zafeiriou, Z.; Rescigno, P.; De Bono, J.S.; et al. Analytical Validation and Clinical Qualification of a New Immunohistochemical Assay for Androgen Receptor Splice Variant-7 Protein Expression in Metastatic Castration-resistant Prostate Cancer. *Eur. Urol.* **2016**, *70*, 599–608. [CrossRef] [PubMed]
12. Kallio, H.M.L.; Hieta, R.; Latonen, L.; Brofeldt, A.; Annala, M.; Kivinummi, K.; Tammela, T.L.; Nykter, M.; Isaacs, W.B.; Lilja, H.G.; et al. Constitutively active androgen receptor splice variants AR-V3, AR-V7 and AR-V9 are co-expressed in castration-resistant prostate cancer metastases. *Br. J. Cancer* **2018**, *119*, 347–356. [CrossRef] [PubMed]
13. Crowley, E.; Di Nicolantonio, F.; Loupakis, F.; Bardelli, A. Liquid biopsy: monitoring cancer-genetics in the blood. *Nat. Rev. Clin. Oncol.* **2013**, *10*, 472–484. [CrossRef] [PubMed]
14. Antonarakis, E.S.; Lu, C.; Luber, B.; Wang, H.; Chen, Y.; Nakazawa, M.; Nadal, R.; Paller, C.J.; Denmeade, S.R.; Carducci, M.A.; et al. Androgen Receptor Splice Variant 7 and Efficacy of Taxane Chemotherapy in Patients With Metastatic Castration-Resistant Prostate Cancer. *JAMA Oncol.* **2015**, *1*, 582–591. [CrossRef] [PubMed]
15. Ma, Y.; Luk, A.; Young, F.P.; Lynch, D.; Chua, W.; Balakrishnar, B.; De Souza, P.; Becker, T.M. Droplet Digital PCR Based Androgen Receptor Variant 7 (AR-V7) Detection from Prostate Cancer Patient Blood Biopsies. *Int. J. Mol. Sci.* **2016**, *17*, 1264. [CrossRef] [PubMed]

16. Hodara, E.; Morrison, G.; Cunha, A.T.; Zainfeld, D.; Xu, T.; Xu, Y.; Dempsey, P.W.; Pagano, P.C.; Bischoff, F.; Khurana, A.; et al. Multiparametric liquid biopsy analysis in metastatic prostate cancer. *JCI Insight* **2019**, *4*, 4. [CrossRef]
17. Todenhöfer, T.; Azad, A.; Stewart, C.; Gao, J.; Eigl, B.J.; Gleave, M.E.; Joshua, A.M.; Black, P.C.; Chi, K.N. AR-V7 Transcripts in Whole Blood RNA of Patients with Metastatic Castration Resistant Prostate Cancer Correlate with Response to Abiraterone Acetate. *J. Urol.* **2017**, *197*, 135–142. [CrossRef]
18. Woo, H.K.; Park, J.; Ku, J.Y.; Lee, C.H.; Sunkara, V.; Ha, H.K.; Cho, Y.K. Urine-based liquid biopsy: non-invasive and sensitive AR-V7 detection in urinary EVs from patients with prostate cancer. *Lab on a chip* **2018**, *19*, 87–97. [CrossRef]
19. Del Re, M.; Biasco, E.; Crucitta, S.; DeRosa, L.; Rofi, E.; Orlandini, C.; Miccoli, M.; Galli, L.; Falcone, A.; Jenster, G.W.; et al. The Detection of Androgen Receptor Splice Variant 7 in Plasma-derived Exosomal RNA Strongly Predicts Resistance to Hormonal Therapy in Metastatic Prostate Cancer Patients. *Eur. Urol.* **2017**, *71*, 680–687. [CrossRef]
20. Antonarakis, E.S.; Lu, C.; Wang, H.; Luber, B.; Nakazawa, M.; Roeser, J.C.; Chen, Y.; Mohammad, T.A.; Chen, Y.; Fedor, H.L.; et al. AR-V7 and Resistance to Enzalutamide and Abiraterone in Prostate Cancer. *N. Eng. J. Med.* **2014**, *371*, 1028–1038. [CrossRef]
21. Bernemann, C.; Schnoeller, T.J.; Luedeke, M.; Steinestel, K.; Boegemann, M.; Schrader, A.J.; Steinestel, J. Expression of AR-V7 in Circulating Tumour Cells Does Not Preclude Response to Next Generation Androgen Deprivation Therapy in Patients with Castration Resistant Prostate Cancer. *Eur. Urol.* **2017**, *71*, 1–3. [CrossRef] [PubMed]
22. Antonarakis, E.S.; Lu, C.; Luber, B.; Wang, H.; Chen, Y.; Zhu, Y.; Silberstein, J.L.; Taylor, M.N.; Maughan, B.L.; Denmeade, S.R.; et al. Clinical Significance of Androgen Receptor Splice Variant-7 mRNA Detection in Circulating Tumor Cells of Men With Metastatic Castration-Resistant Prostate Cancer Treated With First- and Second-Line Abiraterone and Enzalutamide. *J. Clin. Oncol.* **2017**, *35*, 2149–2156. [CrossRef] [PubMed]
23. Onstenk, W.; Sieuwerts, A.M.; Kraan, J.; Van, M.; Nieuweboer, A.J.; Mathijssen, R.H.; Hamberg, P.; Meulenbeld, H.J.; De Laere, B.; Dirix, L.Y.; et al. Efficacy of Cabazitaxel in Castration-resistant Prostate Cancer Is Independent of the Presence of AR-V7 in Circulating Tumor Cells. *Eur. Urol.* **2015**, *68*, 939–945. [CrossRef] [PubMed]
24. Mantalaris, A.; Panoskaltsis, N.; Sakai, Y.; Bourne, P.; Chang, C.; Messing, E.M.; Wu, J.H.D. Localization of androgen receptor expression in human bone marrow. *J. Pathol.* **2001**, *193*, 361–366. [CrossRef]
25. Onstenk, W.; De Klaver, W.; De Wit, R.; Lolkema, M.; Foekens, J.; Sleijfer, S. The use of circulating tumor cells in guiding treatment decisions for patients with metastatic castration-resistant prostate cancer. *Cancer Treat. Rev.* **2016**, *46*, 42–50. [CrossRef] [PubMed]
26. Luk, A.W.S.; Ma, Y.; Ding, P.N.; Young, F.P.; Chua, W.; Balakrishnar, B.; Dransfield, D.T.; De Souza, P.; Becker, T.M. CTC-mRNA (AR-V7) Analysis from Blood Samples—Impact of Blood Collection Tube and Storage Time. *Int. J. Mol. Sci.* **2017**, *18*, 1047. [CrossRef] [PubMed]
27. Hu, R.; Dunn, T.A.; Wei, S.; Isharwal, S.; Veltri, R.W.; Humphreys, E.; Han, M.; Partin, A.W.; Vessella, R.L.; Isaacs, W.B.; et al. Ligand-independent Androgen Receptor Variants Derived from Splicing of Cryptic Exons Signify Hormone Refractory Prostate Cancer. *Cancer Res.* **2009**, *69*, 16–22. [CrossRef]
28. Qu, Y.; Dai, B.; Ye, D.; Kong, Y.; Chang, K.; Jia, Z.; Yang, X.; Zhang, H.; Zhu, Y.; Shi, G. Constitutively Active AR-V7 Plays an Essential Role in the Development and Progression of Castration-Resistant Prostate Cancer. *Sci. Rep.* **2015**, *5*, 7654. [CrossRef]
29. El-Heliebi, A.; Hille, C.; Laxman, N.; Svedlund, J.; Haudum, C.; Ercan, E.; Kroneis, T.; Chen, S.; Smolle, M.; Rossmann, C.; et al. In Situ Detection and Quantification of AR-V7, AR-FL, PSA, and KRAS Point Mutations in Circulating Tumor Cells. *Clin. Chem.* **2018**, *64*, 536–546. [CrossRef]
30. Scher, H.I.; Lu, D.; Schreiber, N.A.; Louw, J.; Graf, R.P.; Vargas, H.A.; Johnson, A.; Jendrisak, A.; Bambury, R.; Danila, D.; et al. Association of AR-V7 on Circulating Tumor Cells as a Treatment-Specific Biomarker With Outcomes and Survival in Castration-Resistant Prostate Cancer. *JAMA Oncol.* **2016**, *2*, 1441–1449. [CrossRef]

© 2019 by the authors. Licensee MDPI, Basel, Switzerland. This article is an open access article distributed under the terms and conditions of the Creative Commons Attribution (CC BY) license (http://creativecommons.org/licenses/by/4.0/).

Article

Prognostic Significance of *TWIST1*, *CD24*, *CD44*, and *ALDH1* Transcript Quantification in EpCAM-Positive Circulating Tumor Cells from Early Stage Breast Cancer Patients

Areti Strati [1], Michail Nikolaou [2], Vassilis Georgoulias [3] and Evi S. Lianidou [1],*

[1] Analysis of Circulating Tumor Cells Lab, Department of Chemistry, University of Athens, 15771 Athens, Greece
[2] Medical Oncology Unit, "Elena Venizelou" Hospital, 11521 Athens, Greece
[3] Metropolitan General Hospital, 15562 Athens, Greece
* Correspondence: lianidou@chem.uoa.gr; Tel.: +30-210-727-4311; Fax: +30-210-727-4750

Received: 31 May 2019; Accepted: 28 June 2019; Published: 29 June 2019

Abstract: (1) Background: The aim of the study was to evaluate the prognostic significance of EMT-associated (*TWIST1*) and stem-cell (SC) transcript (*CD24*, *CD44*, *ALDH1*) quantification in EpCAM+ circulating tumor cells (CTCs) of early breast cancer patients. (2) Methods: 100 early stage breast cancer patients and 19 healthy donors were enrolled in the study. *CD24*, *CD44*, and *ALDH1* transcripts of EpCAM$^+$ cells were quantified using a novel highly sensitive and specific quadraplex RT-qPCR, while *TWIST1* transcripts were quantified by single RT-qPCR. All patients were followed up for more than 5 years. (3) Results: A significant positive correlation between overexpression of *TWIST1* and $CD24^{-/low}/CD44^{high}$ profile was found. Kaplan–Meier analysis revealed that the ER/PR-negative (HR-) patients and those patients with more than 3 positive lymph nodes that overexpressed *TWIST1* in EpCAM$^+$ cells had a significant lower DFI (log rank test; $p < 0.001$, $p < 0.001$) and OS (log rank test; $p = 0.006$, $p < 0.001$). Univariate and multivariate analysis also revealed the prognostic value of *TWIST1* overexpression and $CD24^{-/low}/CD44^{high}$ and $CD24^{-/low}/ALDH1^{high}$ profile for both DFI and OS. (4) Conclusions: Detection of *TWIST1* overexpression and stem-cell (*CD24*, *CD44*, *ALDH1*) transcripts in EpCAM$^+$ CTCs provides prognostic information in early stage breast cancer patients.

Keywords: liquid biopsy; circulating tumor cells; epithelial–mesenchymal transition; stem cells; early breast cancer

1. Introduction

Circulating tumor cells (CTCs) are major players in liquid biopsy [1,2], and their molecular characterization is highly important for rational treatment decisions and for monitoring therapeutic response [3], whereas their analysis at the single cell level has the potential to reveal tumor heterogeneity in real time [4]. In breast cancer, a subpopulation of tumor cells that display stem cell-like properties [5] determines the aggressive characteristics and drug resistance of tumor clonal evolution [6]. Cancer stem cells (CSCs) that mediate tumor metastasis and therapeutic resistance have the capacity to transition between mesenchymal and epithelial-like states [7]. It has already been shown that breast cancer cells with the CD44+CD24−/low phenotype [8] that overexpress aldehyde dehydrogenase 1 (ALDH1+) [9] are able to form tumors in mice with high tumorigenic capacity. It has also been shown that disseminated tumor cells (DTCs) [10] and CTCs express the putative stem cell CD44+/CD24− and/or ALDH1+/CD24− phenotypical profile [11,12]. Moreover, in primary human luminal breast cancer, the metastasis-initiating cells containing CTC that express EPCAM, CD44, CD47, and the proto-oncogene MET are related with reduced overall survival (OS) [13]. In other types of cancer, various stem cell markers have also been identified and correlated with metastatic capacity [14] and poor prognosis [15].

It is now known that breast cancer stem cells exist in distinct mesenchymal-like (epithelial–mesenchymal transition [EMT]) as CD44+/CD24− and epithelial-like (mesenchymal-epithelial transition [MET]) states that express ALDH1. This transition between EMT- and MET-like states is highly important for their capacity to invade, disseminate, and grow at metastatic sites [16]. Many studies have already shown that a major proportion of CTC express both EMT and tumor stem cell characteristics [17–19]. Recently it was shown that an EpCAM-/ALDH1+/HER2+/EGFR+/HPSE+/Notch1+ profile in CTC drives these cells to metastasize to the brain [20]. At the single cell level, it has been shown that CTC that co-express the stem cell marker ALDH1 and the mesenchymal marker *TWIST1* may prevail during disease progression [21]. However, the prognostic significance of EMT and Stem cell (SC) markers in CTC has only been shown up to now in metastatic colorectal cancer [22] and metastatic breast cancer [23].

In early breast cancer, the molecular detection of cytokeratin 19 (CK-19) mRNA-positive cells in peripheral blood before [24], during [25], and after adjuvant therapy [26] is associated with worse prognosis, while their elimination seems to be an efficacy indicator of treatment [27]. The prognostic significance of CTC count using the CellSearch system in neoadjuvant [28] and adjuvant early breast cancer patients [29] has been also shown. Moreover, the administration of "secondary" adjuvant trastuzumab in patients with HER2(−) breast cancer can eliminate chemotherapy-resistant CK19 mRNA-positive CTCs [30], in contrast to the Treat CTC phase II trial that failed to prove the efficacy of trastuzumab in the detection rate of CTC [31]. However, in early breast cancer stages the early detection of recurrence remains a big challenge [32], and until now, there are not solid data proving the prognostic significance of EMT/SC(+) cells. The aim of the current study was to evaluate the prognostic significance of *TWIST1*, *CD24*, *CD44*, and *ALDH1* mRNA quantification in EpCAM-positive circulating tumor cells from early stage breast cancer patients with a long follow-up.

2. Materials and Methods

2.1. Cell Lines

The human mammary carcinoma cell line SKBR-3 was used as a positive control for the development of the quadraplex RT-qPCR assay for *CD24*, *CD44*, *ALDH1*, *HPRT*, while MDA-MB-231 cancer cell line was used as a positive control for the expression of *TWIST1* [33]. Cells were counted in a hemocytometer and their viability was assessed by trypan blue dye exclusion. cDNAs of all cancer cell lines were kept in aliquots at −20 °C and used for the analytical validation of the assay, prior to the analysis of patient's samples.

2.2. Patients

In total, 100 patients with non-metastatic breast cancer from the Medical Oncology Unit "Elena Venizelou" Hospital and IASO General hospital were enrolled in the study from September 2007 until January 2013. Peripheral blood (20 mL) was obtained from all these patients two weeks after the removal of the primary tumor and before the initiation of adjuvant chemotherapy. The chemotherapeutic adjuvant treatment for these patients has been previously reported [34]. The clinical characteristics for these patients at the time of diagnosis are shown in Supplementary Table S1. All patients signed an informed consent to participate in the study, which was approved by the Ethics and Scientific Committees of our Institutions. Peripheral blood (20 mL) was obtained from 19 healthy female blood donors (HD) and was analyzed in the same way as patients' samples (control group).

2.3. Isolation of EpCAM+ CTCs

To reduce blood contamination by epithelial cells from the skin, the first 5 mL of blood were discarded, and the blood collection tube was at the end disconnected before withdrawing the needle.

Peripheral blood (20 mL in EDTA) from (HD) and patients was collected and processed within 3 h in exactly the same manner. After collection, peripheral blood was diluted with 20 mL phosphate buffered saline (PBS, pH 7.3), and peripheral blood mononuclear cells (PBMCs) were isolated by gradient density centrifugation using Ficol-Paque TM PLUS (GE Healthcare, Bio-Sciences AB) at 670 g for 30 min at room temperature. The interface cells were removed and washed twice with 40 mL of sterile PBS (pH 7.3, 4 °C), at 530 g for 10 min. EpCAM+ cells were enriched using immunomagnetic Ber-EP4 coated capture beads (Dynabeads® Epithelial Enrich, Invitrogen, Carlsbad, CA, USA), according to the manufacturer's instructions [33].

2.4. RNA Extraction-cDNA Synthesis

Total RNA isolation was performed using TRIZOL-LS (ThermoFischer, Carlsbad, CA, USA). All RNA preparation and handling steps took place in a laminar flow hood under RNAse-free conditions. The isolated RNA from each fraction was dissolved in 20 µL of RNA storage buffer (Ambion, ThermoFischer, USA) and stored at −70 °C until use. RNA concentration was determined by absorbance readings at 260 nm using the Nanodrop-1000 spectrophotometer (NanoDrop, Technologies, Wilmington, DE, USA). mRNA was isolated from the total RNA using the Dynabeads mRNA Purification kit (ThermoFischer, USA), according to the manufacturer's instructions. cDNA synthesis was performed using the High capacity RNA-to-cDNA kit (ThermoFischer, USA) in a total volume of 20 µL, according to the manufacturer's instructions.

2.5. RT-qPCR

A novel quadraplex RT-qPCR assay was first developed for *CD24*, *CD44*, *ALDH1*, and *HPRT* (reference gene). Primers and dual hybridization probes were de novo in-silico designed, using Primer Premier 5.0 software (Premier Biosoft, Palo Alto, CA, USA). The specificity of all primer and hybridization probe sequences was first tested by homology searches in the nucleotide database (NCBI, nucleotide BLAST). Cross reaction between all oligonucleotide sequences was also examined. Each probe set included a 3′-fluorescein (F) donor probe and a 5′-LC acceptor probe that was different for each gene set: *CD24* (610 nm), *CD44* (640 nm), *ALDH1* (670 nm) and *HPRT* (705 nm). A color compensation test was performed by using pure dye spectra so that spectral overlap between dyes was corrected [35]. Quadraplex RT-qPCR reactions were performed in the LightCycler 2.0 (Roche, Mannheim, Germany). Component concentrations and the cycling conditions for the quadraplex RT-qPCR assay were optimized in detail. The amplification reaction mixture (10 µL) contained 1 µL of the PCR Synthesis Buffer (5X), 2.4 µL of $MgCl_2$ (25 mM), 0.2 µL dNTPs (10 mM), 0.8 µL BSA (10 µg/µL), 0.1 µL Hot Start DNA polymerase (HotStart, 5 U/µL, Promega, Madison, WI, USA), 0.5 µL of a mixture containing all eight primers (10 µM), 0.5 µL of a mixture containing all eight dual hybridization probes (3 µM), and H2O (added to the final volume). Cycling conditions of the *CD24*, *CD44*, *ALDH1*, *HPRT* quadraplex RT-qPCR assay were: 95 °C/2 min; 45 cycles of 95 °C/20 s, annealing at 59 °C/20 s, and extension at 72 °C/20 s. For the development and analytical validation of the novel quadraplex RT-qPCR assay, we generated individual PCR amplicons corresponding to the gene-targets studied that served as quantification calibrators, as we have previously described [33]. RT-qPCR for *TWIST1* was performed as previously described [33,36]. All data were evaluated in respect to *TWIST1*, *CD24*, *CD44*, and *ALDH1* expression by normalizing the EpCAM+ fraction of PBMCs to the expression of *HPRT* and the 2−ΔΔCt approach, as described in detail by Livak and Schmittgen [37]. A cut-off value was calculated as the mean of signals derived by samples of healthy individual analyzed in exactly the same way plus 2SD for *TWIST1*, *CD44*, and *ALDH1* transcripts and as the mean of signals derived by samples of healthy individual minus 2SD for *CD24*.

2.6. Statistical Analysis

Statistical analysis was performed using SPSS (SPSS Statistics 25.0, company, Armonk, NY, USA). The chi-square test of independence or Fisher exact test (SPSS, version 25.0) was used to make comparisons between groups. The DFI and OS rate were calculated by the Kaplan–Meier method and were evaluated by the log-rank test. Cox proportional hazards (PH) models were used to evaluate the relationship between EMT and Stem Cell status and event-time distributions, with tumor size, grade, number of involved lymph nodes, ER, PR, HER2, and age. Parametric and non-parametric tests were used to compare continuous variables between groups. All P-values are two-sided. A level of $p < 0.05$ is considered statistically significant.

3. Results

3.1. Analytical Validation of the Quadraplex RT-qPCR Assay for CD24, CD44, ALDH1, HPRT

The analytical specificity of the developed assay was checked by using all oligonucleotides in a common master mix in four different reactions in the presence of one individual gene target each time. Each primer pair and dual hybridization probe pair amplifies specifically only the corresponding target sequence and is detected only in the corresponding wavelength (Supplementary Figure S1A). The analytical sensitivity was determined for each individual gene target using a calibration curve. These calibration curves were generated using serial dilutions of individual gene-specific external standards in triplicate for each concentration, ranging from 10^5 copies/µL to 10 copies/µL. The analytical detection limit corresponded to 3 copies/µL while the quantification limit was equal to 9 copies/µL (Supplementary Figure S1B). The developed assay showed linearity over the entire quantification range and correlation coefficients greater than 0.99 in all cases, indicating a precise log-linear relationship. Intra and inter-assay variance: Repeatability or intra–assay variance of the quadraplex RT-qPCR was evaluated by repeatedly analyzing four cDNA samples corresponding to 1, 10, 100, and 1000 SKBR-3 cells in the same assay, in three parallel determinations. Reproducibility or interassay variance was evaluated by analyzing the same cDNA sample, representing 1000 SKBR-3 cells on five separate assays performed in five different days (Supplementary Table S2).

3.2. Quantification of CD24, CD44, ALDH1, and TWIST1 mRNA in the EpCAM(+) Fraction in Early Stage BrCa Patients and (HD)

In all EpCAM(+) fractions isolated from 100 early BrCa patient samples and 19 HD *CD24, CD44, ALDH1, HPRT* transcripts were quantified by the developed quadraplex RT-qPCR and *TWIST1* transcripts by the singleplex RT-qPCR assay (Figure 1). Median fold change of *TWIST1* expression in the EpCAM(+) fraction was 0.42 (range: 0–0.95) in HD and 10.06 (range: 2.33–3327) in *TWIST1*^{high} (Mann-Whitney test, $Z = -1.363$, $p = 0.001$) and 0 (range: 0–0) in *TWIST1*$^{low/-}$ early BrCa patient samples (Mann-Whitney test, $Z = -3.634$, $p < 0.001$) (Figure 1A). Median fold change of *CD24* expression in the EpCAM(+) fraction was 2.00 (range: 1.42–3.81) in HD and 1.91 (range: 0.91–15.14) in *CD24*^{high} (Mann-Whitney test, $Z = -0.492$, $p = 0.623$) and 0.62 (range: 0.29–0.88) in *CD24*^{low} early BrCa patients (Mann-Whitney test, $Z = -5.577$, $p < 0.001$) (Figure 1B). Median fold change of *CD44* expression in the EpCAM(+) fraction was 0.71 (range: 0.14–1.06) in HD and 2.33 (range: 1.28–202.75) in *CD44*^{high} (Mann-Whitney test, $Z = -6.084$, $p < 0.001$) and 0.61 (range: 0.01–1.17) in *CD44*^{low} early BrCa patients (Mann-Whitney test, $Z = -1.084$, $p = 0.278$) (Figure 1C). Median fold change of *ALDH1* expression in the EpCAM(+) fraction was 1.32 (range: 0.69–2.19) in HD and 2.97 (range: 2.30–14.72) in *ALDH1*^{high} (Mann-Whitney test, $Z = -5.119$, $p < 0.001$) and 0.84 (range: 0.06–2.16) in *ALDH1*^{low} early BrCa patients (Mann-Whitney test, $Z = -2.190$, $p = 0.029$) (Figure 1D).

In 19/100(19%) breast cancer samples tested, *TWIST1* was overexpressed, while in 15/100(15%) samples the *CD24*$^{-/low}$/*CD44*^{high} profile, and in 9/100(9%) the *CD24*$^{-/low}$/*ALDH1*^{high} profile was detected (Figure 2A). There was a positive correlation between *TWIST1* mRNA overexpression and the *CD24*$^{-/low}$/*CD44*^{high} profile (Fisher's Exact Test; $p = 0.008$), while there was no correlation between *TWIST1* mRNA overexpression and the *CD24*$^{-/low}$/*ALDH1*^{high} profile (Fisher's Exact Test; $p = 0.366$) (Table 1). *TWIST1* overexpression and *CD24*$^{-/low}$/*CD44*^{high} and/or *CD24*$^{-/low}$/*ALDH1*^{high} were detected in 7/100(7%) EpCAM(+) samples. The correlation between these characteristics and the clinical variables of the patients revealed an association between *TWIST1* overexpression with lymph node status (chi-square; $p = 0.036$) and HER2 status of the primary tumor (chi-square; $p = 0.006$) (Supplementary Table S1).

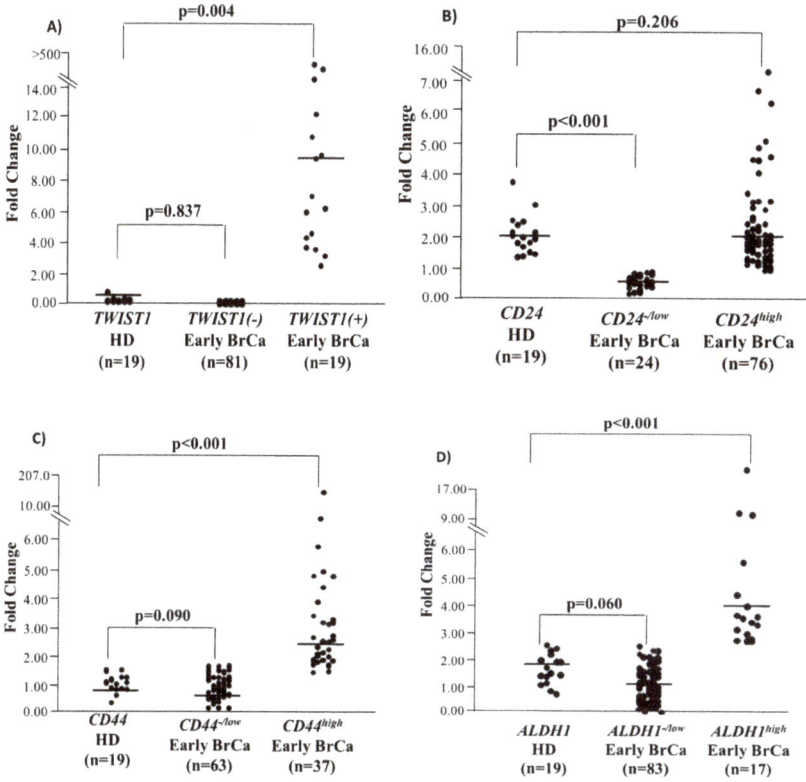

Figure 1. Relative fold change values ($2^{-\Delta\Delta Ct}$) in respect to HPRT expression for: (**A**) *TWIST1* (**B**) *CD24*, (**C**) *CD44*, (**D**) *ALDH1* for early breast cancer patients ($n = 100$) and (HD), ($n = 19$).

3.3. Evaluation of Prognostic Significance

3.3.1. Disease Free Interval

During the follow up period (median: 95 months; range: 4–137 months), 25/100 (25%) patients relapsed and in 9/25 (36%) of them *TWIST1* overexpression was detected in the EpCAM+ CTC fraction (Fisher's Exact Test; $p = 0.019$). Similarly, 6/25 (24%) patients displayed a Stem Cell profile in EpCAM+ CTC fraction (Fisher's Exact Test; $p = 0.194$). In 4/25 (16%) of these patients, both *TWIST1* overexpression and the Stem Cell profile was detected (Fisher's Exact Test; $p = 0.063$) (Supplementary Table S3). The Kaplan–Meier estimates of the cumulative DFI of the patients overexpressing *TWIST1* revealed that these patients had worse survival compared to patients who were negative (83.6mo vs 115.8mo respectively; $p = 0.019$) (Table 2, Figure 3A). However, the stem cell profile alone (86.7mo. vs 113.2mo, respectively in the two groups; log rank test; $p = 0.174$) (Table 2, Supplementary Figure S2A) and both stem cell and mesenchymal characteristics (68.9mo vs 88.8–115.8mo, respectively; $p = 0.087$) (Table 2, Supplementary Figure S2C) failed to show any statistically significant difference even though the mean survival showed a reduced trend. Kaplan–Meier survival analysis of patients with positive axillary lymph nodes and *TWIST1* mRNA overexpression had worst DFI (Table 2, Supplementary Figure S3A) (82.6 mo. vs 88.7–123.3; $p = 0.05$). When all patients were divided into two groups based on the number of positive lymph nodes (1–3, and ≥4 positive nodes) and the overexpression of TWIST1 [(N$_{2-3}$/TWIST1(+), N$_{2-3}$/TWIST1(−), N$_1$/TWIST1(+), and N$_1$/TWIST1(−)], Kaplan–Meier analysis revealed that women harboring more than 3 positive lymph nodes and TWIST1 that was overexpressed in EpCAM+ CTC fraction had a statistically significant shorter DFI (Table 2, Figure 3C) (mean survival: 68.6mo vs. 103.0–114.3mo.; $p = 0.007$). When patients were dichotomized accordingly to the HR status (ER/PR) in the following groups: a) HR(−)/TWIST1(+), b) HR(−)/ TWIST1(−), c) HR(+)/TWIST1(+) and d) HR(+)/TWIST1(−), it was observed that women with HR(−)/TWIST1(+) profile were characterized by statistically significant shorter DFI (36mo. vs 102.3–117.9mo.; $p < 0.001$; Figure 3E). A Univariate analysis (Table 3) also revealed the significance of (a) TWIST1(+), (b) HR(−)/TWIST1(+), (c) TWIST1(+) /N$_{2-3}$, d) SC (+)/ TWIST1(+) (Figure 2B) in the risk of disease progression. Multivariate confirmed the prognostic value of HR(−)/TWIST1(+) and TWIST1(+)/N2-3, in the EpCAM(+) CTC fraction for the prediction of DFI (Table 3) independently from patients' age, tumor T stage, grade, nodal status alone and the HR, and HER2 status of the primary tumor.

Table 1. Correlation between TWIST1 and CD44^{high}/CD24$^{-/low}$ and ALDH1^{high}/CD24$^{-/low}$ expression in early breast cancer EpCAM positive samples ($n = 100$).

	CD44^{high}/CD24$^{-/low}$		p ^{a}	ALDH1^{high}/CD24$^{-/low}$		p ^{a}
TWIST1	Positive	Negative		Positive	Negative	
Positive	7 (46.7%)	12(14.1%)	**0.008**	3 (33.3)	16(17.6%)	0.366
Negative	8 (53.3%)	73(85.9%)		6 (66.7%)	75 (82.4%)	
Concordance	80/100 (80%)			78/100 (78%)		

^{a} Fischer's Exact Test. Bold: highlights the significance of the test.

3.3.2. Overall Survival

Among the 25 patients that relapsed during the follow up period, 14/25 (56.0%) patients died and 11/25 (44.0%) were still alive at the time of the last follow-up. In 6/14 (42.9%) patients that died TWIST1 overexpression was detected in the EpCAM+ fraction (Fisher's Exact Test; $p = 0.024$). Similarly, 4/14 (28.6%) patients displayed a Stem Cell profile in EpCAM+ CTC fraction (Fisher's Exact Test; $p = 0.217$). In 3/14 (21.4%) of these patients, both TWIST1 overexpression and CD24$^{-/low}$/CD44^{high} and/or CD24$^{-/low}$/ALDH1^{high} profiles (Fisher's Exact Test; $p = 0.055$) were detected (Fisher's Exact Test; $p = 0.055$) (Supplementary Table S3). The Kaplan–Meier estimates of the overall survival (OS) of the patients overexpressing TWIST1 were significantly different in favor of patients who were negative for TWIST1 overexpression (106.4 vs 127.2 mo; $p = 0.046$) (Table 2, Figure 3B). Stem Cell profiles (107.3 vs 125.2 mo.; $p = 0.171$) (Table 2, Supplementary Figure S2B) and the co-expression of EMT and SC-associated genes (96.29 vs 109.1–127.3 mo.; $p = 0.118$) (Table 2, Supplementary Figure S2D) failed to show any statistically significant difference. There was no difference in OS in patients with TWIST1 overexpression according to N0 and N+ lymph node involvement (108.8 mo vs 92–129 mo, respectively; $p = 0.194$; Supplementary Figure S3B). However, when the Kaplan–Meier curves for OS for TWIST1 overexpression were additionally stratified according to lymph nodes status (Table 2, Figure 3D) and HR status (Table 2, Figure 3F) our data have shown that patients with >3 LN and TWIST1 overexpression had lower OS (109.8 mo., range: 115–129 mo.; $p = 0.026$); the same was seen for patients that were HR(−) and TWIST1 was overexpressed (65.7 vs 110.2–131.9 mo.; $p < 0.001$). Univariate analysis showed a significantly higher risk of death in the group of patients positive for TWIST1 overexpression that had more than 3 lymph nodes affected or co-expressed the stem cell profile (Figure 2B). Multivariate analysis confirmed the prognostic value of TWIST1 overexpression in combination with N$_{2-3}$, and in combination with HR(−) status in the EpCAM(+) CTC fraction for the prediction of OS, independently from patients' age, tumor T stage, grade, nodal status, and the status of the receptors ER, PR, HER2 of the primary site (Table 3).

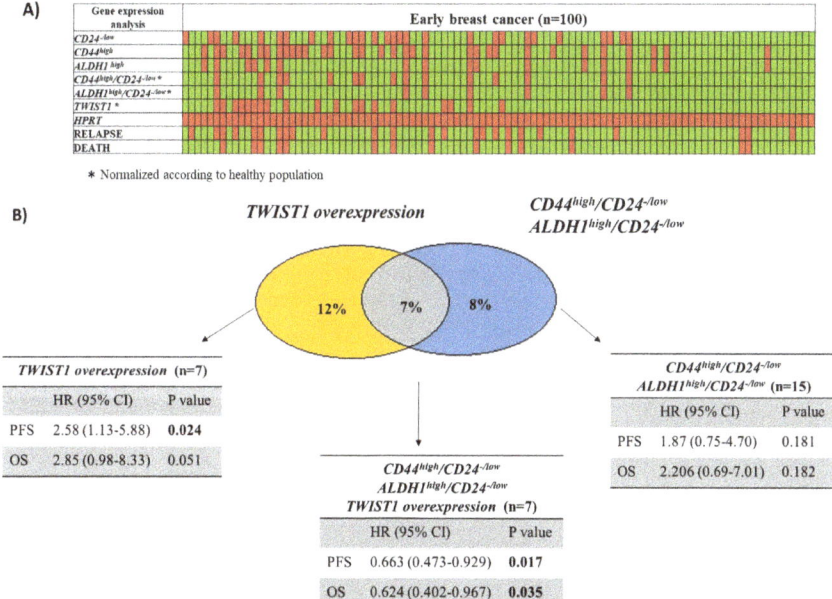

Figure 2. (**A**) Heat map of *TWIST1*, *CD24*, *CD44*, and *ALDH1*-mRNA quantification in the EpCAM+ CTC fraction from early stage breast cancer patients (*n* = 100). Red color represents overexpression, while green color indicates underexpression or lack of expression. Concerning the relapse or death, red color represents the relapse or death, while green color indicates no relapse or alive status. (**B**) Univariate Cox-regression hazard models for TWIST1 overexpression, $CD44^{high}/CD24^{-/low}$, and $ALDH1^{high}/CD24^{-/low}$ and the co-expression of the mesenchymal profile, *TWIST1*, and the stem cell profile, $CD44^{high}/CD24^{-/low}$, and $ALDH1^{high}/CD24^{-/low}$.

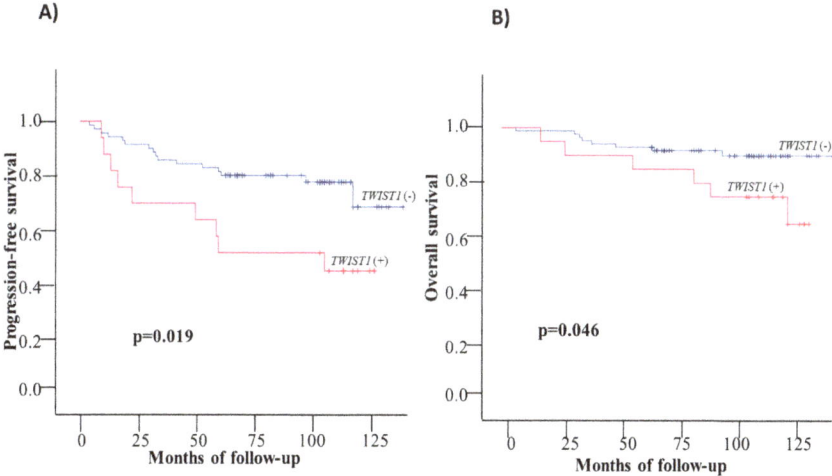

Figure 3. *Cont.*

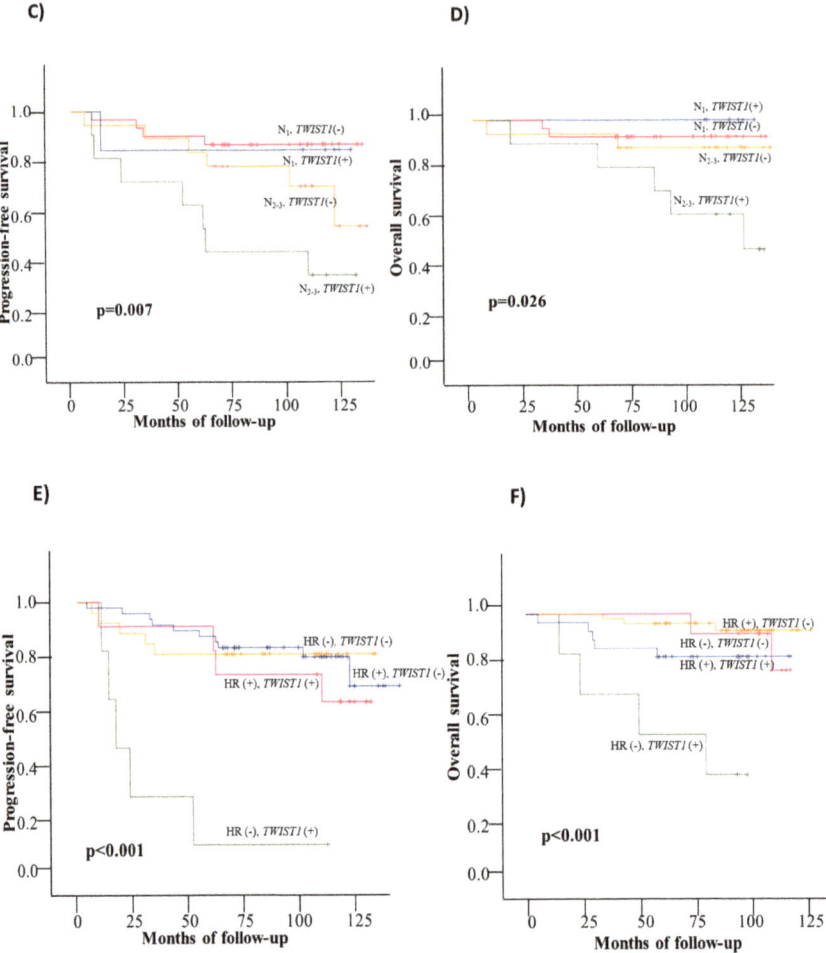

Figure 3. Kaplan–Meier estimates for early BrCa patients: (**A**) DFI: *TWIST1* overexpression, (**B**) OS: *TWIST1* overexpression, (**C**) DFI: *TWIST1* overexpression and number of affected lymph nodes, (**D**) OS: *TWIST1* overexpression and number of affected lymph nodes, (**E**) DFI: *TWIST1* overexpression and HR status, (**F**) OS: *TWIST1* overexpression and HR status.

Table 2. Gene expression in CTCs in respect to DFI and OS.

Gene Expression in CTCs	DFI				OS			
Gene	Mean Survival	95% CI (months)	Range (months)	p	Mean Survival	95% CI (Months)	Range (Months)	p
TWIST1+	83.6	61.9–105.3	9–125	**0.019**	106.4	90.3–122.3	16–127	**0.046**
TWIST1−	115	106.3–125.2	4–137		127.2	120.7–133.7	6–137	
Stem cell profile positive (SC+)	86.7	66.7–106.8	16–118	0.174	107.3	89.8–124.8	26–127	0.171
Stem cell profile negative (SC−)	113.2	103.5–122.8	4–137		125.2	118.3–132.1	6–137	
TWIST1+/SC+	68.9	39.4–98.31	16–112	0.087	96.2	67.4–125.1	26–127	0.118
TWIST1+/SC−	88.8	61.2–116.4	9–125		111.1	93.3–128	16–125	
TWIST1−/SC+	100.1	78.6–121.7	41–118		109.1	92.8–125.3	47–118	
TWIST1−/SC−	115.8	105.8–126	4–137		127.3	120.4–134.2	6–137	
TWIST1+/LN+	82.6	59.3–105.9	9–125	0.05	108.8	93–124.6	16–127	0.194
TWIST1+/LN−	88.7	30.5–146.8	16–125		92	39.1–144.8	26–125	
TWIST1−/LN+	110.3	99.2–121.4	6–130		121.3	113–129.6	6–130	
TWIST1−/LN−	123.3	110.6–135.9	4–137		128.9	120.2–137.7	37–137	
TWIST1+/N_{2-3}	68.6	40.5–96.7	9–125	**0.007**	98.3	75.5–121.2	16–127	**0.026**
TWIST1+/N_1	104.5	71.8–137.5	13–123		112	94–125	101–118	
TWIST1−/N_{2-3}	103.1	84.1–121.9	6–130		118.8	103.4–134.1	6–130	
TWIST1−/N_1	114.4	101.7–126.9	9–128		121.1	111.9–130.3	30–128	
TWIST1+/HR+	102.3	81.2–123.4	9–125	**<0.001**	121.6	113.9–129.3	79–127	**<0.001**
TWIST1+/HR−	36	8.9–63	10–106		65.7	36.7–94.6	76–106	
TWIST1−/HR+	117.9	107.1–128.7	4–137		131.9	126.2–137.5	37–137	
TWIST1−/HR−	107.7	92.3–123.1	6–127		110.2	96.5–123.84	6–127	

Bold: highlights the significance of the test.

Table 3. Univariate and multivariate analyses for DFI and OS in the early breast cancer patients group ($n = 100$).

Covariates	Covariate Value	DFI Univariate Cox Regression Analysis			DFI Multivariate Cox Regression Analysis			OS Univariate Cox Regression Analysis			OS Multivariate Cox Regression Analysis		
		HR[a]	95% CI[b]	p	HR[a]	95% CI[b]	p	HR[a]	95% CI[b]	p	HR[a]	95% CI[b]	p
Age	≥50 vs <50	0.787	0.357–1.734	0.552	0.432	0.169–1.103	0.079	0.718	0.249–2.070	0.539	0.593	0.169–2.080	0.414
ER	Yes vs No	0.647	0.286–1.463	0.295	4.391	1.040–18.549	**0.044**	0.238	0.079–0.721	**0.011**	0.623	0.092–4.215	0.628
PR	Yes vs No	0.492	0.217–1.114	0.089	0.087	0.021–0.362	**0.001**	0.196	0.054–0.707	**0.013**	0.098	0.011–0.851	**0.035**
HER2	Yes vs No	0.500	0.197–1.269	0.145	0.626	0.232–1.693	0.357	0.381	0.106–1.366	0.139	0.247	0.060–1.023	0.054
Lymph nodes	N_0 vs N_1 vs N_{2-3}	2.207	1.261–3.861	**0.006**	2.433	1.272–4.654	**0.007**	1.659	0.817–3.371	0.162	1.351	0.637–2.862	0.433
Size	≥2cm vs <2cm	3.060	1.049–8.922	**0.041**	4.926	1.225–19.811	**0.025**	7.244	0.947–55.432	0.056	17.450	1.464–208.1	**0.024**
Grade	I/II vs III	1.366	0.570–3.273	0.485	0.753	0.228–2.483	0.641	1.953	0.544–7.008	0.305	0.286	0.040–2.028	0.211
TWIST1	Yes vs No	2.582	1.135–5.875	**0.024**	1.382	0.490–3.899	0.540	2.851	0.975–8.33	0.051	1.481	0.382–5.743	0.570
HR and TWIST1 status	HR+TWIST1+ HR+TWIST1- HR-TWIST1+ HR-TWIST1-	0.486	0.302–0.784	**0.003**	0.597	0.360–0.991	**0.046**[c]	0.576	0.313–1.062	0.077	0.666	0.345–1.284	0.225[c]
LN and TWIST1 status	N_{0-1}TWIST1+ N_{0-1}TWIST1- N_{2-3}TWIST1+ N_{2-3}TWIST1-	0.540	0.383–0.762	**<0.001**	0.542	0.371–0.792	**0.002**[d]	0.559	0.357–0.875	**0.011**	0.604	0.373–0.976	**0.040**[d]
Stem cell profile	Yes vs No	1.873	0.746–4.703	0.181	1.755	0.526–5.855	0.360	2.206	0.690–7.050	0.182	3.689	0.806–16.884	0.093
Stem cell profile/TWIST1+	Yes vs No	0.663	0.473–0.929	**0.017**	0.776	0.521–1.146	0.202[e]	0.624	0.402–0.967	**0.035**	0.634	0.378–1.065	0.085[e]

[a] Hazard ratio, estimated from Cox proportional hazard regression mode. [b] Confidence interval of the estimated HR. Results are based on 1000 bootstrap samples and obtained after the Bias corrected and accelerated (BCa) approach. [c] Multivariate model adjusted for age, HER2, LN, Size, Grade, Stem cell. [d] Multivariate model adjusted for age, ER, PR, HER2, Size, Grade, Stem cell. [e] Multivariate model adjusted for age, ER, PR, HER2, LN, Size, Grade. Bold: highlights the significance of the test.

4. Discussion

Molecular characterization of CTCs at the gene expression level has a strong potential to provide novel prognostic and predictive biomarkers. It is now clear through numerous studies that CTCs isolated from breast cancer patients express epithelial markers [38], receptors (ER, PR, HER2, EGFR), stem cell markers [39], and mesenchymal markers [11]. So far, most studies have been performed in the metastatic setting where the number of circulating tumor cells is usually high. However, in the non-metastatic setting of breast cancer, CTCs are not always detected and their numbers are usually very low, thus their molecular characterization is extremely difficult. For this reason, in the early breast cancer setting, a higher volume of peripheral blood used for the analysis of CTCs is very critical. Our group has shown many years ago the prognostic significance of *CK-19* mRNA detection in peripheral blood of early breast cancer patients, using 20 mL of peripheral blood for CTC isolation and further downstream analysis [36,38]. Other groups have also shown that the detection of CTCs in the early breast cancer setting is providing critical prognostic information for these patients [29].

In this study we evaluated for the first time the prognostic significance of *TWIST1*, *CD24*, *CD44*, and *ALDH1* transcript quantification in EpCAM-positive circulating tumor cells isolated from peripheral blood of early stage breast cancer patients. We selected *TWIST1* as this is a very established EMT marker; for this reason, we have developed already in 2011 an RT-qPCR assay for the absolute quantification of *TWIST1*-mRNA expression, and we have validated this assay in EpCAM-positive cells isolated by early and metastatic breast cancer patients [33]. Concerning the selection of stem cell markers, this was based on publications by Al-Hajj M et.al. [8] and Ginestier C et.al. [9], who have shown that the breast cancer stem cell phenotypes of (a) CD44$^+$/CD24$^{-/low}$ phenotype and (b) the overexpression of aldehyde dehydrogenase 1 (ALDH1+) are able to form tumors in mice with high tumorigenic capacity.

Multiplex RT-qPCR assays have many benefits due to their wide dynamic range, the low sample volume required, and the reduced time of analysis [35]. Our study was based on an analytically validated novel multiplex assay for the quantification of *CD24*, *CD44*, *ALDH1*, and *HPRT* and a single RT-qPCR assay for the quantification of *TWIST1* transcripts. The analytical sensitivity and specificity of the novel quadraplexRT-qPCR assay for the simultaneous detection of *CD24*, *CD44*, *ALDH-1*, and *HPRT* transcripts were determined by using calibrators specific for each gene target. Both these assays were validated according to the Minimum Information for Publication of Quantitative Real-Time PCR Experiments (MIQE) guidelines [40].

Relevant prognostic and predictive markers in early breast cancer cohort is of major significance. The SUCCESS A trial has shown that the presence of CTCs, as evaluated in 30 mL of peripheral blood, two years after chemotherapy has been associated with decreased OS and DFS in high-risk early breast cancer patients [41]. Lucci et.al. has also shown that the presence of one or more circulating tumor cells could predict early recurrence and decreased overall survival in chemonaive patients with non-metastatic breast cancer [29]; however, the main limitation of this study is that it was based on CTC enumeration performed in only 7.5 mL of blood. Additionally, molecular characterization of CTC could identify CTC biomarkers that are associated to specific signaling pathways like EMT or CSC. Our findings demonstrate a positive correlation between *TWIST1* overexpression and the CD24$^{-/low}$/CD44^{high} profile in the EpCAM positive CTC fraction. This is in agreement with previous findings showing that the mesenchymal-like breast cancer stem cells are characterized as CD24$^-$/CD44$^+$, while the epithelial-like breast cancer stem cells express high levels of aldehyde dehydrogenase (ALDH) [16]. Univariate analysis revealed a significantly higher risk of relapse and death in the group of patients that expressed both stem cell and mesenchymal characteristics. Mego et al. have shown that patients with *TWIST1*-high tumors had a significantly higher percentage of breast cancer stem cells than patients with *TWIST1*-low tumors [19]. Recently, it was shown that in CTC of NSCLC patients the CD44(+)/CD24(−) population possess epithelial–mesenchymal transition characteristics [42], while another study in metastatic colorectal cancer has shown the prognostic significance of CTC that express both EMT and stem-like genes [22]. At the single CTC-level, Papadaki et.al have shown that CTCs expressing high levels of ALDH1 along with nuclear TWIST expression are more frequently detected in patients with metastatic breast cancer [21] and that these cells represent a chemo-resistant subpopulation with an unfavorable outcome [23]. The main limitation of our study is that we examined the expression of only one EMT marker, *TWIST1* in the EpCAM+ cells of early breast cancer patients. Since there is a high heterogeneity in CTC, it may be possible that we have not detected CTCs that express other mesenchymal markers like Vimentin or Snail. We plan to extend this study by adding more gene expression markers in a biggest sample cohort and correlate our results to the clinical outcome.

According to our results, patients with *TWIST1* overexpression in the EpCAM+ CTCs fraction and more than 3 involved lymph nodes had a significant lower DFI and OS. Similar to our results, recently, Emprou C et al. have shown that in frozen NSCLC tumor samples *TWIST1* is more frequently overexpressed in the N+ group compared to the N0 group showing that partial EMT is involved in lymph node progression in early stages of NSCLC [43], while in primary breast cancer loss of E-cadherin is correlated with more than 3 LN involved in 80% of the patients [44]. Our results also indicate that patients with *TWIST1* overexpression in the EpCAM+ CTCs fraction and a hormone receptor-negative primary tumor had a worse prognosis both for DFI and OS. This is in accordance with previous findings that have shown that the estrogen receptor silencing induces epithelial to mesenchymal transition in breast cancer [45]. It has also been previously shown that in human breast tumors there is an inverse relationship between *TWIST1* and ER expression that may possibly contribute to the generation of hormone-resistant, ER-negative breast cancer [46]. It has also been reported that EMT likely occurs in the basal-like phenotype both in MCF10A cells [47] and in invasive breast cancer carcinomas [48].

5. Conclusions

In conclusion, detection of *TWIST1* overexpression and stem-cell (*CD24, CD44, ALDH1*) transcripts in the EpCAM$^+$ CTC fraction provides prognostic information in early stage breast cancer patients. Overexpression of *TWIST1* in the EpCAM$^+$ CTC fraction in the group of HR negative patients or with >3 positive lymph nodes is associated with worse prognosis.

Supplementary Materials: The following are available online at http://www.mdpi.com/2073-4409/8/7/652/s1. Figure S1: Analytical validation of the multiplex RT-qPCR for *CD24, CD44, ALDH1, HPRT* (all measured in triplicate) (A) analytical specificity, (B) RT-qPCR calibration curves (copies/µL). Figure S2: Kaplan–Meier estimates for early BrCa patients with or without the stem cell profile in respect to (A) DFI, (B) OS and with or without co-expression of *TWIST1* and Stem Cell profile in respect to (C) DFI, (D) OS. Figure S3: Kaplan–Meier estimates for early BrCa patients with positive or negative axillary lymph nodes the (A) DFI and (B) OS. Table S1: Clinical characteristics of the patients with early breast cancer ($n = 100$). Table S2: Quadraplex RT-qPCR for *CD24, CD44, ALDH1, HPRT*, evaluation of intra- ($n = 3$) and inter-assay ($n = 5$) precision. Table S3: Correlation of *TWIST1*, $CD44^{high}/CD24^{-/low}$ and/or $ALDH1^{high}/CD24^{-/low}$ and the co-expression of *TWIST1* and $CD44^{high}/CD24^{-/low}$ and/or $ALDH1^{high}/CD24^{-/low}$ with the patients' clinical outcomes.

Author Contributions: Conceptualization, A.S. and E.S.L.; methodology, A.S.; validation, A.S.; formal analysis, A.S.; investigation, A.S. and E.S.L.; resources, M.N. and V.G.; data curation, A.S.; writing—original draft preparation, A.S.; writing—review and editing, E.S.L. and V.G.; supervision, E.S.L.

Funding: This research received no external funding.

Conflicts of Interest: The authors declare no conflicts of interest.

Abbreviations

EMT	epithelial–mesenchymal transition
MET	mesenchymal-epithelial transition
MICs	metastasis-initiating cells
SC	Stem Cell
HR	Hormone Receptor
BCSC	breast cancer stem cells
CSCs	Cancer stem cells
CTCs	Circulating tumor cells

ER	estrogen receptor
PR	progesterone receptor
HER2	human epidermal growth factor receptor 2
ALDH1	aldehyde dehydrogenase 1
HD	Healthy Donors
RT-qPCR	Quantitative Reverse Transcription Polymerase Chain Reaction
PBMCs	peripheral blood mononuclear cells
PBS	phosphate-buffered saline
LN	lymph nodes

References

1. Lianidou, E.S. Gene expression profiling and DNA methylation analyses of CTCs. *Mol. Oncol.* **2016**, *10*, 431–442. [CrossRef] [PubMed]
2. Bardelli, A.; Pantel, K. Liquid Biopsies, What We Do Not Know (Yet). *Cancer Cell* **2017**, *31*, 172–179. [CrossRef] [PubMed]
3. Boral, D.; Vishnoi, M.; Liu, H.N.; Yin, W.; Sprouse, M.L.; Scamardo, A.; Hong, D.S.; Tan, T.Z.; Thiery, J.P.; Chang, J.C.; et al. Molecular characterization of breast cancer CTCs associated with brain metastasis. *Nat. Commun.* **2017**, *8*, 196. [CrossRef] [PubMed]
4. Jakabova, A.; Bielcikova, Z.; Pospisilova, E.; Matkowski, R.; Szynglarewicz, B.; Staszek-Szewczyk, U.; Zemanova, M.; Petruzelka, L.; Eliasova, P.; Kolostova, K.; et al. Molecular characterization and heterogeneity of circulating tumor cells in breast cancer. *Breast Cancer Res. Treat.* **2017**, *166*, 695–700. [CrossRef] [PubMed]
5. Mansoori, M.; Madjd, Z.; Janani, L.; Rasti, A. Circulating cancer stem cell markers in breast carcinomas: A systematic review protocol. *Syst. Rev.* **2017**, *6*, 262. [CrossRef] [PubMed]
6. Reya, T.; Morrison, S.J.; Clarke, M.F.; Weissman, I.L. Stem cells, cancer, and cancer stem cells. *Nature* **2001**, *414*, 105–111. [CrossRef]
7. Luo, M.; Clouthier, S.G.; Deol, Y.; Liu, S.; Nagrath, S.; Azizi, E.; Wicha, M.S. Breast cancer stem cells: Current advances and clinical implications. *Methods Mol. Biol.* **2015**, *1293*, 1–49.
8. Al-Hajj, M.; Wicha, M.S.; Benito-Hernandez, A.; Morrison, S.J.; Clarke, M.F. Prospective identification of tumorigenic breast cancer cells. *Proc. Natl. Acad. Sci. USA* **2003**, *100*, 3983–3988. [CrossRef]
9. Ginestier, C.; Hur, M.H.; Charafe-Jauffret, E.; Monville, F.; Dutcher, J.; Brown, M.; Jacquemier, J.; Viens, P.; Kleer, C.G.; Liu, S.; et al. ALDH1 is a marker of normal and malignant human mammary stem cells and a predictor of poor clinical outcome. *Cell Stem Cell* **2007**, *1*, 555–567. [CrossRef]
10. Balic, M.; Lin, H.; Young, L.; Hawes, D.; Giuliano, A.; McNamara, G.; Datar, R.H.; Cote, R.J. Most early disseminated cancer cells detected in bone marrow of breast cancer patients have a putative breast cancer stem cell phenotype. *Clin. Cancer Res.* **2006**, *12*, 5615–5621. [CrossRef]
11. Theodoropoulos, P.A.; Polioudaki, H.; Agelaki, S.; Kallergi, G.; Saridaki, Z.; Mavroudis, D.; Georgoulias, V. Circulating tumor cells with a putative stem cell phenotype in peripheral blood of patients with breast cancer. *Cancer Lett.* **2010**, *288*, 99–106. [CrossRef] [PubMed]
12. Bredemeier, M.; Edimiris, P.; Tewes, M.; Mach, P.; Aktas, B.; Schellbach, D.; Wagner, J.; Kimmig, R.; Kasimir-Bauer, S. Establishment of a multimarker qPCR panel for the molecular characterization of circulating tumor cells in blood samples of metastatic breast cancer patients during the course of palliative treatment. *Oncotarget* **2016**, *7*, 41677–41690. [CrossRef] [PubMed]
13. Baccelli, I.; Schneeweiss, A.; Riethdorf, S.; Stenzinger, A.; Schillert, A.; Vogel, V.; Klein, C.; Saini, M.; Bäuerle, T.; Wallwiener, M.; et al. Identification of a population of blood circulating tumor cells from breast cancer patients that initiates metastasis in a xenograft assay. *Nat. Biotechnol.* **2013**, *31*, 539–544. [CrossRef] [PubMed]
14. Schölch, S.; García, S.A.; Iwata, N.; Niemietz, T.; Betzler, A.M.; Nanduri, L.K.; Bork, U.; Kahlert, C.; Thepkaysone, M.-L.; Swiersy, A.; et al. Circulating tumor cells exhibit stem cell characteristics in an orthotopic mouse model of colorectal cancer. *Oncotarget* **2016**, *7*, 27232–27242. [CrossRef] [PubMed]
15. Wang, L.; Li, Y.; Xu, J.; Zhang, A.; Wang, X.; Tang, R.; Zhang, X.; Yin, H.; Liu, M.; Wang, D.D.; et al. Quantified postsurgical small cell size CTCs and EpCAM+ circulating tumor stem cells with cytogenetic abnormalities in hepatocellular carcinoma patients determine cancer relapse. *Cancer Lett.* **2018**, *412*, 99–107. [CrossRef] [PubMed]

16. Liu, S.; Cong, Y.; Wang, D.; Sun, Y.; Deng, L.; Liu, Y.; Martin-Trevino, R.; Shang, L.; McDermott, S.P.; Landis, M.D.; et al. Breast cancer stem cells transition between epithelial and mesenchymal states reflective of their normal counterparts. *Stem Cell Rep.* **2014**, *2*, 78–91. [CrossRef] [PubMed]
17. Aktas, B.; Tewes, M.; Fehm, T.; Hauch, S.; Kimmig, R.; Kasimir-Bauer, S. Stem cell and epithelial-mesenchymal transition markers are frequently overexpressed in circulating tumor cells of metastatic breast cancer patients. *Breast Cancer Res.* **2009**, *11*, R46. [CrossRef]
18. Giordano, A.; Gao, H.; Anfossi, S.; Cohen, E.; Mego, M.; Lee, B.-N.; Tin, S.; De Laurentiis, M.; Parker, C.A.; Alvarez, R.H.; et al. Epithelial-mesenchymal transition and stem cell markers in patients with HER2-positive metastatic breast cancer. *Mol. Cancer* **2012**, *11*, 2526–2534. [CrossRef]
19. Mego, M.; Gao, H.; Lee, B.-N.; Cohen, E.N.; Tin, S.; Giordano, A.; Wu, Q.; Liu, P.; Nieto, Y.; Champlin, R.E.; et al. Prognostic Value of EMT-Circulating Tumor Cells in Metastatic Breast Cancer Patients Undergoing High-Dose Chemotherapy with Autologous Hematopoietic Stem Cell Transplantation. *J. Cancer* **2012**, *3*, 369–380. [CrossRef]
20. Zhang, L.; Ridgway, L.D.; Wetzel, M.D.; Ngo, J.; Yin, W.; Kumar, D.; Goodman, J.C.; Groves, M.D.; Marchetti, D. The identification and characterization of breast cancer CTCs competent for brain metastasis. *Sci. Transl. Med.* **2013**, *5*, 180ra48. [CrossRef]
21. Papadaki, M.A.; Kallergi, G.; Zafeiriou, Z.; Manouras, L.; Theodoropoulos, P.A.; Mavroudis, D.; Georgoulias, V.; Agelaki, S. Co-expression of putative stemness and epithelial-to-mesenchymal transition markers on single circulating tumour cells from patients with early and metastatic breast cancer. *BMC Cancer* **2014**, *14*, 651. [CrossRef] [PubMed]
22. Ning, Y.; Zhang, W.; Hanna, D.L.; Yang, D.; Okazaki, S.; Berger, M.D.; Miyamoto, Y.; Suenaga, M.; Schirripa, M.; El-Khoueiry, A.; et al. Clinical relevance of EMT and stem-like gene expression in circulating tumor cells of metastatic colorectal cancer patients. *Pharm. J.* **2018**, *18*, 29–34. [CrossRef] [PubMed]
23. Papadaki, M.A.; Stoupis, G.; Theodoropoulos, P.A.; Mavroudis, D.; Georgoulias, V.; Agelaki, S. Circulating Tumor Cells with Stemness and Epithelial-to-Mesenchymal Transition Features Are Chemoresistant and Predictive of Poor Outcome in Metastatic Breast Cancer. *Mol. Cancer* **2019**, *18*, 437–447. [CrossRef] [PubMed]
24. Stathopoulou, A.; Vlachonikolis, I.; Mavroudis, D.; Perraki, M.; Kouroussis, C.; Apostolaki, S.; Malamos, N.; Kakolyris, S.; Kotsakis, A.; Xenidis, N.; et al. Molecular Detection of Cytokeratin-19–Positive Cells in the Peripheral Blood of Patients With Operable Breast Cancer: Evaluation of Their Prognostic Significance. *J. Clin. Oncol.* **2002**, *20*, 3404–3412. [CrossRef] [PubMed]
25. Xenidis, N.; Markos, V.; Apostolaki, S.; Perraki, M.; Pallis, A.; Sfakiotaki, G.; Papadatos-Pastos, D.; Kalmanti, L.; Kafousi, M.; Stathopoulos, E.; et al. Clinical relevance of circulating CK-19 mRNA-positive cells detected during the adjuvant tamoxifen treatment in patients with early breast cancer. *Ann. Oncol.* **2007**, *18*, 1623–1631. [CrossRef] [PubMed]
26. Xenidis, N.; Ignatiadis, M.; Apostolaki, S.; Perraki, M.; Kalbakis, K.; Agelaki, S.; Stathopoulos, E.N.; Chlouverakis, G.; Lianidou, E.; Kakolyris, S.; et al. Cytokeratin-19 mRNA-Positive Circulating Tumor Cells After Adjuvant Chemotherapy in Patients With Early Breast Cancer. *J. Clin. Oncol.* **2009**, *27*, 2177–2184. [CrossRef]
27. Xenidis, N.; Perraki, M.; Apostolaki, S.; Agelaki, S.; Kalbakis, K.; Vardakis, N.; Kalykaki, A.; Xyrafas, A.; Kakolyris, S.; Mavroudis, D.; et al. Differential effect of adjuvant taxane-based and taxane-free chemotherapy regimens on the CK-19 mRNA-positive circulating tumour cells in patients with early breast cancer. *Br. J. Cancer* **2013**, *108*, 549–556. [CrossRef]
28. Bidard, F.-C.; Michiels, S.; Riethdorf, S.; Mueller, V.; Esserman, L.J.; Lucci, A.; Naume, B.; Horiguchi, J.; Gisbert-Criado, R.; Sleijfer, S.; et al. Circulating Tumor Cells in Breast Cancer Patients Treated by Neoadjuvant Chemotherapy: A Meta-analysis. *JNCI J. Natl. Cancer Inst.* **2018**, *110*, 560–567. [CrossRef]
29. Lucci, A.; Hall, C.S.; Lodhi, A.K.; Bhattacharyya, A.; Anderson, A.E.; Xiao, L.; Bedrosian, I.; Kuerer, H.M.; Krishnamurthy, S. Circulating tumour cells in non-metastatic breast cancer: A prospective study. *Lancet Oncol.* **2012**, *13*, 688–695. [CrossRef]
30. Georgoulias, V.; Bozionelou, V.; Agelaki, S.; Perraki, M.; Apostolaki, S.; Kallergi, G.; Kalbakis, K.; Xyrafas, A.; Mavroudis, D. Trastuzumab decreases the incidence of clinical relapses in patients with early breast cancer presenting chemotherapy-resistant CK-19mRNA-positive circulating tumor cells: Results of a randomized phase II study. *Ann. Oncol.* **2012**, *23*, 1744–1750. [CrossRef]

31. Ignatiadis, M.; Litière, S.; Rothe, F.; Riethdorf, S.; Proudhon, C.; Fehm, T.; Aalders, K.; Forstbauer, H.; Fasching, P.A.; Brain, E.; et al. Trastuzumab versus observation for HER2 nonamplified early breast cancer with circulating tumor cells (EORTC 90091-10093, BIG 1-12, Treat CTC): A randomized phase II trial. *Ann. Oncol.* **2018**, *29*, 1777–1783. [CrossRef] [PubMed]
32. Schneble, E.J.; Graham, L.J.; Shupe, M.P.; Flynt, F.L.; Banks, K.P.; Kirkpatrick, A.D.; Nissan, A.; Henry, L.; Stojadinovic, A.; Shumway, N.M.; et al. Current approaches and challenges in early detection of breast cancer recurrence. *J. Cancer* **2014**, *5*, 281–290. [CrossRef] [PubMed]
33. Strati, A.; Markou, A.; Parisi, C.; Politaki, E.; Mavroudis, D.; Georgoulias, V.; Lianidou, E. Gene expression profile of circulating tumor cells in breast cancer by RT-qPCR. *BMC Cancer* **2011**, *11*, 422. [CrossRef] [PubMed]
34. Mavroudis, D.; Saloustros, E.; Boukovinas, I.; Papakotoulas, P.; Kakolyris, S.; Ziras, N.; Christophylakis, C.; Kentepozidis, N.; Fountzilas, G.; Rigas, G.; et al. Sequential vs concurrent epirubicin and docetaxel as adjuvant chemotherapy for high-risk, node-negative, early breast cancer: An interim analysis of a randomised phase III study from the Hellenic Oncology Research Group. *Br. J. Cancer* **2017**, *117*, 164–170. [CrossRef] [PubMed]
35. Wittwer, C.T.; Herrmann, M.G.; Gundry, C.N.; Elenitoba-Johnson, K.S.J. Real-Time Multiplex PCR Assays. *Methods* **2001**, *25*, 430–442. [CrossRef]
36. Stathopoulou, A.; Ntoulia, M.; Perraki, M.; Apostolaki, S.; Mavroudis, D.; Malamos, N.; Georgoulias, V.; Lianidou, E.S. A highly specific real-time RT-PCR method for the quantitative determination of CK-19 mRNA positive cells in peripheral blood of patients with operable breast cancer. *Int. J. Cancer* **2006**, *119*, 1654–1659. [CrossRef] [PubMed]
37. Livak, K.J.; Schmittgen, T.D. Analysis of Relative Gene Expression Data Using Real-Time Quantitative PCR and the 2−ΔΔCT Method. *Methods* **2001**, *25*, 402–408. [CrossRef] [PubMed]
38. Ignatiadis, M.; Xenidis, N.; Perraki, M.; Apostolaki, S.; Politaki, E.; Kafousi, M.; Stathopoulos, E.N.; Stathopoulou, A.; Lianidou, E.; Chlouverakis, G.; et al. Different prognostic value of cytokeratin-19 mRNA positive circulating tumor cells according to estrogen receptor and HER2 status in early-stage breast cancer. *J. Clin. Oncol.* **2007**, *25*, 5194–5202. [CrossRef]
39. Aktas, B.; Müller, V.; Tewes, M.; Zeitz, J.; Kasimir-Bauer, S.; Loehberg, C.R.; Rack, B.; Schneeweiss, A.; Fehm, T. Comparison of estrogen and progesterone receptor status of circulating tumor cells and the primary tumor in metastatic breast cancer patients. *Gynecol. Oncol.* **2011**, *122*, 356–360. [CrossRef]
40. Bustin, S.A.; Benes, V.; Garson, J.A.; Hellemans, J.; Huggett, J.; Kubista, M.; Mueller, R.; Nolan, T.; Pfaffl, M.W.; Shipley, G.L.; et al. The MIQE guidelines: Minimum information for publication of quantitative real-time PCR experiments. *Clin. Chem.* **2009**, *55*, 611–622. [CrossRef]
41. Trapp, E.; Janni, W.; Schindlbeck, C.; Jückstock, J.; Andergassen, U.; de Gregorio, A.; Alunni-Fabbroni, M.; Tzschaschel, M.; Polasik, A.; Koch, J.G.; et al. Presence of Circulating Tumor Cells in High-Risk Early Breast Cancer During Follow-Up and Prognosis. *Jnci J. Natl. Cancer Inst.* **2019**, *111*, 380–387. [CrossRef] [PubMed]
42. Mirza, S.; Jain, N.; Rawal, R. Evidence for circulating cancer stem-like cells and epithelial-mesenchymal transition phenotype in the pleurospheres derived from lung adenocarcinoma using liquid biopsy. *Tumour Biol.* **2017**, *39*, 1010428317695915. [CrossRef] [PubMed]
43. Emprou, C.; Le Van Quyen, P.; Jégu, J.; Prim, N.; Weingertner, N.; Guérin, E.; Pencreach, E.; Legrain, M.; Voegeli, A.-C.; Leduc, C.; et al. SNAI2 and TWIST1 in lymph node progression in early stages of NSCLC patients. *Cancer Med.* **2018**, *7*, 3278–3291. [CrossRef] [PubMed]
44. Markiewicz, A.; Wełnicka-Jaśkiewicz, M.; Seroczyńska, B.; Skokowski, J.; Majewska, H.; Szade, J.; Żaczek, A.J. Epithelial-mesenchymal transition markers in lymph node metastases and primary breast tumors - relation to dissemination and proliferation. *Am. J. Transl. Res.* **2014**, *6*, 793–808. [PubMed]
45. Voutsadakis, I.A. Epithelial-Mesenchymal Transition (EMT) and Regulation of EMT Factors by Steroid Nuclear Receptors in Breast Cancer: A Review and in Silico Investigation. *J. Clin. Med.* **2016**, *5*, 11. [CrossRef] [PubMed]
46. Vesuna, F.; Lisok, A.; Kimble, B.; Domek, J.; Kato, Y.; van der Groep, P.; Artemov, D.; Kowalski, J.; Carraway, H.; van Diest, P.; et al. Twist contributes to hormone resistance in breast cancer by downregulating estrogen receptor-α. *Oncogene* **2012**, *31*, 3223–3234. [CrossRef]

47. Sarrió, D.; Rodriguez-Pinilla, S.M.; Hardisson, D.; Cano, A.; Moreno-Bueno, G.; Palacios, J. Epithelial-mesenchymal transition in breast cancer relates to the basal-like phenotype. *Cancer Res.* **2008**, *68*, 989–997. [CrossRef] [PubMed]
48. Choi, Y.; Lee, H.J.; Jang, M.H.; Gwak, J.M.; Lee, K.S.; Kim, E.J.; Kim, H.J.; Lee, H.E.; Park, S.Y. Epithelial-mesenchymal transition increases during the progression of in situ to invasive basal-like breast cancer. *Hum. Pathol.* **2013**, *44*, 2581–2589. [CrossRef]

© 2019 by the authors. Licensee MDPI, Basel, Switzerland. This article is an open access article distributed under the terms and conditions of the Creative Commons Attribution (CC BY) license (http://creativecommons.org/licenses/by/4.0/).

Article

Molecular and Kinetic Analyses of Circulating Tumor Cells as Predictive Markers of Treatment Response in Locally Advanced Rectal Cancer Patients

Bianca C. Troncarelli Flores [1,†], Virgilio Souza e Silva [2,†], Emne Ali Abdallah [1], Celso A.L. Mello [2], Maria Letícia Gobo Silva [3], Gustavo Gomes Mendes [4], Alexcia Camila Braun [1], Samuel Aguiar Junior [5,6] and Ludmilla Thomé Domingos Chinen [1,6,*]

1. International Research Center, A.C. Camargo Cancer Center, São Paulo 01508-010, Brazil
2. Department of Medical Oncology, A.C. Camargo Cancer Center, São Paulo 01509-900, Brazil
3. Department of Radiotherapy, A.C. Camargo Cancer Center, São Paulo 01509-900, Brazil
4. Department of Radiology, A.C. Camargo Cancer Center, São Paulo 01509-900, Brazil
5. Department of Pelvic Surgery, A.C. Camargo Cancer Center, São Paulo 01509-900, Brazil
6. National Institute for Science and Technology in Oncogenomics and Therapeutic Innovation, São Paulo 01509-900, Brazil
* Correspondence: ludmilla.chinen@accamargo.org.br
† These authors have contributed equally to the work.

Received: 10 May 2019; Accepted: 12 June 2019; Published: 26 June 2019

Abstract: Neoadjuvant chemoradiation (NCRT) followed by total mesorectal excision is the standard treatment for locally advanced rectal cancer (LARC). To justify a non-surgical approach, identification of pathologic complete response (pCR) is required. Analysis of circulating tumor cells (CTCs) can be used to evaluate pCR. We hypothesize that monitoring of thymidylate synthase (TYMS) and excision repair protein, RAD23 homolog B (RAD23B), can be used to predict resistance to chemotherapy/radiotherapy. Therefore, the aims of this study were to analyze CTCs from patients with LARC who underwent NCRT plus surgery for expression of TYMS/RAD23B and to evaluate their predictive value. Blood samples from 30 patients were collected prior to NCRT (S1) and prior to surgery (S2). CTCs were isolated and quantified by ISET®, proteins were analyzed by immunocytochemistry, and *TYMS* mRNA was detected by chromogenic in situ hybridization. CTC counts decreased between S1 and S2 in patients exhibiting pCR ($p = 0.02$) or partial response ($p = 0.01$). Regarding protein expression, TYMS was absent in 100% of CTCs from patients with pCR ($p = 0.001$) yet was expressed in 83% of non-responders at S2 ($p < 0.001$). Meanwhile, RAD23B was expressed in CTCs from 75% of non-responders at S1 ($p = 0.01$) and in 100% of non-responders at S2 ($p = 0.001$). Surprisingly, 100% of non-responders expressed *TYMS* mRNA at both timepoints ($p = 0.001$). In addition, TYMS/RAD23B was not detected in the CTCs of patients exhibiting pCR ($p = 0.001$). We found 83.3% of sensitivity for *TYMS* mRNA at S1 ($p = 0.001$) and 100% for TYMS ($p = 0.064$) and RAD23B ($p = 0.01$) protein expression at S2. Thus, *TYMS* mRNA and/or TYMS/RAD23B expression in CTCs, as well as CTC kinetics, have the potential to predict non-response to NCRT and avoid unnecessary radical surgery for LARC patients with pCR.

Keywords: locally advanced rectal cancer; circulating tumor cells; RAD23B; thymidylate synthase; chemoradioresistance

1. Introduction

Colorectal carcinoma is one of the most commonly occurring neoplasms in the Western world. Moreover, despite improvements in treatment, colorectal carcinoma remains an important cause of cancer-related deaths worldwide [1]. Rectal carcinoma, which accounts for approximately 30% of all primary colorectal cancers, is characterized by an anatomy and natural history that are distinct from other colonic tumors [2,3].

Neoadjuvant chemoradiation (NCRT) followed by total mesorectal excision is the standard treatment for locally advanced rectal carcinoma (LARC). NCRT is generally recommended for patients with cT3/T4 disease or lymph node involvement [4].

The absence of viable tumor cells in surgical specimens after NCRT is defined as pathological complete response (pCR). It occurs in 10–30% of patients [5] and it is associated with a better prognosis. Capecitabin and 5-fluorouracil (5-FU) are currently the most widely used radiosensitizers for NCRT. In some patients, chemoradiation induces pCR, although radical surgery is recommended for most patients [6,7].

Accurate identification of patients with pCR is necessary before pursuing a non-surgical approach. Many studies have evaluated the accuracy of the digital rectal exam and images in identifying patients with pCR [8,9]. However, the sensitivity and specificity of these methods are too low to accurately predict pCR. Molecular analyses of gene expression have also been performed and have been shown to be unsuitable for identifying patients with pCR who can be followed without radical surgery [10]. Recently, promising data have supported the use of a "watch and wait" strategy for patients who have no signs of a viable tumor in a digital rectal exam, rectoscopy (with or without biopsy), or imaging. The likelihood of tumor regrowth is minimal for these patients, and most could be cured by salvage surgery [11].

In addition, considering that the majority of patients will not respond to NCRT, it is extremely important to identify these patients, in an attempt to optimize the response to these patients. As there are no randomized phase-III studies to date, comparing this strategy with isolated induction chemoradiotherapy, it becomes fundamental to identify a biomarker capable of selecting patients for these approach [12–14].

For organ-preserving treatment strategies, the ability to identify tissues or blood biomarkers that predict NCRT response/non-response is very important. Circulating tumor cells (CTCs) are a real-time source of biomarkers, which have shown promise in facilitating the detection and monitoring of pCR and non-responder patients. It has been proposed that the identification and analysis of CTCs would facilitate investigations to understand intrinsic tumor features and characteristics of patients with pCR or non-responders so that individualized treatment strategies can be applied [15]. To date, use of CTC counts has been approved by the US Food and Drug Administration as a prognostic tool for metastatic prostate, colon, and breast cancers [16–18]. Additionally, CTC kinetics can be used to monitor tumor response to systemic treatment [19,20]. Meanwhile, molecular characterization of CTCs has facilitated studies of biomarkers, including proteins, gene expression, and chromosomal translocations [21,22].

Since only a small proportion of patients with LARC experience complete response following chemoradiation, an improved understanding of the molecular mechanisms underlying resistance to chemotherapy and radiation therapy resistance is essential. For chemotherapy, the main target of 5-FU is the enzyme thymidylate synthase (TYMS). TYMS expression analysis has been used to predict individual response to NCRT and has exhibited good prognostic value for rectal cancer recurrence [23,24]. Radiation therapy is an effective component of neoadjuvant treatment and it induces genetic damage. Consequently, the ultraviolet excision repair protein, RAD23 homolog B (RAD23B), which is part of the nucleotide excision repair process [25,26], would potentially be induced by the genetic damage introduced by radiation therapy. Recently, RAD23B was found to be associated with breast cancer relapse risk [27].

The aim of the present study was to explore the role of CTCs in patients undergoing NCRT followed by surgery for treatment of LARC. In addition, the predictive values of TYMS and RAD23B before and after NCRT were evaluated.

2. Methods

2.1. Patients and Treatment

This prospective study was conducted at the A. C. Camargo Cancer Center (São Paulo, Brazil) and was approved by the local ethics committee (2141/15C). Written informed consent was obtained from all patients prior to enrolment. Patients met the inclusion criteria for this study if they had a diagnosis of rectal cancer, as confirmed by biopsy pathology; had locally advanced disease staged as cT3–cT4 or N0–N+; and were candidates for chemoradiation therapy per institutional protocol. Patients were excluded if they had evidence of distant metastasis; a history of any surgery (e.g., colostomy) within two weeks prior to the initiation of treatment; or were taking anticoagulants at the time of the study. Cancer stage was determined with pelvic magnetic resonance imaging and chest and abdominal computed tomography.

Blood samples were collected at baseline or prior to NCRT (S1), and then prior to surgery (S2). Venous blood was collected from the antecubital vein and these samples were stored at room temperature for a maximum of 4 h prior to analysis.

Radiation therapy was applied with a three-dimensional (3D) conformal technique. A 45-Gy dose was applied in 25 fractions over the entire pelvis. In addition, a 5.4-Gy radiation boost was administered to the primary tumor and involved lymph nodes in three fractions, for a total of 50.4 Gy applied in 28 fractions. Chemotherapy regimens consisted of either intravenous 5-FU administered at a dose of 1000 mg/m^2 on days 1–5 during weeks 1 and 5 of radiation therapy; or oral capecitabin administered at a dose of 1650 mg/m^2/d during the entire radiation treatment period. Each patient's physician determined which regimen was appropriate.

The evaluation of the response was determined by the comparative analysis of the baseline images in S1 with the images before surgery (S2). In addition, we evaluated the pathologic response in comparison to the clinical staging established by baseline images.

2.2. Isolation and Quantification of CTCs and Protein Analysis of TYMS and RAD

ISET® (Isolation by size of epithelial tumors) was used to quantify and analyze CTCs. Briefly, peripheral blood samples were collected from patients into EDTA tubes (8.0 mL BD Vacutainer, Franklin Lakes, NJ, USA) and then were homogenized at room temperature for up to 4 h. Samples were then prepared as described previously [28]. Briefly, ISET membrane spots were cut out and subjected to immunocytochemistry (ICC) with an anti-RAD23B antibody (1:100 CSB-PA019260LA01HU; CusaBio, Wuhan, People's Republic of China), an anti-TYMS antibody (1:230 WH0007298M1; Sigma-Aldrich, St. Louis, MO, USA), and an anti-CD45 antibody (1:200 HPA000440; Sigma-Aldrich).

Selected spots from the ISET membranes were additionally subjected to a 24-well dual color ICC assay (Polink DS-RR-Hu/Ms A Kit; GBI Labs, Bothell, WA, USA). Briefly, antigen retrieval was performed by using a DakoTM Antigen Retrieval Solution (Dako, Santa Clara, CA, USA). Cells were then hydrated with 1X Tris-buffered saline (TBS) for 10 min and then permeabilized with Triton X-100 for 5 min. Next, cells were rinsed with 1X TBS and incubated with 3% hydrogen peroxide in the dark for 15 min to block endogenous peroxides. Immobilized cells on the membrane spots were incubated overnight with primary antibodies diluted in 10% fetal calf serum in TBS. To amplify primary antibody signals, the spots were incubated for 30 min with rabbit horseradish peroxidase (HRP) polymer (GBI Labs), then with 3,3-diaminobenzidine (DAB) for 10 min. After amplification, the spots were incubated with a second primary antibody for 2 h, a rabbit AP polymer (GBI Labs) for 30 min, and then GBI-Permanent Red (GBI Labs) for 10 min. The latter reagent was freshly prepared according to the manufacturer's instructions. After staining the cells with haematoxylin, specimens were examined with light microscopy (Research System Microscope BX61; Olympus, Waltham, MA, USA). CTCs were counted per 1 mL blood, as previously described by Krebs et al. [29]. CTCs were characterized according to five criteria: Negativity for CD45 staining; nucleus size >12 μm, hyperchromatic and irregular nuclei; visible cytoplasm; and a nuclear to cytoplasm ratio >80% [30]. CTCs were considered positive for TYMS or RAD23B expression if at least one cell in a specimen stained for these markers on ICC analysis.

2.3. CTC Isolation and Immunostaining Control

Negative controls were healthy donor blood filtered by ISET®. Positive controls included healthy donor blood spiked with HCT-8 colorectal carcinoma cells. For the ICC reaction and TYMS antibody control, leukocytes from healthy filtered blood were used. According to the Human Protein Atlas (http://www.proteinatlas.org/), the latter express TYMS. As a positive control for the RAD23B antibody, HCT-8 cells were spiked in healthy donor blood and filtered on ISET®. To create a negative control for ICC, the same cell line was used without primary antibodies in order to ensure exclusion of cross-reactivity. To confirm that analyzed CTCs were not leukocytes, staining with an anti-CD45 antibody was performed.

2.4. Chromogenic In situ Hybridization Assay for TYMS

TYMS mRNA was detected in intact cells by using a chromogenic in situ hybridization (CISH) assay employing RNAscope Technology (ACDbio, Newark, CA, USA), according to the manufacturer's protocol. Cytology methods standardized by our group were also used. Briefly, one spot was cut out for each patient sample and these were placed in individual wells of a 24-well plate. The membrane spots were hydrated with 1X TBS for 5 min before being incubated in 1% formaldehyde solution for 5 min at room temperature. The spots were then rinsed 2× with distilled water before applying 5–8 drops of RNAscope Hydrogen Peroxide (ACDbio) to each spot. After incubating the samples in a humid chamber for 10 min at room temperature, the spots were washed 2× with distilled water and mounted on slides. To each slide, 3 drops of cytology pepsin were applied. After 10 min at room temperature the spots were returned to the 24-well plate, washed 2× with distilled water, then rapidly dehydrated in successive 1-min incubations with 70%, 85%, and 100% ethanol solutions. After the spots were left to air dry on the slides, 3 drops of a TYMS-specific probe were added to each spot. After 2 h at 40 °C in a HybEZ oven hybridizer (ACDbio), drops of Amp1-6 solutions were added, according to the manufacturer's protocol, with washes performed with 1X TBS. Finally, the cells were incubated with chromogen and 50% haematoxylin then placed on slides with aqueous mounting media and coverslips. The samples were inspected with brightfield microscopy (BX61-Olympus; Olympus). To classify TYMS mRNA expression, an absence or presence of staining was classified as negative or positive, respectively.

2.5. Statistical Analysis

Initially, a descriptive analysis was performed to obtain absolute (n) and relative (%) frequency distributions. To evaluate possible associations between the variables of interest, a contingency table was constructed from sample data. Chi-square tests of independence or Fisher's exact test were used, as appropriate. The level of significance was set at 5% and all statistical analyses were performed by using the SPSS program for Windows, version 25.

3. Results

3.1. Patients

Thirty patients with rectal cancer were enrolled in this study between April 2016 and January 2018. Clinical characteristics of these patients are listed in Table 1. The mean age of this cohort was 56 years (range, 34–72) and 60% of the patients were male. For 67% of the patients, their rectal tumors were located 7 cm or less from the anal verge.

The baseline (prior to NCRT) T stage was cT2 for 4 patients (13%), cT3 for 21 patients (70%), and cT4 for 5 patients (17%). Baseline N stage was cN0 for 22 patients (73%) and cN+ for 8 patients (27%). The mean time to surgery completion following NCRT was 77 days. Pathologic T stage was pT0 for 6 patients (20%), pT1 for 5 patients (17%), pT2 for 7 patients (23%), pT3 for 9 patients (30%), and pT4 for 2 patients (7%). Pathologic N stage was pN0 for 21 patients (70%), pN1 for 6 patients (20%), and pN2 for 2 patients (7%). One patient (3%) exhibited disease progression during NCRT. Complete pathologic response after preoperative therapy was detected in 6 patients (20%), while 7 patients (23%) had their tumors down-staged (ypT1-2N0).

Table 1. Patient characteristics.

Characteristics	N (%)
Average age (min–max), years	56 (34–72)
Gender	
Male	18 (60)
Female	12 (40)
Tumor distance from the anal verge	
≤7 cm	20 (67)
>7 cm	10 (33)
Clinical T baseline stage	
T2	04 (13)
T3	21 (70)
T4	05 (17)
Clinical N baseline stage	
N0	22 (73)
N+	08 (27)
Pathological T stage	
T0	06 (20)
T1–T2	12 (40)
T3–T4	11 (37)
DP	1 (3)
Pathological N stage	
N0	21 (70)
N1–N2	08 (27)
DP	01 (3)
Average time (min–max) of completion of RDT for surgery (days)	77 (50–143)

Abbreviations: DP: Disease Progression; RDT: radiotherapy.

All patients were treated with 25 fractions of 45 Gy to the pelvis with a 3D conformal technique. In addition, a 5.4-Gy radiation boost was applied to the primary tumor. There were no treatment delays or interruptions that lasted more than two days. Twenty-six patients (86.7%) received intravenous 5-FU and four patients (13.3%) received capecitabine. The main treatment toxicities were grade 1 or 2 diarrhea in 10 patients (33%) and grade 1 or 2 oral mucositis in 8 patients (27%). No grade 3 or 4 toxicities were reported. There were also no differences in adverse events between the two chemotherapy groups.

3.2. CTCs

CTCs were detected at S1 in all 30 patients of our cohort, and at S2 in 24 patients (Figure 1A,D). The mean CTC concentrations were 6 cells/mL at S1 and 3.5 cells/mL at S2. Kinetic analyses showed that CTC levels were increased in 3 patients, decreased in 22 patients, and remained unchanged in 5 patients. The mean CTC count per mL was 3.1 for those with pCR, 2.5 for those exhibiting a partial response (PR), and 2.9 for those with non-responsive tumors. Patients exhibiting pCR and PR showed a decrease in CTC kinetics (calculated as: CTC baseline [CTC1] × CTC post-CRT [CTC2]) during treatment ($P = 0.02$ and $P = 0.01$, respectively; Table 2).

Table 2. Kinetic counts of circulating tumor cells (CTCs) between baseline (CTC1) and post-neoadjuvant chemoradiation (NCRT) (CTC2) time points.

	Patient ID	CTCs/mL before NCRT	CTCs/mL after NCRT	Kinetics of CTC1 vs. CTC2	
CR	8	3	1	>	
	11	4	1	>	
	18	1	0	>	$P = 0.02$
	21	4	0	>	
	25	4	2	>	
	27	3	2	>	
PR	3	5	5	=	
	4	3	2	>	
	7	3	2	>	
	9	1	0	>	
	10	0	1	<	
	13	1	1	=	$P = 0.01$
	15	2	0	>	
	16	2	0	>	
	23	2	1	>	
	24	6	3	>	
	29	3	0	>	
	30	2	2	=	
NR	1	3	4	<	
	2	3	2	>	
	5	1	1	=	
	6	1	1	=	
	12	1	4	<	
	14	2	1	>	$P = 0.07$
	17	3	1	>	
	19	4	2	>	
	20	5	1	>	
	22	8	4	>	
	26	2	1	>	
	28	2	1	>	

Abbreviations: NCRT: Neoadjuvant chemoradiotherapy; CR: Complete response; PR: Partial response; NR: No response.

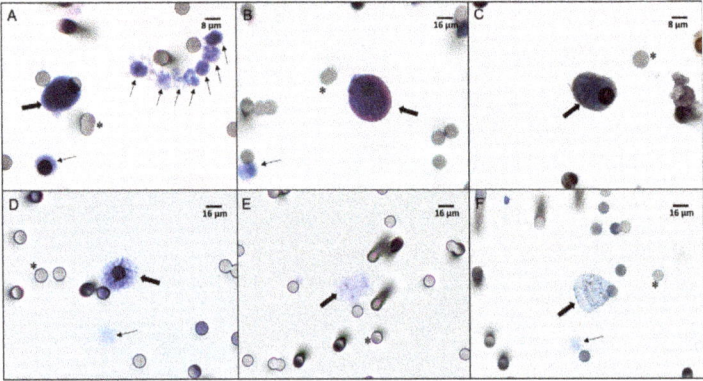

Figure 1. Immunostaining (**A–C**) and chromogenic in situ hybridization (CISH) (**D–F**) of CTCs from locally advanced rectal cancer (LARC) patients. (**A**) CTCs and leukocytes visualized with haematoxylin-eosin staining (×40 magnification). (**B**) CTCs stained with an anti-thymidylate synthase (TYMS) antibody, visualized with Permanent Red, and counterstained with haematoxylin (×40 magnification). (**C**) CTCs stained with an anti-RAD23B antibody, visualized with 3,3-diaminobenzidine (DAB), and counterstained with haematoxylin (×60 magnification). (**D**) CTCs negative for *TYMS* mRNA and counterstained with haematoxylin (×40 magnification). (**E**) CTCs with a low *TYMS* mRNA signal (normal expression) counterstained with haematoxylin (×40 magnification). (**F**) CTCs with a high *TYMS* mRNA signal (overexpression) counterstained with haematoxylin (×40 magnification). All images were analyzed on a Research System Microscope BX61 (Olympus, Tokyo, Japan) coupled to a digital camera (SC100–Olympus). Thick arrows indicate CTCs, thin arrows indicate leukocytes, and asterisks indicate 8 μm pores of the ISET® membranes.

3.3. TYMS and RAD23B

At baseline, TYMS-positive CTCs were detected in 7 patients (23.5%) by ICC (Figure 1B) and in 21 patients (70%) by CISH (Figure 1E,F). RAD23B-positive CTCs were detected in 13 patients (43.3%) by ICC (Figure 1C).

After chemoradiation, TYMS-positive CTCs were detected in 10 patients (41.6%) by ICC, while *TYMS* mRNA was detected in 16 patients (61.5%) by CISH. RAD23B-positive CTCs were detected in 14 patients (58.3%).

Baseline *TYMS* mRNA and RAD23B-positive CTCs were associated with poor clinical response. For example, all 12 patients with non-responsive tumors had *TYMS* mRNA detected in their CTCs by CISH. In contrast, 83.5% of patients with pCR did not have CTCs expressing *TYMS* mRNA ($P = 0.001$; Table 3). At S1, RAD23B-positive CTCs were detected in 33% of patients with pCR, in 16% of patients with PR, and in 75% of patients exhibiting no response ($P = 0.01$). Thus, there was no association between detection of TYMS expression in CTCs by ICC and response type in S1 (Table 3). To confirm the value of these proteins as biomarkers, we compared these found with the most used clinical parameter, which is pathological response of the primary tumor. We found, for TYMS mRNA, 83.3% sensitivity, 83.3% specificity, and 95.2% positive predictive value (PPV) ($p = 0.001$) (supplementary table).

Table 3. Expression profiles of RAD23B and TYMS proteins and *TYMS* mRNA.

			CR	PR	NR	
Before NCRT	CISH (TYMS)	+	16.5	66.5	100	$P = 0.001$
		−	83.5	33.5	0	
	TYMS (protein)	+	16.5	25	25	$P = 1.0$
		−	83.5	75	75	
	RAD (protein)	+	33.5	16.5	75	$P = 0.01$
		−	66.5	83.5	25	
After NCRT	CISH (TYMS)	+	25	30	100	$P = 0.001$
		−	75	70	0	
	TYMS (protein)	+	0	0	83.5	$P = 0.0001$
		−	100	100	16.5	
	RAD (protein)	+	0	25	100	$P = 0.0001$
		−	100	75	0	

Abbreviations: NCRT: Neoadjuvant chemoradiotherapy; CR: Complete response; PR: Partial response; NR: No response.

Post-chemoradiation analyses (S2) showed that expression of TYMS and RAD23B in CTCs strongly correlates with poor response. For example, TYMS-positive CTCs were not detected in the patients with pCR or pPR, yet they were detected in 83% of non-responsive patients ($P < 0.001$; Table 3). Furthermore, *TYMS* mRNA expression at S2 correlated with response, with *TYMS* mRNA detected in all of the patients exhibiting no response ($P = 0.001$). We found, for TYMS protein, 100% sensitivity, 50% specificity, and 100% positive predictive value (PPV) ($p = 0.064$). For RAD23B, the values were: 100% sensitivity, 70% specificity, and 100% positive predictive value (PPV) ($p = 0.01$) (supplementary table).

Lastly, we found that expression of TYMS and RAD23B in CTCs was strongly predictive of response type. For example, after NCRT, TYMS$^-$/RAD23B$^-$ CTCs were detected in 100% of the pCR patients, in 83.5% of the PR patients, and in none of the NR patients. However, TYMS$^+$/RAD23B$^+$ CTCs were not detected in patients with pCR or PR, and yet were detected in 83.5% of patients with no response ($P = 0.001$). At S1, CTCs expressing TYMS and RAD23B did not correlate with response type ($P = 0.1$; Table 4). Meanwhile, expression of *TYMS* mRNA determined which patients did not respond to chemoradiotherapy at S1 and S2 (Figure 2).

Table 4. Correlation between RAD and TYMS protein expression profiles.

Profile	CR %	PR %	NR %	
TYMS−/RAD−	50	66.5	16.5	Before
TYMS+/RAD+	0	8.5	16.5	NCRT
TYMS+/RAD− TYMS−/RAD+	50	25	67	$P = 0.1$
TYMS−/RAD−	100	83.5	0	After
TYMS+/RAD+	0	0	83.5	NCRT
TYMS+/RAD− TYMS−/RAD+	0	16.5	16.5	$P = 0.001$

Abbreviations: NCRT: Neoadjuvant chemoradiotherapy; CR: Complete response; PR: Partial response; NR: No response.

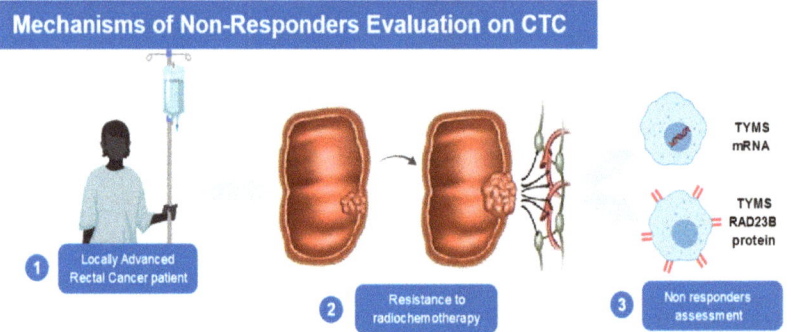

Figure 2. Scheme demonstrating both methodologies used to select responders and non-responders to neoadjuvant chemoradiotherapy: mRNA detected by CISH and protein expression detected by immunocytochemistry (ICC).

4. Discussion

LARC treatment generally consists of NCRT followed by total mesorectal excision. For high-risk patients, postoperative adjuvant chemotherapy may additionally be considered [31]. The ability to identify patients who have undergone complete eradication of a tumor following NCRT is crucial. In the current study, we prospectively demonstrated a strong correlation between expression of TYMS and RAD23B by CTCs in patients with LARC and pCR following NCRT.

Among the 50–60% of patients with LARC who respond to NCRT (i.e., their tumors are down-staged following treatment), many exhibit improved survival. It has been reported that pCR after NCRT is associated with improved cancer outcome and significantly decreased rates of local recurrence. Conversely, the 40% non-response rate after NCRT [31] represents heterogeneity in response to standard treatment [32,33]. For the latter, new therapeutic strategies, avoidance of toxicity associated with ineffective treatment, and novel neoadjuvant chemotherapeutic options are needed [34,35]. In addition, the identification of biomarkers would facilitate the creation of individualized treatment plans.

The ability to identify tumors that do not respond to radiotherapy is useful for helping patients avoid radiation side effects such as fibrosis, fecal and bladder incontinence, diarrhea, dysuria, and myelosuppression [36,37]. This knowledge would also facilitate discussions regarding alternate approaches, such as more potent neoadjuvant multi-agent chemotherapy strategies or a rationale for foregoing radiation therapy. There are several studies that have discussed these considerations, although robust biomarkers that will predict non-responders of NCRT with high accuracy still need to be identified [37–39]. In the present study, we were able to identify NCRT non-responders at S1 by detecting *TYMS* mRNA. Thus, it is possible that detection of *TYMS* mRNA in CTCs could represent a valuable tool in identifying non-responder patients prior to the start of NCRT. Furthermore, detection of

RAD23B expression could make this patient selection process more accurate. It is well-established that RAD23B is involved in DNA repair following radiation damage, and its identification in CTCs can guide treatment plans. In the present study, 75% of non-responding tumors expressed this protein on CTCs before NCRT. Furthermore, when both RAD23B and TYMS protein expression on CTCs were detected after NCRT, 100% of patients with pCR did not express either protein. Meanwhile, 83.5% of patients with non-responsive tumors expressed both proteins and the remaining 16.5% expressed at least one of these proteins. In addition, *TYMS* mRNA expression after NCRT showed high positivity for non-responders (100%) and was not related to protein expression (Figure 3). For the seven patients who presented discordant mRNA/protein positivity at S1 (Supplementary Table S1), they exhibited a correspondence between *TYMS* gene expression and TYMS protein expression at S2. Chemotherapeutic agents have previously been associated with changes in gene expression. In addition, post-transcriptional mechanisms for blocking protein synthesis have been characterized [40–42]. For 5-FU, there are several papers that describe this correlation [43,44]. The present results support the hypothesis that CTC analysis can be a useful tool for identifying patients who will respond to chemoradiotherapy. As a result, a "watch and wait" strategy becomes an option to be considered in addition to radical surgery.

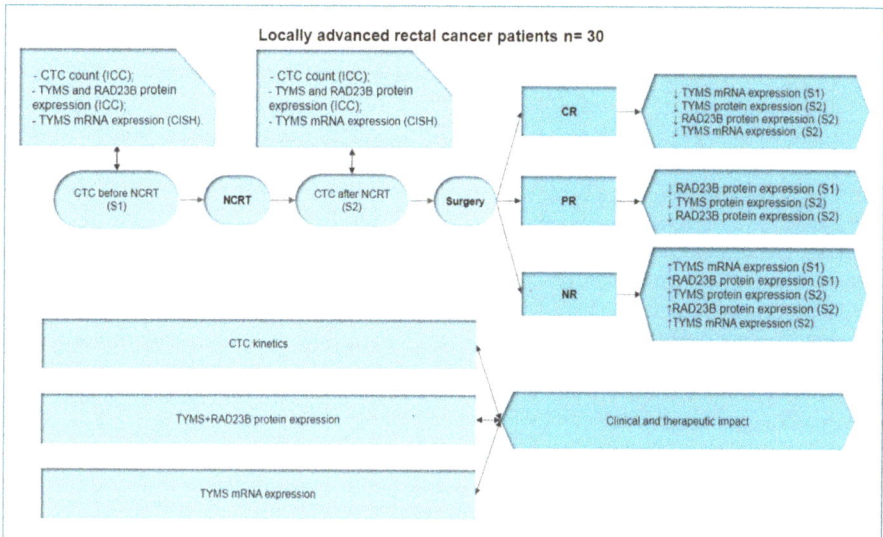

Figure 3. Summary of methodologies, analyses, and results in the present study. Patients were enrolled prior to the start of NCRT. Blood was collected to perform CTC counts and molecular analyses. At baseline, complete response (CR) correlated with low levels of *TYMS* mRNA. In contrast, NR correlated with high levels of *TYMS* mRNA and RAD23B protein expression. Blood samples were collected again during follow-up after NCRT. CTC analyses showed that CR correlated to low levels of *TYMS* mRNA and RAD23B protein expression, while NR correlated to high levels of *TYMS* mRNA and RAD23B/TYMS protein expression. CTC kinetics also correlated to pathological response. Based on these results, *TYMS* mRNA and RAD23B and TYMS protein expression appear to have a clinical and therapeutic impact in LARC patients. Abbreviations: CR: Complete response; LARC: Locally advanced rectal cancer; NCRT: Neoadjuvant chemoradiotherapy; NR: No response; PR: Partial response.

Our analysis of RAD23B and TYMS expression showed similar profiles for these two proteins. Based on the different protein patterns for TYMS in relation to pCR and PR at S1 and S2, we decided to further examine the mRNA expression of *TYMS* with RNA hybridization assays performed in situ with CTCs. All of the CTCs from non-responding patients expressed *TYMS* mRNA at both S1 and

S2. Thus, this assay exhibited high sensitivity and specificity for identifying LARC patients who are predicted to be non-responders to radiochemotherapy.

After NCRT, TYMS expression exhibited a strong correlation with *TYMS* mRNA (kappa value 0.6; $P = 0.003$) in absence of response to treatment. This second assay at S2 cost less to perform besides the possibility of performing more than one protein per time on the CTCs isolated by ISET® methodology. Furthermore, we previously showed that elevated TYMS expression in CTCs was associated with poor prognosis among patients with metastatic colorectal cancer [26].

TYMS, a downstream target molecule of 5-FU, plays a key role in DNA synthesis. The enzyme catalyzes deoxyuridine monophosphate methylation to produce deoxythymidine triphosphate, and subsequently, thymidylate. Increased TYMS expression is thought to be responsible for 5-FU resistance in patients with colorectal cancer [42,43]. Meanwhile, RAD23B is a member of the nucleotide excision repair system, which stabilizes the xeroderma pigmentosum complementation group C (XPC) protein and potentiates its interaction with damaged DNA [45]. XPC subsequently initiates nucleotide excision repair. It is not clear why increased production of RAD23B protein was observed in CTCs from non-responders to NCRT in the present study. A possible hypothesis is that increased levels of this repair protein during treatment make it difficult to eliminate cancer cells by NCRT, a strategy that is based on inducing apoptosis as a result of DNA damage.

A treatment approach that deserves discussion in the context of our study is the "watch and wait" strategy. For the patients whose tumors respond completely to NCRT (as evidenced by radiologic, clinical, and endoscopic evaluations), a "watch and wait" strategy may allow patients to avoid the morbidity, mortality, and functional consequences of radical surgical treatment. A recently published meta-analysis of 880 patients who were monitored with a "watch and wait" strategy showed that the strategy can be safely incorporated into a multidisciplinary management plan for the treatment of patients with rectal cancer who achieve pCR after neoadjuvant treatment [12,46–48]. Therefore, identification of a biomarker that can predict pCR to NCRT treatment prior to treatment, or in the early stages of treatment, would be of great clinical utility in combination with commonly used clinical parameters.

In the present study, it was observed that 100% of patients who responded to NCRT did not express TYMS or RAD23B proteins on their CTCs at the beginning of their follow-up monitoring. This lack of protein expression in CTCs is consistent with imaging and kinetic studies that have showed a reduction or elimination of CTCs post NCRT. Furthermore, we observed that CTC kinetics correlated with disease outcome for our patients with LARC. To the best of our knowledge, this is the first study to demonstrate this result. The observed decrease in CTC levels during treatment in our cases involving complete or partial responses further support the use of CTC analyses to predict response to NCRT.

In summary, our results provide valuable data regarding two potential biomarkers of chemoradiation resistance in patients undergoing neoadjuvant treatment for LARC. Despite the small sample size of our study, CTC kinetics, as well as *TYMS* mRNA and/or RAD23B/TYMS protein expression in CTCs, were found to strongly correlate with pCR. Further studies are needed to validate these findings with a larger patient cohort. If CTC analysis proves useful in predicting pCR with high accuracy, many patients may be spared radical surgery for rectal cancer treatment. In addition, biomarker and kinetic analyses of CTCs may identify potential non-responders to treatment with NCRT, thereby identifying a need to evaluate other forms of therapy for these patients.

Supplementary Materials: The following are available online at http://www.mdpi.com/2073-4409/8/7/641/s1, Table S1: table of sensitivity, specificity and predictive values of biomarkers evaluated.

Author Contributions: B.C.T.F.: conception/design, data analysis and interpretation, collection and/or assembly of data, manuscript writing; V.S.e.S.: conception/design, provision of study material or patients, collection and/or assembly of data, manuscript writing; E.A.A.: data analysis and interpretation; C.A.L.M.: provision of study material or patients; M.L.G.S.: data analysis and interpretation; G.G.M.: data analysis and interpretation; A.C.B.: data analysis and interpretation; S.A.J.: conception/design; L.T.D.C.: conception/design, data analysis and interpretation, manuscript writing, final approval of manuscript.

Funding: Funding was received from the Public Ministry of Brazil (TAC-MP- PAJ no. 000968.2012.10.000/0) and the National Institute for Science and Technology in Oncogenomics and Therapeutic Innovation (INCT FAPESP/CNPq 2014/50943-1).

Acknowledgments: We thank the Public Ministry (TAC-MP- PAJ no. 000968.2012.10.000/0), the National Institute for Science and Technology in Oncogenomics and Therapeutic Innovation (INCT FAPESP/CNPq 2014/50943-1), and the São Paulo Research Foundation (2016/18786-9) for financial support for this study.

Conflicts of Interest: The authors declare no potential conflicts of interest.

References

1. Siegel, R.L.; Miller, K.D.; Jemal, A. Cancer statistics, 2018. *CA Cancer J. Clin.* **2018**, *68*, 7–30. [CrossRef] [PubMed]
2. Cravo, M.; Rodrigues, T.; Ouro, S.; Ferreira, A.; Féria, L.; Maio, R. Management of rectal cancer: Times they are changing. *GE Port. J. Gastroenterol.* **2014**, *21*, 192–200. [CrossRef]
3. Gaertner, W.B.; Kwaan, M.R.; Madoff, R.D.; Melton, G.B. Rectal cancer: An evidence-based update for primary care providers. *World J. Gastroenterol.* **2015**, *21*, 7659–7671. [CrossRef] [PubMed]
4. Li, Y.; Wang, J.; Ma, X.; Tan, L.; Yan, Y.; Xue, C.; Hui, B.; Liu, R.; Ma, H.; Ren, J. A Review of Neoadjuvant Chemoradiotherapy for Locally Advanced Rectal Cancer. *Int. J. Biol. Sci.* **2016**, *12*, 1022–1031. [CrossRef] [PubMed]
5. Ryan, J.E.; Warrier, S.K.; Lynch, A.C.; Ramsay, R.G.; Phillips, W.A.; Heriot, A.G. Predicting pathological complete response to neoadjuvant chemoradiotherapy in locally advanced rectal cancer: a systematic review. *Colorectal Dis.* **2016**, *18*, 234–246. [CrossRef] [PubMed]
6. Zou, X.-C.; Wang, Q.-W.; Zhang, J.-M. Comparison of 5-FU-based and Capecitabine-based Neoadjuvant Chemoradiotherapy in Patients With Rectal Cancer: A Meta-analysis. *Clin. Colorectal Cancer* **2017**, *16*, e123–e139. [CrossRef] [PubMed]
7. Schmoll, H.-J.; Haustermans, K.; Price, T.J.; Nordlinger, B.; Hofheinz, R.; Daisne, J.-F.; Janssens, J.; Brenner, B.; Schmidt, P.; Reinel, H.; et al. Preoperative chemoradiotherapy and postoperative chemotherapy with capecitabine +/- oxaliplatin in locally advanced rectal cancer: Final results of PETACC-6. *JCO* **2018**, *36*, 3500. [CrossRef]
8. Pozo, M.E.; Fang, S.H. Watch and wait approach to rectal cancer: A review. *World J. Gastrointest. Surg.* **2015**, *7*, 306–312. [CrossRef]
9. Vojtíšek, R.; Korčáková, E.; Mařan, J.; Šorejs, O.; Fínek, J. Neoadjuvant chemoradiotherapy of the rectal carcinoma - The correlation between the findings on the restaging multiparametric 3T MRI scanning and the surgical findings. *Rep. Pract. Oncol. Radiother.* **2017**, *22*, 265–276. [CrossRef]
10. Molinari, C.; Matteucci, F.; Caroli, P.; Passardi, A. Biomarkers and Molecular Imaging as Predictors of Response to Neoadjuvant Chemoradiotherapy in Patients With Locally Advanced Rectal Cancer. *Clin. Colorectal Cancer* **2015**, *14*, 227–238. [CrossRef]
11. van der Valk, M.J.M.; Hilling, D.E.; Bastiaannet, E.; Meershoek-Klein Kranenbarg, E.; Beets, G.L.; Figueiredo, N.L.; Habr-Gama, A.; Perez, R.O.; Renehan, A.G.; van de Velde, C.J.H.; et al. Long-term outcomes of clinical complete responders after neoadjuvant treatment for rectal cancer in the International Watch & Wait Database (IWWD): an international multicentre registry study. *Lancet* **2018**, *391*, 2537–2545. [PubMed]
12. Fernandez-Martos, C.; Garcia-Albeniz, X.; Pericay, C.; Maurel, J.; Aparicio, J.; Montagut, C.; Safont, M.J.; Salud, A.; Vera, R.; Massuti, B.; et al. Chemoradiation, surgery and adjuvant chemotherapy versus induction chemotherapy followed by chemoradiation and surgery: long-term results of the Spanish GCR-3 phase II randomized trial†. *Ann. Oncol.* **2015**, *26*, 1722–1728. [CrossRef] [PubMed]
13. Cercek, A.; Roxburgh, C.S.D.; Strombom, P.; Smith, J.J.; Temple, L.K.F.; Nash, G.M.; Guillem, J.G.; Paty, P.B.; Yaeger, R.; Stadler, Z.K.; et al. Adoption of Total Neoadjuvant Therapy for Locally Advanced Rectal Cancer. *JAMA Oncol.* **2018**, *4*, e180071. [CrossRef] [PubMed]
14. Yamashita, K.; Matsuda, T.; Hasegawa, H.; Mukohyama, J.; Arimoto, A.; Tanaka, T.; Yamamoto, M.; Matsuda, Y.; Kanaji, S.; Nakamura, T.; et al. Recent advances of neoadjuvant chemoradiotherapy in rectal cancer: Future treatment perspectives. *Ann. Gastroenterol. Surg.* **2019**, *3*, 24–33. [CrossRef] [PubMed]

15. Sun, W.; Li, G.; Wan, J.; Zhu, J.; Shen, W.; Zhang, Z. Circulating tumor cells: A promising marker of predicting tumor response in rectal cancer patients receiving neoadjuvant chemo-radiation therapy. *Oncotarget* **2016**, *7*, 69507–69517. [CrossRef] [PubMed]
16. Cristofanilli, M.; Hayes, D.F.; Budd, G.T.; Ellis, M.J.; Stopeck, A.; Reuben, J.M.; Doyle, G.V.; Matera, J.; Allard, W.J.; Miller, M.C.; et al. Circulating tumor cells: a novel prognostic factor for newly diagnosed metastatic breast cancer. *J. Clin. Oncol.* **2005**, *23*, 1420–1430. [CrossRef]
17. Cohen, S.J.; Punt, C.J.A.; Iannotti, N.; Saidman, B.H.; Sabbath, K.D.; Gabrail, N.Y.; Picus, J.; Morse, M.; Mitchell, E.; Miller, M.C.; et al. Relationship of circulating tumor cells to tumor response, progression-free survival, and overall survival in patients with metastatic colorectal cancer. *J. Clin. Oncol.* **2008**, *26*, 3213–3221. [CrossRef]
18. León-Mateos, L.; Vieito, M.; Anido, U.; López López, R.; Muinelo Romay, L. Clinical Application of Circulating Tumour Cells in Prostate Cancer: From Bench to Bedside and Back. *Int. J. Mol. Sci.* **2016**, *17*, 1580. [CrossRef]
19. Souza E Silva, V.; Chinen, L.T.D.; Abdallah, E.A.; Damascena, A.; Paludo, J.; Chojniak, R.; Dettino, A.L.A.; de Mello, C.A.L.; Alves, V.S.; Fanelli, M.F. Early detection of poor outcome in patients with metastatic colorectal cancer: tumor kinetics evaluated by circulating tumor cells. *Onco. Targets Ther.* **2016**, *9*, 7503–7513. [CrossRef]
20. Yan, W.-T.; Cui, X.; Chen, Q.; Li, Y.-F.; Cui, Y.-H.; Wang, Y.; Jiang, J. Circulating tumor cell status monitors the treatment responses in breast cancer patients: a meta-analysis. *Sci Rep* **2017**, *7*, 43464. [CrossRef]
21. Benini, S.; Gamberi, G.; Cocchi, S.; Garbetta, J.; Alberti, L.; Righi, A.; Gambarotti, M.; Picci, P.; Ferrari, S. Detection of circulating tumor cells in liquid biopsy from Ewing sarcoma patients. *Cancer Manag. Res.* **2018**, *10*, 49–60. [CrossRef] [PubMed]
22. Pantel, K.; Alix-Panabières, C. Liquid biopsy and minimal residual disease–latest advances and implications for cure. *Nat. Rev. Clin. Oncol.* **2019**. [CrossRef]
23. Kuremsky, J.G.; Tepper, J.E.; McLeod, H.L. Biomarkers for response to neoadjuvant chemoradiation for rectal cancer. *Int. J. Radiat. Oncol. Biol. Phys.* **2009**, *74*, 673–688. [CrossRef] [PubMed]
24. Conradi, L.-C.; Bleckmann, A.; Schirmer, M.; Sprenger, T.; Jo, P.; Homayounfar, K.; Wolff, H.A.; Rothe, H.; Middel, P.; Becker, H.; et al. Thymidylate synthase as a prognostic biomarker for locally advanced rectal cancer after multimodal treatment. *Ann. Surg. Oncol.* **2011**, *18*, 2442–2452. [CrossRef] [PubMed]
25. Watkins, J.F.; Sung, P.; Prakash, L.; Prakash, S. The Saccharomyces cerevisiae DNA repair gene RAD23 encodes a nuclear protein containing a ubiquitin-like domain required for biological function. *Mol. Cell. Biol.* **1993**, *13*, 7757–7765. [CrossRef] [PubMed]
26. Schauber, C.; Chen, L.; Tongaonkar, P.; Vega, I.; Lambertson, D.; Potts, W.; Madura, K. Rad23 links DNA repair to the ubiquitin/proteasome pathway. *Nature* **1998**, *391*, 715–718. [CrossRef] [PubMed]
27. Pérez-Mayoral, J.; Pacheco-Torres, A.L.; Morales, L.; Acosta-Rodríguez, H.; Matta, J.L.; Dutil, J. Genetic polymorphisms in RAD23B and XPC modulate DNA repair capacity and breast cancer risk in Puerto Rican women. *Mol. Carcinog.* **2013**, *52*, E127–E138. [CrossRef]
28. Abdallah, E.A.; Fanelli, M.F.; Buim, M.E.C.; Machado Netto, M.C.; Gasparini Junior, J.L.; Souza E Silva, V.; Dettino, A.L.A.; Mingues, N.B.; Romero, J.V.; Ocea, L.M.M.; et al. Thymidylate synthase expression in circulating tumor cells: a new tool to predict 5-fluorouracil resistance in metastatic colorectal cancer patients. *Int. J. Cancer* **2015**, *137*, 1397–1405. [CrossRef]
29. Krebs, M.G.; Hou, J.-M.; Sloane, R.; Lancashire, L.; Priest, L.; Nonaka, D.; Ward, T.H.; Backen, A.; Clack, G.; Hughes, A.; et al. Analysis of circulating tumor cells in patients with non-small cell lung cancer using epithelial marker-dependent and -independent approaches. *J. Thorac. Oncol.* **2012**, *7*, 306–315. [CrossRef]
30. Khoja, L.; Backen, A.; Sloane, R.; Menasce, L.; Ryder, D.; Krebs, M.; Board, R.; Clack, G.; Hughes, A.; Blackhall, F.; et al. A pilot study to explore circulating tumour cells in pancreatic cancer as a novel biomarker. *Br. J. Cancer* **2012**, *106*, 508–516. [CrossRef]
31. Sung, S.; Son, S.H.; Park, E.Y.; Kay, C.S. Prognosis of locally advanced rectal cancer can be predicted more accurately using pre- and post-chemoradiotherapy neutrophil-lymphocyte ratios in patients who received preoperative chemoradiotherapy. *PLoS ONE* **2017**, *12*, e0173955. [CrossRef] [PubMed]
32. Powell, A.A.; Talasaz, A.H.; Zhang, H.; Coram, M.A.; Reddy, A.; Deng, G.; Telli, M.L.; Advani, R.H.; Carlson, R.W.; Mollick, J.A.; et al. Single Cell Profiling of Circulating Tumor Cells: Transcriptional Heterogeneity and Diversity from Breast Cancer Cell Lines. *PLoS ONE* **2012**, *7*, e33788. [CrossRef] [PubMed]

33. Greenbaum, A.; Martin, D.R.; Bocklage, T.; Lee, J.-H.; Ness, S.A.; Rajput, A. Tumor Heterogeneity as a Predictor of Response to Neoadjuvant Chemotherapy in Locally Advanced Rectal Cancer. *Clin. Colorectal Cancer* **2019**. [CrossRef] [PubMed]
34. Franke, A.J.; Parekh, H.; Starr, J.S.; Tan, S.A.; Iqbal, A.; George, T.J. Total Neoadjuvant Therapy: A Shifting Paradigm in Locally Advanced Rectal Cancer Management. *Clin. Colorectal Cancer* **2018**, *17*, 1–12. [CrossRef] [PubMed]
35. Garcia-Aguilar, J.; Chen, Z.; Smith, D.D.; Li, W.; Madoff, R.D.; Cataldo, P.; Marcet, J.; Pastor, C. Identification of a biomarker profile associated with resistance to neoadjuvant chemoradiation therapy in rectal cancer. *Ann. Surg.* **2011**, *254*, 486–492, discussion 492-493. [CrossRef]
36. Santos, M.D.; Silva, C.; Rocha, A.; Nogueira, C.; Castro-Poças, F.; Araujo, A.; Matos, E.; Pereira, C.; Medeiros, R.; Lopes, C. Predictive clinical model of tumor response after chemoradiation in rectal cancer. *Oncotarget* **2017**, *8*, 58133–58151. [CrossRef]
37. Sun, Y.; Wu, X.; Zhang, Y.; Lin, H.; Lu, X.; Huang, Y.; Chi, P. Pathological complete response may underestimate distant metastasis in locally advanced rectal cancer following neoadjuvant chemoradiotherapy and radical surgery: Incidence, metastatic pattern, and risk factors. *Eur. J. Surg. Oncol.* **2019**. [CrossRef]
38. Gotanda, K.; Hirota, T.; Matsumoto, N.; Ieiri, I. MicroRNA-433 negatively regulates the expression of thymidylate synthase (TYMS) responsible for 5-fluorouracil sensitivity in HeLa cells. *BMC Cancer* **2013**, *13*, 369. [CrossRef]
39. Kwon, M.J.; Soh, J.S.; Lim, S.-W.; Kang, H.S.; Lim, H. HER2 as a limited predictor of the therapeutic response to neoadjuvant therapy in locally advanced rectal cancer. *Pathol.–Res. Pract.* **2019**, *215*, 910–917. [CrossRef]
40. Kaneno, R.; Shurin, G.V.; Kaneno, F.M.; Naiditch, H.; Luo, J.; Shurin, M.R. Chemotherapeutic agents in low noncytotoxic concentrations increase immunogenicity of human colon cancer cells. *Cell Oncol. (Dordr)* **2011**, *34*, 97–106. [CrossRef]
41. Lee, S.-C.; Xu, X.; Lim, Y.-W.; Iau, P.; Sukri, N.; Lim, S.-E.; Yap, H.L.; Yeo, W.-L.; Tan, P.; Tan, S.-H.; et al. Chemotherapy-induced tumor gene expression changes in human breast cancers. *Pharmacogenet. Genomics* **2009**, *19*, 181–192. [CrossRef] [PubMed]
42. Negrei, C.; Hudita, A.; Ginghina, O.; Galateanu, B.; Voicu, S.N.; Stan, M.; Costache, M.; Fenga, C.; Drakoulis, N.; Tsatsakis, A.M. Colon Cancer Cells Gene Expression Signature As Response to 5- Fluorouracil, Oxaliplatin, and Folinic Acid Treatment. *Front. Pharmacol.* **2016**, *7*, 172. [CrossRef] [PubMed]
43. Tang, M.; Lu, X.; Zhang, C.; Du, C.; Cao, L.; Hou, T.; Li, Z.; Tu, B.; Cao, Z.; Li, Y.; et al. Downregulation of SIRT7 by 5-fluorouracil induces radiosensitivity in human colorectal cancer. *Theranostics* **2017**, *7*, 1346–1359. [CrossRef]
44. Liang, X.; Shi, H.; Yang, L.; Qiu, C.; Lin, S.; Qi, Y.; Li, J.; Zhao, A.; Liu, J. Inhibition of polypyrimidine tract-binding protein 3 induces apoptosis and cell cycle arrest, and enhances the cytotoxicity of 5- fluorouracil in gastric cancer cells. *Br. J. Cancer* **2017**, *116*, 903–911. [CrossRef] [PubMed]
45. Ng, J.M.Y.; Vermeulen, W.; van der Horst, G.T.J.; Bergink, S.; Sugasawa, K.; Vrieling, H.; Hoeijmakers, J.H.J. A novel regulation mechanism of DNA repair by damage-induced and RAD23-dependent stabilization of xeroderma pigmentosum group C protein. *Genes Dev.* **2003**, *17*, 1630–1645. [CrossRef] [PubMed]
46. Xu, W.; Jiang, H.; Zhang, F.; Gao, J.; Hou, J. MicroRNA-330 inhibited cell proliferation and enhanced chemosensitivity to 5-fluorouracil in colorectal cancer by directly targeting thymidylate synthase. *Oncol. Lett.* **2017**, *13*, 3387–3394. [CrossRef]
47. Bunick, C.G.; Miller, M.R.; Fuller, B.E.; Fanning, E.; Chazin, W.J. Biochemical and structural domain analysis of xeroderma pigmentosum complementation group C protein. *Biochemistry* **2006**, *45*, 14965–14979. [CrossRef] [PubMed]
48. Rushworth, D.; Mathews, A.; Alpert, A.; Cooper, L.J.N. Dihydrofolate Reductase and Thymidylate Synthase Transgenes Resistant to Methotrexate Interact to Permit Novel Transgene Regulation. *J. Biol. Chem.* **2015**, *290*, 22970–22976. [CrossRef]

© 2019 by the authors. Licensee MDPI, Basel, Switzerland. This article is an open access article distributed under the terms and conditions of the Creative Commons Attribution (CC BY) license (http://creativecommons.org/licenses/by/4.0/).

Article

Bone Marrow Involvement in Melanoma. Potentials for Detection of Disseminated Tumor Cells and Characterization of Their Subsets by Flow Cytometry

Olga Chernysheva *, Irina Markina, Lev Demidov, Natalia Kupryshina, Svetlana Chulkova, Alexandra Palladina, Alina Antipova and Nikolai Tupitsyn

FSBI "N.N. Blokhin Russian Cancer Research Center" of Ministry of Health of the Russian Federation, 115478 Moscow, Russia; irina160771@yandex.ru (I.M.); nntca@yahoo.com (L.D.); natalya-2511@yandex.ru (N.K.); chulkova@mail.ru (S.C.); alexandra.93@mail.ru (A.P.); a.s.antipova@gmail.com (A.A.); nntca@yahoo.com (N.T.)
* Correspondence: dr.chernysheva@mail.ru

Received: 23 May 2019; Accepted: 14 June 2019; Published: 21 June 2019

Abstract: Disseminated tumor cells (DTCs) are studied as a prognostic factor in many non-hematopoietic tumors. Melanoma is one of the most aggressive tumors. Forty percent of melanoma patients develop distant metastases at five or more years after curative surgery, and frequent manifestations of melanoma without an identified primary lesion may reflect the tendency of melanoma cells to spread from indolent sites such as bone marrow (BM). The purpose of this work was to evaluate the possibility of detecting melanoma DTCs in BM based on the expression of a cytoplasmatic premelanocytic glycoprotein HMB-45 using flow cytometry, to estimate the influence of DTCs' persistence in BM on hematopoiesis, to identify the frequency of BM involvement in patients with melanoma, and to analyze DTC subset composition in melanoma. DTCs are found in 57.4% of skin melanoma cases and in as many as 28.6% of stage I cases, which confirms the aggressive course even of localized disease. Significant differences in the groups with the presence of disseminated tumor cells (DTCs$^+$) and the lack thereof (DTC$^-$) are noted for blast cells, the total content of granulocyte cells, and oxyphilic normoblasts of erythroid raw cells.

Keywords: bone marrow; melanoma; disseminated tumor cells; solid cancers; single-cell analysis; enrichment and detection technologies; flow cytometry; tumor stem cells; HMB-45; CD133

1. Introduction

Today oncology is a rapidly developing field of medicine. Every year novel target and immunological agents acting against cancer at the molecular level are added to clinical oncologists' practice, and many such agents are currently at various stages of clinical development. However, notwithstanding significant progress over the last decade and a broad variety of therapeutical options, several fundamental questions remain to be answered: What are causes of cancer development? What are mechanisms of metastasis and recurrence? At what stage of disease development can we influence these processes?

Over the last 150 years there were many theories to explain processes developing both in the tumor and in the patient's body. By the end of the first quarter of the 21st century the world medical community has passed a long way from the first publication by T.R. Ashworth in the *Medical Journal of Australia* in 1869 [1], where the author described for the first time circulating tumor cells in a cancer patient, and the 'seed and soil' theory proposed by Stephen Paget in 1889 [2], through the theory of late dissemination (linear progression) by William Stewart Halsted [3–5] to the theory of early metastasis (parallel progression) by Christophe Klein [6] and the concept of the premetastatic niche by Bethan Psaila and David Lyden in 2009 [7]. The key question in all of these theories was how tumor cells managed to overcome immune surveillance [8], to preserve their proliferative potential and to proliferate in alien environments even after several decades of latency [9,10].

It seems natural that bone marrow (BM) with its advanced capillary network and a cocktail of soluble protein factors, integrins, chemokines, cell adhesion molecules, and a variety of growth factors is the most attractive niche for tumor cells [11,12]. Being basically alien, BM makes its environment appropriate for disseminated tumor cell (DTC) persistence via sophisticated antigenic, immunogenic, and cellular mechanisms [13,14]. DTCs may have different fates in a new microenvironment. Most of them die within several weeks or months [15], while DTCs preserving their vitality without decrease or increase in their total number may enter latency and form so called dormant metastases.

Dormant tumor cells have three main differences from other tumor cells, i.e., the ability to survive in alien and even hostile environments for a long time, temporary but reversible growth arrest, and resistance to target cytostatic agents [16]. These DTC properties make them biologically closer to tumor stem cells, a minor primary tumor subset seeming to play a leading role in the self-maintenance and metastasis of malignancies [17].

BM involvement is described in multiple non-hemopoietic neoplasms and is shown to be an independent poor prognostic factor for overall and disease-free survival [18–23]. Interestingly, these publications mainly address cancers of the breast, stomach, lung, colon, or prostate, while studies of melanoma are few and require further analysis.

Observations of hematogenous metastases from melanoma after 10 [24] or even 40 [25] years after removal of the primary tumor and frequent melanoma manifestations without an identified primary may reflect melanoma cell tendency to spread from indolent sites [26,27] such as BM.

gp100—HMB-45, a cytoplasmatic premelanocytic glycoprotein is a reliable marker of melanoma cells. It was discovered as one of the first melanoma antigens to demonstrate high sensitivity (up to 93%) and specificity (up to 100%) [28] and is usable to identify DTCs.

The Hemopoiesis Immunology Laboratory (N.N. Blokhin Cancer Research Center, Russian Federation Health Ministry) has developed a procedure to identify DTCs by flow cytometry [29]. Flow cytometry has certain advantages as compared to cytology, immunohistochemistry, and molecular biology techniques. For instance, contemporary multicolor flow cytometry can analyze 12 or more parameters in a single cell and accumulate a large number of events with sensitivity close to that of PCR (10^{-4} to 10^{-6}) and allows most complete description of the DTC immunophenotype [30]. Besides pure quantification of DTCs, flow cytometry therefore helps to study DTC subsets such as tumor stem cells or to identify surface molecular targets for drugs (Her2/neu, PDL1).

The purpose of this work was to evaluate the possibility of detecting melanoma DTCs in BM based on the expression of HMB-45 using flow cytometry, to determine the frequency of BM involvement in patients with melanoma, to analyze DTC subset composition in melanoma as to the expression of CD56 and CD57 that were an additional criterion for melanoma immunological diagnosis, and to assess the proportion of tumor stem cells among DTCs based on the presence of CD133.

2. Materials and Methods

A total of 47 patients (23 males and 24 females) aged 20–72 (median 49.8) years managed at the N. N. Blokhin Russian Cancer Research Center for skin melanoma during 2018–2019 were enrolled in the study. The diagnosis was verified histologically in all patients. This study was approved by the institutional ethical committees (Local ethical committee N. N. Blokhin Russian Cancer Research Center of Ministry of Health of the Russian Federation; UDC 616-006, Reg. № AAAA-A16-116122210071-4, Inv. 479.) and was done with the informed consent of the patients. Most of the patients (42.6%) had stage IV disease based on complex examination. BM involvement was assessed by morphology and immunology at diagnosis. Table 1 demonstrates patient distribution by stage.

Table 1. The distribution of patients by disease stage.

Stage	Frequency	Percent (%)
I	7	14.9
IIa	1	2.1
IIb	5	10.6
IIc	3	6.4
III	11	23.4
IV	20	42.6
Total:	47	100

Morphological examination included myelogram count and identification of tumor cells on six Romanovsky-stained bone marrow smears by two morphologists in parallel. Immunological identification of DTCs in BM was done by flow cytometry. Samples were lysed using BD FACS lysing solution (Beckton Dickinson, Franklin Lakes, NJ, USA), then washed in phosphate-buffered saline (PBS), and re-suspended in 100 mL of PBS. Cells were incubated for 15–20 min in the dark at room temperature together with a cocktail of monoclonal antibodies directly conjugated with fluorescein isothiocyanate (FITC), phycoerythrin (PE), allophycocyanin (APC), and Horizon V500 and Horizon V450 fluorochromes (Table 2). All samples were processed within 24 h after collection. Antibody labeling was measured by multiparameter flow cytometry using FACS Canto II (Beckton Dickinson). Twenty million myelokaryocytes (or all cells in the sample) were collected to identify DTCs. Tumor cells were detected by the lack of expression of the common leukocyte antigen CD45 in combination with bright expression of HMB-45. To identify DTC subpopulations expression of CD133, CD56, and CD57 molecules was analyzed among the CD45$^-$HMB-45$^+$ cells.

Table 2. Monoclonal antibodies used in the study.

No.	MoAbs/Fluorochromes	Function	Manufacturer
1	Syto41	Nuclear dye	Thermo Fisher Scientific, Walthem, MA, USA
2	CD45	Leukocyte common antigen	Beckton Dickinson
3	HMB-45	Melanoma cell antigen gp100	Santa Cruz Biotechnology, Dallas, Tx, USA
4	CD56	Neuronal cell adhesion molecule (NCAM)	Beckton Dickinson
5	CD57	NK-cell molecule (HNK1)	Beckton Dickinson
6	CD133	Hematopoietic stem cell antigen	Beckton Dickinson

Results were analyzed using Kaluza Analysis v2.1 (Beckman Coulter, Brea, CA, USA) software. Statistical analysis of data used IBM-SPSS Statistics v.17 package (IBM, Armonk, NY, USA).

3. Results

Morphological analysis of BM biopsies included myelogram count and tumor cell identification.

In the analysis of hematopoiesis, we excluded cases with bone marrow dilution with peripheral blood. Comparison of the average bone marrow parameters according to the myelogram is shown in the Table 3.

Table 3. Comparison of the average bone marrow according to myelogram.

Myelogram Parameters	DTCs	n	Mean Value	Err^{std}mean	p
Cellularity	negative	19	67.0	6.51	NS*
	positive	20	67.3	7.87	
Blasts	negative	20	1.46	0.14	0.026
	positive	25	1.09	0.09	
Promyelocytes	negative	20	0.44	0.11	NS
	positive	25	0.37	0.08	
Neutrophilic myelocytes	negative	20	7.80	0.72	NS
	positive	25	8.95	0.54	
Neutrophilic metamyelocytes	negative	20	8.58	0.65	NS
	positive	25	7.83	0.53	
Band neutrophils	negative	20	16.50	0.91	NS
	positive	25	18.70	1.00	
Segmented neutrophils	negative	20	24.47	1.39	NS
	positive	25	27.266	1.71	
All granulocyte cells	negative	20	60.76	1.45	0.025
	positive	25	65.41	1.38	
Neutrophil maturation index	negative	20	0.43	0.034	NS
	positive	25	0.38	0.034	
Monocytes	negative	20	2.78	0.26	NS
	positive	25	3.30	0.24	
Lymphocytes	negative	20	12.85	0.79	NS
	positive	25	12.02	0.68	
Plasmocytes	negative	20	0.60	0.10	NS
	positive	25	0.77	0.15	
Basophilic normoblasts	negative	20	1.23	0.17	NS
	positive	25	0.97	0.13	
Polychromatophilic normoblasts	negative	20	11.16	0.91	NS
	positive	25	10.19	0.82	
Oxyphilic normoblasts	negative	20	9.08	0.88	0.006
	positive	25	6.25	0.52	
Sum of nucleated erythroid cells	negative	20	21.47	1.44	0.042
	positive	25	17.41	1.29	
Erythroid maturation index	negative	20	0.96	0.01	NS
	positive	25	0.96	0.01	
Leuco–erythroid ratio	negative	20	4.02	0.39	0.034
	positive	25	5.58	0.59	

* NS—not significant.

Significant differences in the groups with the presence of disseminated tumor cells (DTCs+) and the lack thereof (DTC−) were noted for blast cells, the total content of granulocyte cells, and erythroid germ indicators.

The level of blast cells was higher in patients with no DTCs: 1.46% ± 0.14% (n = 20) and 1.1% ± 0.09% (n = 25), p = 0.026.

On the contrary, the total content of granulocyte cells was higher in patients with DTCs in the BM: 65.4% ± 1.4% (n = 25) and 60.8% ± 1.5%: (n = 20), p = 0.025.

The most significant differences were obtained with respect to cells of the erythroid series. Thus, the percentage of oxyphilic normoblasts was significantly higher in patients with no DTCs in the BM: 9.1% ± 0.88% (n = 20) and 6.3% ± 0.52% (n = 25), p = 0.006. It should be noted that, in the group as a whole, the levels of oxyphilic normoblasts were increased compared to the norm in 67% of patients. Accordingly, the sum of nucleated erythroid cells was also higher in melanoma patients with no DTCs in the BM: 21.5% ± 1.4% (n = 20) and 17.4% ± 1.3% (n = 25), p = 0.042. This was reflected in a significantly higher leuco–erythroid ratio in patients with the presence of DTCs in the bone marrow: 5.6% ± 0.6% (n = 25) and 4.0% ± 0.4 (n = 20), p = 0.034.

It is interesting to note that when analyzing according to the tables of conjugacy of characters, only two indicators of the myelogram were reliably associated with the presence of DTCs in the BM—the total content of granulocyte cells and the level of oxyphilic normoblasts.

The relationship of DTCs with the total amount of granulocyte cells consisted in the fact that, in the presence of DTCs, a decrease in the total level of granulocyte cells was observed in only 8% of cases, while in the absence of DTCs a decrease in granulocyte cells was observed in 30% of cases, chi-square = 8.9; $p = 0.012$.

A different situation was noted with respect to oxyphilic normoblasts, whose normal content in the absence of DTCs was observed in 15% of cases, and in the presence of DTCs—three times more often—in 44% of cases. On the contrary, an increase in the level of oxyphilic normoblasts in the absence of DTCs occurred in 85% of cases, in the presence of DTCs, in 56%, chi-square = 4.4; $p = 0.037$. Melanoma cells were identified in BM by morphology in one of 47 cases only (Figure 1).

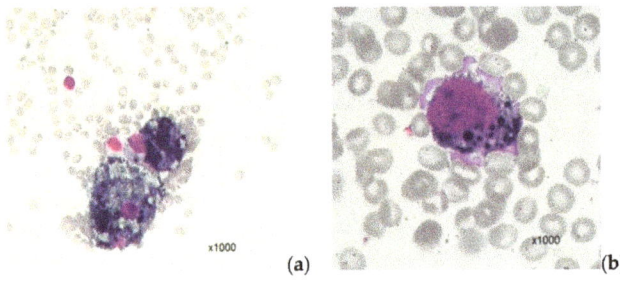

Figure 1. Melanoma disseminated tumor cells in bone marrow. This figure presents a case of detection of skin melanoma cells in bone marrow punctate ((**a**) and (**b**), ×1000 magnification). Punctate bone marrow is poor. Normal lines of myelopoiesis are depressed. Cell complexes of a non-hemopoietic nature are determined. Additionally, there are scattered, separately lying tumor cells. They are represented by cells of a large size, and basophilic colored pigment granules of various sizes are visible in the cytoplasm. The morphological picture of the bone marrow is characteristic of metastatic lesions in melanoma.

Immunological analysis of DTCs in BM was based on a threshold of one tumor cell (Syto41$^+$CD45$^-$HMB-45$^+$) per 10 million myelokaryocytes. A mean of 14,146,987 (±957,728) myelokaryocytes were analyzed in each sample. DTCs were found in 57.4% of BM samples ($n = 27$) based on the threshold level. Interestingly, flow cytometry of melanoma cells has specific features due to morphological characteristics of these cells such as a rather large size and the presence of pigmented inclusions of various diameters (from dust-like to large fused granules of different diameter). For instance, the melanoma DTCs have high direct and side light scatter characteristics and require adequate protocols for flow cytometer tuning.

There were no significant differences in DTC counts with respect to gender, age, or disease stage. What is important is that BM involvement was discovered at all disease stages (Table 4). This means that hematogenic tumor cell dissemination occurs already from clinically localized disease.

Table 4. Frequency of disseminated tumor cell (DTC) detection at various stages of melanoma.

Stage	Number of Patients	Frequency of DTCs + Cases
I	7	28.6%
II	9	55.6%
III	11	63.6%
IV	20	65.0%

DTCs were additionally characterized by CD56 and CD57 expression. In our study, CD56 and CD57 expression was assessed in 23 BM samples. Among them, DTCs were present in 54.2% ($n = 13$) though these cells did not express CD56. CD57 expression on DTCs was found in six cases (46.2%) (Figure 2). Importantly, not all 100% of DTCs in each BM sample demonstrated CD57 expression. On average 87.4% ± 5.8% of DTCs were CD57-positive. Of interest, 50% of CD57+ patients had stage IV, two of six had stage III, and one patient had stage IIc disease.

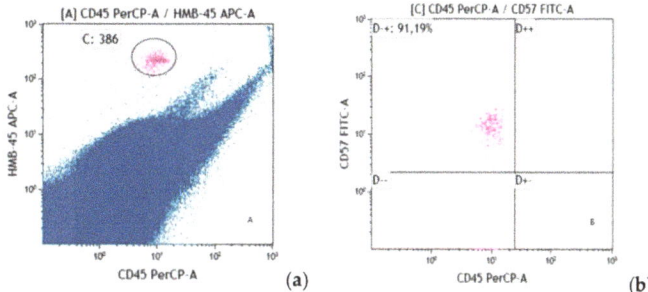

Figure 2. Disseminated tumor cells of skin melanoma as identified by immunological flow cytometry. This figure shows an example of detection of skin melanoma DTCs. On the cytogram (**a**) in gate C, DTCs were observed on the basis of the bright expression of HMB-45 (y-axis) and the absence of CD45 expression (x-axis). On the cytogram (**b**), the analysis of the subpopulation composition of DTCs in melanoma was performed in relation to the expression of the CD57 antigen. Cells are characterized by distinct CD57 expression (y-axis) and lack of CD45 expression (x-axis). Most DTCs (91.19%) are CD57+.

A minor tumor stem cell (TSC) subset with maximum resistance to conventional anticancer therapies plays a special role in metastasis. According to the literature, melanoma TSCs are characterized by expression of antigens such as CD44, CD271, and CD133. We identified TSCs among melanoma DTCs by CD133 expression.

CD133 expression was analyzed in 22 BM samples. Half of these BM samples were DTC-positive. There was a single DTC-positive sample containing a CD133+-DTC subset, which accounted for 1.38% of all DTCs in this case (Figure 3).

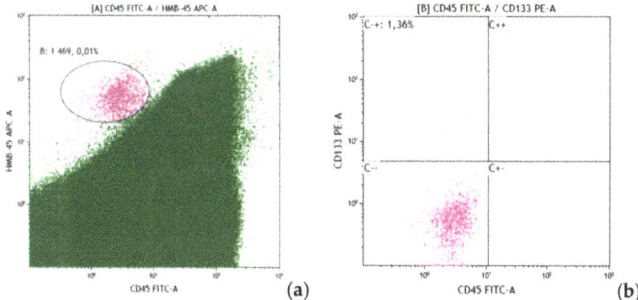

Figure 3. Identification of CD133-positive DTCs. This figure shows an example of the assessment of DTC subpopulations by the tumor stem cell marker CD133. On the cytogram (**a**), 0.01% DTCs was detected by the distinct expression of HMB-45 (*y*-axis) and the lack of expression of the pan-leukocyte antigen CD45 (*x*-axis). Within the DTCs, expression of CD133 evaluated. Cytogram (**b**) (*x*-axis is CD45, *y*-axis is CD133) shows that CD133+ cells make up 1.36% of all DTCs.

We have demonstrated that both the primary tumor and DTCs in BM may have a heterogeneous composition and express various antigens. The significance of this finding for the disease course and prognosis deserves assessment in further studies.

4. Discussion

BM as a niche for micrometastasis plays a key role in hematogeneous dissemination. By creating a unique microenvironment for tumor cells BM maintains their proliferative potential for many years. Disease recurrence decades after treatment of the primary is described for many entities, and skin melanoma is not an exception. Forty percent of skin melanoma patients develop distant metastases at five or more years after curative surgery [27], therefore finding novel factors for disease prognosis and markers of early tumor cell dissemination for personalization of early systemic treatment is of much importance.

As demonstrated in our study, flow cytometry with a specific antibody HMB-45 in combination with CD45 is a useful technique to assess BM involvement in melanoma. DTCs were found in 57.4% of skin melanoma cases. The DTCs were present in 28.6% of stage I disease, which confirms the aggressiveness of skin melanoma even in localized disease. The findings of CD57 and CD133 expression are evidence of DTCs heterogeneity and the complex hierarchical relations between the primary and the DTCs. The prognostic significance of our results will be assessed in further studies.

Thus, we can talk about the complex relationship of hematopoiesis in general and the development of skin melanoma. Both myelo- and erythropoiesis are involved in and reacting to the tumor process occurring in the body. Of particular interest are changes in the bone marrow hematopoiesis arising in the presence of DTCs. Perhaps they are a reflection of the reorganization of the microenvironment of the DTCs, contributing to their long-term persistence in the bone marrow. The role of these changes in the prognosis of the course of the disease remains to be assessed.

The 'seed and soil' theory therefore is still valuable after 150 years and requires further development using up-to-date diagnostic approaches.

Author Contributions: O.C.—conceptualization, methodology, formal analysis, investigation, writing—original draft preparation, visualization, and project administration; I.M.—conceptualization, methodology, data curation, resources, writing—original draft preparation, investigation, and project administration; L.D.—writing—review and editing, supervision, and project administration; N.K.—software, resources and investigation; S.C.—conceptualization and formal analysis; A.P.—resources and formal analysis; A.A.—resources; N.T.—conceptualization, methodology, formal analysis, review and editing, supervision, and project administration.

Funding: This research received no external funding

Conflicts of Interest: The authors declare no conflict of interest.

References

1. Ashworth, T.R. A case of cancer in which cells similar to those in the tumors were seen in the blood after death. *Med. J. Australia* **1869**, *14*, 146–147.
2. Paget, S. The distribution of secondary growths in cancer of the breast. *Lancet* **1889**, *1*, 571–573. [CrossRef]
3. Fidler, I.J.; Kripke, M.L. Metastasis results from preexisting variant cells within a malignant tumor. *Science* **1977**, *197*, 893–895. [CrossRef] [PubMed]
4. Fearon, E.R.; Vogelstein, B. A genetic model for colorectal tumorigenesis. *Cell* **1990**, *61*, 759–767. [CrossRef]
5. Fidler, I.J.; Hart, I.R. Biological diversity in metastatic neoplasms: origins and implications. *Science* **1982**, *217*, 998–1003. [CrossRef] [PubMed]
6. Klein, C.A. Parallel progression of primary tumours and metastases. *Nat. Rev. Cancer* **2009**, *9*, 302–312. [CrossRef] [PubMed]
7. Psaila, B.; Lyden, D. The metastatic niche: Adapting the foreign soil. *Nat. Rev. Cancer* **2009**, *9*, 285–293. [CrossRef] [PubMed]
8. Koebel, C.M.; Vermi, W.; Swann, J.B.; Zerafa, N.; Rodig, S.J.; Old, L.J.; Smyth, M.J.; Schreiber, R.D. Adaptive immunity maintains occult cancer in an equilibrium state. *Nature* **2007**, *450*, 903–907. [CrossRef]

9. Willis, R.A. *The Spread of Tumours in the Human Body*; J. & A. Churchill: London, UK, 1934.
10. Hadfield, G. The dormant cancer cell. *Br. Med. J.* **1954**, *2*, 607–610. [CrossRef]
11. Weilbaecher, K.N.; Guise, T.A.; McCauley, L.K. Cancer to bone: A fatal attraction. *Nat. Rev. Cancer* **2011**, *11*, 411–425. [CrossRef]
12. Jones, D.H.; Nakashima, T.; Sanchez, O.H.; Kozieradzki, I.; Komarova, S.V.; Sarosi, I.; Morony, S.; Rubin, E.; Sarao, R.; Hojilla, C.V.; et al. Regulation of cancer cell migration and bone metastasis by RANKL. *Nature* **2006**, *440*, 692–696. [CrossRef] [PubMed]
13. Aguirre-Ghiso, J.A. Models, mechanisms and clinical evidence for cancer dormancy. *Nat. Rev. Cancer* **2007**, *7*, 834–846. [CrossRef] [PubMed]
14. Sosa, M.S.; Bragado, P.; Aguirre-Ghiso, J.A. Mechanisms of disseminated cancer cell dormancy: An awakening field. *Nat. Rev. Cancer* **2014**, *14*, 611–622. [CrossRef] [PubMed]
15. Luzzi, K.J.; MacDonald, I.C.; Schmidt, E.E.; Kerkvliet, N.; Morris, V.L.; Chambers, A.F.; Groom, A.C. Multistep nature of metastatic inefficiency: Dormancy of solitary cells after successful extravasation and limited survival of early micrometastases. *Am. J. Pathol.* **1998**, *153*, 865–873. [CrossRef]
16. Ghajar, C.M. Metastasis prevention by targeting the dormant niche. *Nat. Rev. Cancer.* **2015**, *15*, 238–247. [CrossRef] [PubMed]
17. Malanchi, I.; Santamaria-Martínez, A.; Susanto, E.; Peng, H.; Lehr, H.A.; Delaloye, J.F.; Huelsken, J. Interactions between cancer stem cells and their niche govern metastatic colonization. *Nature* **2012**, *481*, 85–89. [CrossRef] [PubMed]
18. Janni, W.; Vogl, F.D.; Wiedswang, G.; Synnestvedt, M.; Fehm, T.; Jückstock, J.; Borgen, E.; Rack, B.; Braun, S.; Sommer, H. Persistence of disseminated tumor cells in the bone marrow of breast cancer patients predicts increased risk for relapse—A European pooled analysis. *Clin. Cancer Res.* **2011**, *17*, 2967–2976. [CrossRef]
19. Pantel, K.; Izbicki, J.; Passlick, B.; Angstwurm, M.; Häussinger, K.; Thetter, O.; Riethmüller, G. Frequency and prognostic significance of isolated tumor cells in bone marrow of patients with non-small-cell lung cancer without overt metastases. *Lancet* **1996**, *347*, 649–653. [CrossRef]
20. Lilleby, W.; Stensvold, A.; Mills, I.G.; Nesland, J.M. Disseminated tumor cells and their prognostic significance in nonmetastatic prostate cancer patients. *Int. J. Cancer* **2013**, *133*, 149–155. [CrossRef]
21. Flatmark, K.; Borgen, E.; Nesland, J.M.; Rasmussen, H.; Johannessen, H.O.; Bukholm, I.; Rosales, R.; Hårklau, L.; Jacobsen, H.J.; Sandstad, B.; et al. Disseminated tumour cells as a prognostic biomarker in colorectal cancer. *Br. J. Cancer* **2011**, *104*, 1434–1439. [CrossRef]
22. Besova, N.S.; Obarevich, E.S.; Davydov, M.M.; Beznos, O.A.; Tupitsyn, N.N. Prognostic values of the presence of disseminated tumor cells in the bone marrow in patients with disseminated stomach cancer before start of treatment with antitumor drugs. *Pharmateca* **2017**, *350*, 62–66.
23. Besova, N.S.; Obarevich, E.S.; Beznos, O.A.; Tupitsyn, N.N.; Davydov, M.M. Prognostic value of the dynamics of disseminated tumor cells in the bone marrow in patients with disseminated adenocarcinoma of the stomach or the esophagogastric junction. *Pharmateca* **2017**, *350*, 83–86.
24. Eskelin, S.; Pyrhonen, S.; Summanen, P.; Hahka-Kemppinen, M.; Kivelä, T. Tumor doubling times in metastatic malignant melanoma of the uvea: Tumor progression before and after treatment. *Ophthalmology* **2000**, *107*, 1443–1449. [CrossRef]
25. Coupland, S.E.; Sidiki, S.; Clark, B.J.; McClaren, K.; Kyle, P.; Lee, W.R. Metastatic choroidal melanoma to the contralateral orbit 40 years after enucleation. *Arch. Ophthalmol.* **1996**, *114*, 751–756. [CrossRef] [PubMed]
26. Damsky, W.; Micevic, G.; Meeth, K.; Muthusamy, V.; Curley, D.P.; Santhanakrishnan, M.; Erdelyi, I.; Platt, J.T.; Huang, L.; Theodosakis, N.; et al. mTORC1 activation blocks BrafV600E-induced growth arrest but is insufficient for melanoma formation. *Cancer Cell* **2015**, *27*, 41–56. [CrossRef] [PubMed]
27. Rocken, M. Early tumor dissemination, but late metastasis: Insights into tumor dormancy. *J. Clin. Investig.* **2010**, *120*, 1800–1803. [CrossRef] [PubMed]
28. Wick, M.R.; Swanson, P.E. Recognition of malignant melanoma by monoclonal antibody HMB-45. An immunohistochemical study of 200 paraffin-embedded cutaneous tumors. *J. Cutan. Pathol.* **1988**, *15*, 201–207. [CrossRef] [PubMed]

29. Davydov, M.I.; Tupitsin, N.N. Assessment of minimal bone marrow involvement by flow cytometry. *Hematopoiesis Immunol.* **2014**, *12*, 8–17.
30. Van Dongen, J.J.M.; van der Velden, V.H.J.; Brüggemann, M.; Orfao, A. Minimal residual disease diagnostics in acute lymphoblastic leukemia: Need for sensitive, fast, and standardized technologies. *Blood* **2015**, *125*, 3996–4009. [CrossRef]

© 2019 by the authors. Licensee MDPI, Basel, Switzerland. This article is an open access article distributed under the terms and conditions of the Creative Commons Attribution (CC BY) license (http://creativecommons.org/licenses/by/4.0/).

Article

Fibronectin Regulation of Integrin B1 and SLUG in Circulating Tumor Cells

Jeannette Huaman [1,2], Michelle Naidoo [1,2], Xingxing Zang [3] and Olorunseun O. Ogunwobi [1,2,4,*]

1. Department of Biological Sciences, Hunter College of The City University of New York, New York, NY 10065, USA; JHUAMAN@genectr.hunter.cuny.edu (J.H.); michelle.naidoo86@myhunter.cuny.edu (M.N.)
2. Department of Biology, The Graduate Center of The City University of New York, New York, NY 10016, USA
3. Departments of Microbiology and Immunology, and Medicine (Oncology), Albert Einstein College of Medicine, Bronx, NY 10461, USA; xingxing.zang@einstein.yu.edu
4. Department of Medicine, Weill Cornell Medicine, Cornell University, New York, NY 10065, USA
* Correspondence: ogunwobi@genectr.hunter.cuny.edu

Received: 24 May 2019; Accepted: 19 June 2019; Published: 20 June 2019

Abstract: Metastasis is the leading cause of cancer death worldwide. Circulating tumor cells (CTCs) are a critical step in the metastatic cascade and a good tool to study this process. We isolated CTCs from a syngeneic mouse model of hepatocellular carcinoma (HCC) and a human xenograft mouse model of castration-resistant prostate cancer (CRPC). From these models, novel primary tumor and CTC cell lines were established. CTCs exhibited greater migration than primary tumor-derived cells, as well as epithelial-to-mesenchymal transition (EMT), as observed from decreased E-cadherin and increased SLUG and fibronectin expression. Additionally, when fibronectin was knocked down in CTCs, integrin B1 and SLUG were decreased, indicating regulation of these molecules by fibronectin. Investigation of cell surface molecules and secreted cytokines conferring immunomodulatory advantage to CTCs revealed decreased major histocompatibility complex class I (MHCI) expression and decreased endostatin, C-X-C motif chemokine 5 (CXCL5), and proliferin secretion by CTCs. Taken together, these findings indicate that CTCs exhibit distinct characteristics from primary tumor-derived cells. Furthermore, CTCs demonstrate enhanced migration in part through fibronectin regulation of integrin B1 and SLUG. Further study of CTC biology will likely uncover additional important mechanisms of cancer metastasis.

Keywords: metastasis; circulating tumor cells (CTCs); hepatocellular carcinoma (HCC); castration resistant prostate cancer (CRPC); epithelial-to-mesenchymal transition (EMT); fibronectin; integrin B1; SLUG; major histocompatibility complex class I (MHCI); immunomodulation

1. Introduction

Metastasis is associated with advanced stages of cancer. Resulting in 90% of cancer deaths worldwide [1], metastasis occurs in a series of steps. These steps include the dissociation of cells from the primary tumor, migration through surrounding tissue, intravasation, circulation through blood, followed by extravasation and re-colonization of distant sites throughout the body. At advanced stages of most cancers, there are limited treatment options [1–3]. As such, efforts are increasingly being focused on identification of novel metastasis-related molecular targets.

One way to potentially avoid the need for invasive tissue biopsies when studying cancer metastasis is through the use of circulating tumor cells (CTCs). CTCs are cells which have dissociated from the primary tumor and are found traveling in the blood [4–8]. Some CTCs will eventually form metastatic, secondary lesions. Because CTCs can be obtained from liquid biopsies (from blood), they enable the molecular profiling of potentially unresectable tumors in patients [9] and identification of molecular changes important for progression to advanced cancers [10]. However, there is a challenge with low

CTC numbers frequently found in the blood [11–15]. To address this potential obstacle to studying CTC biology, in this study, we established novel CTC cell lines and primary tumor-derived cells for molecular biological studies.

The two different cancer models used in this study were a syngeneic mouse model of hepatocellular carcinoma (HCC) and a xenograft mouse model of castration-resistant prostate cancer (CRPC). HCC is the most common form of liver cancer and is frequently diagnosed at very late stages. Consequently, it is one of the leading causes of cancer deaths worldwide [16–19]. Moreover, sorafenib, which is the main FDA approved drug to treat advanced HCC, extends life by only six months [20]. As such, better treatment options are needed. Similarly, CRPC is a form of prostate cancer (PCa) that is resistant to both medical and surgical castration [21,22]. However, androgen deprivation therapy (ADT) is the main standard of treatment for localized PCa [23,24]. This makes CRPC particularly challenging to treat. Over one third of CRPC patients will develop bone metastasis for which there is no cure [21,25]. Therefore, finding alternative treatments is critical for this cancer as well.

To this end, we propagated cell lines originating from primary tumors and CTCs. Our aim was to discover differences between these two cell types representing earlier and more advanced stages of cancers. Both HCC and CRPC CTCs demonstrate increased migration and evidence of epithelial-to-mesenchymal transition (EMT). Moreover, we discovered that in CTCs, fibronectin regulates integrin B1 and SLUG, which are known regulators of cell migration. Finally, we identified differences in CTC cell surface marker and cytokine secretion profiles that could have immunomodulatory implications. HCC CTCs had significantly reduced major histocompatibility complex class I (MHCI) expression, as well as significantly decreased secretion of endostatin, CXCL5, and proliferin as compared to primary tumor-derived cells. These findings may have implications for the function of metastatic cells and how they evade the immune system.

2. Materials and Methods

2.1. Cell Lines and Cell Culture

The BNL 1ME A 7R.1 cell line (purchased from ATCC), as well as the newly established primary tumor cell lines (TBOH1 and TBOH9) and circulating tumor cell lines (CBOH4 and CBOH9), were maintained in Dulbecco's Modified Eagle Media (DMEM) media supplemented with 10% fetal bovine serum, 1% penicillin/streptomycin, and L-glutamine. Trypsinization of cells occurred using 0.25% trypsin when 75–80% confluent.

The 22Rv1 cell line (purchased from ATCC), as well as the newly established primary tumor cell line T22OH and circulating tumor cell line C22OH, were maintained in Roswell Park Memorial Institute (RPMI) media supplemented with 10% fetal bovine serum and 1% penicillin/streptomycin. Trypsinization of cells occurred using 0.05% trypsin when 75–80% confluent. All cells lines were cultured in a 5% CO_2, 37 °C atmosphere.

2.2. Mouse Tumor Studies

Figure 1 shows the experimental mouse models used for this study and the subsequent establishment of novel cell lines. For the syngeneic mouse model of HCC, 7-week old, male, BALB/c mice were obtained from Taconic Biosciences Inc. Mice were implanted with 2.5×10^6 BNL 1ME A 7R.1 murine HCC cells. For the NOD scid gamma (NSG) human xenograft mouse model, 7-week old, male NSG mice were obtained from Jackson Laboratory. Mice were implanted with 2.5×10^6 22Rv1 human CRPC cells. Tumors were allowed to grow until reaching a max tumor volume of 2000 cm^3, at which point mice were euthanized, tumors removed, and samples of blood processed for CTCs as previously described by our lab [26]. Briefly, up to 1 mL of blood was obtained from intracardiac blood withdrawal from mice. The blood was spun down and the plasma removed. The rest of the blood sample, most importantly the buffy coat layer, which is where we expect our CTCs to be, was treated with red blood cell lysis buffer. After a series of spins and washes, samples were placed in media. Experiments on CTC cell lines were carried out for as long as forty-five passages (well over 6 months). Features remained consistent among different passages. While efficiency of cell line establishment was moderate to low, once established, the CTC cell lines exhibited high cell viability. In terms of our primary tumor cell lines, tumors were mechanically dissociated in media and given the chance to adhere to the plate to give rise to primary cell culture cell lines. H&E staining was performed to confirm metastasis to lungs using the core facilities at Albert Einstein School of Medicine. All mouse experiments were performed in compliance with Institutional Animal Care and Use Committee (IACUC)–approved protocols at Weill Cornell Medicine.

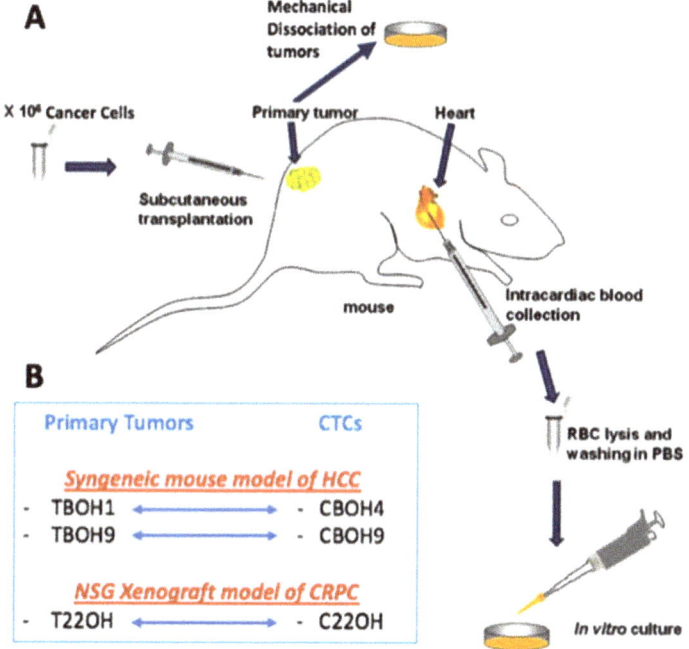

Figure 1. Establishment of novel primary tumor-derived cell lines and circulating tumor cell lines. (A) Schematic diagram summarizing how novel cell lines were established. This figure includes modifications to our previously described work [27]. For the syngeneic hepatocellular carcinoma (HCC) mouse model, BALB/c mice were subcutaneously implanted with the murine hepatocellular carcinoma cell line BNL 1ME A.7R.1. For the xenograft castration-resistant prostate cancer (CRPC) mouse model, NSG mice were subcutaneously implanted with the 22Rv1 human CRPC cell line. Mice were allowed to develop tumors over a period of 3–4 weeks, humanely sacrificed, primary tumors resected and mechanically dissociated, and then put in culture to establish the TBOH series of primary tumor-derived cell lines. The CBOH series of circulating tumor cells (CTCs) was established from cancer cells isolated from the bloodstream of the same mice implanted with either BNL 1ME A.7R.1 or 22Rv1 cells. (B) The newly established cell lines: the TBOH1 and CBOH4 pair and TBOH9 and CBOH9 pair were established from the HCC model. The T22OH and C22OH pair were established from the CRPC model.

2.3. Immunofluorescence Staining

Cells were incubated with coverslips and grown for 48 hr. The coverslips were collected, fixed with 4% paraformaldehyde for 15 min at room temperature, permeabilized with 0.5% Triton X-100 in 1X PBS/1% FBS for 10 min at room temperature. Cells were stained with CREB3L3 antibody (sc-377331; 1:500; Santa Cruz Biotechnology, TX, USA), or endostatin antibody (PA1-601; 1:200; Thermo Fisher

Scientific, Rockford, IL, USA) for 2 h, followed by incubation with Alexa Fluor 635 anti-rabbit secondary antibody (A31577; Thermo Fisher Scientific, Rockford, IL, USA) for 1 h. DAPI was used to counterstain nuclei, and slides were imaged using a Nikon A1 confocal microscope at a 60× magnification.

2.4. Migration Assays

For wound healing migration assays, 5×10^4 murine BNL 1ME A.7R.1 cells were grown in a 6-well tissue culture plate and permitted to reach 90–100% confluency. Using a plastic tip (1 mm thick), wounds were administered to monolayers of cells in each well. Wounded monolayers were washed with 1X PBS and incubated with media. Cells were observed, and images were taken using the Motic AE30 Inverted Microscope.

For transwell migration assays, 1×10^5 22Rv1 CRPC cells were seeded on the top of chambers containing 8 μm pores (Greiner Bio-one, Austria, cat #: 662 638) in serum-free media. The bottom chambers were filled with regular media to serve as a chemoattractant. After 48 hr, the top chambers were rinsed with 1X PBS, fixed with paraformaldehyde, treated with methanol, and stained with trypan blue staining. The chambers were placed on a slide and viewed using the Motic AE30 Inverted Microscope. All migration assays were carried out at least 3 times.

2.5. Protein Extraction and Western Blotting

Whole cell extracts were obtained by treating cells with radioimmunoprecipitation assay (RIPA) lysis buffer (Amresco, Ohio, USA, cat#: N653), supplemented with 10× protease inhibitors (Thermo Fisher Scientific, Rockford, IL, USA, cat#: 88665) and 100 mM PMSF (Amresco, OH, USA, cat#: M145). Protein concentration was calculated via the Bradford Assay using the Bio-Rad Protein Assay Dye Reagent Concentrate. For Western blot analysis, 30 μg of protein was run on precast SDS-PAGE gels and subsequently transferred onto nitrocellulose membranes. Membranes were blocked in 5% BSA in TBS-T for 1 h at room temperature, incubated with primary antibodies overnight at 4 °C, washed with 1× TBS-T, incubated with secondary antibodies for 1 h, washed, and imaged using the LI-COR Odyssey CLx imager with infrared fluorescence. The primary antibodies used were directed against fibronectin (ab2413; 1:1000; Abcam, Cambridge, UK), E-cadherin (3195S; 1:200; Cell Signaling, MA, USA), CREB3L3 (sc-377331; 1:500; Santa Cruz Biotechnology, TX, USA), AR-V7 (ab198394; 1:1000; Abcam, Cambridge, UK), PSA (sc-7316; 1:200; Santa Cruz Biotechnology, TX, USA), integrin B1 (4706S; 1:500; Cell Signaling, MA, USA), GAPDH (5174S; 1:1000; Cell Signaling, MA, USA), alpha-tubulin (sc-32293; 1:500; Santa Cruz Biotechnology, TX, USA), and beta-actin (A5441; 1:5000; Sigma, St.Louis, MO, USA). The secondary antibodies used were anti-mouse (925-32210; 1:15,000; LI-COR, Lincoln, NE, USA) and anti-rabbit (925-32211; 1:15,000; LI-COR, Lincoln, NE, USA). Analysis and quantification of western blots were performed using ImageJ software.

2.6. RNA Extraction and qPCR Analysis

The RNeasy Mini Kit (QIAGEN, Hilden Germany, cat#: 74104) was used to isolate total RNA from each of the cell lines used in this study, according to the protocol specified by the manufacturer. RNA concentration was measured using the spectrophotometer NanoDropTM 2000 (Thermo Fisher Scientific, Inc.). cDNA was obtained using 1 μg of RNA and the QuantiTect Reverse Transcription kit (QIAGEN, Hilden, Germany, cat#: 205311).

Expression of SLUG was measured by quantitative real time qPCR using SYBR-Green Master mix (Life Technologies, CA, USA, cat#: 4309155). Primers for SLUG and GAPDH were created with the OligoPerfect Designer program (ThermoFisher Scientific Inc; Wilmington, DE, USA). The following oligonucleotide sequences were used for primers: murine SLUG-F, 5′-CCTTTCTCTTGCCCTCACTG-3′, and murine SLUG-R, 5′-ACAGCAGCCAGACTCCTCAT-3′; murine GAPDH-F, 5′-TGATGGGTGTGAACCACGAG-3′, and GAPDH-R, 5′-AGTGATGGCATGGACTGTGG-3′; human SLUG-F, 5′-CTTTTTCTTGCCCTCACTGC-3′, and human SLUG-R, 5′-GCTTCGGAGTGAAGAAATGC-3′; human GAPDH-F, 5′-GAGTCAACGGATTTGGTCGT-3′, and human GAPDH-R, 5′-TTGATTTTGGAGGGATCTCG-3′. For each sample, SLUG was normalized with GAPDH expression. The comparative cycle threshold (Ct) method was used to quantify relative target gene expression. The instrument used was the Quantifect Studio System (Applied Biosystems).

2.7. Transfection of siRNAs

Cells were grown in 6-well plates. When a confluency of 60–70% was reached, the cells were transfected according to the manufacturer's instructions. Briefly, the cells were transfected with 10 nM of either fibronectin siRNA (Santa Cruz Biotechnology, TX, USA, cat#: sc-29315) or a non-targeting scramble control (Sigma, St.Louis, MO, USA) using Lipofectamine RNAiMAX (Thermo Fisher Scientific Inc.; Wilmington, DE, USA) diluted in Opti-MEM (ThermoFisher Scientific Inc.; Wilmington, DE, USA). The cells were then incubated for 24 h at 37 °C after which cells were harvested.

2.8. Flow Cytometry

Cells were incubated with trypsin for 3 min in order to harvest them. The pellets were washed, spun at 1200 rpm for 5 min, and re-suspended in 1% BSA in 1X PBS incubation buffer. Cells were spun and treated with mouse Fc block (BD Biosciences, NJ, USA, cat#: 553142) for 20 min. Either FITC anti-mouse MHCI antibody (BD Biosciences, NJ, USA, cat#: 553565) or its corresponding FITC Mouse Isotype control (BD Biosciences, NJ, USA, cat#: 553456) was added to cells and allowed to incubate for 1 h. The samples were fixed with 2% paraformaldehyde for 30 min at 4 °C, washed, and re-suspended in 1X PBS. Analysis was done using the BD FACs instrumentation and software version 7.0 (BD Biosciences, NJ, USA).

2.9. Cytokine Array

TBOH and CBOH cell lines were screened simultaneously for 111 murine cytokines using the Mouse XL Cytokine Array (R&D Biosystems, MN, USA, cat#: ARY028). Duplicate experiments were performed according to the manufacturer's instructions. Cytokine signals were quantified with densitometry using Image J software.

2.10. Statistical Analyses

Data from at least 3 different independent experiments were collected and presented as mean ± standard error of the mean (SEM). Statistical significance was evaluated using Student's t-test. p values of >0.05 were deemed significant.

3. Results

3.1. CTCs Obtained from Blood Express Tissue-Specific Markers

After the establishment of the novel cell lines (as shown in Figure 1), we confirmed tissue specificity of the established CTC lines. To confirm that the CTCs obtained from the bloodstream of the HCC syngeneic mouse models were of liver origin and not other cells potentially isolated from the blood, we performed immunofluorescence staining for CREB3L3, a validated liver specific marker [28,29]. As expected, the primary tumor cell lines TBOH1 and TBOH9 demonstrated strong CREB3L3 expression as shown by the red pigmentation in cells in Figure 2A,C. Similarly, the CTCs CBOH4 and CBOH9 also showed distinct CREB3L3 expression, confirming their liver origin, as well as their derivation from TBOH1 and TBOH9, respectively. No signal was observed in the negative controls in which cells were incubated with only the secondary antibody. Western blotting for CREB3L3 also revealed positive signals for both CBOH4 and CBOH9 as seen in Figure 2B,D.

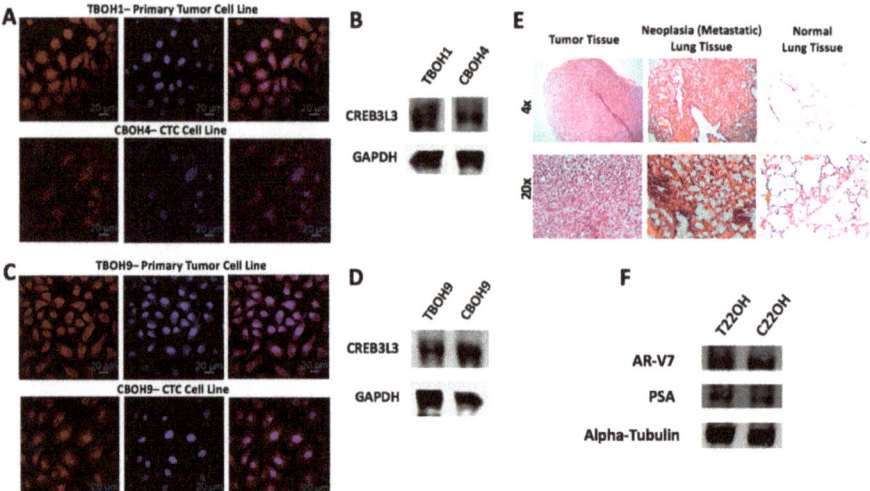

Figure 2. CTCs express tissue-specific markers and were obtained from mice that developed cancer metastasis. (**A,C**) Immunofluorescence staining for CREB3L3, a liver-specific marker; and (**B,D**) western blotting for CREB3L3. DAPI nuclear staining is shown in blue; CREB3L3 cytoplasmic staining is shown in red. (**E**) H&E staining of tumor and lung tissues from implantation of HCC cell line into BALB/c mice demonstrates evidence of cancer metastasis to lungs of cancer-bearing mice. (**F**) Western blotting was carried out to demonstrate prostate cancer origin for both CRPC primary tumor-derived and CTC lines. Experiments were carried out at least three times.

In addition to showing that the CTCs are hepatic-specific, we also performed H&E histological staining on lung tissues obtained from the mice implanted with the HCC cell line. This was to confirm the occurrence of HCC metastasis to the lungs, as shown in Figure 2E by the darker red pigmentation in the lungs of the HCC-implanted mice in comparison to the lungs of the non-cancer bearing mice.

Similarly, to confirm that the cells obtained from the blood of the NSG mice implanted with CRPC cells are of prostate origin, we performed western blotting for prostate specific antigen (PSA) and the AR-V7 variant androgen receptor protein known to be expressed by 22Rv1. These are markers characteristic of prostate cancer cells [23,30,31]. Both the primary tumor-derived cell line and the CTC cell line derived from NSG mice implanted with the 22Rv1 CRPC cell line demonstrated distinct AR-V7 and PSA expression as shown in Figure 2F, confirming their origin from the prostate and the derivation of C22OH from T22OH. Furthermore, the occurrence of metastasis was confirmed by visually finding multiple macroscopic tumors in distant sites during necropsy of the mice.

Our results show that we were successfully able to isolate CTCs from the blood from the syngeneic HCC and NSG CRPC mouse models. Functional assays were subsequently carried out to determine differences between CTCs and their corresponding primary tumor-derived cells.

3.2. CTCs Have a Greater Migratory Capacity than Primary Tumor-Derived Cells

Cancer cell migration is required for cancer metastasis [32,33]. To assess the migratory capability of both primary tumor-derived and CTC HCC cells, we performed wound healing migration assays. The rate at which wounds closed determined the migratory capability of cells. As shown in Figure 3A,C, we observed that CTCs (CBOH4 and CBOH9) were more migratory than their corresponding primary tumor-derived cell lines (TBOH1 and TBOH9, respectively). CBOH4 displayed a 55% increase in migration in comparison with its corresponding primary tumor-derived TBOH1 (Figure 3B), and CBOH9 demonstrated ~30% increased migration in comparison to TBOH9 (Figure 3D). The increased migratory capacity of CTCs is statistically significant.

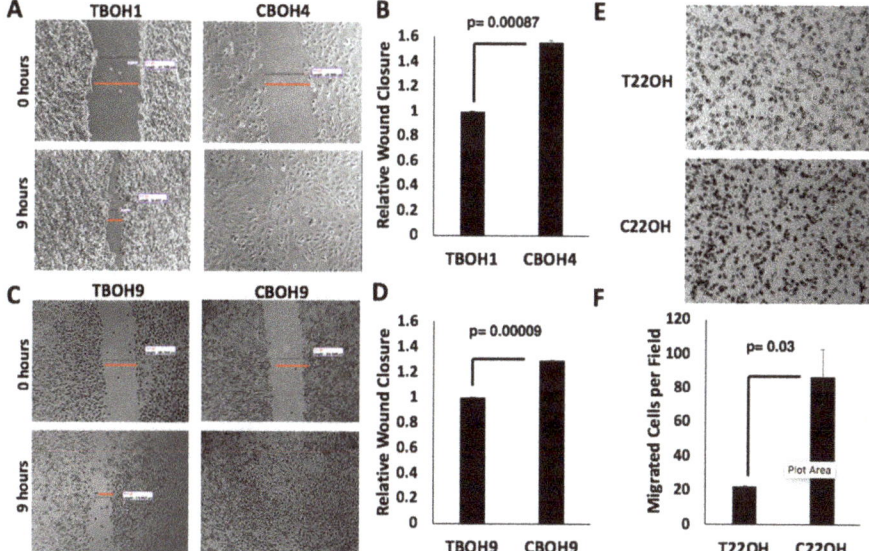

Figure 3. CTCs have greater migratory capacity than cancer cells from primary tumors. (**A,C**) Wound healing migration assays were performed on HCC cell lines. Cells were grown in 6-well plates. When confluent, wounds were made and measured at 0 and 9 h intervals. Wound closure by CTCs, but not by primary tumor-derived cell lines, was complete by 9 h. Images were taken at 10× magnification using Motic AE30 imaging software. (**B,D**) Migration (wound closure) was quantified; N = 3. (**E**) Transwell migration assays were performed on CRPC primary tumor-derived cell lines and CTC lines. Cells that migrated successfully through the pore membrane are represented by dark spots. Images shown were taken at 20× magnification. (**F**) Migration was quantified. Data provided on graphs are presented as mean ± standard error of the mean (SEM); N = 3.

To analyze the migratory capability of CRPC cells, we performed transwell migration assays. The number of cells passing through the chamber pores determined the migratory capability of cells. As shown in Figure 3E,F, the C22OH CTC cell line demonstrated a four-fold increase in migration in comparison to the primary tumor-derived cell line T22OH.

3.3. CTCs Exhibit Epithelial to Mesenchymal Transition (EMT)

Having observed greater migration from CTCs in comparison to primary tumor-derived cells, we wanted to determine whether CTCs were undergoing EMT. This is a phenomenon frequently observed in cancer cells migrating and metastasizing [34–39]. Using western blotting, we examined protein expression of fibronectin, a well-known marker of migratory and mesenchymal cells [40–44]. As observed in Figure 4A–C, CTCs had greater fibronectin protein expression in all three pairs of cell lines. More specifically, CBOH4 showed a 4.81-fold increase in expression of fibronectin when compared to TBOH1; CBOH9 exhibited a 3.95-fold increase in fibronectin expression when compared with TBOH9, and C22OH had a 3.52-fold increase in fibronectin expression in comparison with T22OH. Another marker examined to assess EMT was E-cadherin, a well-known cell adhesion protein characteristic of epithelial cells [45,46]. E-cadherin expression was decreased 11.1-fold in CBOH4 in comparison to TBOH1; decreased 5.5-fold in CBOH9 in comparison with TBOH9, and decreased 2.1-fold in C22OH in comparison with T22OH.

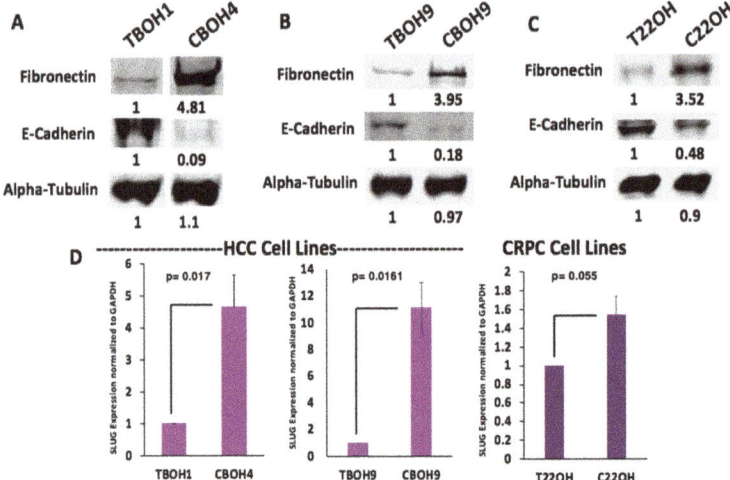

Figure 4. CTCs undergo epithelial-to-mesenchymal transition (EMT) as observed by increased fibronectin, decreased E-cadherin, and increased SLUG expression. (**A,B**) Fibronectin and E-cadherin protein expression by HCC primary tumor-derived cell lines and CTC lines. (**C**) Fibronectin and E-cadherin protein expression by CRPC primary tumor-derived cell line and CTC line. (**D**) SLUG expression was assessed using qPCR. Expression was normalized against GAPDH. Data provided on graphs are presented as mean ± standard error of the mean (SEM); N = 3.

EMT is also made possible by several transcription factors that initiate and maintain it [47]. One such transcription factor observed to be overexpressed in all three CTC cell lines was SLUG. The mRNA expression of this EMT transcription factor was assessed using qPCR. As shown in Figure 4D, CBOH4 exhibited a 4.6-fold increase in SLUG mRNA expression in comparison with TBOH1. CBOH9 demonstrated an 11.1-fold increase in SLUG expression in comparison to the corresponding primary tumor-derived cell line TBOH9. Finally, C22OH exhibited a 1.5-fold increase in SLUG mRNA expression in comparison to the T22OH cell line. These findings demonstrate that EMT is occurring in CTC lines and may be the reason they are more migratory than primary tumor-derived cell lines.

3.4. Fibronectin Expression Regulates Integrin B1 and SLUG Expression in CTCs

Both HCC and CRPC CTCs expressed significantly more fibronectin than primary tumor-derived cells. We therefore investigated the molecular mechanisms of action of fibronectin in CTCs by knocking down fibronectin expression in CTCs. CBOH4 and C22OH were transfected for 24 h with either 25 pmol of fibronectin-specific siRNA or scramble siRNA. In comparison with primary tumor-derived cells, CTCs had higher expression of integrin B1 as shown in Figure 5A,B. Integrins are heterodimeric, transmembrane cell surface receptors that have been linked with metastasis and tumor migration [48–50]. The integrin B1 subunit specifically has been frequently upregulated in tumors [51,52]. Interestingly, knockdown of fibronectin in CTCs resulted in decreased integrin B1 expression (19.4% decrease in fibronectin expression in CBOH4 with knockdown of fibronectin; and 20.3% decrease in C22OH with knockdown of fibronectin).

Like integrin B1, SLUG is also a molecule that has been associated with greater migratory capacity by cancer cells [53–55]. To determine the effect of fibronectin knockdown on SLUG expression in CTCs, we performed qPCR analysis. As shown in Figure 5C, CBOH4 transfected with fibronectin-specific siRNA demonstrated a 40% decrease in SLUG expression in comparison with CBOH4 transfected with scramble siRNA. Similarly, as shown in Figure 5D, C22OH transfected with the fibronectin-specific siRNA demonstrated a 30% decrease in SLUG expression in comparison with C22OH transfected with scramble siRNA. Therefore, fibronectin has significant regulatory effects on integrin B1 and SLUG expression in CTCs.

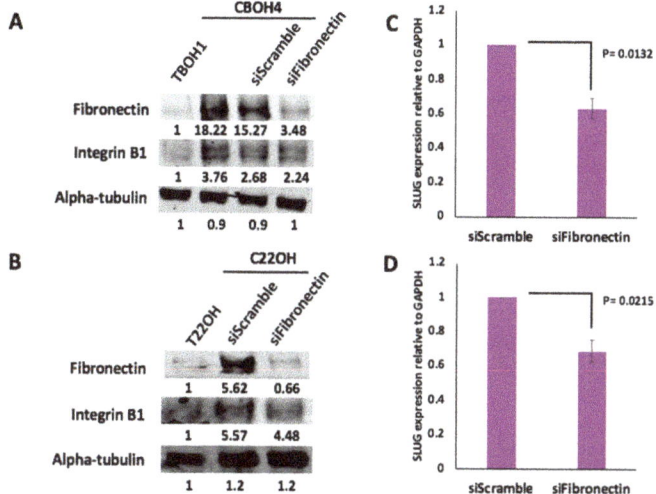

Figure 5. Fibronectin knockdown in CTCs caused decreased integrin B1 and SLUG expression. (A,B) After a 24 h transfection of CBOH4 and C22OH with fibronectin-specific siRNAs, western blotting was performed to determine the effects on integrin B1 expression. Western blotting revealed that fibronectin knockdown in CTCs was successful. The effect of fibronectin knockdown on integrin B1 expression was compared in control untransfected cells, CTCs transfected with scramble siRNAs, and CTCs transfected with fibronectin-specific siRNAs. Western blotting experiments were performed three separate times. (C,D) Effect of Fibronectin knockdown on SLUG expression in CBOH4 and C22OH CTCs was assessed by qPCR. Expression was normalized against GAPDH. Data provided on graphs are presented as mean ± standard error of the mean (SEM); N = 3.

3.5. HCC CTCs have Decreased MHCI Cell Surface Expression

To investigate potential immunomodulatory properties of CTCs, we used the cell lines derived from the syngeneic HCC mouse model. Using flow cytometry, we assayed for MHCI, a well-established cell surface molecule involved in self-recognition and identification of harmful entities for destruction by T-cells [56,57]. As shown in Figure 6A,B, we observed a 1.8-fold (45%) decrease in MHCI expression in the CBOH4 CTC line in comparison to the primary-tumor derived TBOH1 cell line. Similarly, as shown in Figure 6C,D, we observed a 1.5-fold (35%) decrease in MHCI expression in the CBOH9 CTC line in comparison to the TBOH9 primary tumor-derived cell line.

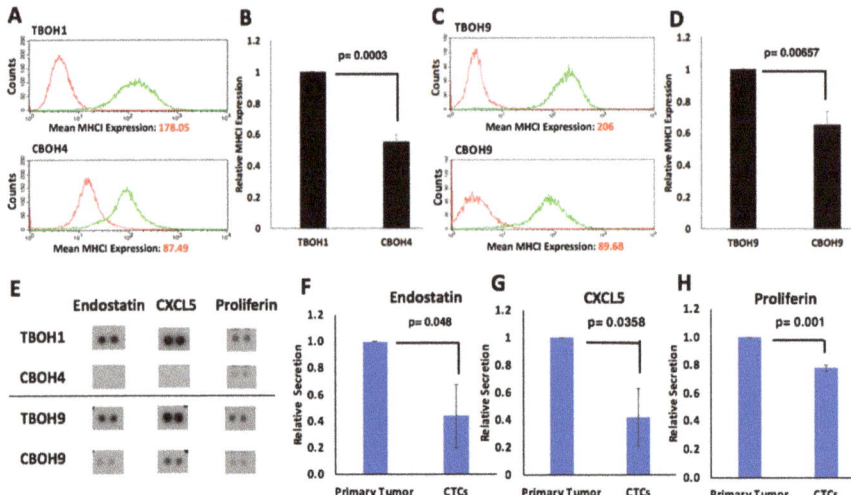

Figure 6. Immunomodulatory mechanisms of CTCs derived from a syngeneic mouse model of HCC. (**A–D**) Major histocompatibility complex class I (MHCI) expression was assessed using flow cytometry. CTCs have decreased MHCI cell surface protein expression in comparison to primary tumor-derived cell lines. Both isotype and MHCI antibodies were conjugated to fluorescein isothiocyanate (FITC). Isotype measuring background is shown in red. MHCI signal is shown in green; N = 5. (**E**) Analysis of 111 different cytokines secreted into cell media reveals consistent and significantly decreased secretion of endostatin, CXCL5, and proliferin in CTCs in comparison to primary tumor-derived cells; N = 2. (**F–H**) Endostatin, CXCL5, and proliferin signals from cytokine array were quantified. Data are presented as mean ± standard error of the mean (SEM).

3.6. HCC CTCs have Significantly Decreased Secretion of Endostatin, CXCL5, and Proliferin

Cytokines are secreted molecules that can activate signaling pathways and have effects on the immune response [58–60]. To determine if there are differences in the cytokine secretion profile between CTCs and primary tumor-derived cells, a cytokine array was used. Of the 111 cytokines assayed, three were consistently and significantly lower in CTCs (see Figure 6E). Endostatin, an anti-angiogenic molecule and inhibitor of tumor growth [61,62], was decreased 2.27-fold in CTCs in comparison to primary tumor-derived cells (Figure 6F). CXCL5, a molecule that plays a role in attracting leukocytes such as neutrophils [63,64], was also downregulated 2.38-fold in CTCs (Figure 6G). Finally, proliferin, a molecule with reported roles in cell growth regulation and differentiation [65–67], was decreased 1.28-fold in CTCs (Figure 6H). Statistical analyses found all three molecules to be significantly decreased with p values of <0.05.

3.7. CTCs, in Comparison to Primary Tumor-Derived Cells, Have Decreased Endostatin Expression

As a follow-up to the observation that CTCs secreted significantly reduced endostatin, we investigated intracellular endostatin expression in both of our HCC and CRPC CTC models using quantitative immunofluorescence. As shown in Figure 7A,B, CBOH4 had a 59% decrease in endostatin expression in comparison to TBOH1, and as shown in Figure 7C,D, CBOH9 had a 48% decrease in endostatin expression compared to TBOH9. Further, as shown in Figure 7E,F, C22OH had a 40% decrease in the expression of endostatin in comparison to T22OH. Thus, we can conclude that CTCs from both HCC and CRPC models examined expressed significantly less intracellular endostatin than their corresponding primary tumor-derived cells.

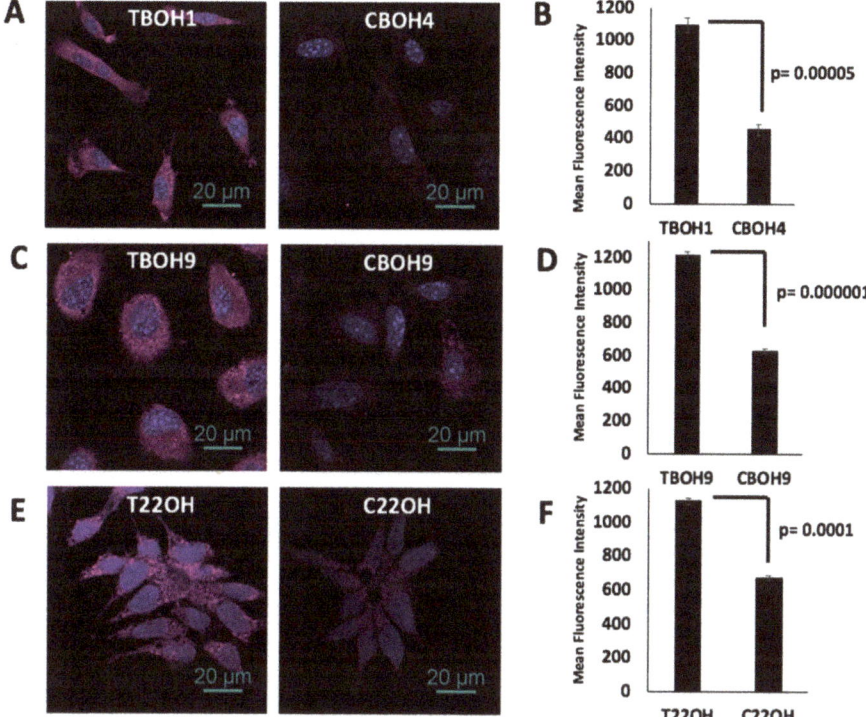

Figure 7. Decreased intracellular endostatin expression by CTCs. (**A,C,E**) Endostatin expression in TBOH1 and CBOH4, TBOH9 and CBOH9, and T22OH and C22OH was determined using immunofluorescence. (**B,D,F**) Mean fluorescence intensity was quantified using the NIS Elements software. Data are presented as mean ± standard error of the mean (SEM).

4. Discussion

Metastasis is the most common cause of death among cancer patients. Molecular mechanisms of cancer metastasis are not yet clear [1–3,33,68]. CTCs are an important step in the metastatic process. Consequently, in this study, we have focused on elucidating the molecular mechanisms of CTCs.

However, CTCs are not always abundant and are usually challenging to obtain in the human clinical setting [12–15]. To address these limitations, we have established an effective method for isolating CTCs, establishing long-term cultures, and propagating them into cell lines that are useful for studying CTC biology. The challenge involved in this task is underscored by the fact that only a few other groups have had success in doing this [69–72]. So far, it has been easier to establish short term CTC cultures, which when coupled with single-cell sequencing or short-term biochemical assays, have resulted in very useful information [73–82]. Further, these techniques can be physically intensive and expensive. Here, we have described an approach for propagating CTC cell lines and using them for functional characterization, identification of novel molecular pathways important to metastasis, as well as to gain insight into molecular receptors and secretions that might have immunomodulatory functions in CTCs.

In this study, we have focused on the use of HCC and CRPC models because they both have significant negative clinical outcomes. At present, there is no effective treatment for advanced HCC [16–18,20]. Similarly, metastatic CRPC currently has no cure [21–24,83].

In the present study, we have successfully propagated three pairs of CTC lines and their corresponding primary tumor-derived cell lines: two pairs of HCC origin and one pair of CRPC origin. Interestingly, all CTCs demonstrated significantly greater migratory capacity than their primary tumor-derived counterparts. Furthermore, we investigated the role of EMT in CTCs. EMT is a phenomenon that is frequently observed in cancer cells during cancer progression [34–36]. Accordingly, we observed evidence of EMT in CTCs as demonstrated by downregulation of E-cadherin and upregulation of fibronectin and SLUG expression. Therefore, our findings indicate that enhanced migratory capacity and EMT are characteristic of CTCs.

Because we observed fibronectin to be overexpressed by CTCs, we were also interested in identifying molecules regulated by fibronectin. Integrin receptors have long been associated with tumor migration [49]. While fibronectin binding and activation of these receptors is established [84,85], we report for the first time our observation that fibronectin has a regulatory role in the expression of the integrin B1 subunit. Fibronectin knockdown in CTCs resulted in decreased expression of integrin B1. Interestingly, another molecule significantly reduced by fibronectin knockdown in CTCs is SLUG. SLUG is a known regulator of EMT, and we observed SLUG overexpression in CTCs in comparison to primary tumor-derived cells. Upon fibronectin knockdown, SLUG expression was significantly decreased in CTCs. Thus, fibronectin expression regulates integrin B1 and SLUG expression in CTCs.

In this study, we also investigated major histocompatibility complex class I (MHCI) expression and secretion of cytokines by CTCs. The major histocompatibility complex class I (MHCI) is an important cell surface molecule that enables T cells to distinguish between "self" vs. "non-self". Upon recognition of harmful entities presented by MHCI, T cells will attack these MHCI-presenting cells [56,57]. We observed significant reduction of MHCI expression in CTCs arising from the syngeneic HCC mouse model. This has interesting implications on how CTCs may circumvent immunosurveillance. Additionally, we sought to identify molecules differentially secreted by CTCs in comparison to primary tumors. To this end, we assayed over 111 different cytokines using media secreted by cells. We found three molecules to be significantly decreased by CTCs: endostatin, CXCL5, and proliferin. We report for the first time that endostatin, CXCL5, and proliferin are significantly secreted less by CTCs than primary tumor-derived cells. Further studies are imperative to clarify the implications of this differential secretion profile of CTCs to cancer metastasis.

Endostatin is known for its anti-angiogenic properties [61,62,86]. The fact that CTCs express less endostatin is intriguing and suggests that CTCs have acquired enhanced angiogenicity, which bodes well for the colonization of secondary sites. Moreover, it is conceivable that decreased endostatin secretion by CTCs may enhance their metastatic capability in other important ways [87–89]. CXCL5 is a chemokine that recruits and activates neutrophils. While some studies associate higher CXCL5 expression with a worse cancer prognosis [63,64,90,91], several reports indicate that a lower CXCL5 expression can also promote metastatic spread [91–94]. Proliferin is a placental growth hormone that has been observed to be pro-angiogenic [65–67]. The implications of our current finding that CTCs secrete reduced amounts of proliferin than primary tumor-derived cells are currently unclear and deserve further study. It is noteworthy, though, that previous studies have found that as tumors grow in volume, intra-tumoral cells become hypoxic due to lack of oxygen within the tumor mass. This, in turn, stimulates the production of pro-angiogenic factors, enabling new blood vessel formation inside the tumor to deliver oxygen [95]. This process may not be necessary for CTCs, and this may explain the reduced secretion of proliferin by CTCs observed.

Interestingly, endostatin expression was found to be downregulated in all our CTCs regardless of tissue of origin. This was even more compelling since fibronectin, which was upregulated in our CTCs, has been previously reported to bind to the same integrin $\alpha5\beta1$ receptor as does endostatin [85,96–100]. It is plausible that there could be a potential autocrine, competitive binding occurring between fibronectin and endostatin to the integrin $\alpha5\beta1$ receptor with resulting implications in a cancer cell's ability to migrate. Further studies are necessary to confirm or refute this.

It is important to also note that while our work highlights a novel methodology and the creation of new cell lines to uncover possible mechanisms of cancer cell action, our cell lines were obtained from murine models. Others have previously isolated CTCs from human patients using microfluidic devices to assess molecular phenotype and drug sensitivity [9,69–72]. We propose our methodology to contribute to the field by adding yet another way in which we can harness the information gained from studying CTCs. In doing so, we may refute or support novel ideas such as CTCs being hybrids between leukocytes and cancer cells [101,102]. Whether this explains the reason our CTCs demonstrate enhanced migration and higher levels of integrin B1, which was previously reported in leukocytes [103,104], remains to be proven.

In summary, we have described the establishment of novel primary tumor-derived cell lines and CTC lines from the same mouse models and have used them to elucidate novel molecular mechanisms of CTCs. This work has demonstrated that EMT, enhanced migration, and decreased endostatin, CXCL5, and proliferin secretion are consistently seen in CTCs. Additionally, we report a novel role for fibronectin in regulating integrin B1 subunit receptor expression in CTCs, and while fibronectin's ability to regulate SLUG activity in renal cell carcinoma cells to promote lung metastasis has been reported [105], we report for the first time an observation of this molecular phenomenon in CTCs of HCC and CRPC origin. Our findings suggest that fibronectin's regulation of SLUG in CTCs may contribute to their role in cancer metastasis. Taken together, our findings demonstrate that CTCs have unique and important molecular mechanisms with implications for cancer metastasis.

5. Conclusions

In conclusion, we have demonstrated the successful establishment of novel primary tumor-derived cell lines and CTC lines from a syngeneic HCC mouse model and a human xenograft CRPC mouse model. The CTCs in these models demonstrate enhanced migration and EMT when compared to primary tumor-derived cells. Here, we report our novel finding of fibronectin's regulation of integrin B1 and SLUG expression in CTCs. This may be a mechanism by which CTCs ensure greater migration and metastasis. Further, we report our observation of decreased MHCI expression and decreased secretion of endostatin, CXCL5, and proliferin by CTCs in comparison to primary tumor-derived cells. These molecular mechanisms of CTCs likely have important implications for cancer metastasis.

Author Contributions: Conceptualization, O.O.; analysis, J.H., M.N., and O.O.; resources, O.O. and X.Z.; funding acquisition, O.O.; methodology, J.H. and O.O; supervision, O.O.; writing—original draft preparation, J.H.; writing—review and editing, O.O.

Funding: Jeannette Huaman was supported by the Research Initiative for Scientific Enhancement (RISE) program at Hunter College, funded by NIH grant GM060665. Dr. Olorunseun Ogunwobi was supported by National Cancer Institute grant number U54CA221704.

Acknowledgments: We would like to thank Kim Ohaegbulam for assistance in facilitating work done using the core facilities at Albert Einstein School of Medicine.

Conflicts of Interest: Olorunseun Ogunwobi is Co-Founder of NucleoBio, Inc, a City University of New York start-up biotechnology company. Other authors declare no conflict of interest. The funders had no role in the design of the study; in the collection, analyses, or interpretation of data; in the writing of the manuscript, or in the decision to publish the results.

References

1. Lee, W.C.; Kopetz, S.; Wistuba, I.I.; Zhang, J. Metastasis of cancer: When and how? *Ann. Oncol.* **2017**, *28*, 2045–2047. [CrossRef]
2. Chaffer, C.L.; Weinberg, R.A. A perspective on cancer cell metastasis. *Science* **2011**, *331*, 1559–1564. [CrossRef] [PubMed]
3. Gupta, G.P.; Massague, J. Cancer metastasis: Building a framework. *Cell* **2006**, *127*, 679–695. [CrossRef] [PubMed]
4. Van de Stolpe, A.; Pantel, K.; Sleijfer, S.; Terstappen, L.W.; Den Toonder, J.M. Circulating tumor cell isolation and diagnostics: Toward routine clinical use. *Cancer Res.* **2011**, *71*, 5955–5960. [CrossRef]
5. Gallerani, G.; Fici, P.; Fabbri, F. Circulating Tumor Cells: Back to the Future. *Front. Oncol.* **2016**, *6*, 275. [CrossRef] [PubMed]

6. Haber, D.A.; Velculescu, V.E. Blood-based analyses of cancer: Circulating tumor cells and circulating tumor DNA. *Cancer Discov.* **2014**, *4*, 650–661. [CrossRef] [PubMed]
7. Pantel, K.; Brakenhoff, R.H.; Brandt, B. Detection, clinical relevance and specific biological properties of disseminating tumour cells. *Nat. Rev. Cancer* **2008**, *8*, 329–340. [CrossRef]
8. Bailey, P.M.; Martin, S.S. Insights on CTC Biology and Clinical Impact Emerging from Advances in Capture Technology. *Cells* **2019**, *8*, 553. [CrossRef]
9. George, T.J., Jr.; Ogunwobi, O.O.; Sheng, W.; Fan, Z.H.; Liu, C. "Tissue is the issue": Circulating tumor cells in pancreatic cancer. *J. Gastrointest. Cancer* **2014**, *45*, 222–225. [CrossRef]
10. Micalizzi, D.S.; Maheswaran, S.; Haber, D.A. A conduit to metastasis: Circulating tumor cell biology. *Genes Dev.* **2017**, *31*, 1827–1840. [CrossRef]
11. Alix-Panabieres, C.; Pantel, K. Challenges in circulating tumour cell research. *Nat. Rev. Cancer* **2014**, *14*, 623–631. [CrossRef] [PubMed]
12. Alvarez Cubero, M.J.; Lorente, J.A.; Robles-Fernandez, I.; Rodriguez-Martinez, A.; Puche, J.L.; Serrano, M.J. Circulating Tumor Cells: Markers and Methodologies for Enrichment and Detection. *Methods Mol. Biol.* **2017**, *1634*, 283–303. [PubMed]
13. Cristofanilli, M.; Budd, G.T.; Ellis, M.J.; Stopeck, A.; Matera, J.; Miller, M.C.; Reuben, J.M.; Doyle, G.V.; Allard, W.J.; Terstappen, L.W.; et al. Circulating tumor cells, disease progression, and survival in metastatic breast cancer. *N. Engl. J. Med.* **2004**, *351*, 781–791. [CrossRef] [PubMed]
14. Miller, M.C.; Doyle, G.V.; Terstappen, L.W. Significance of Circulating Tumor Cells Detected by the CellSearch System in Patients with Metastatic Breast Colorectal and Prostate Cancer. *J. Oncol.* **2010**, *2010*, 617421. [CrossRef] [PubMed]
15. Zieglschmid, V.; Hollmann, C.; Bocher, O. Detection of disseminated tumor cells in peripheral blood. *Crit. Rev. Clin. Lab. Sci.* **2005**, *42*, 155–196. [CrossRef] [PubMed]
16. American Cancer Society. Cancer Facts and Figures 2018. Available online: https://www.cancer.org/content/dam/cancer-org/research/cancer-facts-and-statistics/annual-cancer-facts-and-figures/2018/cancer-facts-and-figures-2018.pdf (accessed on 19 June 2019).
17. Llovet, J.M.; Zucman-Rossi, J.; Pikarsky, E.; Sangro, B.; Schwartz, M.; Sherman, M.; Gores, G. Hepatocellular carcinoma. *Nat. Rev. Dis. Primers* **2016**, *2*, 16018. [CrossRef]
18. Llovet, J.M.; Montal, R.; Sia, D.; Finn, R.S. Molecular therapies and precision medicine for hepatocellular carcinoma. *Nat. Rev. Clin. Oncol.* **2018**, *15*, 599–616. [CrossRef]
19. Ogunwobi, O.H.; Harricharran, T.; Huaman, J.; Galuza, A.; Odumuwagun, O.; Tan, Y.; Ma, G.X.; Nguyen, M.T. Mechanisms of hepatocellular carcinoma progression. *World J. Gastroenterol.* **2019**, *25*, 2279–2293. [CrossRef]
20. Keating, G.M. Sorafenib: A Review in Hepatocellular Carcinoma. *Target Oncol.* **2017**, *12*, 243–253. [CrossRef]
21. Kirby, M.; Hirst, C.; Crawford, E.D. Characterising the castration-resistant prostate cancer population: A systematic review. *Int. J. Clin. Pract.* **2011**, *65*, 1180–1192. [CrossRef]
22. Suzman, D.L.; Antonarakis, E.S. Castration-resistant prostate cancer: Latest evidence and therapeutic implications. *Ther. Adv. Med. Oncol.* **2014**, *6*, 167–179. [CrossRef] [PubMed]
23. Pentyala, S.; Whyard, T.; Pentyala, S.; Muller, J.; Pfail, J.; Parmar, S.; Helguero, C.G.; Khan, S. Prostate cancer markers: An update. *Biomed. Rep.* **2016**, *4*, 263–268. [CrossRef] [PubMed]
24. Litwin, M.S.; Tan, H.J. The Diagnosis and Treatment of Prostate Cancer: A Review. *JAMA* **2017**, *317*, 2532–2542. [CrossRef] [PubMed]
25. Hong, J.H.; Kim, I.Y. Nonmetastatic castration-resistant prostate cancer. *Korean J. Urol.* **2014**, *55*, 153–160. [CrossRef] [PubMed]
26. Das, D.K.; Naidoo, M.K.; Ilboudo, A.; DuBois, P.; Durojaiye, V.; Liu, C.; Ogunwobi, O.O. Isolation and Propagation of Circulating Tumor Cells from a Mouse Cancer Model. *J. Vis. Exp.* **2015**. [CrossRef] [PubMed]
27. Ogunwobi, O.O.; Puszyk, W.; Dong, H.J.; Liu, C. Epigenetic upregulation of HGF and c-Met drives metastasis in hepatocellular carcinoma. *PLoS ONE* **2013**, *8*, e63765. [CrossRef] [PubMed]
28. Omori, Y.; Imai, J.; Watanabe, M.; Komatsu, T.; Suzuki, Y.; Kataoka, K.; Watanabe, S.; Tanigami, A.; Sugano, S. CREB-H: A novel mammalian transcription factor belonging to the CREB/ATF family and functioning via the box-B element with a liver-specific expression. *Nucleic Acids Res.* **2001**, *29*, 2154–2162. [CrossRef]
29. Barbosa, S.; Fasanella, G.; Carreira, S.; Llarena, M.; Fox, R.; Barreca, C.; Andrew, D.; O'Hare, P. An orchestrated program regulating secretory pathway genes and cargos by the transmembrane transcription factor CREB-H. *Traffic* **2013**, *14*, 382–398. [CrossRef]

30. Viswanathan, S.R.; Ha, G.; Hoff, A.M.; Wala, J.A.; Carrot-Zhang, J.; Whelan, C.W.; Haradhvala, N.J.; Freeman, S.S.; Reed, S.C.; Rhoades, J.; et al. Structural Alterations Driving Castration-Resistant Prostate Cancer Revealed by Linked-Read Genome Sequencing. *Cell* **2018**, *174*, 433–447. [CrossRef]
31. Wadosky, K.M.; Koochekpour, S. Androgen receptor splice variants and prostate cancer: From bench to bedside. *Oncotarget* **2017**, *8*, 18550–18576. [CrossRef]
32. Yilmaz, M.; Christofori, G. Mechanisms of motility in metastasizing cells. *Mol. Cancer Res.* **2010**, *8*, 629–642. [CrossRef] [PubMed]
33. Lambert, A.W.; Pattabiraman, D.R.; Weinberg, R.A. Emerging biological principles of metastasis. *Cell* **2017**, *168*, 670–691. [CrossRef] [PubMed]
34. Brabletz, T.; Kalluri, R.; Nieto, M.A.; Weinberg, R.A. EMT in cancer. *Nat. Rev. Cancer* **2018**, *18*, 128–134. [CrossRef] [PubMed]
35. Kalluri, R.; Weinberg, R.A. The basics of epithelial-mesenchymal transition. *J. Clin. Invest.* **2009**, *119*, 1420–1428. [CrossRef] [PubMed]
36. Lamouille, S.; Xu, J.; Derynck, R. Molecular mechanisms of epithelial-mesenchymal transition. *Nat. Rev. Mol. Cell Biol.* **2014**, *15*, 178–196. [CrossRef]
37. Tiwari, N.; Gheldof, A.; Tatari, M.; Christofori, G. EMT as the ultimate survival mechanism of cancer cells. *Semin. Cancer Biol.* **2012**, *22*, 194–207. [CrossRef] [PubMed]
38. Ye, X.; Weinberg, R.A. Epithelial-Mesenchymal Plasticity: A Central Regulator of Cancer Progression. *Trends Cell Biol.* **2015**, *25*, 675–686. [CrossRef]
39. Alix-Panabieres, C.; Mader, S.; Pantel, K. Epithelial-mesenchymal plasticity in circulating tumor cells. *J. Mol. Med.* **2017**, *95*, 133–142. [CrossRef]
40. Meng, X.N.; Jin, Y.; Yu, Y.; Bai, J.; Liu, G.Y.; Zhu, J.; Zhao, Y.Z.; Wang, Z.; Chen, F.; Lee, K.Y.; et al. Characterisation of fibronectin-mediated FAK signalling pathways in lung cancer cell migration and invasion. *Br. J. Cancer* **2009**, *101*, 327–334. [CrossRef]
41. Park, J.; Schwarzbauer, J.E. Mammary epithelial cell interactions with fibronectin stimulate epithelial-mesenchymal transition. *Oncogene* **2014**, *33*, 1649–1657. [CrossRef]
42. Yousif, N.G. Fibronectin promotes migration and invasion of ovarian cancer cells through up-regulation of FAK-PI3K/Akt pathway. *Cell. Biol. Int.* **2014**, *38*, 85–91. [CrossRef] [PubMed]
43. Ramos Gde, O.; Bernardi, L.; Lauxen, I.; Sant'Ana Filho, M.; Horwitz, A.R.; Lamers, M.L. Fibronectin Modulates Cell Adhesion and Signaling to Promote Single Cell Migration of Highly Invasive Oral Squamous Cell Carcinoma. *PLoS ONE* **2016**, *11*, e0151338. [CrossRef] [PubMed]
44. Wang, J.P.; Hielscher, A. Fibronectin: How its Aberrant Expression in Tumors may Improve Therapeutic Targeting. *J. Cancer* **2017**, *8*, 674–682. [CrossRef] [PubMed]
45. Jeanes, A.; Gottardi, C.J.; Yap, A.S. Cadherins and cancer: How does cadherin dysfunction promote tumor progression? *Oncogene* **2008**, *27*, 6920–6929. [CrossRef] [PubMed]
46. Wong, S.H.M.; Fang, C.M.; Chuah, L.H.; Leong, C.O.; Ngai, S.C. E-cadherin: Its dysregulation in carcinogenesis and clinical implications. *Crit. Rev. Oncol. Hematol.* **2018**, *121*, 11–22. [CrossRef] [PubMed]
47. Puisieux, A.; Brabletz, T.; Caramel, J. Oncogenic roles of EMT-inducing transcription factors. *Nat. Cell. Biol.* **2014**, *16*, 488–494. [CrossRef]
48. Desgrosellier, J.S.; Cheresh, D.A. Integrins in cancer: Biological implications and therapeutic opportunities. *Nat. Rev. Cancer* **2010**, *10*, 9–22. [CrossRef]
49. Hamidi, H.; Ivaska, J. Every step of the way: Integrins in cancer progression and metastasis. *Nat. Rev. Cancer* **2018**, *18*, 533–548. [CrossRef]
50. Seguin, L.; Desgrosellier, J.S.; Weis, S.M.; Cheresh, D.A. Integrins and cancer: Regulators of cancer stemness, metastasis, and drug resistance. *Trends Cell Biol.* **2015**, *25*, 234–240. [CrossRef]
51. Blandin, A.F.; Renner, G.; Lehmann, M.; Lelong-Rebel, I.; Martin, S.; Dontenwill, M. beta1 Integrins as Therapeutic Targets to Disrupt Hallmarks of Cancer. *Front. Pharmacol.* **2015**, *6*, 279. [CrossRef]
52. Howe, G.A.; Addison, C.L. beta1 integrin: An emerging player in the modulation of tumorigenesis and response to therapy. *Cell Adh. Migr.* **2012**, *6*, 71–77. [CrossRef] [PubMed]
53. Gu, A.; Jie, Y.; Yao, Q.; Zhang, Y.; Mingyan, E. Slug is associated with tumor metastasis and angiogenesis in ovarian cancer. *Reprod. Sci.* **2017**, *24*, 291–299. [CrossRef] [PubMed]
54. Uygur, B.; Wu, W.S. SLUG promotes prostate cancer cell migration and invasion via CXCR4/CXCL12 axis. *Mol. Cancer* **2011**, *10*, 139. [CrossRef] [PubMed]

55. Yu, M.; Chen, Y.; Li, X.; Yang, R.; Zhang, L.; Huangfu, L.; Zheng, N.; Zhao, X.; Lv, L.; Hong, Y.; et al. YAP1 contributes to NSCLC invasion and migration by promoting Slug transcription via the transcription co-factor TEAD. *Cell Death Dis.* **2018**, *9*, 464. [CrossRef] [PubMed]
56. Bubenik, J. MHC class I down-regulation: Tumour escape from immune surveillance? (review). *Int. J. Oncol.* **2004**, *25*, 487–491. [CrossRef] [PubMed]
57. Shastri, N.; Nagarajan, N.; Lind, K.C.; Kanaseki, T. Monitoring peptide processing for MHC class I molecules in the endoplasmic reticulum. *Curr. Opin. Immunol.* **2014**, *26*, 123–127. [CrossRef]
58. Jang, M.H.; Seoh, J.Y.; Miyasaka, M. Cytokines, chemokines, and their receptors: Targets for immunomodulation. Conference report: International Cytokine Society Conference 2005. *J. Leukoc. Biol.* **2006**, *80*, 217–219. [CrossRef]
59. Turner, M.D.; Nedjai, B.; Hurst, T.; Pennington, D.J. Cytokines and chemokines: At the crossroads of cell signalling and inflammatory disease. *Biochim. Biophys. Acta* **2014**, *1843*, 2563–2582. [CrossRef]
60. Lee, M.; Rhee, I. Cytokine Signaling in Tumor Progression. *Immune Netw.* **2017**, *17*, 214–227. [CrossRef]
61. Cao, Y.; Langer, R. A review of Judah Folkman's remarkable achievements in biomedicine. *Proc. Natl. Acad. Sci. USA* **2008**, *105*, 13203–13205. [CrossRef]
62. Dhanabal, M.; Ramchandran, R.; Volk, R.; Stillman, I.E.; Lombardo, M.; Iruela-Arispe, M.L.; Simons, M.; Sukhatme, V.P. Endostatin: Yeast production, mutants, and antitumor effect in renal cell carcinoma. *Cancer Res.* **1999**, *59*, 189–197. [PubMed]
63. Walz, A.; Burgener, R.; Car, B.; Baggiolini, M.; Kunkel, S.L.; Strieter, R.M. Structure and neutrophil-activating properties of a novel inflammatory peptide (ENA-78) with homology to interleukin 8. *J. Exp. Med.* **1991**, *174*, 1355–1362. [CrossRef] [PubMed]
64. Xia, J.; Xu, X.; Huang, P.; He, M.; Wang, X. The potential of CXCL5 as a target for liver cancer - what do we know so far? *Expert Opin. Ther. Targets* **2015**, *19*, 141–146. [CrossRef] [PubMed]
65. Linzer, D.I.; Lee, S.J.; Ogren, L.; Talamantes, F.; Nathans, D. Identification of proliferin mRNA and protein in mouse placenta. *Proc. Natl. Acad. Sci. USA* **1985**, *82*, 4356–4359. [CrossRef] [PubMed]
66. Jackson, D.; Volpert, O.V.; Bouck, N.; Linzer, D.I. Stimulation and inhibition of angiogenesis by placental proliferin and proliferin-related protein. *Science* **1994**, *266*, 1581–1584. [CrossRef] [PubMed]
67. Toft, D.J.; Rosenberg, S.B.; Bergers, G.; Volpert, O.; Linzer, D.I. Reactivation of proliferin gene expression is associated with increased angiogenesis in a cell culture model of fibrosarcoma tumor progression. *Proc. Natl. Acad. Sci. USA* **2001**, *98*, 13055–13059. [CrossRef] [PubMed]
68. Tabassum, D.P.; Polyak, K. Tumorigenesis: It takes a village. *Nat. Rev. Cancer* **2015**, *15*, 473–483. [CrossRef] [PubMed]
69. Cayrefourcq, L.; Mazard, T.; Joosse, S.; Solassol, J.; Ramos, J.; Assenat, E.; Schumacher, U.; Costes, V.; Maudelonde, T.; Pantel, K.; et al. Establishment and characterization of a cell line from human circulating colon cancer cells. *Cancer Res.* **2015**, *75*, 892–901. [CrossRef]
70. Gao, D.; Vela, I.; Sboner, A.; Iaquinta, P.J.; Karthaus, W.R.; Gopalan, A.; Dowling, C.; Wanjala, J.N.; Undvall, E.A.; Arora, V.K.; et al. Organoid cultures derived from patients with advanced prostate cancer. *Cell* **2014**, *159*, 176–187. [CrossRef]
71. Hamilton, G.; Burghuber, O.; Zeillinger, R. Circulating tumor cells in small cell lung cancer: Ex vivo expansion. *Lung* **2015**, *193*, 451–452. [CrossRef]
72. Yu, M.; Bardia, A.; Aceto, N.; Bersani, F.; Madden, M.W.; Donaldson, M.C.; Desai, R.; Zhu, H.; Comaills, V.; Zheng, Z.; et al. Cancer therapy. Ex vivo culture of circulating breast tumor cells for individualized testing of drug susceptibility. *Science* **2014**, *345*, 216–220. [CrossRef] [PubMed]
73. Bobek, V.; Kacprzak, G.; Rzechonek, A.; Kolostova, K. Detection and cultivation of circulating tumor cells in malignant pleural mesothelioma. *Anticancer Res.* **2014**, *34*, 2565–2569. [PubMed]
74. Bobek, V.; Matkowski, R.; Gurlich, R.; Grabowski, K.; Szelachowska, J.; Lischke, R.; Schutzner, J.; Harustiak, T.; Pazdro, A.; Rzechonek, A.; et al. Cultivation of circulating tumor cells in esophageal cancer. *Folia Histochem. Cytobiol.* **2014**, *52*, 171–177. [CrossRef] [PubMed]
75. Cegan, M.; Kolostova, K.; Matkowski, R.; Broul, M.; Schraml, J.; Fiutowski, M.; Bobek, V. In vitro culturing of viable circulating tumor cells of urinary bladder cancer. *Int. J. Clin. Exp. Pathol.* **2014**, *7*, 7164–7171. [PubMed]
76. Kang, J.H.; Krause, S.; Tobin, H.; Mammoto, A.; Kanapathipillai, M.; Ingber, D.E. A combined micromagnetic-microfluidic device for rapid capture and culture of rare circulating tumor cells. *Lab Chip* **2012**, *12*, 2175–2181. [CrossRef] [PubMed]

77. Kolostova, K.; Matkowski, R.; Gurlich, R.; Grabowski, K.; Soter, K.; Lischke, R.; Schutzner, J.; Bobek, V. Detection and cultivation of circulating tumor cells in gastric cancer. *Cytotechnology* **2016**, *68*, 1095–1102. [CrossRef]
78. Kolostova, K.; Zhang, Y.; Hoffman, R.M.; Bobek, V. In vitro culture and characterization of human lung cancer circulating tumor cells isolated by size exclusion from an orthotopic nude-mouse model expressing fluorescent protein. *J. Fluoresc.* **2014**, *24*, 1531–1536. [CrossRef]
79. Kulasinghe, A.; Perry, C.; Warkiani, M.E.; Blick, T.; Davies, A.; O'Byrne, K.; Thompson, E.W.; Nelson, C.C.; Vela, I.; Punyadeera, C. Short term ex-vivo expansion of circulating head and neck tumour cells. *Oncotarget* **2016**, *7*, 60101–60109. [CrossRef]
80. Ting, D.T.; Wittner, B.S.; Ligorio, M.; Vincent Jordan, N.; Shah, A.M.; Miyamoto, D.T.; Aceto, N.; Bersani, F.; Brannigan, B.W.; Xega, K.; et al. Single-cell RNA sequencing identifies extracellular matrix gene expression by pancreatic circulating tumor cells. *Cell Rep.* **2014**, *8*, 1905–1918. [CrossRef]
81. Zhang, Z.; Shiratsuchi, H.; Lin, J.; Chen, G.; Reddy, R.M.; Azizi, E.; Fouladdel, S.; Chang, A.C.; Lin, L.; Jiang, H.; et al. Expansion of CTCs from early stage lung cancer patients using a microfluidic co-culture model. *Oncotarget* **2014**, *5*, 12383–12397. [CrossRef]
82. Zhang, Z.; Shiratsuchi, H.; Palanisamy, N.; Nagrath, S.; Ramnath, N. Expanded Circulating Tumor Cells from a Patient with ALK-Positive Lung Cancer Present with EML4-ALK Rearrangement Along with Resistance Mutation and Enable Drug Sensitivity Testing: A Case Study. *J. Thorac. Oncol.* **2017**, *12*, 397–402. [CrossRef] [PubMed]
83. Nuhn, P.; De Bono, J.S.; Fizazi, K.; Freedland, S.J.; Grilli, M.; Kantoff, P.W.; Sonpavde, G.; Sternberg, C.N.; Yegnasubramanian, S.; Antonarakis, E.S. Update on Systemic Prostate Cancer Therapies: Management of Metastatic Castration-resistant Prostate Cancer in the Era of Precision Oncology. *Eur. Urol.* **2019**, *75*, 88–99. [CrossRef] [PubMed]
84. Wang, L.; Pan, D.; Yan, Q.; Song, Y. Activation mechanisms of alphaVbeta3 integrin by binding to fibronectin: A computational study. *Protein Sci.* **2017**, *26*, 1124–1137. [CrossRef] [PubMed]
85. Schaffner, F.; Ray, A.M.; Dontenwill, M. Integrin alpha5beta1, the Fibronectin Receptor, as a Pertinent Therapeutic Target in Solid Tumors. *Cancers* **2013**, *5*, 27–47. [CrossRef]
86. Folkman, J. Antiangiogenesis in cancer therapy—Endostatin and its mechanisms of action. *Exp. Cell Res.* **2006**, *312*, 594–607. [CrossRef]
87. Wang, S.; Lu, X.-A.; Liu, P.; Fu, Y.; Jia, L.; Zhan, S.; Luo, Y. Endostatin Has ATPase Activity, Which Mediates Its Antiangiogenic and Antitumor Activities. *Mol. Cancer Ther.* **2015**, *14*, 1192. [CrossRef]
88. Sato, Y. Endostatin as a Biomarker of Basement Membrane Degradation. *J. Atheroscler. Thromb.* **2017**, *24*, 1014–1015. [CrossRef]
89. Walia, A.; Yang, J.F.; Huang, Y.-H.; Rosenblatt, M.I.; Chang, J.-H.; Azar, D.T. Endostatin's Emerging Roles in Angiogenesis, Lymphangiogenesis, Disease, and Clinical Applications. *Biochim. Biophys. Acta* **2015**, *1850*, 2422–2438. [CrossRef]
90. Ocana, A.; Nieto-Jimenez, C.; Pandiella, A.; Templeton, A.J. Neutrophils in cancer: Prognostic role and therapeutic strategies. *Mol. Cancer* **2017**, *16*, 137. [CrossRef]
91. Coffelt, S.B.; Wellenstein, M.D.; De Visser, K.E. Neutrophils in cancer: Neutral no more. *Nat. Rev. Cancer* **2016**, *16*, 431–446. [CrossRef]
92. Granot, Z.; Henke, E.; Comen, E.A.; King, T.A.; Norton, L.; Benezra, R. Tumor entrained neutrophils inhibit seeding in the premetastatic lung. *Cancer Cell* **2011**, *20*, 300–314. [CrossRef] [PubMed]
93. Speetjens, F.M.; Kuppen, P.J.; Sandel, M.H.; Menon, A.G.; Burg, D.; Van de Velde, C.J.; Tollenaar, R.A.; De Bont, H.J.; Nagelkerke, J.F. Disrupted expression of CXCL5 in colorectal cancer is associated with rapid tumor formation in rats and poor prognosis in patients. *Clin. Cancer Res.* **2008**, *14*, 2276–2284. [CrossRef] [PubMed]
94. Blaisdell, A.; Crequer, A.; Columbus, D.; Daikoku, T.; Mittal, K.; Dey, S.K.; Erlebacher, A. Neutrophils Oppose Uterine Epithelial Carcinogenesis via Debridement of Hypoxic Tumor Cells. *Cancer Cell* **2015**, *28*, 785–799. [CrossRef] [PubMed]
95. Liao, D.; Johnson, R.S. Hypoxia: A key regulator of angiogenesis in cancer. *Cancer Metastasis Rev.* **2007**, *26*, 281–290. [CrossRef] [PubMed]

96. Faye, C.; Moreau, C.; Chautard, E.; Jetne, R.; Fukai, N.; Ruggiero, F.; Humphries, M.J.; Olsen, B.R.; Ricard-Blum, S. Molecular interplay between endostatin, integrins, and heparan sulfate. *J. Biol. Chem.* **2009**, *284*, 22029–22040. [CrossRef]
97. Rehn, M.; Veikkola, T.; Kukk-Valdre, E.; Nakamura, H.; Ilmonen, M.; Lombardo, C.; Pihlajaniemi, T.; Alitalo, K.; Vuori, K. Interaction of endostatin with integrins implicated in angiogenesis. *Proc. Natl. Acad. Sci. USA* **2001**, *98*, 1024–1029. [CrossRef]
98. Sudhakar, A.; Sugimoto, H.; Yang, C.; Lively, J.; Zeisberg, M.; Kalluri, R. Human tumstatin and human endostatin exhibit distinct antiangiogenic activities mediated by alpha v beta 3 and alpha 5 beta 1 integrins. *Proc. Natl. Acad. Sci. USA* **2003**, *100*, 4766–4771. [CrossRef]
99. Wickstrom, S.A.; Alitalo, K.; Keski-Oja, J. Endostatin associates with integrin alpha5beta1 and caveolin-1, and activates Src via a tyrosyl phosphatase-dependent pathway in human endothelial cells. *Cancer Res.* **2002**, *62*, 5580–5589.
100. Yokoyama, Y.; Ramakrishnan, S. Binding of endostatin to human ovarian cancer cells inhibits cell attachment. *Int. J. Cancer* **2007**, *121*, 2402–2409. [CrossRef]
101. Gast, C.E.; Silk, A.D.; Zarour, L.; Riegler, L.; Burkhart, J.G.; Gustafson, K.T.; Parappilly, M.S.; Roh-Johnson, M.; Goodman, J.R.; Olson, B.; et al. Cell fusion potentiates tumor heterogeneity and reveals circulating hybrid cells that correlate with stage and survival. *Sci. Adv.* **2018**, *4*, eaat7828. [CrossRef]
102. Laberge, G.S.; Duvall, E.; Haedicke, K.; Pawelek, J. Leukocyte-cancer cell fusion—Genesis of a deadly journey. *Cells* **2019**, *8*, 170. [CrossRef] [PubMed]
103. Sallusto, F.; Baggiolini, M. Chemokines and leukocyte traffic. *Nat. Immunol.* **2008**, *9*, 949–952. [CrossRef] [PubMed]
104. Chakraborty, A.K.; Funasaka, Y.; Ichihashi, M.; Pawelek, J.M. Upregulation of alpha and beta integrin subunits in metastatic macrophage-melanoma fusion hybrids. *Melanoma Res.* **2009**, *19*, 343–349. [CrossRef] [PubMed]
105. Knowles, L.M.; Gurski, L.A.; Engel, C.; Gnarra, J.R.; Maranchie, J.K.; Pilch, J. Integrin alphavbeta3 and fibronectin upregulate Slug in cancer cells to promote clot invasion and metastasis. *Cancer Res.* **2013**, *73*, 6175–6184. [CrossRef] [PubMed]

© 2019 by the authors. Licensee MDPI, Basel, Switzerland. This article is an open access article distributed under the terms and conditions of the Creative Commons Attribution (CC BY) license (http://creativecommons.org/licenses/by/4.0/).

Article

Circulating Tumor Cells and Circulating Tumor DNA Detection in Potentially Resectable Metastatic Colorectal Cancer: A Prospective Ancillary Study to the Unicancer Prodige-14 Trial

François-Clément Bidard [1,2,3], Nicolas Kiavue [1,*], Marc Ychou [4,5], Luc Cabel [1,2,3], Marc-Henri Stern [6], Jordan Madic [2], Adrien Saliou [2], Aurore Rampanou [2], Charles Decraene [2,7], Olivier Bouché [8], Michel Rivoire [9], François Ghiringhelli [10], Eric Francois [11], Rosine Guimbaud [12], Laurent Mineur [13], Faiza Khemissa-Akouz [14], Thibault Mazard [4], Driffa Moussata [15], Charlotte Proudhon [2], Jean-Yves Pierga [1,2,16], Trevor Stanbury [17], Simon Thézenas [18] and Pascale Mariani [19]

1. Department of Medical Oncology, Institut Curie, PSL Research University, 75005 Paris, France; francois-clement.bidard@curie.fr (F.-C.B.); luc.cabel@curie.fr (L.C.); jean-yves.pierga@curie.fr (J.-Y.P.)
2. Circulating Tumor Biomarkers Laboratory, Institut Curie, PSL Research University, 75005 Paris, France; jordanmadic@yahoo.fr (J.M.); saliou.adrien@gmail.com (A.S.); aurore.rampanou@curie.fr (A.R.); charles.decraene@curie.fr (C.D.); charlotte.proudhon@curie.fr (C.P.)
3. UVSQ, Paris Saclay University, 92210 Saint Cloud, France
4. Department of Digestive Oncology, ICM Regional Cancer Institute of Montpellier, 34298 Montpellier, France; marc.ychou@icm.unicancer.fr (M.Y.); t-mazard@chu-montpellier.fr (T.M.)
5. Department of Oncology, Montpellier University, 34000 Montpellier, France
6. INSERM U830, Institut Curie, PSL Research University, 75005 Paris, France; marc-henri.stern@curie.fr
7. CNRS UMR144, Institut Curie, PSL Research University, 75005 Paris, France
8. Department of Medical Oncology, Hôpital Robert Debré, Reims University Hospital, 51100 Reims, France; obouche@chu-reims.fr
9. Department of Digestive Oncology, Centre Léon Bérard, 69008 Lyon, France; michel.rivoire@lyon.unicancer.fr
10. INSERM U866, Centre Georges-François Leclerc, 21000 Dijon, France; fghiringhelli@cgfl.fr
11. Department of Medical Oncology, Centre Antoine Lacassagne, 06189 Nice, France; eric.francois@nice.unicancer.fr
12. Department of Digestive Oncology, CHU de Toulouse, 31059 Toulouse, France; guimbaud.r@chu-toulouse.fr
13. Department of Digestive Oncology, Institut Sainte Catherine, 84000 Avignon, France; l.mineur@isc84.org
14. Department of Gastroenterology, Hôpital Saint Jean, 66000 Perpignan, France; faiza.khemissa@ch-perpignan.fr
15. Department of Gastroenterology, CHRU de Tours, 37044 Tours, France; d.moussata@chu-tours.fr
16. Université Paris Descartes, 75270 Paris, France
17. UCGI Group, R&D UNICANCER, 75654 Paris, France; t-stanbury@unicancer.fr
18. Biometrics Unit, ICM Regional Cancer Institute of Montpellier, 34298 Montpellier, France; simon.thezenas@icm.unicancer.fr
19. Department of Surgical Oncology, Institut Curie, PSL Research University, 75005 Paris, France; pascale.mariani@curie.fr
* Correspondence: nicolas.kiavue@curie.fr

Received: 26 March 2019; Accepted: 20 May 2019; Published: 28 May 2019

Abstract: The management of patients with colorectal cancer (CRC) and potentially resectable liver metastases (LM) requires quick assessment of mutational status and of response to pre-operative systemic therapy. In a prospective phase II trial (NCT01442935), we investigated the clinical validity of circulating tumor cell (CTC) and circulating tumor DNA (ctDNA) detection. CRC patients with potentially resectable LM were treated with first-line triplet or doublet chemotherapy combined with targeted therapy. CTC (Cellsearch®) and Kirsten RAt Sarcoma (KRAS) ctDNA (droplet digital polymerase chain reaction (PCR)) levels were assessed at inclusion, after 4 weeks of therapy and before LM surgery. 153 patients were enrolled. The proportion of patients with high CTC counts (≥3 CTC/7.5mL) decreased during therapy: 19% (25/132) at baseline, 3% (3/108) at week 4 and 0/57 before surgery. ctDNA detection sensitivity at baseline was 91% (N=42/46) and also decreased during treatment. Interestingly, persistently detectable KRAS ctDNA ($p = 0.01$) at 4 weeks was associated with a lower R0/R1 LM resection rate. Among patients who had a R0/R1 LM resection, those with detectable ctDNA levels before liver surgery had a shorter overall survival ($p < 0.001$). In CRC patients with limited metastatic spread, ctDNA could be used as liquid biopsy tool. Therefore, ctDNA detection could help to select patients eligible for LM resection.

Keywords: circulating tumor cells; circulating tumor DNA; liquid biopsy; metastatic colorectal cancer; FOLFIRINOX

1. Introduction

While most patients diagnosed with metastatic colorectal cancer (CRC) have unresectable metastases [1], some can benefit from liver surgery after conversion of unresectable disease to resectable disease by chemotherapy and targeted therapy [1,2]. In this regard, triplet chemotherapy (FOLFOXIRI) may improve the metastasis resection rate and overall survival (OS) [3]. The PRODIGE-14 trial (NCT01442935) was a randomized phase II trial intended to compare prospectively the efficacy of first-line triplet (FOLFIRINOX) versus doublet chemotherapy (FOLFOX: fluorouracil, leucovorin and oxaliplatin or FOLFIRI: fluorouracil, leucovorin and irinotecan), combined with a targeted therapy (bevacizumab in RAt Sarcoma (RAS)-mutated tumors, cetuximab in RAS wild-type tumors), in CRC patients diagnosed with potentially resectable liver metastases (LM). Results of this study have been reported elsewhere [4].

In metastatic CRC, Circulating Tumor Cell (CTC) count by the CellSearch® system is known to be an independent prognostic factor in large studies, using a threshold of ≥3 CTC/7.5 mL of blood [5–7]; these findings were confirmed in a meta-analysis including heterogeneous detection techniques [8] and similar results were reported in other gastro-intestinal cancers [9,10]. Moreover, dynamic changes of CTC levels have been shown to be associated with progression-free survival (PFS) and OS: metastatic CRC patients with persistently elevated CTC levels after one month of chemotherapy had shorter PFS and OS than patients with decreasing CTC counts (PFS: 1.6 vs 6.2 months, $p = 0.02$; OS: 3.7 vs 11.0, $p = 0.0002$) [5].

Similarly, circulating tumor DNA (ctDNA) has proven to be useful for theranostic detection of tumor mutations [11]. While ctDNA analysis has been approved for epidermal growth factor receptor (EGFR) mutation detection in metastatic non-small cell lung cancer [12], it has been suggested as a tool for liquid biopsy in CRC. Preliminary studies addressed the overall concordance between archived tumor tissue and liquid biopsy at any stage of the metastatic disease [13]; more recent results strongly suggested that KRAS mutant subclones may be selected during anti-EGFR therapy, decreasing the overall concordance between nominal (archived tumor tissue-based) and the actual (liquid biopsy-based) KRAS status [14,15]. Furthermore, in RAS-mutated tumors, the RAS mutation may not be detectable in the plasma at first, but may later become detectable under anti-EGFR treatment [16]. In addition to these liquid biopsy applications, ctDNA levels could possibly monitor tumor dynamics [17] with early changes in ctDNA during chemotherapy in CRC associated with tumor response [18]. Detection of a residual disease by ctDNA after surgery in stage II CRC was also associated with early recurrences and poor outcome [19].

While the above-mentioned results were mostly obtained in metastatic CRC patients with heterogeneous clinical settings, we investigated the clinical validity of CTC and ctDNA detection specifically in CRC patients diagnosed with potentially resectable LM and included in the PRODIGE-14 trial. We observed a decrease of CTC and ctDNA detection rates during systemic therapy. We confirmed the prognostic value of CTC detection at baseline and during treatment, and showed that ctDNA detection was associated with a lower R0/R1 LM resection rate.

2. Materials and Methods

2.1. Patients and Treatment

The main trial, identified as NCT01442935, and its ancillary study on circulating tumor biomarkers were approved by a French ethics committee (Comité de Protection des Personnes). All subjects gave their informed consent for inclusion before they participated in the study. The study was conducted in accordance with the Declaration of Helsinki. The patients could accept to participate in the main trial but refuse the ancillary study.

The main inclusion criterion was histologically proven CRC with LM ineligible for curative resection at inclusion and without metastatic spread to other sites (except for up to 3 resectable pulmonary metastases). Other inclusion criteria were: having provided informed consent, good performance status (0–1), known exon 2 KRAS mutational status (as determined locally by standard routine technique on tumor tissue; the clinical trial was later amended to account for other KRAS and Neuroblastoma RAt Sarcoma (NRAS) mutations), adequate hematological, kidney, and liver functions, and no prior therapy for LM. Patients were randomized to either triplet (FOLFIRINOX) [20] or standard doublet (FOLFOX or FOLFIRI) chemotherapy regimens. Chemotherapy was administered in combination with cetuximab in patients with RAS wild-type cancers or with bevacizumab in patients with RAS-mutated cancers.

The main trial objective was to demonstrate the superiority of triplet chemotherapy over doublet chemotherapy in terms of complete (R0/R1) surgical resection of LM and has already been reported [4]. A R0 resection was defined as a microscopically margin-negative resection with a distance between the margins and the tumor ≥ 1 mm. A R1 resection indicates a macroscopically margin-negative resection but with a distance between the margins and the tumor < 1mm. The R0/R1 resection rate was defined as the number of patients who underwent R0/R1 resection divided by the total number of patients included (R0/R1 resection, R2 resection, or no LM surgery).

2.2. Circulating Tumor Biomarker Detection

Three blood draws were required for this ancillary study: before starting treatment, after 1 month of systemic therapy (all patients), and before any surgical resection of LM (in patients referred to surgery after the shrinkage of LM). Blood samples were sent, within 24h, to a central laboratory (Institut Curie, Paris, France).

CTC counts were performed by experienced readers in 7.5 mL of blood (collected in CellSave® tubes) using the CellSearch® system (Menarini Silicon Biosystems), which has previously been reported [21]. The use of different CTC positivity thresholds was planned in order to compare the classical threshold of ≥ 3 CTC [5] to other thresholds and find the optimal cutoff.

For ctDNA analysis, 4 mL of plasma was thawed and cell-free DNA (cfDNA) extracted using the QIAamp® Circulating Nucleic Acid Kit (Qiagen®), after two centrifugations as per routine procedures [22,23]. According to the manufacturer's protocol, digital droplet PCR (ddPCR) reactions were prepared using commercially available primers and TaqMan® probes (Bio-Rad®) with 10 ng of cfDNA. ddPCR mastermix solutions (20 µL) were transferred to a DG8 droplet generator cassette (Bio-Rad®) with 70 µL of oil. Emulsified PCR reactions were then transferred to a 96-well PCR plate and run on a C1000 thermal cycler (Bio-Rad®). Plates were analyzed on a QX-100 droplet reader (Bio-Rad®) with the QuantaSoft v1.7.4 software. Positivity threshold was defined as per manufacturer's instructions, ensuring 0.1% sensitivity. Samples with a variant allele frequency <0.1% were classified as ctDNA-negative. Negative controls were used to minimize the risk of false positive. The assay could detect the G12S, G12R, G12C, G12D, G12A, G12V, and G13D mutations.

In the PRODIGE-14 trial, patients were allocated targeted therapies (cetuximab or bevacizumab) based on the KRAS exon 2 mutational analysis by local assessment on tumor tissue. While extended KRAS exon 3 and 4 and NRAS screening became mandatory in the course of the PRODIGE 14 trial, we confined our ctDNA analysis to KRAS exon 2 mutations. After one month of systemic therapy and before any surgical resection of LM, the ctDNA detection assay was only performed in patients with a known exon 2 KRAS mutation in a tumor tissue sample.

2.3. Statistical Analyses

The main objective of this study was to evaluate CTC and ctDNA detection rates at each time point in mCRC patients. The proportion of patients with detectable ctDNA (using the KRAS exon 2 mutation assay in cfDNA) and with detectable CTC was assessed at baseline, after one month of therapy and before LM surgery, if any. Secondary objectives were to assess the associations of circulating tumor biomarkers and baseline patient characteristics with R0/R1 LM resection and OS. Prespecified analyses were planned accordingly. Analyses conducted with the ctDNA variable (binary: detected or not detected) were also conducted with ctDNA concentration (as a continuous variable: number of mutant KRAS (KRASmut) copies per milliliter), but only for patients with KRAS exon 2 mutated tumors, as determined by routine local assessment on tumor tissues. This hypothesis-generating study had no prespecified power because the detection of circulating tumor biomarkers was done whenever possible, in patients who agreed to participate in the ancillary study. Circulating tumor biomarker detections were blinded to patients and clinicians. Patient characteristics and outcomes were prospectively collected in case report forms for all patients included in the PRODIGE-14 study. OS was defined as time from inclusion to death from any cause. Differences between categorical variables were analyzed by a chi2 test or Fisher's exact test. Continuous variables were analyzed by a Kruskal–Wallis test. Survival curves were plotted according to the Kaplan–Meier method. Statistical significance between survival curves was assessed using the logrank test. Multivariate analysis was done by the Cox proportional hazards model with prognostic factors with a p-value of ≤ 0.10 in univariate analysis. Patients with one or more missing covariable were not included in the multivariate analysis. For all analyses, a p-value of ≤ 0.05 was considered to be statistically significant. This report was written in accordance with the REporting of tumor MARKer studies guidelines.

3. Results

3.1. Patient Characteristics

Between February 2011 and April 2015, 153 patients were enrolled. Patients characteristics are displayed in Table 1. At time of data analysis (01/2017), median follow-up was 37.2 months (IC95% (34–39); range 0–55.3 months); 96 patients were referred to surgery after undergoing a blood draw for circulating biomarker analysis, 91 patients (59%) had a R0/R1 LM resection after chemotherapy and targeted therapy, while 65 deaths (42%) had occurred.

3.2. CTC Detection: Correlation with R0/R1 Resection and Outcome

At baseline, blood samples from 132 patients were available for CTC detection (Figure 1, Supplementary Table S1). ≥ 1 CTC was detected in 7.5 mL of blood in 42% (N=56/132) of patients at baseline and associated with the percentage of liver infiltrated by metastases at baseline ($p = 0.003$). Using the validated ≥ 3 CTC/7.5 mL threshold, elevated CTC counts were observed in 19% (N = 25/132) of patients (Figure 2), and associated with the percentage of liver infiltrated by metastases at baseline ($p = 0.001$) and the synchronicity of LM (p = 0.04). CTC detection at baseline (≥ 1 or ≥ 3 CTC) was not associated with the trial's main objective, the R0/R1 resection of LM ($p = 0.37$ and $p = 0.18$). Associations of CTCs (≥ 3 CTC) or ctDNA detection with baseline clinicopathological characteristics of patients are displayed in Supplementary Tables S2 and S3, respectively.

Table 1. Patients characteristics. N = 153 patients included in the study.

Characteristics	Median Value or Number of Patients
Age, years	Median: 60
	Range: 25–75
Performance Status	
0	95 (63%)
1	57 (37%)
Prior resection of the primary tumor	
No	103 (67%)
Yes	50 (33%)
Synchronous liver metastases	
No	19 (12%)
Yes	134 (88%)
% of liver infiltrated by metastases	
0–25%	41 (45%)
26–50%	28 (30%)
51–75%	15 (16%)
>75%	8 (9%)
CEA	
Normal	17 (11%)
>upper limit of normal	134 (89%)
CA19.9	
Normal	39 (37%)
>upper limit of normal	66 (63%)
KRAS exon 2 mutation in tumor sample	
No	94 (61%)
Yes	59 (39%)
Chemotherapy	
Doublet + targeted therapy	75 (49%)
Triplet + targeted therapy	78 (51%)
R0/R1 resection of liver metastases	
No	62 (41%)
Yes	91 (59%)

At 4 weeks, 108 patients were analyzed for CTC detection. ≥ 1 and ≥ 3 CTC were detected in 11% (N = 12/108) and 3% (N = 3/108) of patients, respectively. CTC counts decreased significantly during therapy ($p < 0.0001$), this decrease being similar in the treatment arms (doublet versus triplet, $p = 0.98$). CTC detection at 4 weeks (≥ 1 or ≥ 3 CTC) was not significantly associated with the eventual R0/R1 resection of LM, although none of the 3 patients with ≥ 3 CTC achieved a R0/R1 resection ($p = 0.06$).

Among patients referred to liver surgery, 57 patients were analyzed for CTC detection. In this selected population, ≥ 1 CTC was detected in 7% (N = 4/57) of patients and no patient had ≥ 3 CTC detected. CTC detection before surgery was not associated with R0/R1 resection ($p = 0.37$).

In regards to the prognostic impact of CTC, ≥ 3 CTC at baseline (HR = 2.2, CI95% [1.2;3.9], $p = 0.01$) and at 4 weeks (HR = 10.9, 95%CI [3.2;36.9]; $p < 0.001$) were correlated with shorter OS (Figure 3). ≥ 1 CTC was significantly associated with shorter OS at 4 weeks ($p = 0.04$), but not at baseline ($p = 0.38$) or before liver surgery ($p = 0.71$). In multivariate analysis, ≥ 3 CTC was found to be an independent prognostic factor for OS at both baseline and at 4 weeks (Supplementary Table S4).

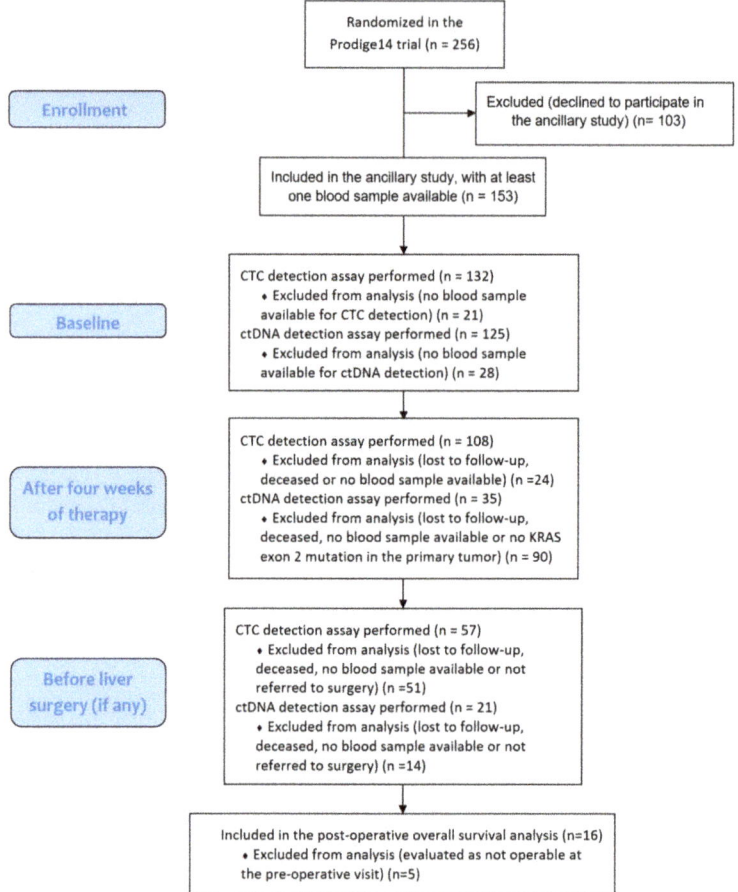

Figure 1. Flow chart of patients included in the analyses at the different time points.

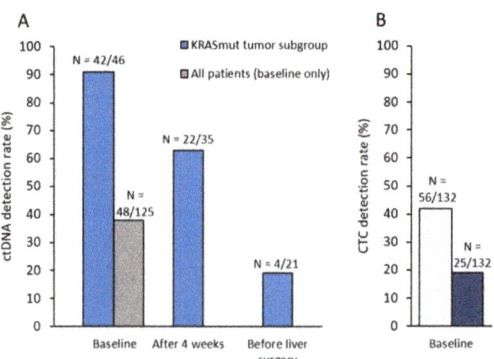

Figure 2. (**A**) ctDNA detection rate (KRAS exon 2 mutation with a variant allele frequency ≥ 0.1%) in all patients at baseline, and in the subgroup of patients with a KRAS exon 2 mutation as determined by routine local assessment on tumor tissues, at baseline, after 4 weeks and before liver surgery (if any). (**B**) CTC detection rate at each timepoint, with the ≥1CTC or the ≥3CTC/7.5mL of blood.

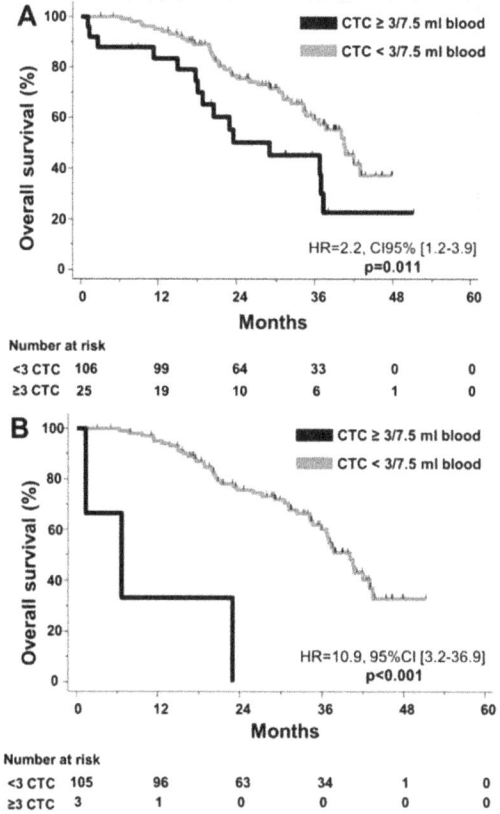

Figure 3. Kaplan–Meier curves for Overall Survival according to CTC detection (**A**) at baseline. (**B**) at 4 weeks.

3.3. KRAS Mutation: Correlation between Liquid and Solid Biopsy

At baseline, blood samples from 125 patients were available for KRAS exon 2 status assessment on plasma as part of our study; 46 of these 125 patients had a KRAS exon 2 mutated tumor according to their medical files (i.e., determined by routine local assessment; Table S1). Among these 46 patients, KRASmut ctDNA was detected at baseline in 42 patients (sensitivity of the liquid biopsy = 0.91, 95%CI [0.79;0.96]). The median number of KRASmut copies/mL plasma in all 46 patients was 378 (range [0;25380]). Among the 79 patients with KRAS wild-type tumors per local assessment, 6 patients (8%) had detectable KRASmut ctDNA (nominal specificity=92%, 95%CI [0.84;0.96]). However, all 6 patients displayed high levels of ctDNA (>150 KRASmut copies/mL plasma), suggesting the actual presence of a KRAS mutation rather than a lack of specificity of the liquid biopsy.

3.4. Dynamic Changes of ctDNA Levels, Correlation with R0/R1 Resection and Outcome

The following analyses were performed in the subgroup of patients with KRAS exon 2 mutated tumors, as determined by routine local assessment on tumor tissues (except at baseline for ctDNA detection as a dichotomized variable, because all patients underwent the ctDNA detection assay at this timepoint).

At baseline, we found that KRASmut copies in plasma (continuous variable) were significantly associated with CTC positivity ($\geq$1 CTC; Kruskal–Wallis test, $p = 0.04$) but not with serum markers

(CA19.9: p = 0.24; CEA: p = 0.25). Baseline ctDNA concentration was however correlated with a lower R0/R1 resection rate (p = 0.05, Figure 4).

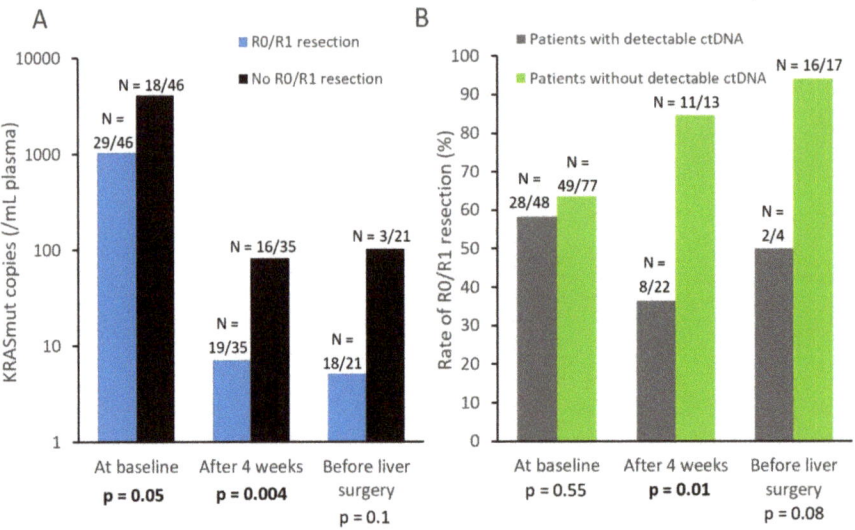

Figure 4. (**A**) Mean number of KRASmut copies per mL of plasma (continuous variable) at baseline, after four weeks, and before LM resection. N indicates the number of patients who achieved or did not achieve R0/R1 resection, among patients (with a KRAS mutated tumor) available for KRASmut assessment at each time point. (**B**) Rate of R0/R1 resection for patients with or without detectable ctDNA (dichotomized variable). N indicates the number of patients who achieved R0/R1 resection according to their ctDNA detection status, among patients who underwent the ctDNA detection assay at each timepoint.

KRAS ctDNA levels significantly decreased during therapy (p = 0.0001). 63% (N = 22/35) of patients with KRAS mutated tumors displayed detectable ctDNA at 4 weeks, while this ctDNA positivity rate dropped to 19% (N = 4/21) before surgery (Figure 2). At 4 weeks, lower ctDNA levels (as a continuous variable) were significantly correlated with eventual R0/R1 resection (p = 0.004, Figure 4). Similar results were observed with ctDNA detection as a dichotomized variable: patients with still detectable ctDNA after 4 weeks of systemic therapy had a lower R0/R1 resection rate than those with no ctDNA detected (36% vs 85%, p = 0.01, Figure 4).

In terms of OS, the 4 patients with no detectable ctDNA levels at baseline had an excellent prognosis (p = 0.05, HR not available; Figure 5A). ctDNA detection at 4 weeks had no prognostic impact (p = 0.31, Figure 5B, Supplementary Figure S1). However, among patients referred to LM resection, the detection of residual ctDNA levels before surgery was significantly associated with a short OS (HR = 31 CI95% [3.2;317], p < 0.001) (Figure 5C). A similar association was found with a short post-operative OS (Figure 5D).

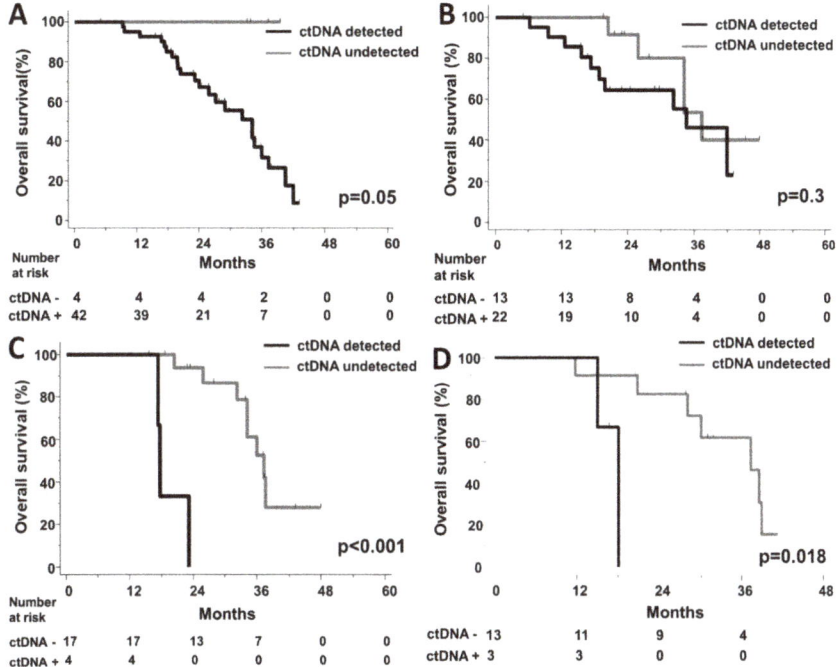

Figure 5. Kaplan–Meier curves for Overall Survival according to ctDNA detection (**A**) at baseline, (**B**) at 4 weeks, (**C**) before liver surgery (**D**) Kaplan–Meier curve for post-operative Overall Survival according to ctDNA detection before liver surgery.

4. Discussion

This is the first study to investigate the clinical validity of both CTC and ctDNA in patients with potentially resectable LM of CRC in a prospective clinical trial. These patients should be treated with intensive first-line systemic therapy combining poly-chemotherapy and the most appropriate targeted therapy (anti-EGFR antibodies in RAS wild-type tumors). Our study found that before the start of systemic therapy, the results of the assessment of tumor mutation status using ctDNA were closely correlated with those of local testing. Even more interestingly, we identified a few patients considered KRAS wild-type by tumor tissue sequencing that had significant KRAS mutant levels in their blood. A similar discrepancy was observed, but at much higher rates, in studies that focused on heavily pre-treated patients [15], suggesting a role of prior anti-EGFR therapies in the emergence of KRAS mutants subclones. Importantly, the probable benefit of anti-EGFR therapy is very limited in such cases [14]. In chemotherapy-naïve patients, the recent RAS Mutation Testing in the Circulating Blood of Patients With Metastatic Colorectal Cancer (RASANC) study [24] found that 8 of 412 patients had a RAS mutation in plasma but not in the primary tumor by local assessment. However, the authors performed central re-analysis on 6 of 8 tumor samples, by next-generation sequencing (NGS) or ddPCR, and found RAS mutations in all six samples. These results, and the shorter testing time, strongly suggest that ctDNA analysis might become a valuable theranostic tool in patients diagnosed with potentially resectable LM.

In addition to liquid biopsy applications at baseline, the clinical validity of ctDNA quantification was investigated at different time points during therapy. First, systemic therapy induced a significant decrease in ctDNA levels, highlighting that liquid biopsy has a very limited sensitivity once therapy has been initiated. We also found that ctDNA levels at different time points yielded significant prognostic information: undetectable ctDNA levels at baseline tended to be a prognostic factor, as demonstrated in other cancers such as metastatic lung cancer [25], ctDNA being correlated with tumor burden in various cancer types [26]. More interestingly, the absence of ctDNA at 4 weeks was correlated with a very high R0/R1 resection rate of LM (85%), suggesting that this biomarker could help decide whether liver surgery is appropriate for patients.

Finally, in patients referred to surgery for LM resection, persistently detectable ctDNA levels before surgery was associated with short post-surgical OS, suggesting that LM were not fully responding to therapy and/or that extra-hepatic micro-metastases were present. A study by Narayan and colleagues [27] in 59 metastatic CRC patients who underwent LM resection found an association between worse disease-specific survival and the detection of circulating mutant *TP53* copies during surgery (but not with ctDNA). However, blood samples were only obtained during and after surgery. Recent studies have found an association between dynamic changes in ctDNA detection and outcome in CRC, in the adjuvant setting, or in the metastatic setting. In the adjuvant setting [28], change of ctDNA status (as a dichotomized variable: detected or not detected) from positive to negative or from negative to positive was associated with respectively superior or lower recurrence-free survival. In the metastatic setting, a recent study [29], using a composite marker evaluating the decrease of ctDNA levels during chemotherapy, demonstrated that it could be used to predict response, progression-free survival, and OS.

If confirmed by further studies, we hypothesize that the absence of detectable ctDNA might become an important criterion prior to any LM resection in this patient population.

Regarding ctDNA analyses, limitations of our study include the limited number of KRAS mutated tumors enrolled and the focus on KRAS exon 2 mutations, as predefined in the study protocol at time of initiation, with no assessment of other KRAS, NRAS, and Rapidly Accelerated Fibrosarcoma homolog B (BRAF) mutations. Of note, while assessing several mutation hotspots in a single assay is usually achieved by NGS; multiplex ddPCR [30] and, more recently, drop-off ddPCR [31] may allow the screening of several hotspots in a single reaction. Larger mutation panel or methylation patterns can be used to detect and quantify ctDNA in a larger proportion of patients [18,24,32]. In the RASANC study [24], plasma samples from chemotherapy-naïve metastatic CRC patients were analyzed by NGS combined with methylation ddPCR, which allowed for a high detection rate of ctDNA (329/425, 77%).

Regarding CTC detection, our study showed its correlation with ctDNA levels, as already reported in patients with uveal melanoma LM [33]. However, the CTC detection rate was lower in our patient population than in prior studies in non-resectable metastatic CRC patients [5,7], probably because of the limited tumor burden in patients included in this study. While our study confirmed the prognostic impact of CTC count at baseline, the number of patients with persistently elevated CTC counts during therapy appeared very limited and prevents any clinical utility in this clinical setting, despite a proven clinical validity. We propose that more sensitive CTC detection techniques [34] be investigated in metastatic CRC to assess the clinical utility of CTC level, such as those relying on microfluidics [35], on EpCAM-independent CTC detection [36] and/or on the screening of larger blood volume [37].

Lastly, newly developed circulating biomarkers such as free serum amino acids [38] could be compared to CTC or ctDNA detection for their prognostic value. Similarly, circulating extracellular matrix components have been evaluated as biomarkers for cancer diagnosis and prognosis in various tumor types [39].

This prospective study showed that CTC and ctDNA had different detection profiles in mCRC patients with potentially resectable LM, the latter demonstrating interesting validity with regards to liquid biopsy and pre-operative prognostic applications.

Supplementary Materials: The following are available online at http://www.mdpi.com/2073-4409/8/6/516/s1, Table S1: blood samples available. Table S2: associations of CTCs (as a dichotomous variable: ≥3 or <3 CTC/7.5 mL) and clinicopathological characteristics of patients at baseline (when available), treatment arm, or primary endpoint. Table S3: associations between ctDNA detection (as a dichotomous variable) and clinicopathological characteristics of patients at baseline (when available), treatment arm, or primary endpoint. Table S4: multivariate Cox regression with CTC detection at baseline and at 4 weeks (Overall Survival). Figure S1: post-operative overall survival according to ctDNA detection at 4 weeks.

Author Contributions: Conceptualization, F.-C.B., J.-Y.P., T.S. and P.M.; methodology F.-C.B., J.-Y.P. and P.M.; software, S.T.; validation, F.-C.B., J.M., A.S., A.R., C.D., C.P. and J.-Y.P.; formal analysis, S.T.; investigation, F.-C.B., M.Y., M.H.S., J.M., A.S., A.R., C.D., O.B., M.R., F.G., E.F., R.G, L.M., F.K.A., T.M., D.M., W.C., C.P., J.-Y.P. and P.M.; resources, F.-C.B., C.D., C.P. and J.-Y.P.; data curation, S.T.; writing—original draft preparation, F.-C.B., L.C. and N.K.; writing—review and editing, F.-C.B. and N.K.; visualization, F.-C.B., L.C. and N.K.; supervision, F.-C.B.; project administration, F.-C.B.; funding acquisition, F.-C.B., M.Y., T.S. and P.M.

Funding: This research was funded by the French Ministry of Health (Programme hospitalier de recherche clinique) and the pharmaceutical company Merck.

Conflicts of Interest: The authors declare no conflict of interest. The funders had no role in the design of the study; in the collection, analyses, or interpretation of data; in the writing of the manuscript, or in the decision to publish the results.

References

1. Adam, R.; Delvart, V.; Pascal, G.; Valeanu, A.; Castaing, D.; Azoulay, D.; Giacchetti, S.; Paule, B.; Kunstlinger, F.; Ghémard, O.; et al. Rescue surgery for unresectable colorectal liver metastases downstaged by chemotherapy: A model to predict long-term survival. *Ann. Surg.* **2004**, *240*, 644–657. [CrossRef] [PubMed]
2. Delaunoit, T.; Alberts, S.R.; Sargent, D.J.; Green, E.; Goldberg, R.M.; Krook, J.; Fuchs, C.; Ramanathan, R.K.; Williamson, S.K.; Morton, R.F.; et al. Chemotherapy permits resection of metastatic colorectal cancer: Experience from Intergroup N9741. *Ann. Oncol.* **2005**, *16*, 425–429. [CrossRef] [PubMed]
3. Falcone, A.; Ricci, S.; Brunetti, I.; Pfanner, E.; Allegrini, G.; Barbara, C.; Crinò, L.; Benedetti, G.; Evangelista, W.; Fanchini, L.; et al. Phase III trial of infusional fluorouracil, leucovorin, oxaliplatin, and irinotecan (FOLFOXIRI) compared with infusional fluorouracil, leucovorin, and irinotecan (FOLFIRI) as first-line treatment for metastatic colorectal cancer: The gruppo oncologico nord ovest. *J. Clin. Oncol.* **2007**, *25*, 1670–1676.
4. Ychou, M.; Rivoire, M.; Thezenas, S.; Guimbaud, R.; Ghiringhelli, F.; Mercier-Blas, A.; Mineur, L.; Francois, E.; Khemissa, F.; Moussata, D.; et al. FOLFIRINOX combined to targeted therapy according RAS status for colorectal cancer patients with liver metastases initially non-resectable: A phase II randomized Study—Prodige 14–ACCORD 21 (METHEP-2), a unicancer GI trial. *J. Clin. Oncol.* **2016**, *34*, 3512. [CrossRef]
5. Cohen, S.J.; Punt, C.J.A.; Iannotti, N.; Saidman, B.H.; Sabbath, K.D.; Gabrail, N.Y.; Picus, J.; Morse, M.; Mitchell, E.; Miller, M.C.; et al. Relationship of circulating tumor cells to tumor response, progression-free survival, and overall survival in patients with metastatic colorectal cancer. *J. Clin. Oncol.* **2008**, *26*, 3213–3221. [CrossRef]
6. Sastre, J.; Vidaurreta, M.; Gómez, A.; Rivera, F.; Massutí, B.; López, M.R.; Abad, A.; Gallen, M.; Benavides, M.; Aranda, E.; et al. Prognostic value of the combination of circulating tumor cells plus kras in patients with metastatic colorectal cancer treated with chemotherapy plus bevacizumab. *Clin. Colorectal Cancer* **2013**, *12*, 280–286. [CrossRef]
7. Tol, J.; Koopman, M.; Miller, M.C.; Tibbe, A.; Cats, A.; Creemers, G.J.M.; Vos, A.H.; Nagtegaal, I.D.; Terstappen, L.W.M.M.; Punt, C.J.A. Circulating tumour cells early predict progression-free and overall survival in advanced colorectal cancer patients treated with chemotherapy and targeted agents. *Ann. Oncol.* **2010**, *21*, 1006–1012. [CrossRef] [PubMed]
8. Huang, X.; Gao, P.; Song, Y.; Sun, J.; Chen, X.; Zhao, J.; Xu, H.; Wang, Z. Meta-analysis of the prognostic value of circulating tumor cells detected with the CellSearch System in colorectal cancer. *BMC Cancer* **2015**, *15*, 202. [CrossRef]
9. Bidard, F.C.; Huguet, F.; Louvet, C.; Mineur, L.; Bouche, O.; Chibaudel, B.; Artru, P.; Desseigne, F.; Bachet, J.B.; Mathiot, C.; et al. Circulating tumor cells in locally advanced pancreatic adenocarcinoma: The ancillary CirCe 07 study to the LAP 07 trial. *Ann. Oncol.* **2013**, *24*, 2057–2061. [CrossRef]
10. Bidard, F.C.; Ferrand, F.R.; Huguet, F.; Hammel, P.; Louvet, C.; Malka, D.; Boige, V.; Ducreux, M.; Andre, T.; de Gramont, A.; et al. Disseminated and circulating tumor cells in gastrointestinal oncology. *Crit. Rev. Oncol. Hematol.* **2012**, *82*, 103–115. [CrossRef]
11. Bidard, F.-C.; Weigelt, B.; Reis-Filho, J.S. Going with the flow: From circulating tumor cells to DNA. *Sci. Transl. Med.* **2013**, *5*, 207ps14. [CrossRef]

12. Douillard, J.-Y.; Ostoros, G.; Cobo, M.; Ciuleanu, T.; Cole, R.; McWalter, G.; Walker, J.; Dearden, S.; Webster, A.; Milenkova, T.; et al. Gefitinib treatment in EGFR mutated caucasian NSCLC: Circulating-free tumor dna as a surrogate for determination of egfr status. *J. Thorac. Oncol.* **2014**, *9*, 1345–1353. [CrossRef]
13. Thierry, A.R.; Mouliere, F.; El Messaoudi, S.; Mollevi, C.; Lopez-Crapez, E.; Rolet, F.; Gillet, B.; Gongora, C.; Dechelotte, P.; Robert, B.; et al. Clinical validation of the detection of KRAS and BRAF mutations from circulating tumor DNA. *Nat. Med.* **2014**, *20*, 430–435. [CrossRef]
14. Siravegna, G.; Mussolin, B.; Buscarino, M.; Corti, G.; Cassingena, A.; Crisafulli, G.; Ponzetti, A.; Cremolini, C.; Amatu, A.; Lauricella, C.; et al. Clonal evolution and resistance to EGFR blockade in the blood of colorectal cancer patients. *Nat. Med.* **2015**, *21*, 795. [CrossRef]
15. Tabernero, J.; Lenz, H.-J.; Siena, S.; Sobrero, A.; Falcone, A.; Ychou, M.; Humblet, Y.; Bouché, O.; Mineur, L.; Barone, C.; et al. Analysis of circulating DNA and protein biomarkers to predict the clinical activity of regorafenib and assess prognosis in patients with metastatic colorectal cancer: A retrospective, exploratory analysis of the CORRECT trial. *Lancet Oncol.* **2015**, *16*, 937–948. [CrossRef]
16. Raimondi, C.; Nicolazzo, C.; Belardinilli, F.; Loreni, F.; Gradilone, A.; Mahdavian, Y.; Gelibter, A.; Giannini, G.; Cortesi, E.; Gazzaniga, P. Transient disappearance of RAS mutant clones in plasma: A counterintuitive clinical use of EGFR inhibitors in RAS mutant metastatic colorectal cancer. *Cancers* **2019**, *11*, 42. [CrossRef]
17. Diehl, F.; Schmidt, K.; Choti, M.A.; Romans, K.; Goodman, S.; Li, M.; Thornton, K.; Agrawal, N.; Sokoll, L.; Szabo, S.A.; et al. Circulating mutant DNA to assess tumor dynamics. *Nat. Med.* **2008**, *14*, 985–990. [CrossRef] [PubMed]
18. Tie, J.; Kinde, I.; Wang, Y.; Wong, H.L.; Roebert, J.; Christie, M.; Tacey, M.; Wong, R.; Singh, M.; Karapetis, C.S.; et al. Circulating tumor DNA as an early marker of therapeutic response in patients with metastatic colorectal cancer. *Ann. Oncol.* **2015**, *26*, 1715–1722. [CrossRef] [PubMed]
19. Tie, J.; Wang, Y.; Tomasetti, C.; Li, L.; Springer, S.; Kinde, I.; Silliman, N.; Tacey, M.; Wong, H.-L.; Christie, M.; et al. Circulating tumor DNA analysis detects minimal residual disease and predicts recurrence in patients with stage II colon cancer. *Sci. Transl. Med.* **2016**, *8*, 346ra92. [CrossRef] [PubMed]
20. Ychou, M.; Rivoire, M.; Thezenas, S.; Quenet, F.; Delpero, J.-R.; Rebischung, C.; Letoublon, C.; Guimbaud, R.; Francois, E.; Ducreux, M.; et al. A randomized phase II trial of three intensified chemotherapy regimens in first-line treatment of colorectal cancer patients with initially unresectable or not optimally resectable liver metastases. The METHEP trial. *Ann. Surg. Oncol.* **2013**, *20*, 4289–4297. [CrossRef]
21. Allard, W.J. Tumor cells circulate in the peripheral blood of all major carcinomas but not in healthy subjects or patients with nonmalignant diseases. *Clin. Cancer Res.* **2004**, *10*, 6897–6904. [CrossRef]
22. Lebofsky, R.; Decraene, C.; Bernard, V.; Kamal, M.; Blin, A.; Leroy, Q.; Rio Frio, T.; Pierron, G.; Callens, C.; Bieche, I.; et al. Circulating tumor DNA as a non-invasive substitute to metastasis biopsy for tumor genotyping and personalized medicine in a prospective trial across all tumor types. *Mol. Oncol.* **2015**, *9*, 783–790. [CrossRef]
23. Madic, J.; Kiialainen, A.; Bidard, F.-C.; Birzele, F.; Ramey, G.; Leroy, Q.; Frio, T.R.; Vaucher, I.; Raynal, V.; Bernard, V.; et al. Circulating tumor DNA and circulating tumor cells in metastatic triple negative breast cancer patients: ctDNA and CTC in metastatic triple negative breast cancer. *Int. J. Cancer* **2015**, *136*, 2158–2165. [CrossRef]
24. Bachet, J.B.; Bouché, O.; Taieb, J.; Dubreuil, O.; Garcia, M.L.; Meurisse, A.; Normand, C.; Gornet, J.M.; Artru, P.; Louafi, S.; et al. RAS mutation analysis in circulating tumor DNA from patients with metastatic colorectal cancer: The AGEO RASANC prospective multicenter study. *Ann. Oncol.* **2018**, *29*, 1211–1219. [CrossRef]
25. Pécuchet, N.; Zonta, E.; Didelot, A.; Combe, P.; Thibault, C.; Gibault, L.; Lours, C.; Rozenholc, Y.; Taly, V.; Laurent-Puig, P.; et al. Base-position error rate analysis of next-generation sequencing applied to circulating tumor dna in non-small cell lung cancer: A prospective study. *PLoS Med.* **2016**, *13*, e1002199. [CrossRef] [PubMed]
26. Bettegowda, C.; Sausen, M.; Leary, R.J.; Kinde, I.; Wang, Y.; Agrawal, N.; Bartlett, B.R.; Wang, H.; Luber, B.; Alani, R.M.; et al. Detection of circulating tumor dna in early- and late-stage human malignancies. *Sci. Transl. Med.* **2014**, *6*, 224ra24. [CrossRef] [PubMed]
27. Narayan, R.R.; Goldman, D.A.; Gonen, M.; Reichel, J.; Huberman, K.H.; Raj, S.; Viale, A.; Kemeny, N.E.; Allen, P.J.; Balachandran, V.P.; et al. Peripheral circulating tumor dna detection predicts poor outcomes after liver resection for metastatic colorectal cancer. *Ann. Surg. Oncol.* **2019**, *26*, 1824–1832. [CrossRef] [PubMed]

28. Tie, J.; Cohen, J.; Wang, Y.; Lee, M.; Wong, R.; Kosmider, S.; Ananda, S.; Cho, J.H.; Faragher, I.; McKendrick, J.J.; et al. Serial circulating tumor DNA (ctDNA) analysis as a prognostic marker and a real-time indicator of adjuvant chemotherapy (CT) efficacy in stage III colon cancer (CC). *J. Clin. Oncol.* **2018**, *36*, 3516. [CrossRef]
29. Garlan, F.; Laurent-Puig, P.; Sefrioui, D.; Siauve, N.; Didelot, A.; Sarafan-Vasseur, N.; Michel, P.; Perkins, G.; Mulot, C.; Blons, H.; et al. Early evaluation of circulating tumor dna as marker of therapeutic efficacy in metastatic colorectal cancer patients (PLACOL Study). *Clin. Cancer Res.* **2017**, *23*, 5416–5425. [CrossRef]
30. Taly, V.; Pekin, D.; Benhaim, L.; Kotsopoulos, S.K.; Le Corre, D.; Li, X.; Atochin, I.; Link, D.R.; Griffiths, A.D.; Pallier, K.; et al. Multiplex picodroplet digital PCR to detect KRAS mutations in circulating DNA from the plasma of colorectal cancer patients. *Clin. Chem.* **2013**, *59*, 1722–1731. [CrossRef]
31. Decraene, C.; Silveira, A.B.; Bidard, F.-C.; Vallée, A.; Michel, M.; Melaabi, S.; Vincent-Salomon, A.; Saliou, A.; Houy, A.; Milder, M.; et al. Multiple hotspot mutations scanning by single droplet digital PCR. *Clin. Chem.* **2018**, *64*, 317–328. [CrossRef]
32. Garrigou, S.; Perkins, G.; Garlan, F.; Normand, C.; Didelot, A.; Le Corre, D.; Peyvandi, S.; Mulot, C.; Niarra, R.; Aucouturier, P.; et al. A Study of hypermethylated circulating tumor DNA as a universal colorectal cancer biomarker. *Clin. Chem.* **2016**, *62*, 1129–1139. [CrossRef]
33. Bidard, F.-C.; Madic, J.; Mariani, P.; Piperno-Neumann, S.; Rampanou, A.; Servois, V.; Cassoux, N.; Desjardins, L.; Milder, M.; Vaucher, I.; et al. Detection rate and prognostic value of circulating tumor cells and circulating tumor DNA in metastatic uveal melanoma: Circulating tumor DNA in uveal melanoma. *Int. J. Cancer* **2014**, *134*, 1207–1213. [CrossRef]
34. Ferreira, M.M.; Ramani, V.C.; Jeffrey, S.S. Circulating tumor cell technologies. *Mol. Oncol.* **2016**, *10*, 374–394. [CrossRef]
35. Saliba, A.-E.; Saias, L.; Psychari, E.; Minc, N.; Simon, D.; Bidard, F.-C.; Mathiot, C.; Pierga, J.-Y.; Fraisier, V.; Salamero, J.; et al. Microfluidic sorting and multimodal typing of cancer cells in self-assembled magnetic arrays. *Proc. Natl. Acad. Sci. USA* **2010**, *107*, 14524–14529. [CrossRef]
36. Kuhn, P.; Keating, S.; Baxter, G.; Thomas, K.; Kolatkar, A.; Sigman, C. Lessons learned: Transfer of the high-definition circulating tumor cell assay platform to development as a commercialized clinical assay platform. *Clin. Pharmacol. Ther.* **2017**, *102*, 777–785. [CrossRef]
37. Andree, K.C.; Mentink, A.; Zeune, L.L.; Terstappen, L.W.M.M.; Stoecklein, N.H.; Neves, R.P.; Driemel, C.; Lampignano, R.; Yang, L.; Neubauer, H.; et al. Toward a real liquid biopsy in metastatic breast and prostate cancer: Diagnostic LeukApheresis increases CTC yields in a European prospective multicenter study (CTCTrap): Toward a real liquid biopsy in metastatic breast and prostate cancer. *Int. J. Cancer* **2018**, *143*, 2584–2591. [CrossRef]
38. Vsiansky, V.; Svobodova, M.; Gumulec, J.; Cernei, N.; Sterbova, D.; Zitka, O.; Kostrica, R.; Smilek, P.; Plzak, J.; Betka, J.; et al. Prognostic significance of serum free amino acids in head and neck cancers. *Cells* **2019**, *8*, 428. [CrossRef]
39. Giussani, M.; Triulzi, T.; Sozzi, G.; Tagliabue, E. Tumor extracellular matrix remodeling: new perspectives as a circulating tool in the diagnosis and prognosis of solid tumors. *Cells* **2019**, *8*, 81. [CrossRef]

© 2019 by the authors. Licensee MDPI, Basel, Switzerland. This article is an open access article distributed under the terms and conditions of the Creative Commons Attribution (CC BY) license (http://creativecommons.org/licenses/by/4.0/).

Review

Are Circulating Tumor Cells (CTCs) Ready for Clinical Use in Breast Cancer? An Overview of Completed and Ongoing Trials Using CTCs for Clinical Treatment Decisions

Fabienne Schochter *, Thomas W. P. Friedl *, Amelie deGregorio, Sabrina Krause, Jens Huober, Brigitte Rack and Wolfgang Janni

Department of Obstetrics and Gynecology, University Hospital Ulm, University of Ulm, 89075 Ulm, Germany; Amelie.deGregorio@uniklinik-ulm.de (A.d.G.); sabrina.krause@uniklinik-ulm.de (S.K.); jens.huober@uniklinik-ulm.de (J.H.); brigitte.rack@uniklinik-ulm.de (B.R.); wolfgang.janni@uniklinik-ulm.de (W.J.)
* Correspondence: fabienne.schochter@uni-ulm.de (F.S.); Thomas.Friedl@uniklinik-ulm.de (T.W.P.F.); Tel.: +49-731-500-58688 (F.S.); +49-731-500-58598 (T.W.P.F)

Received: 23 October 2019; Accepted: 6 November 2019; Published: 8 November 2019

Abstract: In recent years, breast cancer treatment has become increasingly individualized. Circulating tumor cells (CTCs) have the potential to move personalized medicine another step forward. The prognostic relevance of CTCs has already been proven both in early and metastatic breast cancer. In addition, there is evidence that changes in CTC numbers during the course of therapy can predict treatment response. Thus, CTCs are a suitable tool for repeated treatment monitoring through noninvasive liquid biopsy. The next step is to evaluate how this information can be used for clinical decision making with regard to the extension, modification, or abandonment of a treatment regimen. This review will summarize the completed and ongoing clinical trials using CTC number or phenotype for treatment decisions. Based on current knowledge, CTCs can be regarded as a useful prognostic and predictive marker that is well suited for both risk stratification and treatment monitoring in breast cancer patients. However, there is still the need to provide sufficient and unequivocal evidence for whether CTCs may indeed be used to guide treatment decisions in everyday clinical practice. The results of the ongoing trials described in this review are eagerly awaited to answer these important questions.

Keywords: circulating tumor cells (CTCs); clinical trials; breast cancer; CTC-based treatment decisions

1. Introduction

In recent years, breast cancer treatment has become increasingly individualized. New targeted therapies have improved survival for many patients. With the aim of becoming even more patient-specific, circulating tumor cells (CTCs) are an interesting topic in translational oncology. CTCs represent rare cancer cells in the peripheral blood that have disseminated from the primary tumor (or metastatic sites) and play an important role in tumor progression and the formation of (new) metastases. Several recent reviews describe the biology of CTCs and discuss both their potential of being used as a prognostic and/or predictive marker and the challenges involved in incorporating CTCs in clinical practice [1,2].

CTCs have already proven their prognostic relevance in early breast cancer (EBC) and metastatic breast cancer (MBC) [3–5]. In addition, the SUCCESS-A and the ECOG-ACRIN study E5103 showed that the detection of persisting CTCs two years and even five years after (neo)adjuvant chemotherapy is related to an increased risk of recurrence [6–8]. In a multivariate analysis presented by Sparano

2018, CTC-positivity assessed after five years was the strongest predictor of late disease recurrence in patients with hormone receptor-positive breast cancer that had no signs of disease recurrence in the first five years after primary diagnosis [8].

If CTC-positivity predicts a worse clinical outcome [9] and CTC dynamics is a predictor of therapy response [10], the question remains whether this information can be used for forming treatment decisions. Trials are attempting to find a way to answer this challenging question using either CTC number or CTC phenotype as a criterion for therapy decisions. This article gives an overview of the current status (see Table 1) and—if available—results of all clinical breast cancer trials that involve interventions based on CTC number or phenotype.

Table 1. Clinical trials with CTC-based treatment decisions.

Trail	Status	Enrollment	Condition	Intervention	Primary Endpoints
Treat-CTC, NCT01548677 (phase II)	Closed after interim analysis	1317 enrolled 63 randomized (CTC-pos)	HER2-neg EBC with CTCs after CT	Trastuzumab iv 6 cycles vs. observation	CTC detection rate at week 18
SWOG S0500, NCT00382018 (phase III)	Completed	595 enrolled; 123 randomized (persisting high CTC count)	CT-resistant, CTC-pos MBC	Early switch in therapy vs. treatment until progression	OS, PFS
CirCe01, NCT01349842 (phase III)	Recruitment completed,	568 planed	CT-resistant, CTC-pos MBC	Early switch in therapy vs. treatment until progression	OS
STIC-CTC, NCT01710605 (phase III)	Completed	778 randomized	HR-pos and HER2-neg MBC	Decision CT or ET by clinical choice vs. CTC count	PFS, economic value
Circe TDM-1 NCT01975142 (phase II)	Closed after interim analysis	155 screened; 11 treated	HER2-neg MBC and HER2-pos CTCs	T-DM1	TRR
DETECT III, NCT01619111 (phase III)	Recruiting	120 planned, up to date: 105	HER2-neg MBC and HER2-pos CTCs	Standard treatment vs. Standard treatment + lapatinib	CTC clearance
DETECT IV, NCT02035813 (phase II)	Recruiting	Group A: 180 planned Up to date: 103 Group B: 120 planned Up to date: 107	HER2-neg MBC and HER2-neg CTCs	A: ET + ribociclib or everolimus B: eribulin	A: CTC clearance B: PFS
DETECT V, NCT02344472 (phase III)	Recruiting	270 planned Up to date 151	HR-pos, HER2-pos, MBC	trastuzumab/ pertuzumab + CT or ET with ribociclib	Tolerability, safety, and quality of life

CT = chemotherapy, CTC = circulating tumor cells, ET = endocrine treatment, HER2 = human epidermal growth factor receptor 2, HR = hormone receptor, EBC = early breast cancer, MBC = metastatic breast cancer, OS = overall survival, PFS = progression-free survival, TRR = tumor response rate, T-DM1 = trastuzumab emtansine, pos = positive, neg = negative

2. Material and Methods

On the database site ClinicalTrials.gov, 59 studies were retrieved by searching using the keywords "circulating tumor cells" and "breast cancer" at the time the search was performed (end of March 2019). Overall, 26 trials were initiated in Europe, 20 in the United States, 11 in Asian countries, and 2 in Brazil. In total, 31 interventional trials are registered, from which 10 are either recruiting or active but not recruiting. However, only 7 trials use either the presence, amount, or phenotype of CTCs in determining an intervention for the chosen treatment. One is an adjuvant trial, while the others describe trials in a metastatic setting. In the following, we will give an overview of these studies.

All clinical trials used the only FDA-cleared and current gold standard method for the enrichment and detection of CTCs, the CELLSEARCH® system (Menarini Silicon Biosystems, Bologna, Italy), which has been described in detail by Park Y et al. [11]. Briefly, the first step during analysis is the immunomagnetic enrichment of EpCAM-positive cells using antibody-coated ferrofluid nanoparticles. The EpCAM-enriched cells are then stained with antibodies specific for cytokeratins 8, 18, and 19 (epithelial cell markers) and for CD45 (leukocyte marker), as well as with the fluorescent nucleic acid dye 4,6-diamidino-2-phenylindole dihydrochloride (DAPI) for labeling of cell nuclei. CTCs are identified and counted using a semiautomated fluorescence-based microscope system that generates images of the stained cells, with CTCs being defined as cytokeratin-positive nucleated cells that lack CD45 expression.

3. CTC-Based Clinical Trials in EBC

Treat-CTC

The TREAT-CTC study (NCT01548677) is the only study in the (neo)adjuvant setting in which treatment decisions are based on the presence of CTCs. This trial tried to address the question of an additional treatment possibility to eliminate CTCs persistent after (neo)adjuvant chemotherapy. A total of 1317 patients with HER2-negative EBC were screened for CTCs after completion of (neo)adjuvant chemotherapy and patients with at least 1 CTC/15 mL blood were randomized to either an additional treatment with trastuzumab (6 cycles of trastuzumab i.v.) or observation. In 95 (7.2%) of the patients, CTCs could be detected; 31 patients were randomized to trastuzumab treatment, while 32 patients were randomized to the observational control arm. The CTC-positivity rate was similar, and not significantly different in both arms after 18 weeks of treatment (17.2% vs. 13.8%); furthermore, no difference in disease-free survival could be seen [12]. Following the recommendation of the independent Data Monitoring Committee to stop the trial for futility, study recruitment was not continued after the first interim analysis. A possible explanation for this negative result might be that, while the HER2 status of CTCs was determined, HER2-positivity of CTCs was not required for study inclusion. In the majority of the patients (76%), the detected CTCs were HER2-negative. This is in accordance with the results of the NSABP-B47 trial that failed to show improved disease-free survival if trastuzumab is added to chemotherapy in patients with HER2-low (IHC 1+ or 2+ staining intensity) breast cancer [13]. Thus, both the NSABP B47 and the Treat CTC trial failed to confirm the hypothesis that women with early breast cancer showing low *HER2* expression might benefit from treatment with trastuzumab following adjuvant chemotherapy. Taken together, these results suggest that the failure of the Treat CTC trial was due to choosing an inappropriate treatment intervention for the targeted patient population rather than indicating a general failure of the concept of CTC-based intervention decisions.

4. CTC-Based Clinical Trials in MBC

In the MBC setting, trials that are based on CTC number usually use the cutoff of ≥5 CTCs which, initially, was not meant to be used for treatment decisions but was presented as a tool to separate the patients into two groups with different survival prospects. A recent retrospective pooled analysis including 2436 MBC patients confirmed the utility of the cutoff of ≥5 CTCs for risk stratification, as MBC patients could be separated into categories of either stage IV indolent (<5 CTCs) or stage IV aggressive (≥5 CTCs) with significantly longer overall survival in the group with <5 CTCs independently of clinical and molecular variables [14].

4.1. SWOG S0500

Since the first knowledge that a high count of CTCs predicts a worse clinical outcome and that changes in CTCs reflect therapy response, the question has been raised of whether MBC patients can be monitored and treated based on CTC dynamics. The first clinical phase III trial to investigate this hypothesis was initiated by the Southwest Oncology Group (SWOG). The S0500 study (NCT00382018) included 595 patients with MBC scheduled for first-line chemotherapy in the advanced setting. Before the start of first-line chemotherapy, patients were tested for CTCs. If patients did not have an increased CTC count (defined as less than 5 CTCs per 7.5 mL blood) at baseline (n = 276), they were treated according to physician's choice and no additional blood draws or interventions were performed. A total of 319 patients had an elevated CTC count (defined as five or more CTCs per 7.5 mL blood) before the start of first-line chemotherapy. A total of 288 of these high-risk patients were re-tested after the first cycle of first-line therapy, which was approximately 22 days after the first chemotherapy administration. Patients with a follow-up CTC count of less than 5 CTCs per 7.5 mL blood (n = 165) continued on the initially chosen chemotherapy regimen until progression. If the follow-up CTC level was persistently high (i.e., the CTC count remained at a level of five or more CTCs per 7.5 mL blood), the patients (n = 123) were included in the interventional part and randomly assigned to either continue the treatment until clinical and/or radiographic evidence of progression (n = 64) or switch early to another chemotherapy of the physician's choice (n = 59). The results showed no significant improvement in survival for an early change in treatment regime (median progression-free survival (PFS) 4.6 vs. 3.5 months, HR = 0.92; 95%CI = 0.64–1.32; median overall survival (OS) 1.5 vs. 10.7 months, HR = 1.00, 95%CI = 0.69–1.47). Nevertheless, the results confirmed the prognostic impact of CTCs: the median OS reached 35 months in patients with a low CTC count before treatment, 23 months in patients with a high CTC count before treatment but a low follow-up CTC count, and 13 months in patients with a persistently high CTC count. These differences were highly significant after adjusting for hormone receptor and HER2 status in a multivariable Cox model ($p < 0.001$). Overall, the study showed that for patients with persistently increased CTCs after one cycle of first-line chemotherapy, and that early switching to another standard chemotherapy regimen was not effective in terms of prolonged survival. Thus, persisting CTCs might represent a chemoresistant population of tumor cells which requires an alternative approach [15].

4.2. CirCe01

Another similar trial evaluating the response to chemotherapy by the CTC decrease after the first cycle is the CirCe01 study (NCT01349842). This French, multicenter, randomized phase III study included MBC patients that were CTC-positive (≥5 CTCs/7.5 mL) after two lines of chemotherapy. Patients were either followed by the conventional clinical and radiological assessments or by the determination of CTCs. Patients in the interventional (CTC-driven) arm were switched to another chemotherapy if CTC count did not decrease after one cycle of chemotherapy [16]. The aim was to switch non-responding patients early to avoid ineffective but toxic treatments. The recruitment has been completed, but the results are still pending.

4.3. STIC CTC

The French study group also initiated the STIC CTC Trial (NCT01710605). This phase III trial tries to answer the question of whether the choice between chemotherapy and endocrine therapy in first-line, HER2-negative, hormone receptor-positive MBC patients can be driven by the amount of CTCs. The patients were randomized to clinician's choice in the standard arm (i.e., the decision for chemotherapy or endocrine therapy was based on clinical assessment) or CTC-based decision in the interventional arm. The trial used the common cutoff of ≥5 CTCs in 7.5 mL peripheral blood; patients in the CTC-driven interventional arm with less than 5 CTCs received endocrine treatment, while patients with 5 or more CTCs were treated with chemotherapy (by the physician's choice). The results were presented at the San Antonio Breast Cancer Meeting 2018 [17]. Overall, 778 patients were randomized. In the clinician's choice arm, 72.6% and 27.4% of patients respectively received hormone therapy and chemotherapy, while in the CTC-driven arm, 62.6% and 37.4% of patients were respectively treated with endocrine therapy and chemotherapy. The primary endpoint of the study was met, as PFS of patients in the CTC decision-based arm (median 15.6 months) was not inferior to PFS of patients in the clinician's choice arm (median 14.0 months). The analysis focused further on the planned subgroup analysis of the discordant groups (CTCs high/clinical low risk; CTCs low/clinical high risk). Patients in the CTCs high/clinical low-risk subgroup who received a CTC-driven chemotherapy discordant to the physicians choice had a significantly longer PFS than the patients in the clinical-decision arm receiving endocrine therapy (HR = 0.62, 95%CI = 0.45–0.84; p = 0.002), with a non-significant trend toward longer OS (HR = 0.69, 95%CI = 0.43–1.11; p = 0.12). On the other side, patients in the CTCs low/clinical high-risk subgroup who were deescalated to a hormonal therapy from the clinically chosen chemotherapy due to having a low CTC count showed no significantly worse PFS compared to the patients treated with chemotherapy according to clinician's choice. In addition, when the patients from the two discordant subgroups were pooled, patients receiving chemotherapy showed significantly better PFS (HR = 0.66, 95%CI = 0.51–0.85; p = 0.001) and OS (HR = 0.65, 95%CI = 0.43–0.98; p = 0.04) than patients receiving endocrine therapy, challenging current treatment standards. The results of this trial are promising and indicate that including a CTC count in the decision algorithm for HER2-negative, hormone receptor-positive MBC patients might improve patient outcome in some cases. However, the results need to be confirmed for the new era of endocrine treatment regimen with CDK4/6 inhibitors.

4.4. CirCe T-DM1

The CirCe T-DM1 study (NCT01975142) is the first clinical trial not using the number but the phenotype of CTCs as a decision criterion. HER2-negative MBC patients needing a third or fourth line of therapy and with detected HER2-positive CTCs in baseline-screening were treated with the antibody–drug conjugate trastuzumab emtansine (TDM-1). The study was closed after the first interim analysis and presented at the ESMO 2017 [18]. A total of 155 patients were screened, 14 (9.0%) had HER2-positive CTCs and 11 patients were treated with TDM-1. Partial response was observed in only one patient; median PFS was 4.9 months (range: 1.8–10.1). Due to these results and the very low prevalence of HER2-positive CTCs (1.6% of the detected CTCs), the authors conclude that the tested therapeutic approach was not promising.

4.5. DETECT Study Program

The DETECT study concept is a comprehensive trial observing CTCs for liquid biopsy in MBC patients with different biologic tumor features. The study concept includes DETECT III, DETECT IV, and DETECT V. It investigates the efficacy of treatment decisions based on the presence and phenotype of CTCs.

Patients with HER2-negative MBC and HER2-positive CTCs are eligible for the DETECT III study (NCT01619111). In this phase III trial, 120 patients were randomized (1:1) to a standard endocrine (letrozole, anastrozole, or exemestane) or standard chemotherapeutic (docetaxel, paclitaxel, capecitabine, vinorelbine) treatment according to the physician's choice, with or without an additional HER2-targeted treatment with lapatinib. The primary endpoint of the DETECT III study is the efficacy as assessed by the CTC clearance rate, i.e., the proportion of patients with no evidence of CTCs in the blood after treatment.

Patients with HER2-negative MBC and only HER2-negative CTCs are being treated in the DETECT IV trial (NCT02035813), which is divided into two separate treatment and observation cohorts. In cohort A, patients with hormone receptor-positive breast cancer are treated with endocrine therapy (anastrozole, letrozole, exemestane, fulvestrant or tamoxifen; tamoxifen only for patients in combination with ribociclib) combined with the mTOR inhibitor everolimus or (after an amendment) the CDK4/6 inhibitor ribociclib. Cohort B includes patients with either triple-negative breast cancer or patients with hormone receptor-positive breast cancer that need a more aggressive treatment. Here, patients receive the mitotic inhibitor eribulin (halichondrin B analogue) as treatment in single-arm observation. The primary endpoints for cohort A and B are efficacy as assessed by the CTC clearance rate and PFS, respectively.

With DETECT V/CHEVENDO (NCT02344472), the study concept also offers a therapeutic option for patients with HER2-positive and hormone receptor-positive breast cancer, even though interventions are not based on CTCs. In this two-arm randomized phase III trial, all patients are treated with the dual HER2-targeted therapy with pertuzumab and trastuzumab, and are randomized to either in combination with chemotherapy (docetaxel, paclitaxel, capecitabine, vinorelbine) or endocrine therapy (fulvestrant, tamoxifen, letrozole, anastrozole, or exemestane) combined with the CDK4/6 inhibitor ribociclib. A modified adverse event score and the "quality-adjusted time without symptoms and toxicity" (Q-TWiST) method are used for assessing safety, tolerability and quality of life. In fact, the DETECT V trail is the only part in which the therapeutic decision is not driven by the number or phenotype of CTCs. However, one aim of the DETECT V trial is the development of an "endocrine responsiveness score" (ERS), based on the estrogen receptor and *HER2* expression of CTCs, to derive a predictive tool for the hormone receptor-positive, HER2-positive disease. The first results coming from the COMETI-2 study suggest that high rates of estrogen receptor are associated with a better response to endocrine treatment, while high rates of *HER2* expression are associated with a worse response [19]. The general goal of the endocrine responsiveness score is to identify patients with a predicted good response to endocrine therapy to avoid unnecessary chemotherapeutic treatments, which are generally associated with more adverse events and decreased quality of life compared to endocrine therapies.

In all trials of the DETECT study concept, CTCs are measured repetitively during treatment to obtain data on CTC dynamics and their possible role as a tool for treatment monitoring and early response assessment. Furthermore, the DETECT study concept is accompanied by a comprehensive translational research project ("DETECT-CTC") which is funded by the Deutsche Krebshilfe (German Cancer Aid). The main aim of DETECT-CTC is to apply innovative biomarkers and assays focusing on molecular characteristics of CTCs and circulating nucleic acids to analyze their potential function as a liquid biopsy tool for assessing the biological status of the advanced disease, and to determine their relevance for predicting treatment response and therapy monitoring in order to optimize treatment for patients with metastatic breast cancer. Various subprojects of DETECT-CTC focus on the molecular characterization of CTCs, circulating free DNA and microRNA in blood, the evaluation of DNA damage and repair markers on CTCs, the evaluation of the origins and molecular causes of resistance to endocrine therapy at the level of individual CTCs, the comparison of phenotypic expression of biomarkers between CTCs, disseminated tumor cells from bone marrow, primary tumor and metastases from the same patients, and the study of microevolution of resistant subclones in metastatic breast cancer through single-cell analysis of CTCs [20].

In a comprehensive analysis of a large cohort of DETECT screening patients with HER2-negative primary tumors, the rate of discordance in the HER2 phenotype between primary tumor and CTCs was assessed and, as such, a discordance might have far-reaching implications in terms of follow-up treatment and the addition of targeted therapies [21]. The analysis included data from 1123 patients with HER2-negative MBC, and CTC screening was performed on average 56 months after primary diagnosis. In blood samples from 711 (63.3%) of 1123 patients, at least one CTC was detected (median 7 CTCs). To assess the HER2 status of CTCs, cells were labeled with HER2 antibodies, and only CTCs with a strong IHC staining intensity for HER2 (IHC score 3+) were considered as HER2-positive. In 134 of the 711 CTC-positive patients, at least one HER2-positive CTC was detected, yielding a HER2 discordance between primary tumor and CTCs in 18.8% of patients. A multivariable logistic regression with presence of at least one HER2-positive CTC (yes vs. no) as binary response variable revealed that lobular tumor histology, positive hormone receptor status, younger age, and the presence of 5 or more CTCs significantly predicted discordance in HER2 phenotype between primary tumor and CTCs. The authors concluded that in view of the well-known tumor heterogeneity in MBC, CTC-based liquid biopsy might better reflect total tumor burden and heterogeneity of tumor biology than biopsy of a single metastatic site.

Interestingly, the rates of patients with HER2-positive CTCs seem to be lower in the French CirCe T-DM1 trial (9.0% of screened patients) as compared to the DETECT screening (18.8% of screened patients). Both trials used the CELLSEARCH® System for the detection of CTCs, and HER2-positivity as a criterion for patient inclusion was defined as the presence of at least one HER2-positive CTC both for the CirCe T-DM1 and the DETECT III trial. However, HER2-positivity was assessed based on single-cell FISH-analysis in CirCe T-DM1 and by an immunohistochemical antibody staining procedure in DETECT. Another difference between CirCe T-DM1 and DETECT is the fact that MBC patients could be included in the DETECT III trial regardless of treatment line, while for the CirCe T-DM1 trial only patients starting a third- or fourth-line systemic therapy were eligible. In CirCe T-DM1, the prevalence rate of HER2-positive CTCs out of all detected CTCs in included patients was very low with a median of 1.6%, and the French group concluded that the low prevalence rate of HER2-positive CTCs was an important reason for the failure of the study [18]. Results of the DETECT III trial are still pending, and it will be interesting to see whether the rate of HER2-positive CTCs out of all detected CTCs will be as low as the rate observed in the CirCe T-DM1 study.

5. Conclusions

A general finding of all trials evaluating the clinical utility for CTCs described here is that CTCs are rare cells which are not present in every patient, particularly so in patients with EBC. Nevertheless, there is ample evidence that CTCs—once detected—are a strong prognostic factor for reduced survival. However, at present, it is not clear how this knowledge can be transferred to a prediction of therapy response and an improved clinical outcome. The STIC CTC trial showed the first data for a positive effect of a CTC-based decision in a subgroup with a high count of CTCs before starting treatment. Thus, CTCs might represent a helpful early treatment monitoring tool and can be used in situations with uncertain therapy response. The awaited results of the ongoing trials described here will hopefully provide much needed data that help to answer the question regarding the clinical utility of CTCs. Furthermore, the identification of potential targets for more individualized treatment options might improve the use of CTCs in clinical practice.

Funding: This research received no external funding.

Conflicts of Interest: Dr. Schochter: honoraria by Roche, Novartis and Pfizer; Dr. Friedl: honoraria by Novartis; Dr. deGregorio: honoraria from Roche, Pfizer, Novartis, Celgene, Tesaro, Daiichi; Dr. Krause: no relevant financial relationsships; Prof. Huober: research grants from Celgene, Novartis; Speaker fees by Lilly, Novartis, Roche, Pfizer, AstraZeneca, MSD, Celgene, Eisai; consultant by Lilly, Novartis, Roche, Pfizer, Hexal, AstraZeneca, MSD, Celgene; Travel expenses by Roche, Pfizer, Novartis, Celgene, Daiichi; Prof Rack: research grants and honoraria Aventis, Pharmacia, Amgen, Novartis, AstraZenica, Roche, Pfizer, Chugai; Prof. Janni has received research grants and honoraria from Menarini Silicon Biosystems, Johnson&Johnson. Institutional founding by the Deutsche Krebshilfe (German Cancer Aid), Menarini Silicon Biosystems, Roche, Novartis, GlaxoSmithKline, Amgen, Celgene, Eisai, and TEVA.

References

1. Joosse, S.A.; Gorges, T.M.; Pantel, K. Biology, detection, and clinical implications of circulating tumor cells. *EMBO Mol. Med.* **2015**, *7*, 1–11. [CrossRef] [PubMed]
2. Pantel, K.; Alix-Panabieres, C. Liquid biopsy and minimal residual disease—Latest advances and implications for cure. *Nat. Rev. Clin. Oncol.* **2019**, *16*, 409–424. [CrossRef] [PubMed]
3. Bidard, F.-C.; Peeters, D.J.; Fehm, T.; Nolè, F.; Gisbert-Criado, R.; Mavroudis, D.; Grisanti, S.; Generali, D.; Garcia-Saenz, J.A.; Stebbing, J.; et al. Clinical validity of circulating tumour cells in patients with metastatic breast cancer: A pooled analysis of individual patient data. *Lancet Oncol.* **2014**, *15*, 406–414. [CrossRef]
4. Janni, W.J.; Rack, B.; Terstappen, L.W.M.M.; Pierga, J.-Y.; Taran, F.-A.; Fehm, T.; Hall, C.; De Groot, M.R.; Bidard, F.-C.; Friedl, T.W.; et al. Pooled Analysis of the Prognostic Relevance of Circulating Tumor Cells in Primary Breast Cancer. *Clin. Cancer Res.* **2016**, *22*, 2583–2593. [CrossRef] [PubMed]

5. Bidard, F.-C.; Michiels, S.; Riethdorf, S.; Mueller, V.; Esserman, L.J.; Lucci, A.; Naume, B.; Horiguchi, J.; Gisbert-Criado, R.; Sleijfer, S.; et al. Circulating Tumor Cells in Breast Cancer Patients Treated by Neoadjuvant Chemotherapy: A Meta-analysis. *J. Natl. Cancer Inst.* **2018**, *110*, 560–567. [CrossRef] [PubMed]
6. Janni, W.; Rack, B.K.; Fasching, P.A.; Haeberle, L.; Tesch, H.; Lorenz, R.; Schochter, F.; Tzschaschel, M.; De Gregorio, A.; Fehm, T.N.; et al. Persistence of circulating tumor cells in high risk early breast cancer patients five years after adjuvant chemotherapy and late recurrence: Results from the adjuvant SUCCESS A trial. *J. Clin. Oncol.* **2018**, *36*, 515. [CrossRef]
7. Trapp, E.; Janni, W.; Schindlbeck, C.; Jückstock, J.; Andergassen, U.; de Gregorio, A.; Alunni-Fabbroni, M.; Tzschaschel, M.; Polasik, A.; Koch, J.G.; et al. Presence of Circulating Tumor Cells in High-Risk Early Breast Cancer During Follow-Up and Prognosis. *J. Natl. Cancer Inst.* **2019**, *111*, 380–387. [CrossRef] [PubMed]
8. Sparano, J.; O'Neill, A.; Alpaugh, K.; Wolff, A.C.; Northfelt, D.W.; Dang, C.T.; Sledge, G.W.; Miller, K.D. Association of Circulating Tumor Cells with Late Recurrence of Estrogen Receptor-Positive Breast Cancer: A Secondary Analysis of a Randomized Clinical Trial. *JAMA Oncol.* **2018**, *4*, 1700–1706. [CrossRef] [PubMed]
9. Rack, B.; Schindlbeck, C.; Jückstock, J.; Andergassen, U.; Hepp, P.; Zwingers, T.; Friedl, T.W.P.; Lorenz, R.; Tesch, H.; Fasching, P.A.; et al. Circulating tumor cells predict survival in early average-to-high risk breast cancer patients. *J. Natl. Cancer Inst.* **2014**, *106*. [CrossRef] [PubMed]
10. Pierga, J.Y.; Hajage, D.; Bachelot, T.; Delaloge, S.; Brain, E.; Campone, M.; Dieras, V.; Rolland, E.; Mignot, L.; Mathiot, C.; et al. High independent prognostic and predictive value of circulating tumor cells compared with serum tumor markers in a large prospective trial in first-line chemotherapy for metastatic breast cancer patients. *Ann. Oncol.* **2012**, *23*, 618–624. [CrossRef] [PubMed]
11. Park, Y.; Kitahara, T.; Urita, T.; Yoshida, Y.; Kato, R. Expected clinical applications of circulating tumor cells in breast cancer. *World J. Clin. Oncol.* **2011**, *2*, 303–310. [CrossRef] [PubMed]
12. Ignatiadis, M.; Litière, S.; Rothé, F.; Riethdorf, S.; Proudhon, C.; Fehm, T.; Aalders, K.; Forstbauer, H.; Fasching, P.A.; Brain, E.; et al. Trastuzumab versus observation for HER2 nonamplified early breast cancer with circulating tumor cells (EORTC 90091-10093, BIG 1-12, Treat CTC): A randomized phase II trial. *Ann. Oncol.* **2018**, *29*, 1777–1783. [CrossRef] [PubMed]
13. Fehrenbacher, L.; Cecchini, R.S.; Geyer, C.E.; Rastogi, P.; Costantino, J.P.; Atkins, J.N.; Polikoff, J.; Boileau, J.F.; Provencher, L.; Stokoe, C.; et al. Abstract GS1-02: NSABP B-47 (NRG oncology): Phase III randomized trial comparing adjuvant chemotherapy with adriamycin (A) and cyclophosphamide (C) â†' weekly paclitaxel (WP), or docetaxel (T) and C with or without a year of trastuzumab (H) in women with node-positive or high-risk node-negative invasive breast cancer (IBC) expressing HER2 staining intensity of IHC 1+ or 2+ with negative FISH (HER2-Low IBC). *Cancer Res.* **2018**, *78*, GS1–GS02.
14. Cristofanilli, M.; Pierga, J.-Y.; Reuben, J.; Rademaker, A.; Davis, A.A.; Peeters, D.J.; Fehm, T.; Nolé, F.; Gisbert-Criado, R.; Mavroudis, D.; et al. The clinical use of circulating tumor cells (CTCs) enumeration for staging of metastatic breast cancer (MBC): International expert consensus paper. *Crit. Rev. Oncol.* **2019**, *134*, 39–45. [CrossRef] [PubMed]
15. Smerage, J.B.; Barlow, W.E.; Hortobagyi, G.N.; Winer, E.P.; Leyland-Jones, B.; Srkalovic, G.; Tejwani, S.; Schott, A.F.; O'Rourke, M.A.; Lew, D.L.; et al. Circulating tumor cells and response to chemotherapy in metastatic breast cancer: SWOG S*J. Clin. Oncol.* **2014**, *32*, 3483–3489. [CrossRef] [PubMed]
16. Bidard, F.-C.; Fehm, T.; Ignatiadis, M.; Smerage, J.B.; Alix-Panabières, C.; Janni, W.; Messina, C.; Paoletti, C.; Müller, V.; Hayes, D.F.; et al. Clinical application of circulating tumor cells in breast cancer: Overview of the current interventional trials. *Cancer Metastasis Rev.* **2013**, *32*, 179–188. [CrossRef] [PubMed]
17. Bidard, F.; Jacot, W.; Dureau, S.; Brain, E.; Bachelot, T.; Bourgeois, H.; Goncalves, A.; Ladoire, S.; Naman, H.; Dalenc, F.; et al. Abstract GS3-07: Clinical utility of circulating tumor cell count as a tool to chose between first line hormone therapy and chemotherapy for ER+ HER2- metastatic breast cancer: Results of the phase III STIC CTC trial. *Cancer Res.* **2019**, *79*, GS3–GS07.
18. Bidard, F.-C.; Cottu, P.; Dubot, C.; Venat-Bouvet, L.; Lortholary, A.; Bourgeois, H.; Bollet, M.; Servent Hanon, V.; Luporsi-Gely, E.; Espie, M.; et al. Anti-HER2 therapy efficacy in HER2-negative metastatic breast cancer with HER2-amplified circulating tumor cells: Results of the CirCe T-DM1 trial. *Ann. Oncol.* **2017**, *28*. [CrossRef]

19. Paoletti, C.; Regan, M.M.; Liu, M.C.; Marcom, P.K.; Hart, L.L.; Smith, J.W.; Tedesco, K.L.; Amir, E.; Krop, I.E.; DeMichele, A.M.; et al. Abstract P1-01-01: Circulating tumor cell number and CTC-endocrine therapy index predict clinical outcomes in ER positive metastatic breast cancer patients: Results of the COMETI Phase 2 trial. *Cancer Res.* **2017**, *77*. [CrossRef]
20. Polasik, A.; Tzschaschel, M.; Schochter, F.; de Gregorio, A.; Friedl, T.W.; Rack, B.; Hartkopf, A.; Fasching, P.A.; Schneeweiss, A.; Mueller, V.; et al. Circulating Tumour Cells, Circulating Tumour DNA and Circulating MicroRNA in Metastatic Breast Carcinoma—What is the Role of Liquid Biopsy in Breast Cancer? *Geburtshilfe Frauenheilkd.* **2017**, *77*, 1291–1298. [CrossRef] [PubMed]
21. De Gregorio, A.; Friedl, T.W.; Huober, J.; Scholz, C.; De Gregorio, N.; Rack, B.; Trapp, E.; Alunni-Fabbroni, M.; Riethdorf, S.; Mueller, V.; et al. Discordance in Human Epidermal Growth Factor Receptor 2 (HER2) Phenotype Between Primary Tumor and Circulating Tumor Cells in Women With HER2-Negative Metastatic Breast Cancer. *JCO Precis. Oncol.* **2017**, 1–12. [CrossRef]

© 2019 by the authors. Licensee MDPI, Basel, Switzerland. This article is an open access article distributed under the terms and conditions of the Creative Commons Attribution (CC BY) license (http://creativecommons.org/licenses/by/4.0/).

Review

Cooperative and Escaping Mechanisms between Circulating Tumor Cells and Blood Constituents

Carmen Garrido-Navas [1], Diego de Miguel-Pérez [1,2], Jose Exposito-Hernandez [3], Clara Bayarri [1,4], Victor Amezcua [3], Alba Ortigosa [1], Javier Valdivia [3], Rosa Guerrero [3], Jose Luis Garcia Puche [1], Jose Antonio Lorente [1,2] and Maria José Serrano [1,3,*]

1. GENYO, Centre for Genomics and Oncological Research (Pfizer/University of Granada/Andalusian Regional Government), PTS Granada Av. de la Ilustración, 114, 18016 Granada, Spain; carmen.garrido@genyo.es (C.G.-N.); diego.miguel@genyo.es (D.d.M.-P.); ci.bayarri@gmail.com (C.B.); albaortigosa@correo.ugr.es (A.O.); jlpuche@ugr.es (J.L.G.P.); jlorente@ugr.es (J.A.L.)
2. Laboratory of Genetic Identification, Department of Legal Medicine, University of Granada, Av. de la Investigación, 11, 18071 Granada, Spain
3. Integral Oncology Division, Virgen de las Nieves University Hospital, Av. Dr. Olóriz 16, 18012 Granada, Spain; jose.exposito.sspa@juntadeandalucia.es (J.E.-H.); victor.amezcua.md@gmail.com (V.A.); jvaldib@gmail.com (J.V.); mjs@ugr.es (R.G.)
4. Department of Thoracic Surgery, Virgen de las Nieves University Hospital, Av. de las Fuerzas Armadas, 2, 18014 Granada, Spain
* Correspondence: mjose.serrano@genyo.es; Tel.: +34-958-715-500

Received: 2 September 2019; Accepted: 1 November 2019; Published: 3 November 2019

Abstract: Metastasis is the leading cause of cancer-related deaths and despite measurable progress in the field, underlying mechanisms are still not fully understood. Circulating tumor cells (CTCs) disseminate within the bloodstream, where most of them die due to the attack of the immune system. On the other hand, recent evidence shows active interactions between CTCs and platelets, myeloid cells, macrophages, neutrophils, and other hematopoietic cells that secrete immunosuppressive cytokines, which aid CTCs to evade the immune system and enable metastasis. Platelets, for instance, regulate inflammation, recruit neutrophils, and cause fibrin clots, which may protect CTCs from the attack of Natural Killer cells or macrophages and facilitate extravasation. Recently, a correlation between the commensal microbiota and the inflammatory/immune tone of the organism has been stablished. Thus, the microbiota may affect the development of cancer-promoting conditions. Furthermore, CTCs may suffer phenotypic changes, as those caused by the epithelial–mesenchymal transition, that also contribute to the immune escape and resistance to immunotherapy. In this review, we discuss the findings regarding the collaborative biological events among CTCs, immune cells, and microbiome associated to immune escape and metastatic progression.

Keywords: circulating tumor cells; tumor cell dissemination; immune system; microbiome

1. Introduction

The presence of circulating tumor cells (CTCs) in the peripheral blood has been largely associated with reduced disease-free and overall survival [1–3]. Even though metastasis is a highly inefficient process (tumor cell survival is less than 0.01%) [4], it is responsible for the majority of cancer-associated deaths [5]. In fact, it is accepted that CTCs are the initiator factor of metastatic relapse and their presence identifies patients with a higher risk of developing metastasis [6,7]. However, the complex biological processes enabling CTCs to survive and disseminate is not yet well understood and little is known about the cellular and genetic events involved both in the metastatic initiation and in its progression.

The success of the metastatic process is conditioned by the established relationship between tumor cells and the surrounding microenvironment. During the metastatic process, tumor cells interact with the immune system, which modulates this process [8]. The immune system has a dual role, both repressing but also promoting cancer progression. In fact, formation of CTC clusters or microemboli, not only composed of CTCs but also leukocytes, cancer-associated fibroblasts, endothelial cells, and platelets, was shown to facilitate the metastatic process and thus be related to poorer outcome in patients with breast [9] and gastric cancer [10], among others.

In this review, we will focus our interest in the "intimate friendship" between CTCs and the immune system. This private alliance benefits tumor progression through CTCs survival in this hostile microenvironment, the blood.

However, we cannot forget that the microenvironment is not only composed by immune system cells, stromal cells, and components of the extracellular matrix (ECM). CTCs and microbes co-evolve inside the ecosystem within our bodies [11,12] as will be further described in Section 3. This interaction influences the activity of the immune system on cell survival and expansion of CTCs [13].

In this review, we evaluate the current literature on interactions among CTCs, immune system cells, and microbiome in the tumor progression. We discuss how immune cells–CTC interactions contribute to the survival of these CTCs and how the microbiome can promote this positive association, finally supporting the metastatic process.

2. Promotion of Circulating Tumor Cells through the Immune System

The immune system is educated to eliminate the foreign and to respect the innate [14]. However, in the case of cancer, tumor cells are able to use the immune system to facilitate their own survival and migration. This phenomenon is known as concomitant immunity (CI) [15,16].

The plasticity of the immune system is well known and thus, according with the tumor type, the functional contribution of each immune cell can also be different [17,18]. However, some immune events are intimately associated with promotion of cancer, independently on the tumor type. Inflammation is one of these events and it is recognized as one of the "hallmarks" of cancer [19]. This process involves different types of immune cells, among which platelets, macrophages, and neutrophils can be highlighted [20,21].

Platelets are anucleated blood cells with a diameter of 2–4 μm originated during megakaryocytes maturation in the bone marrow and circulate in large numbers (1.5–4.0×10^9/L) in the bloodstream [22,23]. Platelets are the main cells involved in thrombosis and hemostasis, thus, related with the physiological and pathophysiological processes occurring during inflammation [24]. Interestingly, several studies have reported their role in cancer progression, especially during cancer metastasis [25] as they actively promote the metastatic process. Metastasis-promoting mechanisms affected by platelets are related to both migration of tumor cells and cancer cell survival in circulation [26].

Regarding the migration process, platelets store large amounts of transforming growth factor β (TGFβ), which is associated with an increase of the invasion potential of tumor cells. Thus, tumor cells-conjugated platelets release mediators to modify blood vessels permeability, including dense granule-release, histamine, eicosanoid metabolites, or serotonin [23]. These mediators induce endothelial cell retraction, exposing the basement membrane, and thus facilitate cancer cell extravasation [27]. In addition, platelets activation by cancer cells lead to the generation of platelet-derived microparticles (PDMPs), which can also release mediators like TxA2 and 12-HETE. These metabolites may enhance cell migration and invasion, eventually increasing the metastatic potential of cancer cells [28].

However, self-migration ability of CTCs is not enough to complete a successful metastatic process, survival of these cells, once in the blood, depends on the formation of circulating microemboli [9] as well as the acquisition of resistant phenotypes to the surrounding microenvironment. The acquisition of these phenotypes involves a biological process known as epithelial to mesenchymal transition (EMT) [29]. The EMT process explains how tumor cells change their phenotype, allowing them to detach, invade, and metastasize through the blood or lymphatic systems. Among others, the EMT involves loss of E-cadherin, disrupting cell-to-cell adhesions and altering gene expression by increasing β-catenin nuclear localization [30]. In contrast, N-cadherin, which is highly expressed in mesenchymal cells, fibroblasts, neural tissue, and cancer cells, is elevated during EMT. This cadherin switch, from E-cadherin to N-cadherin, is closely associated with the increased invasiveness, motility and metastasis potential of tumor cells. Moreover, activated platelets induce EMT through secretion of growth factors and cytokines (e.g., TNFα and TGFβ) [31]. Interestingly, these cytokines are also associated with the inflammatory process as previously explained [32].

Furthermore, CTCs-conjugated platelets also coordinate the engagement of other immune cells during the dissemination process [33]. In fact, CTCs-conjugated platelets recruit neutrophils, macrophages, and other immune cells through release of chemokines, such as CXCL5 or CXCL7. Among white blood cells (WBC), neutrophils are recognized as the mediators of metastasis initiation [34,35]. Neutrophils promote tumor development by initiating an angiogenic switch and facilitating colonization of CTCs. In fact, some groups support the idea that WBC shape a protective cover around CTCs, avoiding their recognition and destruction by other immune cells [36]. It has long been known that circulating platelet–neutrophil complexes are present in a wide range of inflammatory conditions including cancer. In this interaction, neutrophils are responsible to activate platelets and it was shown that the neutrophils–platelets interaction initiates inflammatory responses [37]. Platelets interact with neutrophils by multiple intermediates including platelet P-selectin binding to neutrophil P-selectin glycoprotein ligand-1 (PSGL-1) [38]. In addition, it has been suggested that the neutrophils-platelets complexes interacting to CTCs bring the latter to the endothelium, which is an essential step in hematogenous dissemination metastasis [34]. Thus, platelets prime tumor cells to promote neutrophil extracellular traps (NETs) formation, which are also involved in endothelial activation [39].

However, the interaction between the platelets–tumor cells complex and immune cells is not only restricted to neutrophils. The release of CXCL12, which is highly present in platelets, allows recruitment of CXCR4-positive cells such as macrophages to prepare the metastatic niche for CTCs [40]. Neutrophils are the first leukocytes to be recruited in response to chemotactic signaling and are responsible for stimulating the repair process and initiating inflammation. This influx is followed by monocytes, which, upon entry into the tissue, differentiate into macrophages. These macrophages promote invasion and metastasis from the primary tumor site through their ability to engage cancer cells in an autocrine loop that promotes cancer cell [41]. This autocrine signaling triggers cancer cells to produce CSF-1, which promote epidermal growth factor (EGF) production by macrophages. Finally, cancer cells and macrophages co-migrate towards tumor blood vessels, where macrophage-derived VEGF-A promotes cancer cell intravasation [42]. In addition, tumor migration is upregulated by macrophage-derived cathepsins, SPARC, or CCL18, that enhance tumor cell adhesion to extracellular matrix proteins [41]. Finally, CTCs produce CCL2 that recruits inflammatory monocytes, which in turn increase vascular permeability and allow migration of these tumor cells [16].

Nevertheless, migration and survival of CTCs belong together, so the promotion of CTC migration alone is not enough to allow metastasis. Anoikis is a programmed cell death induced by cell detachment [43] and essential for CTC survival. Another effect of the collaboration between immune cells and CTCs includes the protection of CTCs from anoikis [44]. Platelets are involved in this protective mechanism as it was observed that they induce RhoA-(myosin phosphatase targeting subunit 1) and MYPT1-protein phosphatase (PP1)-mediated Yes-associated protein 1 (YAP1) dephosphorylation and nuclear translocation, resulting in apoptosis resistance [45]. Apoptosis signal-regulating kinase 1 (Ask1) is a stress-responsive Ser/Thr mitogen-activated protein kinase kinase kinase (MAP3K) in the Jun N-terminal kinases (JNK) and p38 pathways. Once Ask1 levels are reduced in platelets, active phosphorylation of protein kinase B (Akt), JNK and p38 is downregulated, and thus tumor metastasis is attenuated [46].

In conclusion, the fate of CTCs is not to survive alone but with help of their mates within the immune system and thus, survival of CTCs depends on their ability to interact with immune cells. However, this favorable interaction between CTC and immune cells depends also on the status of our gut microbiota that is intimately linked with the nature of the immune system.

3. Survival of Circulating Tumor Cells Through the Interaction of Microbiota with the Immune System

The evolution of any disease, including cancer, depends highly on the physiological status of the host. The gut microbiota has emerged as an important factor of health and disease [47]. Likewise, our microbiota conditions the status of our immune system [48]. In fact, gut microorganisms are involved in the immune system development and in the response of the host against different pathologies, like cancer. Taken together, tumor cells-microbiome-immune system interactions may improve the likelihood of cell survival and induce tumor cell migration (Figure 1) [49].

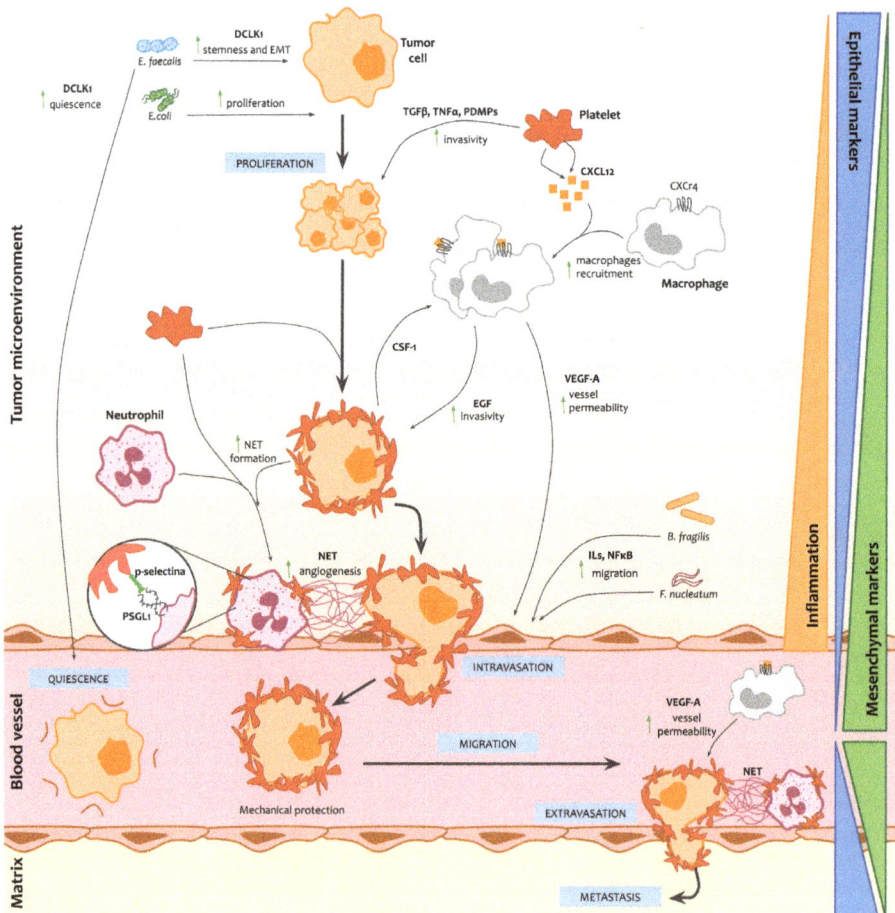

Figure 1. Interactions between circulating tumor cells (CTCs), immune system cells, and microbiome. Metabolites and cytokines produced by bacteria such as *Bacterioides fragilis*, *Enterococcus faecium*, *Escherichia coli*, and *Fusobacterium nucleatum* facilitate proliferation and migration of circulating tumor cells (CTCs), promote stemness and epithelial to mesenchymal transition (EMT), and help CTCs to enter quiescence. Furthermore, platelets interact with proliferating tumor cells directly, by formation of CTCs-platelet complexes allowing CTCs to escape the immune system but also indirectly, through three different ways: secretion of growth factors such as TFGβ, TNFα either alone or enclosed in platelets-derived microparticles (PDMPs) that increase invasivity of CTCs; secretion of chemokines such as CXCL12, increasing macrophages recruitment, what ultimately impact on invasivity and vessel permeability through epidermal growth factor (EGF) and VEFG-A, respectively; and formation of platelet-neutrophil

complexes (through P-selectin and PSGL1) that eventually generate neutrophil extracellular traps (NET) promoting angiogenesis and facilitating CTC intravasation to blood vessels. Finally, macrophages and NET also facilitate CTC extravasation from blood vessels to the extracellular matrix to produce metastasis.

As it has been mentioned before, the inflammatory process is an essential step in the development and progression of cancer. Microbes have a critical role in the initiation and maintenance of chronic inflammatory conditions [50,51]. However, how do microbes influence on the inflammation process and on the migration and survival of CTCs?

The gut microbiota contributes to cancer progression through different mechanisms. Recently, it was demonstrated that both, DNA-damaging superoxide radicals or genotoxins produced by the gut bacteria could initiate colon cancer. In addition, bacteria may induce cell proliferation through interactions with T-helper cells or Toll-like receptors, respectively [52]. In colon cancer patients, an increase of the *Escherichia coli* population was observed to induce colitis and colibactin synthesis and thereby, to promote inflammation.

Furthermore, it has been demonstrated that microbes as *Bacillus* sp., *Enterococcus faecium*, and *E. coli* produce peptides which alter host epithelial growth factor, activating intracellular pathways associated to migration. In a pioneer work, Wynendaele, E et al. [53] discovered that certain quorum sensing peptides produced by bacteria (molecules that microbes use to coordinate their gene expression and behavior) interact with cancer cells. This study demonstrated that Phr0662 (*Bacillus* sp.), EntF-metabolite (*E. faecium*), and EDF-derived (*E. coli*) peptides can initiate HCT-8/E11 colon cancer cell invasion. According to results of this group, the Phr0662 peptide targets epidermal growth factor receptors (EGFR and ErbB2). Upregulation of EGFR induces activation of the Ras/raf/MEK/MAPK, PI3K/Akt, and STAT intracellular signaling cascades [54] altering gene transcription and allowing migration of tumor cells. However, despite this work being extremely interesting, it is only a preliminary and exploratory in vitro assay and more exhaustive analyses including cancer patients should be carried out to validate these results.

Microbes can also alter cancer cell epigenetics through production of metabolites affecting gene expression [55]. This is the case of *Bifidobacterium* spp., which produces folate, one of the most powerful methyl donors involved in gene silencing. Thus, the gut microbiome is also involved in chromatin remodeling via acetylation and deacetylation of histones through butyrate production. Butyrate is a common metabolite of the microbiome, inducing cell differentiation via histone acetylation of the intestinal T reg cells [56].

Moreover, the microbiome plays an important role in the epithelial mesenchymal transition (EMT), an essential step for CTC migration and survival. In fact, microbes produce toxins that contribute to EMT [57]. Some of those microbes, as *Bacteroides fragilis*, *Fusobacterium nucleatum*, and *E. faecalis*, clear E-cadherin from epithelial cells, a transmembrane adhesion protein, leading to colonic epithelial proliferation [58]. Most of the studies on the interaction of the microbiome with cancer cells have been developed on colon cancer, murine models, or in vitro assays. Likewise, in a recent study including a murine model, colonic epithelial cells were transformed to express Ly6A/E, a stem cell marker implicating mesenchymal features, and Doublecortin-like kinase 1 (DCLK1), a marker of cancer, by the presence of *E. faecalis* [59]. DCLK1 is a member of the protein kinase super family and the doublecortin family, which is overexpressed in many cancers, including colon, pancreas, liver, esophageal, and kidney cancers. It is now suggested to be a master regulator of pluripotency factors, including Nanog, Oct4, Sox2, Klf4, and Myc, that are critical for stemness of cancer cells (CSC, cancer stem cells) and EMT transcriptional factors, including Snail, Slug, Twist, and Zeb 1 [60]. Interestingly, all these markers are involved in regulation of both EMT and CSCs and are controlled by DCLK1 expression in cancer models [61]. Furthermore, Westphalen, CB et al. [62], reported that DCLK1 induces quiescence of tumor cells. Quiescence is a common property of CSCs that is associated with the EMT process as a critical step for the migration and progression of tumor cells [63]. In consequence, EMT allows not only CTCs migration and tumor relapse, but also, induces the ability of CTCs to escape the immune system cancer treatments (Figure 1) [64].

Another biological mechanism used by the microbiome to enhance cancer progression includes the modulation of the immune system. Among all the microbes involved in this process, the enterotoxin *Bacteriodes fragilis* (ETBF) stands up due to the activation of STATA3 and T helper cells, both with an important role in the inflammatory process [65]. In fact, Chung, L et al. [65], demonstrated that *Bacteroides fragilis* toxin (BFT) can activate a pro-carcinogenic inflammatory cascade, related to IL-17R, NF-κB, and Stat3 signaling, in colonic epithelial cells (CECs). Likewise, the activation of NF-κB in these cells, induces other chemokines as CXCL1 that mediates recruitment of CXCR2-expressing polymorphonuclear immature myeloid cells, promoting ETBF-mediated distal colon tumorigenesis. Another bacterium associated with poor oncological outcomes is *Fuscobacterium nucleatum*. It has been suggested that *F. nucleatum* promotes tumorigenesis through both pro-inflammatory and immunosuppressive effects. Furthermore, *F. nucleatum* is associated with activation of cytokines IL-6, IL-12, IL-17, and TNF-α, which cooperatively upregulate NFκB, a critical regulator of cellular proliferation [66]. Some studies associated the presence of high levels of *F. nucleatum* with the EMT process [67]. In this context, Mima, K et al. [68], identified that *F. nucleatum* adheres to and invades epithelial cells mainly through the virulence factors, including Fusobacterium adhesin A, Fusobacterium autotransporter protein 2, and fusobacterial outer membrane protein. To the contrary, other studies raise their skepticism about the role of *F. nucleatum* on EMT, as it still remains unclear whether *F. nucleatum* triggers the colonic EMT process. Ma, CT, et al., showed that *F. nucleatum* infection did not affect expression levels of E-cadherin and β-catenin [69]. However, it was associated with proliferation and invasion of colon cancer cells as it significantly increased phosphorylation of p65 (a subunit of nuclear factor-κB), as well as expression of interleukin (IL)-6, IL-1β and matrix metalloproteinase (MMP)-13. Regardless of the fact that there are not any explicit studies evaluating the direct action of the microbiota on CTCs, we here reviewed some of the biological processes in which microbes alter tumorigenic pathways. As they are involved in inflammation or inducing EMT, both biological processes intimately associated with the ability of CTCs to migrate and to survive, we suggest the potential interaction between them.

4. Conclusions and Perspectives

In conclusion, the complex interactions between the microbiome, the immune system and CTCs may allow us to grasp the insights of the dissemination process occurring in cancer and the immune system's mechanisms involved in this process. Therefore, the interactions among microbiome, immune system, and CTCs could aid the rational design of interventions that strengthen the antimetastatic potential of combined treatments to prevent appearance of metastasis. Moreover, emerging evidences may provide new mechanisms to control the dissemination process through the development of new therapeutic strategies with the microbiota as target. However, this topic is still an incipient area of research and further investigation is needed to clarify the association of the microbiome with the immune system and the dissemination process.

Author Contributions: Design and writing by M.J.S., C.G.-N., and J.E.-H.; Figures designed by A.O., C.B., J.A.L., and D.d.M.-P.; Discussion J.L.G.P., V.A., J.V., and R.G.

Funding: This research received no external funding.

Acknowledgments: We would like to extend our gratitude to Hugh Ilyine for the English revision.

Conflicts of Interest: The authors declare no conflict of interest.

References

1. Bayarri-Lara, C.; Ortega, F.G.; De Guevara, A.C.L.; Puche, J.L.; Zafra, J.R.; De Miguel-Pérez, D.; Ramos, A.S.-P.; Giraldo-Ospina, C.F.; Gómez, J.A.N.; Delgado-Rodríguez, M.; et al. Circulating Tumor Cells Identify Early Recurrence in Patients with Non-Small Cell Lung Cancer Undergoing Radical Resection. *PLoS ONE* **2016**, *11*, e0148659. [CrossRef] [PubMed]
2. Delgado-Ureña, M.; Ortega, F.G.; De Miguel-Pérez, D.; Rodriguez-Martínez, A.; García-Puche, J.L.; Ilyine, H.; Lorente, J.A.; Exposito-Hernandez, J.; Garrido-Navas, M.C.; Delgado-Ramirez, M.; et al. Circulating tumor cells criteria (CyCAR) versus standard RECIST criteria for treatment response assessment in metastatic colorectal cancer patients. *J. Transl. Med.* **2018**, *16*, 251. [CrossRef] [PubMed]
3. Mamdouhi, T.; Twomey, J.D.; McSweeney, K.M.; Zhang, B. Fugitives on the run: Circulating tumor cells (CTCs) in metastatic diseases. *Cancer Metastasis Rev.* **2019**, *38*, 297–305. [CrossRef] [PubMed]
4. Rejniak, K.A. Circulating Tumor Cells: When a Solid Tumor Meets a Fluid Microenvironment. *Adv. Exp. Med. Biol.* **2016**, *936*, 93–106. [PubMed]
5. Chaffer, C.L.; Weinberg, R.A. A Perspective on Cancer Cell Metastasis. *Science* **2011**, *331*, 1559–1564. [CrossRef]
6. Tsai, W.-S.; Chen, J.-S.; Shao, H.-J.; Wu, J.-C.; Lai, J.-M.; Lu, S.-H.; Hung, T.-F.; Chiu, Y.-C.; You, J.-F.; Hsieh, P.-S.; et al. Circulating Tumor Cell Count Correlates with Colorectal Neoplasm Progression and Is a Prognostic Marker for Distant Metastasis in Non-Metastatic Patients. *Sci. Rep.* **2016**, *6*, 24517. [CrossRef]
7. Al-Mehdi, A.; Tozawa, K.; Fisher, A.; Shientag, L.; Lee, A.; Muschel, R. Intravascular origin of metastasis from the proliferation of endothelium-attached tumor cells: A new model for metastasis. *Nat. Med.* **2000**, *6*, 100–102. [CrossRef]
8. Balkwill, F.; Mantovani, A. Inflammation and cancer: Back to Virchow? *Lancet* **2001**, *357*, 539–545. [CrossRef]
9. Aceto, N.; Bardia, A.; Miyamoto, D.T.; Donaldson, M.C.; Wittner, B.S.; Spencer, J.A.; Yu, M.; Pely, A.; Engstrom, A.; Zhu, H.; et al. Circulating Tumor Cell Clusters Are Oligoclonal Precursors of Breast Cancer Metastasis. *Cell* **2014**, *158*, 1110–1122. [CrossRef]
10. Abdallah, E.A.; Braun, A.C.; Flores, B.C.; Senda, L.; Urvanegia, A.C.; Calsavara, V.; De Jesus, V.H.F.; Almeida, M.F.A.; Begnami, M.D.; Coimbra, F.J.; et al. The Potential Clinical Implications of Circulating Tumor Cells and Circulating Tumor Microemboli in Gastric Cancer. *Oncologist* **2019**, *24*, e854–e863. [CrossRef]
11. Whisner, C.M.; Athena Aktipis, C. The Role of the Microbiome in Cancer Initiation and Progression: How Microbes and Cancer Cells Utilize Excess Energy and Promote One Another's Growth. *Curr. Nutr. Rep.* **2019**, *8*, 42–51. [CrossRef] [PubMed]
12. Contreras, A.V.; Cocom-Chan, B.; Hernandez-Montes, G.; Portillo-Bobadilla, T.; Resendis-Antonio, O. Host-Microbiome Interaction and Cancer: Potential Application in Precision Medicine. *Front. Physiol.* **2016**, *7*, 606. [CrossRef] [PubMed]
13. Bose, M.; Mukherjee, P. Role of Microbiome in Modulating Immune Responses in Cancer. *Mediat. Inflamm.* **2019**, *2019*, 1–7. [CrossRef] [PubMed]
14. Mellman, I.; Coukos, G.; Dranoff, G. Cancer immunotherapy comes of age. *Nature* **2011**, *480*, 480–489. [CrossRef] [PubMed]
15. Janssen, L.M.; Ramsay, E.E.; Logsdon, C.D.; Overwijk, W.W. The immune system in cancer metastasis: Friend or foe? *J. Immunother. Cancer* **2017**, *5*, 79. [CrossRef] [PubMed]
16. Kitamura, T.; Qian, B.-Z.; Pollard, J.W. Immune cell promotion of metastasis. *Nat. Rev. Immunol* **2015**, *15*, 73–86. [CrossRef]
17. Blomberg, O.S.; Spagnuolo, L.; de Visser, K.E. Immune regulation of metastasis: Mechanistic insights and therapeutic opportunities. *Dis. Model. Mech.* **2018**, *11*, dmm036236. [CrossRef]
18. Leone, K.; Poggiana, C.; Zamarchi, R. The Interplay between Circulating Tumor Cells and the Immune System: From Immune Escape to Cancer Immunotherapy. *Diagnostics* **2018**, *8*, 59. [CrossRef]
19. Hanahan, D.; Weinberg, R.A. Hallmarks of Cancer: The Next Generation. *Cell* **2011**, *144*, 646–674. [CrossRef]
20. Morrell, C.N.; Aggrey, A.A.; Chapman, L.M.; Modjeski, K.L. Emerging roles for platelets as immune and inflammatory cells. *Blood* **2014**, *123*, 2759–2767. [CrossRef]
21. Hamilton, G.; Rath, B. Circulating tumor cell interactions with macrophages: Implications for biology and treatment. *Transl. Lung Cancer Res.* **2017**, *6*, 418–430. [CrossRef] [PubMed]
22. Machlus, K.R.; Italiano, J.E. The incredible journey: From megakaryocyte development to platelet formation. *J. Cell Biol.* **2013**, *201*, 785–796. [CrossRef] [PubMed]

23. Li, N. Platelets in cancer metastasis: To help the "villain" to do evil. *Int. J. Cancer* **2016**, *138*, 2078–2087. [CrossRef] [PubMed]
24. Margetic, S. Inflammation and haemostasis. *Biochem. Med.* **2012**, *22*, 49–62. [CrossRef]
25. Micalizzi, D.S.; Maheswaran, S.; Haber, D.A. A conduit to metastasis: Circulating tumor cell biology. *Genes Dev.* **2017**, *31*, 1827–1840. [CrossRef]
26. Yu, L.-X.; Yan, L.; Yang, W.; Wu, F.-Q.; Ling, Y.; Chen, S.-Z.; Tang, L.; Tan, Y.-X.; Cao, D.; Wu, M.-C.; et al. Platelets promote tumour metastasis via interaction between TLR4 and tumour cell-released high-mobility group box1 protein. *Nat. Commun.* **2014**, *5*, 5256. [CrossRef]
27. Van Zijl, F.; Krupitza, G.; Mikulits, W. Initial steps of metastasis: Cell invasion and endothelial transmigration. *Mutat. Res. Mutat. Res.* **2011**, *728*, 23–34. [CrossRef]
28. Plantureux, L.; Mège, D.; Crescence, L.; Dignat-George, F.; Dubois, C.; Panicot-Dubois, L. Impacts of Cancer on Platelet Production, Activation and Education and Mechanisms of Cancer-Associated Thrombosis. *Cancers (Basel)* **2018**, *10*, 441. [CrossRef]
29. Lim, J.; Thiery, J.P. Epithelial-mesenchymal transitions: Insights from development. *Development* **2012**, *139*, 3471–3486. [CrossRef]
30. Serrano, M.J.; Ortega, F.G.; Alvarez-Cubero, M.J.; Nadal, R.; Sánchez-Rovira, P.; Salido, M.; Rodriguez, M.; García-Puche, J.L.; Delgado-Rodríguez, M.; Sole, F.; et al. EMT and EGFR in CTCs cytokeratin negative non-metastatic breast cancer. *Oncotarget* **2014**, *5*, 7486–7497. [CrossRef]
31. Tsubakihara, Y.; Moustakas, A. Epithelial-Mesenchymal Transition and Metastasis under the Control of Transforming Growth Factor β. *Int. J. Mol. Sci.* **2018**, *19*, 3672. [CrossRef] [PubMed]
32. Wojdasiewicz, P.; Poniatowski, Ł.A.; Szukiewicz, D. The role of inflammatory and anti-inflammatory cytokines in the pathogenesis of osteoarthritis. *Mediat. Inflamm.* **2014**, *2014*, 561459. [CrossRef] [PubMed]
33. Gruber, I.; Landenberger, N.; Staebler, A.; Hahn, M.; Wallwiener, D.; Fehm, T. Relationship between circulating tumor cells and peripheral T-cells in patients with primary breast cancer. *Anticancer. Res.* **2013**, *33*, 2233–2238. [PubMed]
34. Tao, L.; Zhang, L.; Peng, Y.; Tao, M.; Li, L.; Xiu, D.; Yuan, C.; Ma, Z.; Jiang, B. Neutrophils assist the metastasis of circulating tumor cells in pancreatic ductal adenocarcinoma: A new hypothesis and a new predictor for distant metastasis. *Medicine (Baltimore)* **2016**, *95*, e4932. [CrossRef] [PubMed]
35. Zhang, J.; Qiao, X.; Shi, H.; Han, X.; Liu, W.; Tian, X.; Zeng, X. Circulating tumor-associated neutrophils (cTAN) contribute to circulating tumor cell survival by suppressing peripheral leukocyte activation. *Tumor Biol.* **2016**, *37*, 5397–5404. [CrossRef]
36. Uppal, A.; Wightman, S.C.; Ganai, S.; Weichselbaum, R.R.; An, G. Investigation of the essential role of platelet-tumor cell interactions in metastasis progression using an agent-based model. *Theor. Biol. Med. Model.* **2014**, *11*, 17. [CrossRef] [PubMed]
37. Sreeramkumar, V.; Adrover, J.M.; Ballesteros, I.; Cuartero, M.I.; Rossaint, J.; Bilbao, I.; Nácher, M.; Pitaval, C.; Radovanovic, I.; Fukui, Y.; et al. Neutrophils scan for activated platelets to initiate inflammation. *Science* **2014**, *346*, 1234–1238. [CrossRef]
38. Moore, K.L.; Patel, K.D.; Bruehl, R.E.; Li, F.; Johnson, D.A.; Lichenstein, H.S.; Cummings, R.D.; Bainton, D.F.; McEver, R.P. P-selectin glycoprotein ligand-1 mediates rolling of human neutrophils on P-selectin. *J. Cell Biol.* **1995**, *128*, 661–671. [CrossRef]
39. Abdol Razak, N.; Elaskalani, O.; Metharom, P. Pancreatic Cancer-Induced Neutrophil Extracellular Traps: A Potential Contributor to Cancer-Associated Thrombosis. *Int. J. Mol. Sci.* **2017**, *18*, 487. [CrossRef]
40. Sun, X.; Cheng, G.; Hao, M.; Zheng, J.; Zhou, X.; Zhang, J.; Taichman, R.S.; Pienta, K.J.; Wang, J.; Zhou, X. CXCL12/CXCR4/CXCR7 chemokine axis and cancer progression. *Cancer Metastasis Rev.* **2010**, *29*, 709–722. [CrossRef]
41. Sangaletti, S.; Di Carlo, E.; Gariboldi, S.; Miotti, S.; Cappetti, B.; Parenza, M.; Rumio, C.; Brekken, R.A.; Chiodoni, C.; Colombo, M.P. Macrophage-Derived SPARC Bridges Tumor Cell-Extracellular Matrix Interactions toward Metastasis. *Cancer Res.* **2008**, *68*, 9050–9059. [CrossRef] [PubMed]
42. Nielsen, S.R.; Schmid, M.C. Macrophages as Key Drivers of Cancer Progression and Metastasis. *Mediat. Inflamm.* **2017**, *2017*, 9624760. [CrossRef] [PubMed]
43. Kim, Y.-N.; Koo, K.H.; Sung, J.Y.; Yun, U.-J.; Kim, H. Anoikis Resistance: An Essential Prerequisite for Tumor Metastasis. *Int. J. Cell Biol.* **2012**, *2012*, 1–11. [CrossRef] [PubMed]

44. Heeke, S.; Mograbi, B.; Alix-Panabières, C.; Hofman, P. Never Travel Alone: The Crosstalk of Circulating Tumor Cells and the Blood Microenvironment. *Cells* **2019**, *8*, 714. [CrossRef]
45. Abylkassov, R.; Xie, Y. Role of Yes-associated protein in cancer: An update. *Oncol. Lett.* **2016**, *12*, 2277–2282. [CrossRef] [PubMed]
46. Kamiyama, M.; Shirai, T.; Tamura, S.; Suzuki-Inoue, K.; Ehata, S.; Takahashi, K.; Miyazono, K.; Hayakawa, Y.; Sato, T.; Takeda, K.; et al. ASK1 facilitates tumor metastasis through phosphorylation of an ADP receptor P2Y12 in platelets. *Cell Death Differ.* **2017**, *24*, 2066–2076. [CrossRef] [PubMed]
47. Mohajeri, M.H.; Brummer, R.J.M.; Rastall, R.A.; Weersma, R.K.; Harmsen, H.J.M.; Faas, M.; Eggersdorfer, M. The role of the microbiome for human health: From basic science to clinical applications. *Eur. J. Nutr.* **2018**, *57*, 1–14. [CrossRef]
48. Cianci, R.; Pagliari, D.; Piccirillo, C.A.; Fritz, J.H.; Gambassi, G. The Microbiota and Immune System Crosstalk in Health and Disease. *Mediat. Inflamm.* **2018**, *2018*, 1–3. [CrossRef]
49. Li, W.; Deng, Y.; Chu, Q.; Zhang, P. Gut microbiome and cancer immunotherapy. *Cancer Lett.* **2019**, *447*, 41–47. [CrossRef]
50. Ong, H.S.; Yim, H.C.H. Microbial Factors in Inflammatory Diseases and Cancers. In *Advances in Experimental Medicine and Biology*; Springer: Singapore, 2017; pp. 153–174.
51. Sussman, D.A.; Santaolalla, R.; Strobel, S. Cancer in inflammatory bowel disease. *Curr. Opin. Gastroenterol.* **2012**, *28*, 327–333. [CrossRef]
52. Sieling, P.A.; Chung, W.; Duong, B.T.; Godowski, P.J.; Modlin, R.L. Toll-like receptor 2 ligands as adjuvants for human Th1 responses. *J. Immunol.* **2003**, *170*, 194–200. [CrossRef] [PubMed]
53. Wynendaele, E.; Verbeke, F.; D'Hondt, M.; Hendrix, A.; Van De Wiele, C.; Burvenich, C.; Peremans, K.; De Wever, O.; Bracke, M.; De Spiegeleer, B. Crosstalk between the microbiome and cancer cells by quorum sensing peptides. *Peptides* **2015**, *64*, 40–48. [CrossRef] [PubMed]
54. Wee, P.; Wang, Z. Epidermal Growth Factor Receptor Cell Proliferation Signaling Pathways. *Cancers (Basel)* **2017**, *9*, 52.
55. Gerhauser, C. Impact of dietary gut microbial metabolites on the epigenome. *Philos. Trans. R. Soc. Lond. B Biol. Sci.* **2018**, *373*, 20170359. [CrossRef]
56. Ji, J.; Shu, D.; Zheng, M.; Wang, J.; Luo, C.; Wang, Y.; Guo, F.; Zou, X.; Lv, X.; Li, Y.; et al. Microbial metabolite butyrate facilitates M2 macrophage polarization and function. *Sci. Rep.* **2016**, *6*, 24838. [CrossRef] [PubMed]
57. Gaines, S.; Williamson, A.J.; Kandel, J. How the microbiome is shaping our understanding of cancer biology and its treatment. *Semin. Colon Rectal Surg.* **2018**, *29*, 12–16. [CrossRef]
58. Sears, C.L.; Geis, A.L.; Housseau, F. Bacteroides fragilis subverts mucosal biology: From symbiont to colon carcinogenesis. *J. Clin. Investig.* **2014**, *124*, 4166–4172. [CrossRef] [PubMed]
59. Wang, X.; Yang, Y.; Huycke, M.M. Commensal bacteria drive endogenous transformation and tumour stem cell marker expression through a bystander effect. *Gut* **2015**, *64*, 459–468. [CrossRef]
60. Chandrakesan, P.; Weygant, N.; May, R.; Qu, D.; Chinthalapally, H.R.; Sureban, S.M.; Ali, N.; Lightfoot, S.A.; Umar, S.; Houchen, C.W. DCLK1 facilitates intestinal tumor growth via enhancing pluripotency and epithelial mesenchymal transition. *Oncotarget* **2014**, *5*, 9269–9280. [CrossRef]
61. Chandrakesan, P.; Panneerselvam, J.; Qu, D.; Weygant, N.; May, R.; Bronze, M.; Houchen, C. Regulatory Roles of Dclk1 in Epithelial Mesenchymal Transition and Cancer Stem Cells. *J. Carcinog. Mutagen.* **2016**, *7*, 1–8.
62. Westphalen, C.B.; Asfaha, S.; Hayakawa, Y.; Takemoto, Y.; Lukin, D.J.; Nuber, A.H.; Brandtner, A.; Setlik, W.; Remotti, H.; Muley, A.; et al. Long-lived intestinal tuft cells serve as colon cancer–initiating cells. *J. Clin. Investig.* **2014**, *124*, 1283–1295. [CrossRef]
63. Thiery, J.P.; Acloque, H.; Huang, R.Y.; Nieto, M.A. Epithelial-Mesenchymal Transitions in Development and Disease. *Cell* **2009**, *139*, 871–890. [CrossRef] [PubMed]
64. Alvarez-Cubero, M.J.; Vázquez-Alonso, F.; Puche-Sanz, I.; Ortega, F.G.; Martin-Prieto, M.; Garcia-Puche, J.L.; Pascual-Geler, M.; Lorente, J.A.; Cozar-Olmo, J.M.; Serrano, M.J. Dormant Circulating Tumor Cells in Prostate Cancer: Therapeutic, Clinical and Biological Implications. *Curr. Drug Targets* **2016**, *17*, 693–701. [CrossRef] [PubMed]
65. Chung, L.; Orberg, E.T.; Geis, A.L.; Chan, J.L.; Fu, K.; Shields, C.E.D.; Dejea, C.M.; Fathi, P.; Chen, J.; Finard, B.B.; et al. Bacteroides fragilis Toxin Coordinates a Pro-carcinogenic Inflammatory Cascade via Targeting of Colonic Epithelial Cells. *Cell Host Microbe* **2018**, *23*, 203–214.e5. [CrossRef] [PubMed]

66. Yang, Y.; Weng, W.; Peng, J.; Hong, L.; Yang, L.; Toiyama, Y.; Gao, R.; Liu, M.; Yin, M.; Pan, C.; et al. Fusobacterium nucleatum Increases Proliferation of Colorectal Cancer Cells and Tumor Development in Mice by Activating Toll-Like Receptor 4 Signaling to Nuclear Factor−κB, and Up-regulating Expression of MicroRNA-21. *Gastroenterology* **2017**, *152*, 851–866.e24. [CrossRef] [PubMed]
67. Rubinstein, M.R.; Wang, X.; Liu, W.; Hao, Y.; Cai, G.; Han, Y.W. Fusobacterium nucleatum Promotes Colorectal Carcinogenesis by Modulating E-Cadherin/β-Catenin Signaling via its FadA Adhesin. *Cell Host Microbe* **2013**, *14*, 195–206. [CrossRef] [PubMed]
68. Mima, K.; Nishihara, R.; Qian, Z.R.; Cao, Y.; Sukawa, Y.; Nowak, J.A.; Yang, J.; Dou, R.; Masugi, Y.; Song, M.; et al. *Fusobacterium nucleatum* in colorectal carcinoma tissue and patient prognosis. *Gut* **2016**, *65*, 1973–1980. [CrossRef]
69. Ma, C.; Luo, H.; Gao, F.; Tang, Q.; Chen, W. Fusobacterium nucleatum promotes the progression of colorectal cancer by interacting with E-cadherin. *Oncol. Lett.* **2018**, *16*, 2606–2612. [CrossRef]

© 2019 by the authors. Licensee MDPI, Basel, Switzerland. This article is an open access article distributed under the terms and conditions of the Creative Commons Attribution (CC BY) license (http://creativecommons.org/licenses/by/4.0/).

Review

New Frontiers in Diagnosis and Therapy of Circulating Tumor Markers in Cerebrospinal Fluid In Vitro and In Vivo

Olga A. Sindeeva [1], Roman A. Verkhovskii [1], Mustafa Sarimollaoglu [2], Galina A. Afanaseva [1,3], Alexander S. Fedonnikov [1,3], Evgeny Yu. Osintsev [1,3], Elena N. Kurochkina [1,3], Dmitry A. Gorin [4], Sergey M. Deyev [5], Vladimir P. Zharov [1,2] and Ekaterina I. Galanzha [1,6,*]

[1] Laboratory of Biomedical Photoacoustics, Saratov State University, 83 Astrakhanskaya St, 410012 Saratov, Russia; mouse-oa@rambler.ru (O.A.S.); r.a.verhovskiy@mail.ru (R.A.V.); gafanaseva@yandex.ru (G.A.A.); fedonnikov@mail.ru (A.S.F.); dr_osintsev@mail.ru (E.Y.O.); e.katamadze@yandex.ru (E.N.K.); zharovvladimirp@uams.edu (V.P.Z.)
[2] Arkansas Nanomedicine Center & Winthrop P. Rockefeller Cancer Institute, University of Arkansas for Medical Sciences, Little Rock, AR 72205, USA; msarimollaoglu@uams.edu
[3] Saratov State Medical University, 112 Bolshaya Kazachia St., 410012 Saratov, Russia
[4] Laboratory of Biophotonics, Skolkovo Institute of Science and Technology, 3 Nobelya Str., 121205 Moscow, Russia; gorinda@mail.ru
[5] Shemyakin-Ovchinnikov Institute of Bioorganic Chemistry, Russian Academy of Sciences, Miklukho-Maklaya St., 16/10, Moscow 117997, Russia; biomem@mail.ru
[6] Laboratory of Lymphatic Research, Diagnosis and Therapy (LDT), University of Arkansas for Medical Sciences, Little Rock, AR 72205, USA
* Correspondence: egalanzha@uams.edu

Received: 28 June 2019; Accepted: 26 September 2019; Published: 3 October 2019

Abstract: One of the greatest challenges in neuro-oncology is diagnosis and therapy (theranostics) of leptomeningeal metastasis (LM), brain metastasis (BM) and brain tumors (BT), which are associated with poor prognosis in patients. Retrospective analyses suggest that cerebrospinal fluid (CSF) is one of the promising diagnostic targets because CSF passes through central nervous system, harvests tumor-related markers from brain tissue and, then, delivers them into peripheral parts of the human body where CSF can be sampled using minimally invasive and routine clinical procedure. However, limited sensitivity of the established clinical diagnostic cytology in vitro and MRI in vivo together with minimal therapeutic options do not provide patient care at early, potentially treatable, stages of LM, BM and BT. Novel technologies are in demand. This review outlines the advantages, limitations and clinical utility of emerging liquid biopsy in vitro and photoacoustic flow cytometry (PAFC) in vivo for assessment of CSF markers including circulating tumor cells (CTCs), circulating tumor DNA (ctDNA), microRNA (miRNA), proteins, exosomes and emboli. The integration of in vitro and in vivo methods, PAFC-guided theranostics of single CTCs and targeted drug delivery are discussed as future perspectives.

Keywords: cerebrospinal liquid biopsy; in vivo flow cytometry; tumor biomarkers; circulating tumor cells; ctDNA; miRNA; exosomes; emboli; targeted therapy

1. Introduction

Leptomeningeal and brain metastasis (LM and BM) as a result of metastatic dissemination of solid tumors (e.g., melanoma, breast cancer, lung cancer and colorectal cancer) and hematological neoplasms as well as primary brain tumors (BTs, e.g., glioma) are commonly fatal with minimum treatment options [1–11]. Relatively high number of underdiagnosed LM, BM and BT and often ineffective therapy are the major challenges. For example, autopsy data demonstrate that BM contribute to death in ~75% of melanoma patients but they are clinically diagnosed in only 37% cases [8].

Among other parts of central nervous system (CNS), cerebrospinal fluid (CSF) is the easiest accessible medium that can directly uptake tumor markers from different parts of CNS [12–17]. Normally, CSF is a colorless liquid (a total volume of 130–150 mL for human) that contains up to 5 cells/µL, mainly leukocytes (white blood cells [WBCs]) [18–20]. CSF is produced by the choroidal plexus of the ventricular system and ependymal brain cells from blood [18,20,21].

In tumor patients with CNS involvement, CSF contains various markers associated with disease progression and responses to therapy [2–4,13–17,22–30]. Among others, circulating tumor cells (CTCs) are direct seeds of metastasis and, therefore, their diagnostic significance encourages high attention of researchers and clinicians. Furthermore, multiple recent reports suggested that detection of tumor-derived markers such as exosomes, circulating tumor DNA (ctDNA), micro-RNA (miRNA) and proteins is relevant to LM, BM and BT. The diagnostic significance of these markers seems especially important for BT because some BTs are not metastatic and do not typically shed CTCs but may release tumor-derived markers in CSF. CTC aggregates (so-called clusters or emboli) in CSF may also have diagnostic value. This speculation is based on: (1) finding CTC emboli in CSF samples of patients with lung cancer and LM [30]; (2) detection of CTC clusters in blood of patients with BT (e.g., glioblastoma) assuming their leaving CNS through the compromised blood-brain barrier (BBB) [31]; and (3) experimental and clinical evidences that multicellular CTC aggregates in peripheral blood represent the aggressive cell subset responsible for initiating and promoting metastasis [31–40].

Based on the physiology of CNS and mechanisms of tumor development (e.g., compromising BBB to penetrate tumor cells [41]), CTCs, their aggregates and other tumor-derived markers may invade CSF through different mechanisms that include (1) crossing the compromised BBB by blood and lymphatic CTCs and/or (2) shedding tumor cells by existing BM and BT. The latter mechanism provides a solid basis for using CSF tumor markers to diagnose progression of BM and BT, and to estimate responses to therapy. The first mechanism likely works for LM and BM and suggests the origin of CSF tumor biomarkers from blood or/and lymph and their possible entry to CSF before colonization of brain tissue and meninges.

Thus, testing CSF might predict deadly LM, BM and BT; and advanced methods to assess CSF tumor markers in CSF are urgently needed to prolong life of patients suffering from CNS tumor lesions.

2. In Vitro Detection of CSF Tumor Markers

The gold standard for routine clinical examination of CSF is cytology after lumbar puncture [9–11,24,42,43]. The detection approach is based on cytomorphology of tumor cells after staining samples with Wright-Giemsa or Papanicolaou dyes. However, the sensitivity of CSF cytology is estimated as low as 50% [9]. Furthermore, cytology is a relatively subjective method since its results depend on the ability of a laboratory technician to correctly identify types of cells, for example, to distinguish tumor cells from normal leukocytes [24–26]. This may lead to delaying of therapeutic interventions until other diagnostic criteria (e.g., abnormal magnetic resonance imaging [MRI]) and/or strong clinical symptoms emerge. As a result, involvement of CNS in some patients is found at autopsy only.

The limitations of cytology and deadly nature of LM, BM and BT encouraged researchers and clinicians to develop more sensitive and accurate markers using modern technologies. During the past decade, substantial efforts have been made to assess CSF samples using new concept of liquid biopsy (Figure 1) [2,15–17,23,26–30,44–48].

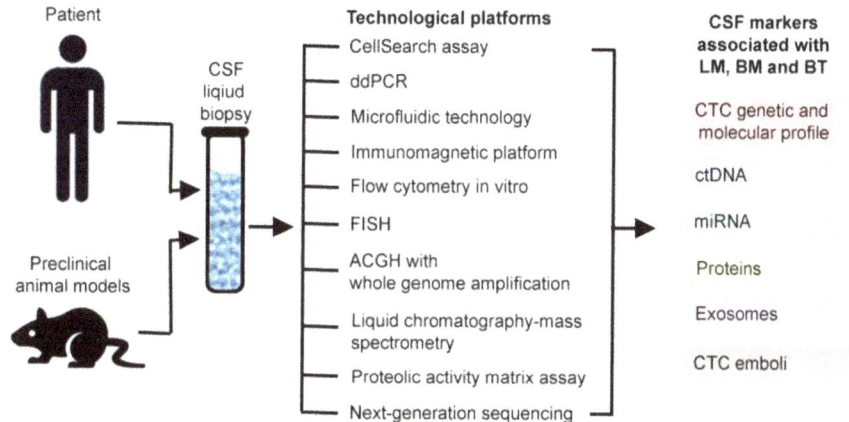

Figure 1. Cerebrospinal fluid (CSF) liquid biopsy detection of tumor markers in vitro.

CSF Liquid Biopsy

Several years ago, Patel et al. showed that FDA-approved CellSearch method can be used to identify CTCs in 7.5 mL CSF samples of breast cancer patients [22]. Compared to traditional cytology, the CellSearch assay has been demonstrated significantly higher number of CTCs [22,28,30,49,50]. Despite promise, this technological platform is limited in detection of only a few tumor markers, typically EpCam, for patients with epithelial cancers (e.g., breast cancer) and CD 146 and HMW-MMA for patients with melanoma [21,49]. Thus, CellSearch obviously cannot identify a highly heterogeneous population of CTCs and not suitable for diagnosis of many tumors such as glioblastoma. These limitations somewhat reduced enthusiasm to recommend this method in routine clinical practice.

Using real-time polymerase chain reaction (RT-PCR) for examination of patients with BM and LM has demonstrated higher sensitivity than conventional cytology [51]. However, relatively high rate of false-negatives during RT-PCR analysis make it a suboptimal method for CSF testing [9]. To solve this problem, cancer researchers and clinical oncologists recently explored the use of high-sensitive droplet-digital PCR (ddPCR) [52–56]. It was shown that ddPCR provides accurate and reliable CSF analysis. It can work with poor DNA quality and measure multiple parameters including absolute allele quantification, rare mutation, copy number variations, DNA methylation and gene rearrangements [52]. In a few clinical studies, ddPCR of CSF was able to detect ctDNA in patients with melanoma and CNS metastasis; and the obtained results were strongly correlated with cytology results and detection of abnormalities in MRI [52,56]. It is interesting that some patients with high level of ctDNA showed negative cytological results [56]. The small volume of CSF fluid required for testing ctDNA is definitely an additional advantage but high level of false results is a challenge. Overall, to date, it is too early to make conclusions on diagnostic value of ctDNA.

Another promising emerging data of CSF liquid biopsy have been obtained using immunofluorescence in situ hybridization (FISH) technology [24,57–60]. The published results hold promise to provide more accurate diagnosis of CSF CTCs than cytology. The main advantage of FISH is phenotypic and karyotypic identification and characterization of the highly heterogeneous CTCs, which can be assessed by both chromosome ploidy and the expression of various tumor markers [57]. However, FISH is not currently standardized for liquid biopsy and requires future development and research to clarify whether or not this method is reliable for identification of CTCs.

Integration of array comparative genomic hybridization (ACGH) analysis and whole genome amplification provided achieving the genomic characterization of rare CSF CTCs [61,62]. The clonal similarity between CSF CTCs and primary tumor genomic profiles with more copy number alterations in CTCs was demonstrated using samples of CSF and primary tumor from breast cancer patients with LM [62].

Analysis of CSF samples with conventional flow cytometry in vitro has been reported to diagnose CTCs in CSF [63–65]. Flow cytometry immunophenotypic testing of bulk breast cancer receptors, cancer stem cell markers and various WBC subpopulations looks interesting and suggests interplay of CSF and lymph fluid during CTC migration [63]. However, well-known limitation of flow cytometry to detect rare events might reduce enthusiasm for its use of assessment of CTCs which is supposed to be rare (up to 1-5 CTCs per sample) at early stage of CNS involvement.

In the past few years, the clinical potential of some other technological platforms including microfluidic technology, immunomagnetic platform, high performance liquid chromatography-mass spectrometry, next generation sequencing (NGS) and proteolytic activity matrix assay (PrAMA) has also been demonstrated [25,50,55,66–69]. Despite interest and promises, the singularity of these reports does not allow yet making conclusions on suitability of these methods to improve prognosis in patients.Overall, despite CSF liquid biopsy is expected to yield clinically significant biomarkers and assays, the main drawback to all aforementioned approaches in vitro is that their sensitivity is substantially limited by the volume of the sample [70,71]. Typically, up to 10 mL of CSF is used for examination, which is estimated to be less than 6–7% of the total 130–150 mL volume of human CSF. It means that in vitro testing misses up to 93–94% of CTCs [71]. A simple recalculation of the results in vitro, which detected minimum 1–2 CTCs per CSF patient sample (5–10 mL) with the existing LM and BM [21,49], shows that the real number of CTCs at the time of diagnosis was more than 15–20 cells in the total CSF volume. Serial analysis of multiple samples from repeated punctures increases sensitivity [28]. However, repeated punctures are a challenge because it can be performed over several days and may lead to delaying of therapies. In addition, the existing methods in vitro are burdened with: (1) low throughput, which may require many hours (if not, days) to assess a typical CSF sample and (2) multiple time-consuming sample-processing steps including staining, immunomagnetic capture, isolation and washing, which result in loss of many CTCs [21,23,30,49,51]. As a result, CTCs in small quantities may escape detection, which also contributes to late diagnosis and poor outcomes.

Based on this, liquid biopsy in vitro can provide advanced molecular and genetic analysis of tumor associated markers in CSF but it cannot detect rare CTCs at early stage of LM and BM and possibly, before LM and BM initiation (Table 1). The rarity of CSF CTCs definitely demands a new strategy. An attractive solution to these problems is to monitor almost entire CSF volume in vivo (Table 1).

Table 1. New and emerging technologies for detection of tumor biomarkers in CSF.

Detection Method	Biomarker	Disease	Approach In Vitro	Approach In Vivo	Significant Advantages	Main Limitations	Application Research	Application Clinical	Refs
CellSearch	CTCs, Cell emboli	BM, LM	+		FDA-approved technology, Higher sensitivity than cytology	Small sample volume, Processing delay, Limited number of detected markers	+	+	[22,28,30,49,50]
Microfluidic technologies	CTCs	BM, BT	+		Single CTC capture in sub-nanoliter trap, Relatively quick (~1 h) analysis	Early stage of research using cell lines	+		[69]
Immuno-magnetic platform	CTCs	LM	+		Capable to detect and separate rare cells	Low sensitivity due to small sample volume, Limited number of detected markers		+	[25]
FC in vitro	CTCs	LM	+		Standardized technology, Higher sensitivity than cytology	Impossibility to detect rare cells	+	+	[63–65]
ddPCR	ctDNA, miRNA, CTCs	BT, BM, LM	+		High specificity	False-positivity, Not standardized	+	+	[52–56]
FISH	CTCs	LM, BT	+		Analysis poor DNA, Relatively high resolution	Early stage of research, Not standardized	+	+	[24,57–60]
ACGH	CTCs		+		Whole genome sequencing, High resolution compared to conventional CGH	Inability to detect aberrations that do not result in copy number changes		+	[61,62]
NGS	ctDNA, miRNA	LM	+		High-throughput whole genome sequencing, High specificity	High price, Complex data analysis		+	[50,56]
PrAMA	Proteases	LM	+		Detection of protease activity as indicator of BBB degradation	Early stage of research		+	[66]
PAFC in vivo	CTCs, Cell emboli	BM		+	Extremely high sensitivity, Theranostic capability	Detection of surface CTC receptors	+		[71]

CTCs—circulating tumor cells, LM—leptomeningeal metastasis; BM—brain metastasis; BT—brain tumor; FC—flow cytometry; ddRCP—droplet-digital polymerase chain reaction; ctDNA—cell free DNA; miRNA—microRNA; FISH—immunofluorescence in situ hybridization; ACGH—array comparative genomic hybridization; CGH—comparative genomic hybridization; NGS—next generation sequencing; PrAMA—proteolytic activity matrix assay; PAFC—photoacoustic flow cytometry.

3. In Vivo Diagnosis of CSF

Despite significant progress in neuroimaging in vivo (e.g., MRI, computed tomography [CT], radiography) [9,11,50,64,72–74], existing diagnosis, even advanced multi-modal imaging is not sufficient to make judgments about early LM and BM. The low spatial and temporal resolution of CT and MRI allows identification of only macroscopic changes in the CNS (e.g., metastases ≥10 mm by CT). Therefore, the diagnosis is typically based on the indirect signs of LM including pathological meningeal contrast enhancement at the MRI examination, which are often equivocal. In addition, a recent study has found that immunotherapy might be a source of MRI false positivity ('pseudomeningeosis') [73]. New generations of MRI, such as phase-contrast MRI, enable quantitative measurements of CSF flow but not suitable for detection of relatively fast moving single CTCs and particles due to slow time response [75]. The same limitation applies to intravital fluorescence microscopy which has been used for imaging CSF plasma (so-called, cisternography) but not single cells in CSF [76]. Furthermore, the translation of fluorescent neuroimaging to humans in vivo is problematic due to (1) cytotoxicity of fluorophores, (2) undesirable immune responses to tags and (3) assessing only superficial fluid flows due to strong influence of autofluorescent and scattering background.

Photoacoustic Flow Cytometry In Vivo

The most promising method for detecting CTCs in CSF is photoacoustic (PA) flow cytometry (PAFC), which compared to other in vivo diagnostic techniques, demonstrated ultra-sensitive molecular detection and counting of single cells in different body fluids (e.g., blood, lymph and CSF) [40,70,77–83]. The principle of multicolor PAFC is based on noninvasive (i.e., through intact skin) irradiation of the selected fluid with short laser pulses at different wavelengths followed by the detection of laser-induced acoustic waves (referred to as PA signals) using an ultrasound transducer placed on the skin (Figure 2a). PA methods provide higher sensitivity and resolution in deeper tissues (up to 2–3 cm, with potential up to 5–7 cm [70,84]) than other optical modalities. These benefits make possible detection of CTCs in CSF through the atlanto-occipital membrane. In PAFC, this allows distinguishing signals from single fast-moving particles (e.g., CTCs, exosomes, and emboli) at laser energies within the safety standards for humans [70,77,81,83,85,86]. In regards of CSF detection, PAFC has advantages compared to other in vivo methods. Specifically, the colorlessness and optical transparency of CSF, commonly accepted as a diagnostic limitation, provides low absorbance and, therefore, extremely low PA background signal, which significantly improves distinguishing stronly light absorbing objects [71]. It means that CTCs, exosomes or emboli with strong absorbing molecules (e.g., natural melanin or nanoparticles) are predominated over the absorption of CSF by a few orders of magnitude, especially in the near-infrared window of transparency for biotissues ("first window": 700–1100 nm). Based on this, some strong absorbing cells such as melanoma CTCs with natural intracellular high absorbing melanin as intrinsic non-toxic PA contrast agents, can be easily detected by PAFC in label-free mode. To detect low absorbing tumor-related CSF markers (e.g., breast cancer CTCs), they should be labeled by exogenous PA contrast agents conjugated with ligands (e.g., antibodies, peptides, or folic acid) against specific surface receptor(s). The key requirements for in vivo use of contrast agents include low toxicity and high PA contrast. Some of the best candidates are gold and magnetic nanoparticles [77,87,88].

The first successful demonstration of PAFC's capability to diagnose CSF tumor markers was reported using preclinical models of breast metastatic cancer (Figure 2b–d) [71]. It was shown that PAFC was able to detect CSF CTCs with 10–20 times higher sensitivity compared to in vitro methods. The most important finding is that some tumor-bearing mice without histologically detectable BM exhibited rare CSF CTCs (e.g., 1–3 signals every 40–60 min). The presence of blood CTCs in these mice suggests the possible origin of CSF CTCs to be from blood CTCs and indicates the potential of CSF CTCs as a predictive biomarker of BM. The obtained experimental evidence is in line with the aforementioned suggestion that blood and lymphatic CTCs might pass the compromised BBB and enter brain tissue, meninges and CSF to form BM and LM. This may serve as a scientific foundation for prognosis and prediction of LM and BM in patients.

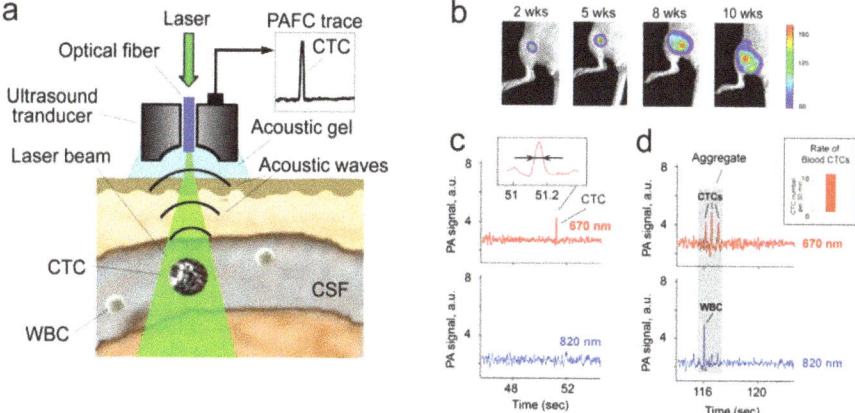

Figure 2. Assessment of circulating tumor markers in CSF in vivo with multicolor PAFC. (**a**) Principle of diagnosis with PAFC. (**b**) Intravital luminescence imaging of metastatic breast cancer progression in orthotopic xenograft mouse model after inoculation of human MDA-MB-231-luc2-GFP cells. (**c**) Two-color PAFC of the spontaneous CSF CTCs in vivo; inset: the photoacoustic signal width (indicated by arrows), which is associated with a single circulating tumor cell (CTC). (**d**) PAFC of circulating CTC-containing embolus in tumor-bearing mice; gray rectangle: aggregate of CSF-CTCs and leukocyte (WBC); insert: the blood CTC rate at the time of CSF monitoring.

Another interesting finding is the existence of CTC-containing emboli in CSF in vivo (Figure 2d). Identification of embolus is based on the width and shape of PA signal, assuming that embolus' multicellular structure produces a relatively wider PA signal containing a set of narrower peaks.

Overall, the success of preclinical studies together with the simplicity and safety of PAFC give confidence to rapidly translate this method into clinical practice. PAFC diagnosis of CSF in human subarachnoid space and spinal canal at a depth of 1–3 cm seems possible and was supported by the reports on high sensitivity and resolution of PA methods in deeper tissues. Recently, the clinical relevance of PAFC was successfully demonstrated in clinical trials with melanoma patients by detecting blood CTCs in 1–2 mm hand vessels at depth of 1-3 mm with a detection limit of 1 CTC/1000 mL (i.e.,10^3 –fold increased sensitivity compared to existing CTC assays) [40].

4. Future Directions

To date, crucial steps in increasing the survival of patients with LM and BM are (1) early diagnosis; (2) initiating preventive therapy such as targeted therapy of single CSF CTCs and their emboli and (3) assessing therapeutic efficacy in order to optimize an individual course of therapy.

4.1. Advance Diagnosis

Although many promising technologies to detect various CSF tumor markers during liquid biopsy have been reported, there is no standardized and validated assay that is currently ready to introduce for daily clinical practice as an advanced alternative or supplement of conventional cytology.

Novel approaches integrating unprecedented high sensitivity of in vivo flow cytometry and comprehensive molecular and genetic characterization of tumor markers in CSF in vitro are highly desired for clinical needs.

In addition, one of the possible future alternatives is CSF diagnosis in vivo using updated GILUPI CellCollector. This method was introduced in 2016 for EpCam-based detection of CTCs in blood by introduction of EpCAM-coated wire into a vein of the patient [89]. However, the invasive nature of the method and possibility of missing CTCs, which transit outside the wire, somewhat reduce enthusiasm of using GILUPI device for CSF assessment.

The new looks are also suggesting continuous cell exchange between CSF, blood, lymph and brain tissue [90–92] that should be considered at the diagnosis. The prognostic value of CTCs, if they are simultaneously tested in blood, lymph and CSF, would provide a new, highly sensitive and accurate prognostic biomarker of metastasis progression and therapy efficacy.

4.2. Therapeutic Perspectives

Minimal treatment options in current management of LM and BM lead to poor prognosis for patients due to low efficacy, late therapy initiation, use of common (i.e., not-personalized) therapeutic schematics and high toxicity. From this, one of the top future priorities is development of novel targeted and immune therapies. The molecular-targeted nanotechnology platform is highly promising for targeted drug delivery. For this purpose, nanoparticles should have high sensitivity, specificity and selectivity as well as safety, multifunctionality, multimodality, ability to penetrate BBB and high efficiency of drug delivery to tumor. Among existing nanoparticle-based drug cargoes, the most promising candidates include low toxic individual nanoparticles, high-contrast spasers, liposomes, polymer micelles, lipid micelles packaged with semiconducting polymer dots as simultaneous MRI and PA imaging and photodynamic and photothermal dual-modal therapeutic agents, layer by layer based composite structures (core-shells) and microcapsules (shells) and biocompatible natural magnetic nanoparticles [87,88,93–98]. The targeting could be achieved by surface modification using targeted molecules specific to CTCs, exosomes and emboli [88,99]. For example, a single injection of core-shells in CSF has shown the effectiveness of their use for the long-term delivery of painkillers in the treatment of persistent pain [100]. Potentially, these drug delivery systems may be effective for treating CTCs in CSF.

There is a high therapeutic potential of modern technologies for creating synthetic truncated antibodies [101] and scaffolds [102]. The revolutionary progress in genetic and protein engineering methods make it possible to directionally modify the molecular size, affinity, specificity and immunogenicity of an antibody, their derivatives and analogues, oriented to the use in the diagnosis and targeted therapy of cancer. Today, rational design and molecular engineering allow modelling of the compounds with preprogrammed properties and to create biotechnological producers of therapeutic medicines [102–106]. A promising direction is conjugation of these unique theranostic agents with nanoparticles. The advantages of using nanoparticles in these conjugates include developed surface of nanoparticles, which can be decorated with biocompatible functional moieties for targeted delivery; and diagnosis that guides and monitors effects of the nanoparticle-assisted therapy [107–110]. Recently, the design of a hybrid nanocomplex based on an upconversion nanoparticle (UCNP) was reported [111]. Owing to their unique photophysical properties, UCNPs are high-potential platform for theranostics complexes. Conversion of near-infrared light, which can deeply penetrate in biological tissue, to the higher photon energy visible, ultra-violet and near-infrared light is among UCNP's most useful properties. Two toxic agents—beta-emitting radionuclide yttrium-90 and a highly efficient targeted toxin DARPin-exotoxin A from Pseudomonas aeruginosa—were coupled to UCNP core to exert toxicity to cancer cells. As a result, on the one hand, the photophysical properties of hybrid nanocomplex enable background-free imaging of its distribution in cells and animals. On the other hand, specific delivery of UCNP complexes to cancer cells results in combined therapy by two toxic agents with markedly increased synergetic effect [111]. The design of the hybrid multifunctional nanoheterocomplex proves the principle "when the whole is greater than the sum of the parts."

The novel targeted CSF therapy may also use the advanced design of heterostructures based on the barnase:barstar pair [112]. The ribonuclease barnase and its inhibitor, barstar, are small stable proteins. They form extremely tight complex with a $K_d \sim 10^{-14}$ M. The strategy is applicable to any proteins or nanoparticles that can be functionally attached to the barstar and barnase, especially for production of heterooligomeric constructs because the extremely specific barnase barstar interaction eliminates reliably the mispairing problems. This universal platform is a promising alternative to commonly used chemical conjugation techniques in nanobiotechnology, theranostics and clinical applications. It provides a straightforward technology to design wide range of multifunctional nanoheterostructures for the highly efficient delivery of active agents to tumor cells for theranostics [112–117].

A very exciting future direction is the possibility of integration of in vivo molecular diagnosis, targeted therapy and estimation of therapeutic efficacy in one technological platform of PAFC [40,79,118]. PAFC's capability to identify a single high-absorbing CTC and immediately "kill" it through photothermal-indiced nanobubbles with photomechanical action on CTC membranes and vital intracellular structures was demonstrated for blood CTCs in experiments and, recently, in clinical research in blood circulation [40]. Furthermore, the following disappearance of the CTC-associated PA signals might serve as the criterion of effective therapy. These data bring hope that earliest rare CTCs might be identified and "killed" directly in CSF before colonization of brain tissue and formation of BM and LM.

It is expected that technological innovations and ongoing clinical trials would contribute to the finding of novel approaches to provide advances in BM and LM theranostics at the earliest possible stages before development of overt deadly lesions, to select patients with high risk of BM and LM for personalized therapy, to identify early disease progression and thereby improve survival rates of cancer patients.

Funding: This work was supported in part by grants CA131164, EB009230, EB022698 and R21CA230059 from the National Institute of Health (NIH), grants from the Arkansas Biosciences Institute and the Translational Research Institute at UAMS, the Arkansas EPSCoR-NSF Program and RF 14.Z50.31.0044.

Acknowledgments: We thank V.T., M.J., D.N. and many other colleagues who participated in this work and co-authored the cited joint papers.

Conflicts of Interest: The authors declare no conflict of interest. The funders had no role in the design of the study; in the collection, analyses or interpretation of data; in the writing of the manuscript or in the decision to publish the results.

References

1. Siegel, R.L.; Miller, K.D.; Jemal, A. Cancer statistics, 2019. *CA Cancer J. Clin.* **2019**, *69*, 7–34. [CrossRef] [PubMed]
2. Taillibert, S.; Chamberlain, M.C. Leptomeningeal metastasis. *Handb. Clin. Neurol.* **2018**, *149*, 169–204. [PubMed]
3. Steeg, P.S.; Camphausen, K.A.; Smith, Q.R. Brain metastases as preventive and therapeutic targets. *Nat. Rev. Cancer* **2011**, *11*, 352–363. [CrossRef] [PubMed]
4. Langley, R.R.; Fidler, I.J. The biology of brain metastasis. *Clin. Chem.* **2013**, *59*, 180–189. [CrossRef] [PubMed]
5. Figura, N.B.; Rizk, V.T.; Armaghani, A.J.; Arrington, J.A.; Etame, A.B.; Han, H.S.; Czerniecki, B.J.; Forsyth, P.A.; Ahmed, K.A. Breast leptomeningeal disease: A review of current practices and updates on management. *Breast Cancer Res. Treat.* **2019**, *177*, 277–294. [CrossRef] [PubMed]
6. Taylor, G.; Karlin, N.; Halfdanarson, T.R.; Coppola, K.; Grothey, A. Leptomeningeal Carcinomatosis in Colorectal Cancer: The Mayo Clinic Experience. *Clin. Colorectal. Cancer* **2018**, *17*, e183–e187. [CrossRef] [PubMed]
7. Blakeley, J.O.; Coons, S.J.; Corboy, J.R.; Kline Leidy, N.; Mendoza, T.R.; Wefel, J.S. Clinical outcome assessment in malignant glioma trials: Measuring signs, symptoms, and functional limitations. *Neuro Oncol.* **2016**, *18* (Suppl. 2), ii13–ii20. [CrossRef]
8. Sloan, A.E.; Nock, C.J.; Einstein, D.B. Diagnosis and treatment of melanoma brain metastasis: A literature review. *Cancer Control* **2009**, *16*, 248–255. [CrossRef] [PubMed]
9. Chamberlain, M.C.; Glantz, M.; Groves, M.D.; Wilson, W.H. Diagnostic tools for neoplastic meningitis: Detecting disease, identifying patient risk, and determining benefit of treatment. *Semin. Oncol.* **2009**, *36*, S35–S45. [CrossRef]

10. Nayar, G.; Ejikeme, T.; Chongsathidkiet, P.; Elsamadicy, A.A.; Blackwell, K.L.; Clarke, J.M.; Lad, S.P.; Fecci, P.E. Leptomeningeal disease: Current diagnostic and therapeutic strategies. *Oncotarget* **2017**, *8*, 73312–73328. [CrossRef] [PubMed]
11. Pellerino, A.; Bertero, L.; Rudà, R.; Soffietti, R. Neoplastic meningitis in solid tumors: From diagnosis to personalized treatments. *Ther. Adv. Neurol. Disord.* **2018**, *11*, 1756286418759618. [CrossRef] [PubMed]
12. Boon, J.M.; Abrahams, P.H.; Meiring, J.H.; Welch, T. Lumbar puncture: Anatomical review of a clinical skill. *Clin. Anat.* **2004**, *17*, 544–553. [CrossRef] [PubMed]
13. Verheul, C.; Kleijn, A.; Lamfers, M.L.M. Cerebrospinal fluid biomarkers of malignancies located in the central nervous system. *Handb. Clin. Neurol.* **2017**, *146*, 139–169. [PubMed]
14. Hepnar, D.; Adam, P.; Žáková, H.; Krušina, M.; Kalvach, P.; Kasík, J.; Karpowicz, I.; Nasler, J.; Bechyně, K.; Fiala, T.; et al. Recommendations for cerebrospinal fluid analysis. *Folia Microbiol.* **2019**, *64*, 443–452. [CrossRef] [PubMed]
15. Nevel, K.S.; Wilcox, J.A.; Robell, L.J.; Umemura, Y. The Utility of Liquid Biopsy in Central Nervous System Malignancies. *Curr. Oncol. Rep.* **2018**, *20*, 60. [CrossRef]
16. Fontanilles, M.; Duran-Peña, A.; Idbaih, A. Liquid Biopsy in Primary Brain Tumors: Looking for Stardust! *Curr. Neurol. Neurosci. Rep.* **2018**, *18*, 13. [CrossRef]
17. Shankar, G.M.; Balaj, L.; Stott, S.L.; Nahed, B.; Carter, B.S. Liquid biopsy for brain tumors. *Expert Rev. Mol. Diagn.* **2017**, *17*, 943–947. [CrossRef]
18. Johanson, C.E.; Duncan, J.A.; Klinge, P.M.; Brinker, T.; Stopa, E.G.; Silverberg, G.D. Multiplicity of cerebrospinal fluid functions: New challenges in health and disease. *Cerebrospinal Fluid Res.* **2008**, *5*, 10. [CrossRef]
19. Sweetman, B.; Linninger, A.A. Cerebrospinal fluid flow dynamics in the central nervous system. *Ann. Biomed. Eng.* **2011**, *39*, 484–496. [CrossRef]
20. Huff, T.; Tadi, P.; Varacallo, M. Neuroanatomy, Cerebrospinal Fluid. In *StatPearls [Internet]*; StatPearls Publishing: Treasure Island, FL, USA, 2019.
21. Brinker, T.; Stopa, E.; Morrison, J.; Klinge, P. A new look at cerebrospinal fluid circulation. *Fluids Barriers CNS* **2014**, *11*, 10. [CrossRef]
22. Patel, A.S.; Allen, J.E.; Dicker, D.T.; Peters, K.L.; Sheehan, J.M.; Glantz, M.J.; El-Deiry, W.S. Identification and enumeration of circulating tumor cells in the cerebrospinal fluid of breast cancer patients with central nervous system metastases. *Oncotarget* **2011**, *2*, 752–760. [CrossRef] [PubMed]
23. Burns, T.F.; Wolff, A.C. Detection of circulating tumor cells in the cerebrospinal fluid: A new frontier. *Cell Cycle* **2012**, *11*, 203–204. [CrossRef]
24. Li, X.; Zhang, Y.; Ding, J.; Wang, M.; Li, N.; Yang, H.; Wang, K.; Wang, D.; Lin, P.P.; Li, M.; et al. Clinical significance of detecting CSF-derived tumor cells in breast cancer patients with leptomeningeal metastasis. *Oncotarget* **2017**, *9*, 2705–2714. [CrossRef] [PubMed]
25. Lin, X.; Fleisher, M.; Rosenblum, M.; Lin, O.; Boire, A.; Briggs, S.; Bensman, Y.; Hurtado, B.; Shagabayeva, L.; DeAngelis, L.M.; et al. Cerebrospinal fluid circulating tumor cells: A novel tool to diagnose leptomeningeal metastases from epithelial tumors. *Neuro Oncol.* **2017**, *19*, 1248–1254. [CrossRef]
26. Zorofchian, S.; Iqbal, F.; Rao, M.; Aung, P.P.; Esquenazi, Y.; Ballester, L.Y. Circulating tumour DNA, microRNA and metabolites in cerebrospinal fluid as biomarkers for central nervous system malignancies. *J. Clin. Pathol.* **2019**, *72*, 271–280. [CrossRef] [PubMed]
27. Touat, M.; Duran-Peña, A.; Alentorn, A.; Lacroix, L.; Massard, C.; Idbaih, A. Emerging circulating biomarkers in glioblastoma: Promises and challenges. *Expert Rev. Mol. Diagn.* **2015**, *15*, 1311–1323. [CrossRef]
28. Lee, J.S.; Melisko, M.E.; Magbanua, M.J.; Kablanian, A.T.; Scott, J.H.; Rugo, H.S.; Park, J.W. Detection of cerebrospinal fluid tumor cells and its clinical relevance in leptomeningeal metastasis of breast cancer. *Breast Cancer Res. Treat.* **2015**, *154*, 339–349. [CrossRef] [PubMed]
29. Ning, M.; Chunhua, M.; Rong, J.; Yuan, L.; Jinduo, L.; Bin, W.; Liwei, S. Diagnostic value of circulating tumor cells in cerebrospinal fluid. *Open Med.* **2016**, *11*, 21–24. [CrossRef] [PubMed]
30. Tu, Q.; Wu, X.; Le Rhun, E.; Blonski, M.; Wittwer, B.; Taillandier, L.; De Carvalho Bittencourt, M.; Faure, G.C. CellSearch technology applied to the detection and quantification of tumor cells in CSF of patients with lung cancer leptomeningeal metastasis. *Lung Cancer* **2015**, *90*, 352–357. [CrossRef]

31. Krol, I.; Castro-Giner, F.; Maurer, M.; Gkountela, S.; Szczerba, B.M.; Scherrer, R.; Coleman, N.; Carreira, S.; Bachmann, F.; Anderson, S.; et al. Detection of circulating tumour cell clusters in human glioblastoma. *Br. J. Cancer* **2018**, *119*, 487–491. [CrossRef]
32. Young, A.; Chapman, O.; Connor, C.; Poole, C.; Rose, P.; Kakar, A.K. Thrombosis and cancer. *Nat. Rev. Clin. Oncol.* **2012**, *9*, 437–449. [CrossRef] [PubMed]
33. Au, S.H.; Storey, B.D.; Moore, J.C.; Tang, Q.; Chen, Y.L.; Javaid, S.; Sarioglu, A.F.; Sullivan, R.; Madden, M.W.; O'Keefe, R.; et al. Clusters of circulating tumor cells traverse capillary-sized vessels. *Proc. Natl. Acad. Sci. USA* **2016**, *113*, 4947–4952. [CrossRef] [PubMed]
34. Mu, Z.; Wang, C.; Ye, Z.; Austin, L.; Civan, J.; Hyslop, T.; Palazzo, J.P.; Jaslow, R.; Li, B.; Myers, R.E.; et al. Prospective assessment of the prognostic value of circulating tumor cells and their clusters in patients with advanced-stage breast cancer. *Breast Cancer Res. Treat.* **2015**, *154*, 563–571. [CrossRef] [PubMed]
35. Wang, C.; Mu, Z.; Chervoneva, I.; Austin, L.; Ye, Z.; Rossi, G.; Palazzo, J.P.; Sun, C.; Abu-Khalaf, M.; Myers, R.E.; et al. Longitudinally collected CTCs and CTC-clusters and clinical outcomes of metastatic breast cancer. *Breast Cancer Res. Treat.* **2017**, *161*, 83–94. [CrossRef] [PubMed]
36. Au, S.H.; Edd, J.; Stoddard, A.E.; Wong, K.H.K.; Fachin, F.; Maheswaran, S.; Haber, D.A.; Stott, S.L.; Kapur, R.; Toner, M. Microfluidic Isolation of Circulating Tumor Cell Clusters by Size and Asymmetry. *Sci. Rep.* **2017**, *7*, 2433. [CrossRef]
37. Kulasinghe, A.; Schmidt, H.; Perry, C.; Whitfield, B.; Kenny, L.; Nelson, C.; Warkiani, M.E.; Punyadeera, C. A Collective Route to Head and Neck Cancer Metastasis. *Sci. Rep.* **2018**, *8*, 746. [CrossRef] [PubMed]
38. Gkountela, S.; Castro-Giner, F.; Szczerba, B.M.; Vetter, M.; Landin, J.; Scherrer, R.; Krol, I.; Scheidmann, M.C.; Beisel, C.; Stirnimann, C.U.; et al. Circulating Tumor Cell Clustering Shapes DNA Methylation to Enable Metastasis Seeding. *Cell* **2019**, *176*, 98–112. [CrossRef]
39. Kulasinghe, A.; Kapeleris, J.; Cooper, C.; Warkiani, M.E.; O'Byrne, K.; Punyadeera, C. Phenotypic Characterization of Circulating Lung Cancer Cells for Clinically Actionable Targets. *Cancers* **2019**, *11*, 380. [CrossRef]
40. Galanzha, E.I.; Menyaev, Y.A.; Yadem, A.C.; Sarimollaoglu, M.; Juratli, M.A.; Nedosekin, D.A.; Foster, S.R.; Jamshidi-Parsian, A.; Siegel, E.R.; Makhoul, I.; et al. In vivo liquid biopsy using Cytophone platform for photoacoustic detection of circulating tumor cells in patients with melanoma. *Sci. Transl. Med.* **2019**, *11*, eaat5857. [CrossRef]
41. Wrobel, J.K.; Toborek, M. Blood–brain Barrier Remodeling during Brain Metastasis Formation. *Mol. Med.* **2016**, *22*, 32–40. [CrossRef]
42. Deeken, J.F.; Loscher, W. The Blood-Brain Barrier a Overview of cerebrospinal fluid cytology and Cancer: Transporters, Treatment, and Trojan Horses. *Clin. Cancer Res.* **2007**, *13*, 1663–1674. [CrossRef] [PubMed]
43. Rahimi, J.; Woehrer, A. Overview of cerebrospinal fluid cytology. *Handb. Clin. Neurol.* **2017**, *145*, 563–571. [PubMed]
44. Pantel, K.; Alix-Panabières, C. The potential of circulating tumor cells as a liquid biopsy to guide therapy in prostate cancer. *Cancer Discov.* **2012**, *2*, 974–975. [CrossRef] [PubMed]
45. Pantel, K.; Alix-Panabières, C. Liquid biopsy: Potential and challenges. *Mol. Oncol.* **2016**, *10*, 371–373. [CrossRef] [PubMed]
46. Connolly, I.D.; Li, Y.; Gephart, M.H.; Nagpal, S. The "Liquid Biopsy": The Role of Circulating DNA and RNA in Central Nervous System Tumors. *Curr. Neurol. Neurosci. Rep.* **2016**, *16*, 25. [CrossRef]
47. Best, M.G.; Sol, N.; Zijl, S.; Reijneveld, J.C.; Wesseling, P.; Wurdinger, T. Liquid biopsies in patients with diffuse glioma. *Acta Neuropathol.* **2015**, *129*, 849–865. [CrossRef]
48. Pantel, K.; Alix-Panabières, C. Liquid biopsy and minimal residual disease—Latest advances and implications for cure. *Nat. Rev. Clin. Oncol.* **2019**, *16*, 409–424. [CrossRef]
49. Le Rhun, E.; Tu, Q.; De Carvalho Bittencourt, M.; Farre, I.; Mortier, L.; Cai, H.; Kohler, C.; Faure, G.C. Detection and quantification of CSF malignant cells by the CellSearch technology in patients with melanoma leptomeningeal metastasis. *Med. Oncol.* **2013**, *30*, 538. [CrossRef]
50. Jiang, B.Y.; Li, Y.S.; Guo, W.B.; Zhang, X.C.; Chen, Z.H.; Su, J.; Zhong, W.Z.; Yang, X.N.; Yang, J.J.; Shao, Y.; et al. Detection of Driver and Resistance Mutations in Leptomeningeal Metastases of NSCLC by Next-Generation Sequencing of Cerebrospinal Fluid Circulating Tumor Cells. *Clin. Cancer Res.* **2017**, *23*, 5480–5488. [CrossRef]

51. Hoon, D.S.; Kuo, C.T.; Wascher, R.A.; Fournier, P.; Wang, H.J.; O'Day, S.J. Molecular detection of metastatic melanoma cells in cerebrospinal fluid in melanoma patients. *J. Investig. Dermatol.* **2001**, *117*, 375–378. [CrossRef]
52. Olmedillas-López, S.; García-Arranz, M.; García-Olmo, D. Current and Emerging Applications of Droplet Digital PCR in Oncology. *Mol. Diagn. Ther.* **2017**, *21*, 493–510. [CrossRef] [PubMed]
53. Momtaz, P.; Pentsova, E.; Abdel-Wahab, O.; Diamond, E.; Hyman, D.; Merghoub, T.; You, D.; Gasmi, B.; Viale, A.; Chapman, P.B. Quantification of tumor-derived cell free DNA(cfDNA) by digital PCR (DigPCR) in cerebrospinal fluid of patients with BRAFV600 mutated malignancies. *Oncotarget* **2016**, *7*, 85430–85436. [CrossRef] [PubMed]
54. Hiemcke-Jiwa, L.S.; Minnema, M.C.; Radersma-van Loon, J.H.; Jiwa, N.M.; de Boer, M.; Leguit, R.J.; de Weger, R.A.; Huibers, M.M.H. The use of droplet digital PCR in liquid biopsies: A highly sensitive technique for MYD88 p.(L265P) detection in cerebrospinal fluid. *Hematol. Oncol.* **2018**, *36*, 429–435. [CrossRef] [PubMed]
55. Zhao, J.; Ye, X.; Xu, Y.; Chen, M.; Zhong, W.; Sun, Y.; Yang, Z.; Zhu, G.; Gu, Y.; Wang, M. EGFR mutation status of paired cerebrospinal fluid and plasma samples in EGFR mutant non-small cell lung cancer with leptomeningeal metastases. *Cancer Chemother. Pharmacol.* **2016**, *78*, 1305–1310. [CrossRef] [PubMed]
56. Ballester, L.Y.; Glitza Oliva, I.C.; Douse, D.Y.; Chen, M.M.; Lan, C.; Haydu, L.E.; Huse, J.T.; Roy-Chowdhuri, S.; Luthra, R.; Wistuba, I.I.; et al. Evaluating Circulating Tumor DNA From the Cerebrospinal Fluid of Patients with Melanoma and Leptomeningeal Disease. *J. Neuropathol. Exp. Neurol.* **2018**, *77*, 628–635. [CrossRef] [PubMed]
57. Ge, F.; Zhang, H.; Wang, D.D.; Li, L.; Lin, P.P. Enhanced detection and comprehensive in situ phenotypic characterization of circulating and disseminated heteroploid epithelial and glioma tumor cells. *Oncotarget* **2015**, *6*, 27049–27064. [CrossRef]
58. Jiao, X.D.; Ding, C.; Zang, Y.S.; Yu, G. Rapid symptomatic relief of HER2-positive gastric cancer leptomeningeal carcinomatosis with lapatinib, trastuzumab and capecitabine: A case report. *BMC Cancer* **2018**, *18*, 206. [CrossRef]
59. Lv, Y.; Mu, N.; Ma, C.; Jiang, R.; Wu, Q.; Li, J.; Wang, B.; Sun, L. Detection value of tumor cells in cerebrospinal fluid in the diagnosis of meningeal metastasis from lung cancer by immuno-FISH technology. *Oncol. Lett.* **2016**, *12*, 5080–5084. [CrossRef]
60. Ma, C.; Lv, Y.; Jiang, R.; Li, J.; Wang, B.; Sun, L. Novel method for the detection and quantification of malignant cells in the CSF of patients with leptomeningeal metastasis of lung cancer. *Oncol. Lett.* **2016**, *11*, 619–623. [CrossRef]
61. Magbanua, M.J.; Melisko, M.; Roy, R.; Sosa, E.V.; Hauranieh, L.; Kablanian, A.; Eisenbud, L.E.; Ryazantsev, A.; Au, A.; Scott, J.H.; et al. Molecular profiling of tumor cells in cerebrospinal fluid and matched primary tumors from metastatic breast cancer patients with leptomeningeal carcinomatosis. *Cancer Res.* **2013**, *73*, 7134–7143. [CrossRef]
62. Magbanua, M.J.; Roy, R.; Sosa, E.V.; Hauranieh, L.; Kablanian, A.; Eisenbud, L.E.; Ryazantsev, A.; Au, A.; Scott, J.H.; Melisko, M.; et al. Genome-wide copy number analysis of cerebrospinal fluid tumor cells and their corresponding archival primary tumors. *Genom. Data* **2014**, *2*, 60–62. [CrossRef] [PubMed]
63. Cordone, I.; Masi, S.; Summa, V.; Carosi, M.; Vidiri, A.; Fabi, A.; Pasquale, A.; Conti, L.; Rosito, I.; Carapella, C.M.; et al. Overexpression of syndecan-1, MUC-1, and putative stem cell markers in breast cancer leptomeningeal metastasis: A cerebrospinal fluid flow cytometry study. *Breast Cancer Res.* **2017**, *19*, 46. [CrossRef] [PubMed]
64. Milojkovic Kerklaan, B.; Pluim, D.; Bol, M.; Hofland, I.; Westerga, J.; van Tinteren, H.; Beijnen, J.H.; Boogerd, W.; Schellens, J.H.; Brandsma, D. EpCAM-based flow cytometry in cerebrospinal fluid greatly improves diagnostic accuracy of leptomeningeal metastases from epithelial tumors. *Neuro Oncol.* **2016**, *18*, 855–862. [CrossRef] [PubMed]
65. Gong, X.; Lin, D.; Wang, H.; Wang, Y.; Liu, B.; Wei, H.; Zhou, C.; Liu, K.; Wei, S.; Gong, B.; et al. Flow cytometric analysis of cerebrospinal fluid in adult patients with acute lymphoblastic leukemia during follow-up. *Eur. J. Haematol.* **2018**, *100*, 279–285. [CrossRef] [PubMed]
66. Conrad, C.; Dorzweiler, K.; Miller, M.A.; Lauffenburger, D.A.; Strik, H.; Bartsch, J.W. Profiling of metalloprotease activities in cerebrospinal fluids of patients with neoplastic meningitis. *Fluids Barriers CNS* **2017**, *14*, 22. [CrossRef] [PubMed]

67. Warkiani, M.E.; Khoo, B.; Wu, L.; Tay, A.K.; Bhagat, A.A.; Han, J.; Lim, C.T. Ultra-fast, label-free isolation of circulating tumor cells from blood using spiral microfluidics. *Nat. Protoc.* **2016**, *11*, 134–148. [CrossRef]
68. Zhou, J.; Kulasinghe, A.; Bogseth, A.; O'Byrne, K.; Punyadeera, C.; Papautsky, I. Isolation of circulating tumor cells in non-small-cell-lung-cancer patients using a multi-flow microfluidic channel. *Microsyst. Nanoeng.* **2019**, *5*, 8. [CrossRef]
69. Turetsky, A.; Lee, K.; Song, J.; Giedt, R.J.; Kim, E.; Kovach, A.E.; Hochberg, E.P.; Castro, C.M.; Lee, H.; Weissleder, R. On chip analysis of CNS lymphoma in cerebrospinal fluid. *Theranostics* **2015**, *5*, 796–804. [CrossRef]
70. Galanzha, E.I.; Shashkov, E.V.; Spring, P.; Suen, J.Y.; Zharov, V.P. In vivo noninvasive label-free detection and eradication of circulating metastatic melanoma cells using two-color photoacoustic flow cytometry with a diode laser. *Cancer Res.* **2009**, *69*, 7926–7934. [CrossRef]
71. Nedosekin, D.A.; Juratli, M.A.; Sarimollaoglu, M.; Moore, C.L.; Rusch, N.J.; Smeltzer, M.S.; Zharov, V.P.; Galanzha, E.I. Photoacoustic and photothermal detection of circulating tumor cells, bacteria and nanoparticles in cerebrospinal fluid in vivo and ex vivo. *J. Biophotonics* **2013**, *6*, 523–533. [CrossRef]
72. Barajas, R.F., Jr.; Cha, S. Imaging diagnosis of brain metastasis. *Prog. Neurol. Surg.* **2012**, *25*, 55–73. [PubMed]
73. Bier, G.; Klumpp, B.; Roder, C.; Garbe, C.; Preibsch, H.; Ernemann, U.; Hempel, J.M. Meningeal enhancement depicted by magnetic resonance imaging in tumor patients: Neoplastic meningitis or therapy-related enhancement? *Neuroradiology* **2019**, *61*, 775–782. [CrossRef] [PubMed]
74. Salehi Ravesh, M.; Huhndorf, M.; Moussavi, A. Non-contrast enhanced molecular characterization of C6 rat glioma tumor at 7 T. *Magn. Reason. Imaging* **2019**, *61*, 175–186. [CrossRef] [PubMed]
75. Sakhare, A.R.; Barisano, G.; Pa, J. Assessing test-retest reliability of phase contrast MRI for measuring cerebrospinal fluid and cerebral blood flow dynamics. *Magn. Reason. Med.* **2019**, *82*, 658–670. [CrossRef] [PubMed]
76. Shibata, Y.; Kruskal, J.B.; Palmer, M.R. Imaging of cerebrospinal fluid space and movement of hydrocephalus mice using near infrared fluorescence. *Neurol. Sci.* **2007**, *28*, 87–92. [CrossRef]
77. Galanzha, E.I.; Shashkov, E.V.; Kelly, T.; Kim, J.-W.; Yang, L.; Zharov, V.P. In vivo magnetic enrichment and multiplex photoacoustic detection of circulating tumour cells. *Nat. Nanotechnol.* **2009**, *4*, 855–860. [CrossRef]
78. Zharov, V.P. Ultrasharp nonlinear photothermal and photoacoustic resonances and holes beyond the spectral limit. *Nat. Photonics* **2011**, *5*, 110–116. [CrossRef]
79. Galanzha, E.I.; Zharov, V.P. Circulating tumor cell detection and capture by photoacoustic flow cytometry in vivo and ex vivo. *Cancers* **2013**, *5*, 1691–1738. [CrossRef]
80. Galanzha, E.I.; Viegas, M.G.; Malinsky, T.I.; Melerzanov, A.V.; Juratli, M.A.; Sarimollaoglu, M.; Nedosekin, D.A.; Zharov, V.P. In vivo acoustic and photoacoustic focusing of circulating cells. *Sci. Rep.* **2016**, *6*, 21531. [CrossRef]
81. Nolan, J.; Sarimollaoglu, M.; Nedosekin, D.A.; Jamshidi-Parsian, A.; Galanzha, E.I.; Kore, R.A.; Griffin, R.J.; Zharov, V.P. In Vivo Flow Cytometry of Circulating Tumor-Associated Exosomes. *Anal. Cell. Pathol.* **2016**, *2016*, 1628057. [CrossRef]
82. Tuchin, V.V.; Tarnok, A.; Zharov, V.P. In vivo flow cytometry: A horizon of opportunities. *Cytom. Part A* **2011**, *79A*, 737–745. [CrossRef] [PubMed]
83. Galanzha, E.I.; Zharov, V.P. Photoacoustic flow cytometry. *Methods* **2012**, *57*, 280–296. [CrossRef] [PubMed]
84. Wang, L.V.; Hu, S. Photoacoustic tomography: In vivo imaging from organelles to organs. *Science* **2012**, *335*, 1458–1462. [CrossRef] [PubMed]
85. Juratli, M.A.; Menyaev, Y.A.; Sarimollaoglu, M.; Siegel, E.R.; Nedosekin, D.A.; Suen, J.Y.; Melerzanov, A.V.; Juratli, T.A.; Galanzha, E.I.; Zharov, V.P. Real-Time Label-Free Embolus Detection Using In Vivo Photoacoustic Flow Cytometry. *PLoS ONE* **2016**, *11*, e0156269. [CrossRef] [PubMed]
86. Juratli, M.A.; Sarimollaoglu, M.; Nedosekin, D.A.; Melerzanov, A.V.; Zharov, V.P.; Galanzha, E.I. Dynamic Fluctuation of Circulating Tumor Cells during Cancer Progression. *Cancers* **2014**, *6*, 128–142. [CrossRef] [PubMed]
87. Kim, J.-W.; Galanzha, E.I.; Shashkov, E.V.; Moon, H.-M.; Zharov, V.P. Golden carbon nanotubes as multimodal photoacoustic and photothermal molecular agents. *Nat. Nanotechnol.* **2009**, *4*, 688–694. [CrossRef]
88. Kim, J.W.; Galanzha, E.I.; Zaharoff, D.A.; Griffin, R.J.; Zharov, V.P. Nanotheranostics of circulating tumor cells, infections and other pathological features in vivo. *Mol. Pharm.* **2013**, *10*, 813–830. [CrossRef]

89. Gorges, T.M.; Penkalla, N.; Schalk, T.; Joosse, S.A.; Riethdorf, S.; Tucholski, J.; Lücke, K.; Wikman, H.; Jackson, S.; Brychta, N.; et al. Enumeration and Molecular Characterization of Tumor Cells in Lung Cancer Patients Using a Novel In Vivo Device for Capturing Circulating Tumor Cells. *Clin. Cancer Res.* **2016**, *22*, 2197–2206. [CrossRef]
90. Garzia, L.; Kijima, N.; Morrissy, A.S.; De Antonellis, P.; Guerreiro-Stucklin, A.; Holgado, B.L.; Wu, X.; Wang, X.; Parsons, M.; Zayne, K.; et al. A Hematogenous Route for Medulloblastoma Leptomeningeal Metastases. *Cell* **2018**, *172*, 1050–1062. [CrossRef]
91. Tuchin, V.V.; Zharov, V.P.; Galanzha, E.I. Biophotonics for lymphatic theranostics in animals and humans. *J. Biophotonics* **2018**, *11*, e201811001. [CrossRef]
92. Louveau, A.; Smirnov, I.; Keyes, T.J.; Eccles, J.D.; Rouhani, S.J.; Peske, J.D.; Derecki, N.C.; Castle, D.; Mandell, J.W.; Lee, K.S.; et al. Structural and functional features of central nervous system lymphatic vessels. *Nature* **2015**, *523*, 337–341. [CrossRef] [PubMed]
93. Beziere, N.; Lozano, N.; Nunes, A.; Salichs, J.; Queiros, D.; Kostarelos, K.; Ntziachristos, V. Dynamic imaging of PEGylated indocyanine green (ICG) liposomes within the tumor microenvironment using multi-spectral optoacoustic tomography (MSOT). *Biomaterials* **2015**, *37*, 415–424. [CrossRef] [PubMed]
94. Galanzha, E.I.; Weingold, R.; Nedosekin, D.A.; Sarimollaoglu, M.; Nolan, J.; Harrington, W.; Kuchyanov, A.S.; Parkhomenko, R.G.; Watanabe, F.; Nima, Z.; et al. Spaser as a biological probe. *Nat. Commun.* **2017**, *8*, 15528. [CrossRef] [PubMed]
95. Zhang, D.; Wu, M.; Zeng, Y.; Liao, N.; Cai, Z.; Liu, G.; Liu, J. Lipid micelles packaged with semiconducting polymer dots as simultaneous MRI/photoacoustic imaging and photodynamic/photothermal dual-modal therapeutic agents for liver cancer. *J. Mater. Chem. B* **2016**, *4*, 589–599. [CrossRef]
96. Yashchenok, M.; Jose, J.; Trochet, P.; Sukhorukov, G.B.; Gorin, D.A. Multifunctional polyelectrolyte microcapsules as a contrast agent for photoacoustic imaging in blood. *J. Biophotonics* **2016**, *9*, 792–799. [CrossRef] [PubMed]
97. Novoselova, M.V.; Bratashov, D.N.; Sarimollaoglu, M.; Nedosekin, D.A.; Harrington, W.; Watts, A.; Han, M.; Khlebtsov, B.N.; Galanzha, E.I.; Gorin, D.A.; et al. Photoacoustic and fluorescent effects in multilayer plasmon-dye interfaces. *J. Biophotonics* **2019**, *12*, e201800265. [CrossRef]
98. Nima, Z.A.; Watanabe, F.; Jamshidi-Parsian, A.; Sarimollaoglu, M.; Nedosekin, D.A.; Han, M.; Watts, J.A.; Biris, A.S.; Zharov, V.P.; Galanzha, E.I. Bioinspired magnetic nanoparticles as multimodal photoacoustic, photothermal and photomechanical contrast agents. *Sci. Rep.* **2019**, *9*, 887. [CrossRef] [PubMed]
99. Zhang, H.K.; Chen, Y.; Kang, J.; Lisok, A.; Minn, I.; Pomper, M.G.; Boctor, E.M. Prostate-specific membrane antigen-targeted photoacoustic imaging of prostate cancer in vivo. *J. Biophotonics* **2018**, *11*, e201800021. [CrossRef]
100. Kopach, O.; Zheng, K.; Dong, L.; Sapelkin, A.; Voitenko, N.; Sukhorukov, G.B.; Rusakov, D.A. Nano-engineered microcapsules boost the treatment of persistent pain. *Drug Deliv.* **2018**, *25*, 435–447. [CrossRef]
101. Mironova, K.E.; Proshkina, G.M.; Ryabova, A.V.; Stremovskiy, O.A.; Lukyanov, S.A.; Petrov, R.V.; Deyev, S.M. Genetically encoded immunophotosensitizer 4D5scFV-miniSOG is a highly selective agent for targeted photokilling of tumor cells in vitro. *Theranostics* **2013**, *3*, 831–840. [CrossRef]
102. Deyev, S.M.; Lebedenko, E.N.; Petrovskaya, L.E.; Dolgikh, D.A.; Gabibov, A.G.; Kirpichnikov, M.P. Man-made antibodies and immunoconjugates with desired properties: Function optimization using structural engineering. *Rus. Chem. Rev.* **2015**, *84*, 1–26. [CrossRef]
103. Martsev, S.P.; Chumanevich, A.A.; Vlasov, A.P.; Dubnovitsky, A.P.; Tsybovsky, Y.I.; Deyev, S.M.; Cozzi, A.; Arosio, P.; Kravchuk, Z.I. Antiferritin single-chain Fv fragment is a functional protein with properties of a partially structured state: Comparison with the completely folded V(L) domain. *Biochemistry* **2000**, *39*, 8047–8057. [CrossRef] [PubMed]
104. Vorobyeva, A.; Bragina, O.; Altai, M.; Mitran, B.; Orlova, A.; Shulga, A.; Proshkina, G.; Chernov, V.; Tolmachev, V.; Deyev, S. Comparative Evaluation of Radioiodine and Technetium-Labeled DARPin 9_29 for Radionuclide Molecular Imaging of HER2 Expression in Malignant Tumors. *Contrast Media Mol. Imaging* **2018**, *2018*. [CrossRef] [PubMed]
105. Proshkina, G.M.; Shilova, O.N.; Ryabova, A.V.; Stremovskiy, O.A.; Deyev, S.M. A new anticancer toxin based on HER2/neu-specific DARPin and photoactive flavoprotein miniSOG. *Biochimie* **2015**, *118*, 116–122. [CrossRef] [PubMed]

106. Shipunova, V.O.; Nikitin, M.P.; Nikitin, P.I.; Deyev, S.M. MPQ-cytometry: A magnetism-based method for quantification of nanoparticle-cell interactions. *Nanoscale* **2016**, *8*, 12764–12772. [CrossRef] [PubMed]
107. Deyev, S.; Proshkina, G.; Ryabova, A.; Tavanti, F.; Menziani, M.C.; Eidelshtein, G.; Avishai, G.; Kotlyar, A. Synthesis, Characterization, and Selective Delivery of DARPin-Gold Nanoparticle Conjugates to Cancer Cells. *Bioconjug. Chem.* **2017**, *28*, 2569–2574. [CrossRef]
108. Balalaeva, I.V.; Zdobnova, T.A.; Krutova, I.V.; Brilkina, A.A.; Lebedenko, E.N.; Deyev, S.M. Passive and active targeting of quantum dots for whole-body fluorescence imaging of breast cancer xenografts. *J. Biophotonics* **2012**, *5*, 860–867. [CrossRef]
109. Zdobnova, T.A.; Stremovskiy, O.A.; Lebedenko, E.N.; Deyev, S.M. Self-Assembling Complexes of Quantum Dots and scFv Antibodies for Cancer Cell Targeting and Imaging. *PLoS ONE* **2012**, *7*, e48248. [CrossRef]
110. Deyev, S.; Proshkina, G.; Baryshnikova, O.; Ryabova, A.; Avishai, G.; Katrivas, L.; Giannini, C.; Levi-Kalisman, Y.; Kotlyar, A. Selective staining and eradication of cancer cells by protein-carrying DARPin-functionalized liposomes. *Eur. J. Pharm. Biopharm.* **2018**, *130*, 296–305. [CrossRef]
111. Guryev, E.L.; Volodina, N.O.; Shilyagina, N.Y.; Gudkov, S.V.; Balalaeva, N.Y.; Volovetskiy, A.B.; Lyubeshkin, A.V.; Sen', A.V.; Ermilov, S.A.; Vodeneev, V.A.; et al. Radioactive ($_{90}$Y) upconversion nanoparticles conjugated with recombinant targeted toxin for synergistic nanotheranostics of cancer. *Proc. Natl. Acad. Sci. USA* **2018**, *11*, 9690–9695.
112. Nikitin, M.P.; Zdobnova, T.A.; Lukash, S.V.; Stremovskiy, O.A.; Deyev, S.M. Protein-assisted self-assembly of multifunctional nanoparticles. *Proc. Natl. Acad. Sci. USA* **2010**, *107*, 5827–5832. [CrossRef] [PubMed]
113. Aghayeva, U.F.; Nikitin, M.P.; Lukash, S.V.; Deyev, S.M. Denaturation-Resistant Bifunctional Colloidal Superstructures Assembled via the Proteinaceous Barnase-Barstar Interface. *ACS Nano* **2013**, *7*, 950–961. [CrossRef] [PubMed]
114. Shipunova, V.O.; Zelepukin, I.V.; Stremovskiy, O.A.; Nikitin, M.P.; Care, A.; Sunna, A.; Zvyagin, A.V.; Deyev, S.M. Versatile Platform for Nanoparticle Surface Bioengineering Based on SiO$_2$-Binding Peptide and Proteinaceous Barnase*Barstar Interface. *ACS Appl. Mater. Interfaces* **2018**, *10*, 17437–17447. [CrossRef]
115. Shipunova, V.O.; Kotelnikova, P.A.; Aghayeva, U.F.; Stremovskiy, O.A.; Novikov, I.A.; Schulga, A.A.; Nikitin, M.P.; Deyev, S.M. Self-assembling nanoparticles biofunctionalized with magnetite-binding protein for the targeted delivery to HER2/neu overexpressing cancer cells. *J. Magn. Magn. Mater.* **2019**, *469*, 450–455. [CrossRef]
116. Sreenivasan, V.K.A.; Kelf, T.A.; Grebenik, E.A.; Stremovskiy, O.A.; Say, J.M.; Rabeau, J.R.; Zvyagin, A.V.; Deyev, S.M. A modular design of low-background bioassays based on a high-affinity molecular pair barstar:barnase. *Proteomics* **2013**, *13*, 1437–1443. [CrossRef] [PubMed]
117. Grebenik, E.A.; Nadort, A.; Generalova, A.N.; Nechaev, A.V.; Sreenivasan, V.K.; Khaydukov, E.V.; Semchishen, V.A.; Popov, A.P.; Sokolov, V.I.; Akhmanov, A.S.; et al. Feasibility study of the optical imaging of a breast cancer lesion labeled with upconversion nanoparticle biocomplexes. *J. Biomed. Opt.* **2013**, *18*, 76004. [CrossRef] [PubMed]
118. Kim, J.-W.; Shashkov, E.V.; Galanzha, E.I.; Kotagiri, N.; Zharov, V.P. Photothermal antimicrobial nanotherapy and nanodiagnostics with self-assembling carbon nanotube clusters. *Lasers Surgery Med.* **2007**, *39*, 622–634. [CrossRef]

© 2019 by the authors. Licensee MDPI, Basel, Switzerland. This article is an open access article distributed under the terms and conditions of the Creative Commons Attribution (CC BY) license (http://creativecommons.org/licenses/by/4.0/).

Review

CTC-Derived Models: A Window into the Seeding Capacity of Circulating Tumor Cells (CTCs)

Tala Tayoun [1,2,3], Vincent Faugeroux [1,2], Marianne Oulhen [1,2], Agathe Aberlenc [1,2], Patrycja Pawlikowska [2] and Françoise Farace [1,2,*]

1. "Circulating Tumor Cells" Translational Platform, CNRS UMS3655 – INSERM US23AMMICA, Gustave Roussy, Université Paris-Saclay, F-94805 Villejuif, France; tala.tayoun@gustaveroussy.fr (T.T.); vincent.faugeroux@laposte.net (V.F.); marianne.oulhen@gustaveroussy.fr (M.O.); agathe.aberlenc@gustaveroussy.fr (A.A.)
2. INSERM, U981 "Identification of Molecular Predictors and new Targets for Cancer Treatment", F-94805 Villejuif, France; patrycjamarta.pawlikowska@gustaveroussy.fr
3. Faculty of Medicine, Université Paris Sud, Université Paris-Saclay, F-94270 Le Kremlin-Bicetre, France
* Correspondence: francoise.farace@gustaveroussy.fr; Tel.: +33-1-42-11-51-98

Received: 11 July 2019; Accepted: 24 September 2019; Published: 25 September 2019

Abstract: Metastasis is the main cause of cancer-related death owing to the blood-borne dissemination of circulating tumor cells (CTCs) early in the process. A rare fraction of CTCs harboring a stem cell profile and tumor initiation capacities is thought to possess the clonogenic potential to seed new lesions. The highest plasticity has been generally attributed to CTCs with a partial epithelial-to-mesenchymal transition (EMT) phenotype, demonstrating a large heterogeneity among these cells. Therefore, detection and functional characterization of these subclones may offer insight into mechanisms underlying CTC tumorigenicity and inform on the complex biology behind metastatic spread. Although an in-depth mechanistic investigation is limited by the extremely low CTC count in circulation, significant progress has been made over the past few years to establish relevant systems from patient CTCs. CTC-derived xenograft (CDX) models and CTC-derived ex vivo cultures have emerged as tractable systems to explore tumor-initiating cells (TICs) and uncover new therapeutic targets. Here, we introduce basic knowledge of CTC biology, including CTC clusters and evidence for EMT/cancer stem cell (CSC) hybrid phenotypes. We report and evaluate the CTC-derived models generated to date in different types of cancer and shed a light on challenges and key findings associated with these novel assays.

Keywords: metastasis; tumor-initiating cells (TICs); circulating tumor cells (CTCs); CTC-derived xenografts; CTC-derived ex vivo models

1. Introduction

Metastatic spread and its resistance to treatment remain the leading cause of death in cancer patients. This process is fueled by malignant cells that dissociate from the primary tumor and travel through the bloodstream to colonize distant organs. These cells are referred to as "circulating tumor cells" (CTCs) and are able to enter vasculature during the early course of disease. Nonetheless, the majority of the tumor cell population dies during transit as a result of biological and physical constraints such as shear stress and immune surveillance, and only a minor subset of the surviving CTCs (0.01%) acquires the capacity of tumor-initiating cells (TICs) [1–4]. The outcome of tumor dissemination is dependent on a selection process that favors the survival of a small proportion of cancer cells holding the self-renewal ability of stem cells along with TIC properties, which enables them to seed tumors and reconstitute tumor heterogeneity [5–7]. These cells are termed "cancer stem cells"

(CSCs), and CTCs holding a CSC phenotype have been detected and associated with high invasiveness and tumorigenicity in many cancers including breast cancer (BC), colorectal cancer, and glioma [8–11].

An important aspect of CTC research is to study the mechanistic basis behind their TIC properties and explore new CTC-based biomarkers and targeting strategies. The generation of CTC-derived xenografts (CDXs) or CTC-derived cell lines at relevant time points during disease progression is therefore crucial to achieve a longitudinal and functional characterization of these cells, along with in vivo and in vitro pharmacological testing. Although this task remains challenging owing to CTCs scarcity in peripheral blood and technical hurdles related to their enrichment strategies, significant efforts have been made in the establishment of clinically relevant systems to study CTC biology in different cancer types. In this review, we briefly cover basic knowledge of TIC-related properties in CTCs and evaluate the existing CTC-derived models, including both in vivo CDXs and in vitro functional culture assays in different cancers. We also highlight the important findings which have helped unveil new insights into CTC biology and novel therapeutic strategies.

2. Brief Glimpse into TIC-Related Properties of CTCs

CTC profile evolves as the initial events of the metastatic cascade take place. Indeed, CTCs undergo reversible phenotypic alterations to achieve intravasation, survival in vasculature and extravasation, known as epithelial-to-mesenchymal transition (EMT). During EMT—a key phenomenon in embryonic development—cancer cells undergo cytoskeletal changes and typically lose their cell–cell adhesion proteins as well as their polarity to become motile cells and intravasate [12,13]. EMT signatures were detectable in CTCs of BC patients [14–17]. Increasing experimental evidence draws a potential link between EMT and acquisition of stemness [12,13,18,19]. In fact, several EMT-inducing transcription factors have been shown to confer malignancy in neoplastic cells, leading to the emergence of highly aggressive clones with combined EMT/CSC traits [20–23]. Nevertheless, this association is not universal. Indeed, it has been suggested that the loss of the EMT-inducing factor *Prxx1* is required for cancer cells to colonize organs in vivo, which revert to the epithelial state and acquire CSC traits, thus uncoupling EMT and stemness [19,24,25]. Moreover, the requirement of EMT for CTC dissemination has long been subject to debate. Several studies have shown that mesenchymal features in tumor cells may indeed be dispensable for their migratory activity but could contribute molecularly and phenotypically to chemoresistance [26–28]. It is currently hypothesized that CTC subclones displaying an intermediate phenotype between epithelial and mesenchymal have the highest plasticity to adapt to the microenvironment and generate a more aggressive CTC population resistant to conventional chemotherapy and capable of metastatic outgrowth. Our group showed the existence of a hybrid epithelial/mesenchymal (E/M) phenotype in CTCs from patients with non-small cell lung cancer (NSCLC) [29]. Heterogeneous expression of EMT markers within SCLC and NSCLC patient cohorts was described by Hou et al., while Hofman et al. reported the presence of proportions of NSCLC CTCs which expressed the mesenchymal marker vimentin and correlated with shorter disease-free survival [30,31]. Recent data in metastatic BC patients showed the enrichment of CTC subpopulations with a CSC$^+$/partial EMT$^+$ signature in patients post-treatment, which correlated with worse clinical outcome [32]. Indeed, the CTC population is described as a highly heterogeneous pool of tumor cells with low numbers of metastasis-initiating cells (MICs) that are sometimes prone to apoptosis [33]. The different factors influencing MIC properties of CTCs and their survival underlie the complexity and inefficiency of organ invasion and macro-metastases formation, relevant both clinically and in experimental mouse models [4,34,35]. Recent advances in single-cell technologies have unraveled CTC-specific genetic mutations and profiling of the CTC population thus points out the emergence of subclones with dynamic phenotypes that contribute to the evolution of the tumor genome during disease progression and treatment [36–39]. CTCs are less frequently found in clusters, also termed "circulating tumor microemboli" (CTM), which travel as 2–50 cells in vasculature and present extremely enhanced metastatic competency [40]. This can be explained by the survival advantage they hold over single CTCs, as CTM were shown to escape anoikis as well as stresses in circulation [30,41]. A recent report showed that these characteristics are due to CSC properties of CTM, notably a CD44-directed cell aggregation mechanism that forms these clusters, promotes their survival and favors polyclonal metastasis [42]. Another group also investigated the factors behind CTM metastatic potential: Gkountela et al. reported that CTC clusters from BC patients and CTC cell lines exhibit a DNA methylation pattern distinct from that of single CTCs and which represents targetable vulnerabilities [43]. Moreover, CTC-neutrophils clusters are occasionally formed in the bloodstream and in vivo evidence shows that this association triggers cell cycle progression and thus drives metastasis formation in BC [44].

3. Brief Introduction to CTC Enrichment and Detection Strategies

A plethora of technologies have been developed over the last decade to respond to specific CTC applications. CTC identification remains a technically challenging task due to the extreme phenotypic heterogeneity and rarity of these cells in the bloodstream and therefore requires methods with high sensitivity and specificity. Enrichment strategies can be based on either biological properties (i.e., cell-surface markers) or physical characteristics (i.e., size, density, electric charge) and are usually combined with detection techniques (e.g., immunofluorescence, immunohistochemistry, FISH) to identify CTCs. CTC capture relies on a positive selection among normal blood cells or a negative selection by leukocyte depletion. Among biologically-based technologies is the CellSearch system (Menarini-Silicon Biosystem, Bologna, Italy). It is the most commonly applied assay for CTC enumeration in which CTCs are captured in whole blood by EpCAM (epithelial cell adhesion molecule)-coated immunomagnetic beads followed by fluorescent detection using anti-cytokeratins (CK 8, CK 18, CK 19), anti-CD45 (leukocyte marker), and a nuclear stain (DAPI). It is the only technology cleared by the US Food and Drug Administration to aid in prognosis for patients with metastatic breast, prostate, and colorectal cancer [45–49]. Although standardized and reproducible, this method has a limited sensitivity most likely due to failure in recognizing cells undergoing EMT and thus inevitably misses an aggressive and clinically relevant CTC subpopulation. Platforms relying on the depletion of leukocytes (negative selection) are being investigated and used to overcome this bias. One example is the widely used RosetteSep technique which enriches CTCs without phenotypic a priori by excluding $CD45^+$ and $CD36^+$ cells in rosettes and eliminating them in a Ficoll-Paque PLUS density-gradient centrifugation. Physical property-based methods including filtration systems have been developed to capture CTCs based on their large size compared to leukocytes, notably the ISET® (*Isolation by Size of Tumor Cells*) (RareCells Diagnostics, Paris, France) and the ScreenCell® (Paris, France) methods, which are able to detect CTCs as well as CTM using microporous polycarbonate filters [50,51]. In line with this notion, we and others have reported an overall higher recovery rate using ISET compared to CellSearch for CTC enumeration in NSCLC and prostate cancer patients [31,52]. Our lab developed a novel CTC detection approach combining ISET filtration with a FISH assay, optimized for the detection of *ALK*- or *ROS1*-rearranged pattern of NSCLC CTCs on filters [53,54]. To ensure a wider coverage of CTC heterogeneity, new devices are being developed (and some commercially) such as the CTC-iChip which relies on both biological and physical properties of CTCs: it applies size-based filtration using microfluidics processing, followed by positive selection of CTCs with EpCAM-conjugated beads or negative selection with $CD45^-$-coated beads to deplete hematopoietic cells [55]. Different technologies have been implemented to isolate live CTCs (without a fixation step) and perform subsequent functional studies. Some strategies have integrated isolation protocols for molecular analysis of single CTCs. One example is the DEPArray™ (Silicon Biosystems S.p.A., Bologna, Italy), a microfluidic system which sorts live single CTCs based on image selection followed by entrapment of CTCs inside dielectrophoretic cages [56–58]. FACS has also been adapted for molecular characterization of CTCs as well as their isolation in the aim of xenograft establishment [59].

At this point, none of the technologies fully respond to the phenotypic heterogeneity of CTCs. Indeed, each method has its own advantages and limitations and researchers have based the development of capture strategies on the specific aim of further CTC characterization studies. New insights in CTC biology should be integrated into current enrichment, detection, and isolation techniques to optimize the process and improve their reliability. As shown in Table 1, RosetteSep and FACS have been used for CDX establishment. Enrichment using RosetteSep may be advantageous owing to the lack of phenotypic a priori on tumorigenic CTCs and a higher recovery rate.

4. CTC-Derived Xenografts

Patient-derived xenograft (PDX) technology has rapidly emerged as a standard translational research platform to improve understanding of cancer biology and test novel therapeutic strategies [60]. PDXs are generated by implantation of surgically-removed tumor tissue (primary or metastasis) into immunodeficient mice. Although these models have proven utility as a preclinical tool in many cancers, their feasibility remains challenged by limited tumor tissue availability, as single-site biopsies may be impossible or detrimental in some malignancies [61]. This limitation can be overcome by the generation of CDX models after enrichment of CTCs collected from a readily accessible blood draw and subsequent injection into immunodeficient mice [62–64]. Nevertheless, it is noteworthy that CDX development still presents an enormous challenge due to low CTC prevalence in several cancers. Until now, CDXs have been established in breast, melanoma, lung and prostate cancer and are discussed in this section (Table 1).

In 2013, Baccelli et al. reported the first experimental proof that primary human luminal BC CTC populations contain MICs in a xenograft assay. Injection of CTCs from 110 patients was performed. Six recipient mice developed bone, lung, and liver metastases within 6–12 months after CTC transplant (~1000 CTCs) from three patients with advanced metastatic BC. Cell sorting analysis of the MIC-containing population shared a common $EpCAM^+CD44^+MET^+CD47^+$ phenotype, highlighting a CSC characteristic of CTCs. The authors also showed that the number of CTCs positive for these markers strongly correlated with decreased progression-free survival of metastatic BC patients. This study has therefore revealed a first phenotypic identification of luminal BC CTCs with MIC properties, making them an attractive tool to track and potentially target metastatic development in BC [59]. A second group derived a CDX model from a metastatic triple-negative BC (TNBC) patient for the first time. The patient selected for CDX establishment had advanced TNBC with a very high CTC count obtained with CellSearch analysis (969 CTCs and 74 CTC clusters/7.5 mL). Enriched cells were injected subcutaneously into nude mice and a palpable tumor was observed five months later. The authors carried out a longitudinal study and samples were collected at two different time points (metastasis and progression) during the course of the disease, which allowed real-time assessment of molecular changes between patient tumor, CTCs, and CDXs samples. The obtained CDX phenocopied the patient tumor. Most importantly, RNA sequencing of the CDX tumor disclosed key mechanisms relevant in TNBC biology such as the WNT pathway, which is necessary for the maintenance of CSCs and was shown to correlate with metastasis and poor clinical outcome in TNBC subtypes. CTC analysis also deciphered a panel of potential tumor biomarkers [65]. An additional TNBC CDX model of liver metastasis was established very recently by Vishnoi et al. Similar genomic profiling of metastatic tissue was obtained in four sequential CDX generations, representing the recapitulation of liver metastasis in all the models. Notably, the authors deciphered a first 597-gene CTC signature related to liver metastasis in TNBC which, despite small sample size bias, can provide insight into the mechanistic basis of TNBC disease progression in the liver [66].

In melanoma, Girotti et al. demonstrated the tumorigenicity of advanced-disease CTCs in immunocompromised mice. The authors resorted to CDX development when tumor material was inaccessible for PDX generation. They reported a success rate of 13% with six CDX established, 15 failed attempts, and 26 additional models followed at the time of publication. CDX tumor growth was detectable as of one month after CTC implantation and was sustainable in secondary hosts. Moreover, the CDXs were representative of patient tumors and mirrored therapy response. This proof-of-principle was developed along with PDX technology and circulating tumor DNA analysis as part of a platform to optimize precision medicine for melanoma patients. It explored the TIC properties of melanoma CTCs but did not achieve a biological characterization of these cells [67].

In lung cancer, Hodgkinson et al. showed that CTCs in chemosensitive or chemorefractory SCLC are tumorigenic. CTCs were isolated from six late-stage SCLC patients having never received chemotherapy and were subsequently injected into NSG mice. Each patient presented with more than 400 CTCs and four out of six CTC samples gave rise to CDX tumors detected as of 2.4 months post-implantation. CDXs recapitulated the genomic profile of CellSearch-enriched CTCs and mimicked donor patients response to standard of care chemotherapy (platinum and etoposide), proving the clinical relevance of these models [68]. CDX tumors were subsequently dissociated and expanded into short-term in vitro CDX cultures (Table 2). These cells maintained the genomic landscape of donor tumors as well as their drug sensitivity profiles. CDX-derived cells were also labeled in vitro with the GFP lentivirus and successfully implanted into mice, where they can serve as a tracking tool to study tumor dissemination patterns in vivo [69]. Additional 16 SCLC CDX models were recently generated by Drapkin et al. from CTCs collected at initial diagnosis or at progression, with 38% efficiency. Somatic mutations were maintained between patient tumors and CDX as shown by whole-exome sequencing (WES) and the genomic landscape remained stable throughout early CDX passages showing clonal homogeneity. The authors also developed serial CDX models from one patient at baseline of the combination olaparib and temozolomide and at relapse. Interestingly, the models accurately reflected the evolving drug sensitivity profiles of the patient's malignancy, which highlights the potential utility of serial CDXs to study the evolution of resistance to treatment in SCLC [70]. One CDX model was also described in NSCLC. In this study, CTC samples were retrieved at two different time points: Baseline and post-brain radiotherapy. No CDX was developed at baseline. Notably, no EpCAM$^+$ CTCs were detected during CellSearch analysis at disease progression, yet injection of post-radiotherapy CTCs gave rise to a palpable tumor 95 days after engraft. Phenotypic and molecular characterizations showed no epithelial CTCs, but revealed a sizeable population of phenotypically heterogeneous CTCs mostly expressing the mesenchymal marker vimentin. This study suggests that the absence of EpCAM$^+$ CTCs in NSCLC does not preclude the existence of CTCs with TIC potential in patients and underlines the importance of investigating CTCs undergoing EMT in this malignancy [71].

Our group generated the first CDX model of castration-resistant prostate cancer (CRPC) and derived a permanent ex vivo culture from CDX tumor cells. A total of 22 samples from metastatic CRPC patients were collected, among which seven were obtained from diagnostic leukapheresis (DLA). DLA products were generated as part of the European FP7 program CTCTrap which aimed for an increased CTC yield to perform molecular characterization of the tumor [72]. One patient with a very high CTC count (~20,000 CTCs) obtained by DLA gave rise to a palpable tumor within 5.5 months. Acquisition of key genetic drivers (i.e., *TP53*, *PTEN*, and *RB1*) that govern the trans-differentiation of CRPC into CRPC-neuroendocrine (CRPC-NE) malignancy was detected in CTCs, highlighting the role of tumorigenic CRPC-NE CTCs in this transformation. Moreover, the obtained in vitro CDX-derived cell line faithfully recapitulated the genetic characteristics and tumorigenicity of the CDX and mimicked patient response to standard of care treatments for CRPC (i.e., enzalutamide and docetaxel) (Table 2) [73,74].

Table 1. Overview of in vivo circulating tumor cell (CTC)-derived models established to date.

CTC-Derived Xenografts								
Type of Cancer	Stage	Live CTC Isolation Technique	# of CTCs	Injection Procedure	Take Rate	Passaging	Main Findings	Ref
Breast cancer	Metastatic luminal	FACS isolation (PI⁻CD45⁻EpCAM⁺) or RosetteSep	≥1109 CTCs EpCAM⁺ (CellSearch)	- Dilution in matrigel - Injection in femoral medullar cavity	5%	N/A	- Specific CTC MIC signature EpCAM⁺CD44⁺MET⁺CD47⁺ - Recapitulation of patient metastases phenotype in CDX metastases - No drug sensitivity study	[59]
	Metastatic triple-negative	Density gradient centrifugation: Histopaque®	969 CTCs EpCAM⁺ (CellSearch)	- Dilution in matrigel - Subcutaneous inj.	3%	Piece of tumor explant or injection of explant culture	- RT-qPCR for genomic profiling of CTC/CDX samples before and after injection - WNT pathway upregulation as a potential therapeutic target in TNBC identified by RNAseq - No drug sensitivity study	[65]
	Metastatic triple-negative	FACS (CD45⁻/CD34⁻/CD105⁻/CD90⁻CD73⁻)	N/A	Intracardiac injection	33%	Minced metastatic liver tissue	- Identification of a TNBC liver metastasis CTC-specific signature (whole-transcriptome) - Survival analyses for signature transcripts	[66]
Melanoma	Stage IV	RosetteSep	N/A	- Dilution in matrigel - Subcutaneous inj.	13%	Tumor fragments	- recapitulation of patient response to dabrafenib in the CDX - concordance in SNV profiles (WES/RNAseq)	[67]
	Metastatic	RosetteSep	>400 CTCs EpCAM⁺ (CellSearch)	Dilution in matrigel/subcutaneous	67%	Tumor fragments	- Recapitulation of CTC genomic profile by CDX tumors - CDX mimicked donor's response to chemotherapy	[68]
SCLC	Limited or extensive stage	CTC-iChip + RosetteSep Ficoll	N/A	Dilution in matrigel/subcutaneous	38%	Tumor fragments	- Faithful recapitulation of the tumor genome - Reflection of evolving treatment sensitivities of patient tumor	[70]

Table 1. Cont.

Type of Cancer	Stage	Live CTC Isolation Technique	# of CTCs	Injection Procedure	Take Rate	Passaging	Main Findings	Ref
CTC-Derived Xenografts								
NSCLC	Metastatic	RosetteSep	>150 CTCs by FACS (CD45/CD144/vimentin/CK)	Dilution in matrigel/subcutaneous	100%	Disaggregation of tumor	- Importance of mesenchymal CTCs with tumorigenic capacity	[71]
CRPC	Metastatic	DLA/RosetteSep	~20,000 CTCs EpCAM+ (CellSearch)	Dilution in matrigel/subcutaneous	14%	Tumor fragments	- Recapitulation of genome characteristics in CTC, patient tumor and CDX (WES) - Tumorigenic CTCs with acquired CRPC-NE features	[73]

* N/A: not available; FACS: Fluorescent-activated cell sorting; CDX: CTC-derived xenograft; MIC: Metastasis-initiating cell; TIC: Tumor-initiating cell; TNBC: Triple-negative breast cancer; SCLC: Small-cell lung cancer; SNV: Single nucleotide variant; NSCLC: Non-small cell lung cancer; CRPC: Castration-resistant prostate cancer; NE: Neuroendocrine; WES: Whole-exome sequencing.

Table 2. Overview of CDX-derived ex vivo cultures established to date.

CDX-Derived Ex Vivo Cultures				
Type of Cancer	Stage	Culturing Conditions	Main Findings	Ref
SCLC	Metastatic	HITES medium with ROCK inhibitor—non-adherent cell clusters—short-term	Recapitulate genomic landscape and in vivo drug response Tumorigenic in vivo Lentiviral transduction of one cell line	[69]
CRPC	Metastatic	DMEM/F12 medium—adherent conditions—permanent	Recapitulation of genomic characteristics and standard of care drug response	[73]

5. CTC-Derived Ex Vivo Models

Although CDXs represent classical preclinical mouse models that are relatively easy to handle, they cannot be derived from every patient depending on tumor type and the process could take several months, a time frame that would not provide proper aid for the clinical guidance of donor patients. Expansion of viable CTCs ex vivo may offer an attractive alternative allowing both molecular analysis and high-throughput drug screening in a shorter time, but with CTC scarcity remaining, a fortiori, a significant limitation. In vitro CTC cultures were reported in colon, breast, prostate, and lung cancer and are evaluated in this section (Table 3).

The first long-term colon cancer CTC cell line was derived by Cayrefourcq et al. from a metastatic colon cancer patient who had 302 EpCAM$^+$ CTCs detected by the CellSearch platform. Importantly, the characterized CTC-MCC-41 cell line shared the main genomic features of both the donor patient primary tumor and lymph node metastasis [9]. In a second study, the authors established and characterized eight additional cell lines from the same patient with CTCs collected at different time points during his follow-up. Transcriptomics analyses in the nine cell lines revealed an intermediate epithelial/mesenchymal phenotype promoting their metastatic potential, as well as stem cell-like properties that increased in cell lines isolated at later stages of progression. This may highlight the selection mechanism of treatment-resistant clones with specific phenotypes that drive disease progression. Functional experiments showed that these cells favor angiogenesis in vitro, which was concordant with the secretion of potent angiogenesis inducers such as VEGF and FGF2 as well as the tumorigenicity of these cells in vivo [9,10].

In BC, Zhang et al. presented the characterization of EpCAM$^-$ CTCs and revealed a shared protein signature HER2$^+$/EGFR$^+$/HPSE$^+$/Notch1$^+$ in CTCs competent for brain metastasis. Indeed, the three established CTC lines expressing this signature promoted brain and lung localization after xenotransplantation into nude mice. The authors therefore deciphered a preliminary signature which provides insight into metastatic competency of BC CTCs and pushes towards using CTC research to explore new potential biomarkers [75]. Another study reported the establishment of non-adherent CTC lines under hypoxic conditions (4% O_2) with CTCs issued from six patients with metastatic luminal-subtype BC. Three out of five tested cell lines were tumorigenic in vivo, giving rise to tumors with histological and immunohistochemical similarities with the primary patient tumor. This proof-of-concept study also identified targetable mutations acquired de novo in CTC cell lines, elucidating the importance of monitoring the mutational evolution of the tumor throughout the disease. To explore this, the authors performed sensitivity assays on the CTC lines with large panels of single drug and drug combinations targeting the different mutations identified [76]. In vitro phenotypic analysis of these cell lines and patient CTCs was recently performed. A CTM-specific DNA methylation status was revealed in which binding sites for stemness and proliferation transcription factors were hypomethylated, suggesting potential targets. This pattern correlated with poor prognosis in patients and targeting of clusters with Na+/K+ ATPase inhibitors shed them into single cell and enabled DNA methylation remodeling, leading to suppression of metastasis. These data therefore highlight a key connection between phenotypic properties of CTCs and DNA methylation patterns at specific stemness- and proliferation-related sites [43].

Table 3. Overview of ex vivo CTC-derived models established to date.

CTC-Derived Ex Vivo Models

Type of Cancer	Stage	Live CTC Isolation Technique	# of CTCs (CellSearch)	Culturing Conditions	Success Rate	Main Findings	Ref
Colon cancer	Nonresectable metastatic	RosetteSep	≥300	- Hypoxic in medium 1 DMEM/F12 to normoxic conditions in medium 2 RPMI1640 - 2D, sustained for >6months	1%	- Recapitulation of main genomic features - Tumorigenic in vivo - Intermediate EMT + stem cell properties	[9,10]
Breast cancer	Metastatic	FACS	0	- Normoxic stem cell culture medium - 2D	8%	- Tumorigenic in vivo, brain metastasis signature (EpCAM⁻HER2⁺/EGFR⁺/HPSE⁺/Notch1⁺)	[75]
	Metastatic luminal	CTC-iChip	3–3000	- Hypoxic, nonadherent - 2D, Sustained for >6 months	83%	- Tumorigenic in vivo - Drug sensitivity panels and CTM-specific methylation profile	[43,76]
CRPC	Metastatic	RosetteSep-Ficoll	>100	- Growth factors reduced Matrigel/Advanced DMEM/F12 - 3D, sustained for >6 months	6%	- Tumorigenic in vivo	[77]
NSCLC	Early stage	Microfluidic CTC-capture device	1–11	- Matrigel + collagen - 3D, sustained for ~1 month	73%	- Common mutations between cultured CTCs and primary tumor	[78]

271

Despite successful in vitro expansion of patient CTCs in several cancer types as reported above, important limitations should be noted when handling 2D cultures, including cell morphology alterations due to adherence to plastic and lack of tumor microenvironment. Moreover, cell-cell and spatial interactions in vitro are not fully representative of the setting in the tumor mass in vivo [79]. These constraints can thus interfere with physiological functions and molecular responses of the tumor cells, making them less representative of the actual malignancy. To circumvent this problem, 3D models were proposed in prostate and lung cancer [77,78]. In prostate cancer, Gao et al. generated the first seven fully characterized organoid lines from a CRPC patient including a CTC-derived 3D organoid system from a patient who had more than 100 CTCs in 8 mL of blood. Success rate for the establishment of the CTC-derived organoid was not provided. Whole-exome sequencing (WES) analysis showed that all the 3D models recapitulated the molecular diversity of prostate cancer subtypes and were amenable to pharmacological assays. Engraftment of the CTC-derived organoid in vivo gave rise to tumors with a histological pattern similar to that of the primary cancer. This research, therefore, contributes a patient-derived model of CRPC which, with further optimization, may respond to the pressing need of in vitro models that faithfully recapitulate CRPC [77]. In lung cancer, Zhang et al. developed a novel ex vivo CTC-derived model using a 3D co-culture system which stimulated a microenvironment to sustain tumor development. CTCs were enriched and expanded for a short period of time from 14 to 19 early lung cancer patients. Next-generation sequencing detected several mutations including *TP53* found in both cultured CTCs and matched patient primary tumors [78].

6. Discussion

During the last decade, tremendous technological progress has been made to reliably detect, quantify and characterize CTCs at phenotypic, genomic, and functional levels. The characterization of CTC-derived models has paved the way toward an improved understanding of tumor dissemination by these cells (Figure 1). As depicted in Table 1, procedures for developing CDXs can vary from one study to another. Subcutaneous (SC) injection of cells in mice is the simplest method for tumor engrafts which has been used for decades and was most recently applied for PDX establishment. It facilitates tumor growth monitoring as it does not require fluorescent labeling or imaging. Most CDX models published to date have been developed through SC injection of CTCs. SC tumors do not usually metastasize probably due to the absence of the human microenvironment and the impact of murine angiogenesis, which influence dissemination of primary human tumors. Moreover, as the time-frame needed for tumor growth extends to several months, ethical regulations may not allow waiting for metastatic spread. To this end, these studies were limited to the characterization of the CDX primary tumor. Injection in mouse bone marrow as done by Baccelli et al. may also be an appropriate way to investigate MICs as this microenvironment has been previously described as a reservoir for disseminating tumor cells [35,59]. Conversely, studies aiming to assess metastatic and not only tumorigenic competency of CTCs have resorted to intracardiac injection [66,75]. This method, similarly to tail vein (TV) injection, allows a more rapid spread of the cells as they directly enter the bloodstream and thus mimics CTCs in their original setting. Propagation of CDX models through intracardiac or TV injection is less common or completely lacking, most likely due to potential dissemination bias. Indeed, organ metastasis could be influenced by the injection site of CTCs and defined by the first capillary bed encountered by cells post-injection. TV has been observed to induce lung metastases, thus generating false-positive results [80].

Another important challenge is ensuring the CDX consistently maintains its clinical relevance and serves as a patient surrogate. To this end, stringent validation is required and several aspects must be addressed. Firstly, it is crucial to verify the human origin of the CDX, as spontaneous tumors could grow in immunocompromised mouse models. Secondly, confirming cancer type and comparing the CDX tumor to the donor patient's biopsy through histopathology, followed by genomic studies to assess CDX genomic fidelity with patient tumor. Moreover, in the context of establishing preclinical models for precision medicine, functional drug sensitivity assays are needed to evaluate recapitulation of patient response to therapy in the CDX [68].

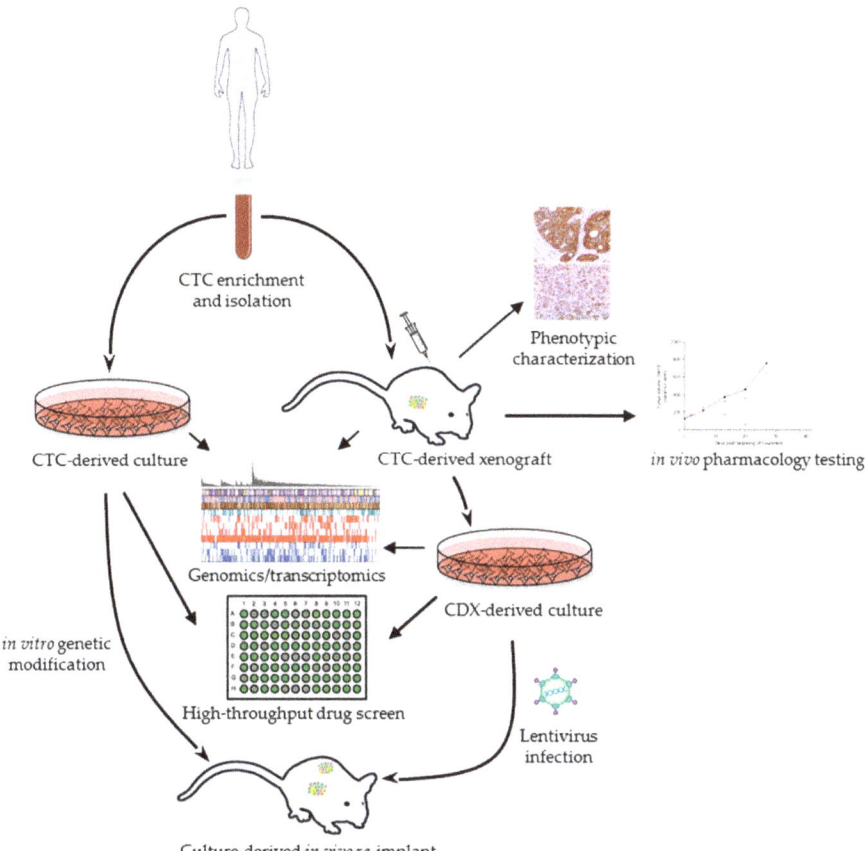

Figure 1. CTC-derived models as tractable systems to explore tumor-initiating cells (TICs) and new therapeutic strategies. CTCs isolated from late-stage cancer patients are used to generate CTC-derived xenografts (CDXs) to perform functional characterizations and pharmacology studies. CDX tumors can be isolated and dissociated into ex vivo cultures for drug screening and genome-wide analyses. CDX-derived cultures are amenable to lentiviral infection and can be re-injected into mice and used as tools to track tumor dissemination. In parallel, CTCs can be expanded in vitro and used as readouts of drug sensitivity. CTC = circulating tumor cell. CDX = CTC-derived xenograft.

Although PDX models serve as reliable tools for tumor modeling, CDXs offer added value for the understanding of tumor biology and metastasis. Detection and characterization of metastasis-competent CTCs using in vivo models offer a more representative molecular snapshot of the disease, as they serve as easily accessible "surrogates" of metastatic tissue, which is otherwise unobtainable in many cancer organs (e.g., bones or lungs) [81]. Indeed, CDX models could help showcase tumor heterogeneity in the metastatic setting in contrast to a localized biopsy in the case of PDX and are attainable at different time points throughout disease progression [63]. Most importantly, CDX models established to date reveal the high tumorigenic capacity of CTCs—even at a low number of cells (as low as 400 CTCs [68]). As reviewed above, CTCs with survival and MIC properties are assumed to be selected for seeding CDX tumors, similar to what has been observed in PDXs [82]. It is expected that the proportion of tumorigenic CTCs may vary between cancer types and patients as well as under selective pressure of treatment, which highlights a potential selective process for the acquisition of minor metastasis-competent CTC subclones [73]. CTC clusters and hybrid E/M CTCs have been described as the most aggressive cells with a high propensity for tumorigenesis. However, it is currently difficult to evaluate the impact each subpopulation could have on CDX tumor take rate. Indeed, as detailed before, Aceto et al. have shown an increased metastatic competency in CTM vs. single CTCs but this remains limited to murine models and is difficult to translate to human subjects [40]. It is worth noting however that, although in vivo models are sustained by the host tissue microenvironment and can faithfully recapitulate the tumor genome, the absence of immune components constitutes an important bias.

On the other hand, CTC expansion ex vivo is promising but is still very far from routine applications as culturing conditions are still under investigation and need further optimization. Therefore, CDX-derived cultures represent an attractive intermediate model to characterize this aggressive population in vitro. In the event of molecular similarities between the two models, CDX and CDX-derived cell lines offer complementary, tractable systems for CTC functional characterization and therapy testing. Re-injection of the CDX-derived cell line in immunodeficient mice could allow the identification of candidate genes in metastasis and chemoresistance mechanisms [64,69]. Additional model systems such as the chick embryo chorioallantoic membrane have also opened up new promising avenues in the in vivo studies of tumor metastasis, as the highly vascularized setting sustains tumor formation and dissemination rapidly after engraft [83]. Moreover, organoids have recently emerged as novel robust 3D models optimized to propagate in vitro and reminiscent of tumoral heterogeneity, with amenability to genetic modifications and drug screening assays [77,84]. One can hypothesize that the establishment of several CTC-derived organoid lines from the same patient could be useful in modeling metastatic disease and acquired CTC mutational profiles to monitor disease progression. However, these models lack in vivo host complexity and recent efforts have been put into the generation of 3D co-cultures in microfluidic devices to model ex vivo tumor microenvironments by the integration of different cell populations (e.g., immune cells, fibroblasts) [78,85].

7. Concluding Remarks

CDX models have shown unprecedented opportunities to provide insight into the complex biology of the metastatic process. However, at the present time, these functional models serve as proof-of-principle tools as their development is limited to late-stage disease settings and high CTC counts. The main goal of functional CTC studies being the identification and characterization of MICs and candidate target genes among CTCs, it is crucial to expand analyses to earlier stages of cancer [63]. Unfortunately, these rare preclinical models are derived from patients in exceptional clinical situations and we are currently unable to predict if the limitation caused by CTC scarcity could be circumvented. Nevertheless, the establishment of CTC-derived models from only a few CTCs is a major achievement today and an invaluable opportunity to decipher new biomarkers, which are urgently needed for novel therapeutic strategies in advanced cancers.

Funding: Our CDX project was supported by ANR-15-CE17-0006-01, CTCTrap FP7 HEALTH #305341 and the Innovative Medicines Joint Undertaking CANCER ID (IMI-JU-11- 2013, grant no. 115749. TT is supported by La Ligue Nationale Contre le Cancer.

Acknowledgments: We are grateful to the patients and their families.

Conflicts of Interest: The authors declare no conflict of interest.

References

1. Fidler, I.J. Metastasis: Quantitative Analysis of Distribution and Fate of Tumor Emboli Labeled with ^{125}I-5-Iodo-2′-deoxyuridine23. *JNCI J. Natl. Cancer Inst.* **1970**, *45*, 773–782. [PubMed]
2. Chambers, A.F.; Groom, A.C.; MacDonald, I.C. Dissemination and growth of cancer cells in metastatic sites. *Nat. Rev. Cancer* **2002**, *2*, 563–572. [CrossRef] [PubMed]

3. Luzzi, K.J.; MacDonald, I.C.; Schmidt, E.E.; Kerkvliet, N.; Morris, V.L.; Chambers, A.F.; Groom, A.C. Multistep Nature of Metastatic Inefficiency. *Am. J. Pathol.* **1998**, *153*, 865–873. [CrossRef]
4. Massagué, J.; Obenauf, A.C. Metastatic colonization by circulating tumour cells. *Nature* **2016**, *529*, 298–306. [CrossRef] [PubMed]
5. Fidler, I.J. Tumor heterogeneity and the biology of cancer invasion and metastasis. *Cancer Res.* **1978**, *38*, 2651–2660. [CrossRef]
6. Al-Hajj, M.; Wicha, M.S.; Benito-Hernandez, A.; Morrison, S.J.; Clarke, M.F. Prospective identification of tumorigenic breast cancer cells. *Proc. Natl. Acad. Sci. USA* **2003**, *100*, 3983–3988. [CrossRef] [PubMed]
7. Ricci-Vitiani, L.; Lombardi, D.G.; Pilozzi, E.; Biffoni, M.; Todaro, M.; Peschle, C.; De Maria, R. Identification and expansion of human colon-cancer-initiating cells. *Nature* **2007**, *445*, 111–115. [CrossRef] [PubMed]
8. Agnoletto, C.; Corrà, F.; Minotti, L.; Baldassari, F.; Crudele, F.; Cook, W.; Di Leva, G.; d'Adamo, A.; Gasparini, P.; Volinia, S. Heterogeneity in Circulating Tumor Cells: The Relevance of the Stem-Cell Subset. *Cancers* **2019**, *11*, 483. [CrossRef] [PubMed]
9. Cayrefourcq, L.; Mazard, T.; Joosse, S.; Solassol, J.; Ramos, J.; Assenat, E.; Schumacher, U.; Costes, V.; Maudelonde, T.; Pantel, K.; et al. Establishment and Characterization of a Cell Line from Human Circulating Colon Cancer Cells. *Cancer Res.* **2015**, *75*, 892–901. [CrossRef] [PubMed]
10. Soler, A.; Cayrefourcq, L.; Mazard, T.; Babayan, A.; Lamy, P.-J.; Assou, S.; Assenat, E.; Pantel, K.; Alix-Panabières, C. Autologous cell lines from circulating colon cancer cells captured from sequential liquid biopsies as model to study therapy-driven tumor changes. *Sci. Rep.* **2018**, *8*, 15931. [CrossRef]
11. Liu, T.; Xu, H.; Huang, M.; Ma, W.; Saxena, D.; Lustig, R.A.; Alonso-Basanta, M.; Zhang, Z.; O Rourke, D.M.; Zhang, L.; et al. Circulating glioma cells exhibit stem cell-like properties. *Cancer Res.* **2018**. [CrossRef] [PubMed]
12. Nieto, M.A.; Huang, R.Y.-J.; Jackson, R.A.; Thiery, J.P. EMT: 2016. *Cell* **2016**, *166*, 21–45. [CrossRef] [PubMed]
13. Shibue, T.; Weinberg, R.A. EMT, CSCs, and drug resistance: The mechanistic link and clinical implications. *Nat. Rev. Clin. Oncol.* **2017**, *14*, 611–629. [CrossRef] [PubMed]
14. Kasimir-Bauer, S.; Hoffmann, O.; Wallwiener, D.; Kimmig, R.; Fehm, T. Expression of stem cell and epithelial-mesenchymal transition markers in primary breast cancer patients with circulating tumor cells. *Breast Cancer Res.* **2012**, *14*. [CrossRef] [PubMed]
15. Yu, M.; Bardia, A.; Wittner, B.S.; Stott, S.L.; Smas, M.E.; Ting, D.T.; Isakoff, S.J.; Ciciliano, J.C.; Wells, M.N.; Shah, A.M.; et al. Circulating Breast Tumor Cells Exhibit Dynamic Changes in Epithelial and Mesenchymal Composition. *Science* **2013**, *339*, 580–584. [CrossRef] [PubMed]
16. Kallergi, G.; Papadaki, M.A.; Politaki, E.; Mavroudis, D.; Georgoulias, V.; Agelaki, S. Epithelial to mesenchymal transition markers expressed in circulating tumour cells of early and metastatic breast cancer patients. *Breast Cancer Res.* **2011**, *13*. [CrossRef] [PubMed]
17. Giordano, A.; Gao, H.; Anfossi, S.; Cohen, E.; Mego, M.; Lee, B.-N.; Tin, S.; Laurentiis, M.D.; Parker, C.A.; Alvarez, R.H.; et al. Epithelial–Mesenchymal Transition and Stem Cell Markers in Patients with HER2-Positive Metastatic Breast Cancer. *Mol. Cancer Ther.* **2012**, *11*, 2526–2534. [CrossRef]
18. Tam, W.L.; Weinberg, R.A. The epigenetics of epithelial-mesenchymal plasticity in cancer. *Nat. Med.* **2013**, *19*, 1438–1449. [CrossRef]
19. Puisieux, A.; Brabletz, T.; Caramel, J. Oncogenic roles of EMT-inducing transcription factors. *Nat. Cell Biol.* **2014**, *16*, 488–494. [CrossRef]
20. Mani, S.A.; Guo, W.; Liao, M.-J.; Eaton, E.N.; Ayyanan, A.; Zhou, A.Y.; Brooks, M.; Reinhard, F.; Zhang, C.C.; Shipitsin, M.; et al. The Epithelial-Mesenchymal Transition Generates Cells with Properties of Stem Cells. *Cell* **2008**, *133*, 704–715. [CrossRef]
21. Morel, A.-P.; Lièvre, M.; Thomas, C.; Hinkal, G.; Ansieau, S.; Puisieux, A. Generation of Breast Cancer Stem Cells through Epithelial-Mesenchymal Transition. *PLoS ONE* **2008**, *3*, e2888. [CrossRef] [PubMed]
22. Polyak, K.; Weinberg, R.A. Transitions between epithelial and mesenchymal states: Acquisition of malignant and stem cell traits. *Nat. Rev. Cancer* **2009**, *9*, 265–273. [CrossRef] [PubMed]
23. Creighton, C.J.; Li, X.; Landis, M.; Dixon, J.M.; Neumeister, V.M.; Sjolund, A.; Rimm, D.L.; Wong, H.; Rodriguez, A.; Herschkowitz, J.I.; et al. Residual breast cancers after conventional therapy display mesenchymal as well as tumor-initiating features. *Proc. Natl. Acad. Sci. USA* **2009**, *106*, 13820–13825. [CrossRef] [PubMed]

24. Ocaña, O.H.; Córcoles, R.; Fabra, Á.; Moreno-Bueno, G.; Acloque, H.; Vega, S.; Barrallo-Gimeno, A.; Cano, A.; Nieto, M.A. Metastatic Colonization Requires the Repression of the Epithelial-Mesenchymal Transition Inducer Prrx1. *Cancer Cell* **2012**, *22*, 709–724. [CrossRef] [PubMed]
25. Beerling, E.; Seinstra, D.; de Wit, E.; Kester, L.; van der Velden, D.; Maynard, C.; Schäfer, R.; van Diest, P.; Voest, E.; van Oudenaarden, A.; et al. Plasticity between Epithelial and Mesenchymal States Unlinks EMT from Metastasis-Enhancing Stem Cell Capacity. *Cell Rep.* **2016**, *14*, 2281–2288. [CrossRef] [PubMed]
26. Fischer, K.R.; Durrans, A.; Lee, S.; Sheng, J.; Li, F.; Wong, S.T.C.; Choi, H.; El Rayes, T.; Ryu, S.; Troeger, J.; et al. Epithelial-to-mesenchymal transition is not required for lung metastasis but contributes to chemoresistance. *Nature* **2015**, *527*, 472–476. [CrossRef] [PubMed]
27. Zheng, X.; Carstens, J.L.; Kim, J.; Scheible, M.; Kaye, J.; Sugimoto, H.; Wu, C.-C.; LeBleu, V.S.; Kalluri, R. Epithelial-to-mesenchymal transition is dispensable for metastasis but induces chemoresistance in pancreatic cancer. *Nature* **2015**, *527*, 525–530. [CrossRef]
28. Bailey; Martin Insights on CTC Biology and Clinical Impact Emerging from Advances in Capture Technology. *Cells* **2019**, *8*, 553. [CrossRef]
29. Lecharpentier, A.; Vielh, P.; Perez-Moreno, P.; Planchard, D.; Soria, J.C.; Farace, F. Detection of circulating tumour cells with a hybrid (epithelial/mesenchymal) phenotype in patients with metastatic non-small cell lung cancer. *Br. J. Cancer* **2011**, *105*, 1338–1341. [CrossRef]
30. Hou, J.-M.; Krebs, M.; Ward, T.; Sloane, R.; Priest, L.; Hughes, A.; Clack, G.; Ranson, M.; Blackhall, F.; Dive, C. Circulating tumor cells as a window on metastasis biology in lung cancer. *Am. J. Pathol.* **2011**, *178*, 989–996. [CrossRef]
31. Hofman, V.; Ilie, M.I.; Long, E.; Selva, E.; Bonnetaud, C.; Molina, T.; Vénissac, N.; Mouroux, J.; Vielh, P.; Hofman, P. Detection of circulating tumor cells as a prognostic factor in patients undergoing radical surgery for non-small-cell lung carcinoma: Comparison of the efficacy of the CellSearch Assay^{TM} and the isolation by size of epithelial tumor cell method. *Int. J. Cancer* **2011**, *129*, 1651–1660. [CrossRef] [PubMed]
32. Papadaki, M.A.; Stoupis, G.; Theodoropoulos, P.A.; Mavroudis, D.; Georgoulias, V.; Agelaki, S. Circulating Tumor Cells with Stemness and Epithelial-to-Mesenchymal Transition Features Are Chemoresistant and Predictive of Poor Outcome in Metastatic Breast Cancer. *Mol. Cancer Ther.* **2019**, *18*, 437–447. [CrossRef] [PubMed]
33. Wong, C.W.; Lee, A.; Shientag, L.; Yu, J.; Dong, Y.; Kao, G.; Al-Mehdi, A.B.; Bernhard, E.J.; Muschel, R.J. Apoptosis: An early event in metastatic inefficiency. *Cancer Res.* **2001**, *61*, 333–338. [PubMed]
34. Bednarz-Knoll, N.; Alix-Panabières, C.; Pantel, K. Clinical relevance and biology of circulating tumor cells. *Breast Cancer Res. BCR* **2011**, *13*, 228. [CrossRef] [PubMed]
35. Braun, S.; Vogl, F.D.; Naume, B.; Janni, W.; Osborne, M.P.; Coombes, R.C.; Schlimok, G.; Diel, I.J.; Gerber, B.; Gebauer, G.; et al. A Pooled Analysis of Bone Marrow Micrometastasis in Breast Cancer. *N. Engl. J. Med.* **2005**, *353*, 793–802. [CrossRef]
36. Faugeroux, V.; Lefebvre, C.; Pailler, E.; Pierron, V.; Marcaillou, C.; Tourlet, S.; Billiot, F.; Dogan, S.; Oulhen, M.; Vielh, P.; et al. An Accessible and Unique Insight into Metastasis Mutational Content Through Whole-exome Sequencing of Circulating Tumor Cells in Metastatic Prostate Cancer. *Eur. Urol. Oncol.* **2019**. [CrossRef] [PubMed]
37. Fernandez, S.V.; Bingham, C.; Fittipaldi, P.; Austin, L.; Palazzo, J.; Palmer, G.; Alpaugh, K.; Cristofanilli, M. TP53 mutations detected in circulating tumor cells present in the blood of metastatic triple negative breast cancer patients. *Breast Cancer Res. BCR* **2014**, *16*, 445. [CrossRef]
38. Jordan, N.V.; Bardia, A.; Wittner, B.S.; Benes, C.; Ligorio, M.; Zheng, Y.; Yu, M.; Sundaresan, T.K.; Licausi, J.A.; Desai, R.; et al. HER2 expression identifies dynamic functional states within circulating breast cancer cells. *Nature* **2016**, *537*, 102–106. [CrossRef]
39. Heitzer, E.; Auer, M.; Gasch, C.; Pichler, M.; Ulz, P.; Hoffmann, E.M.; Lax, S.; Waldispuehl-Geigl, J.; Mauermann, O.; Lackner, C.; et al. Complex Tumor Genomes Inferred from Single Circulating Tumor Cells by Array-CGH and Next-Generation Sequencing. *Cancer Res.* **2013**, *73*, 2965–2975. [CrossRef]
40. Aceto, N.; Bardia, A.; Miyamoto, D.T.; Donaldson, M.C.; Wittner, B.S.; Spencer, J.A.; Yu, M.; Pely, A.; Engstrom, A.; Zhu, H.; et al. Circulating Tumor Cell Clusters Are Oligoclonal Precursors of Breast Cancer Metastasis. *Cell* **2014**, *158*, 1110–1122. [CrossRef]
41. Aceto, N.; Toner, M.; Maheswaran, S.; Haber, D.A. En Route to Metastasis: Circulating Tumor Cell Clusters and Epithelial-to-Mesenchymal Transition. *Trends Cancer* **2015**, *1*, 44–52. [CrossRef] [PubMed]

42. Liu, X.; Taftaf, R.; Kawaguchi, M.; Chang, Y.-F.; Chen, W.; Entenberg, D.; Zhang, Y.; Gerratana, L.; Huang, S.; Patel, D.B.; et al. Homophilic CD44 Interactions Mediate Tumor Cell Aggregation and Polyclonal Metastasis in Patient-Derived Breast Cancer Models. *Cancer Discov.* **2019**, *9*, 96–113. [CrossRef] [PubMed]
43. Gkountela, S.; Castro-Giner, F.; Szczerba, B.M.; Vetter, M.; Landin, J.; Scherrer, R.; Krol, I.; Scheidmann, M.C.; Beisel, C.; Stirnimann, C.U.; et al. Circulating Tumor Cell Clustering Shapes DNA Methylation to Enable Metastasis Seeding. *Cell* **2019**, *176*, 98–112.e14. [CrossRef] [PubMed]
44. Szczerba, B.M.; Castro-Giner, F.; Vetter, M.; Krol, I.; Gkountela, S.; Landin, J.; Scheidmann, M.C.; Donato, C.; Scherrer, R.; Singer, J.; et al. Neutrophils escort circulating tumour cells to enable cell cycle progression. *Nature* **2019**, *566*, 553–557. [CrossRef] [PubMed]
45. Allard, W.J.; Matera, J.; Miller, M.C.; Repollet, M.; Connelly, M.C.; Rao, C.; Tibbe, A.G.J.; Uhr, J.W. Tumor Cells Circulate in the Peripheral Blood of All Major Carcinomas but not in Healthy Subjects or Patients with Nonmalignant Diseases. *Clin. Cancer Res* **2004**, *10*, 6897–6904. [CrossRef] [PubMed]
46. Cristofanilli, M.; Budd, G.T.; Ellis, M.J.; Stopeck, A.; Matera, J.; Miller, M.C.; Reuben, J.M.; Doyle, G.V.; Allard, W.J.; Terstappen, L.W.M.M.; et al. Circulating Tumor Cells, Disease Progression, and Survival in Metastatic Breast Cancer. *N. Engl. J. Med.* **2004**, *351*, 781–791. [CrossRef] [PubMed]
47. Cristofanilli, M.; Hayes, D.F.; Budd, G.T.; Ellis, M.J.; Stopeck, A.; Reuben, J.M.; Doyle, G.V.; Matera, J.; Allard, W.J.; Miller, M.C.; et al. Circulating Tumor Cells: A Novel Prognostic Factor for Newly Diagnosed Metastatic Breast Cancer. *J. Clin. Oncol.* **2005**, *23*, 1420–1430. [CrossRef]
48. De Bono, J.S.; Scher, H.I.; Montgomery, R.B.; Parker, C.; Miller, M.C.; Tissing, H.; Doyle, G.V.; Terstappen, L.W.W.M.; Pienta, K.J.; Raghavan, D. Circulating Tumor Cells Predict Survival Benefit from Treatment in Metastatic Castration-Resistant Prostate Cancer. *Clin. Cancer Res.* **2008**, *14*, 6302–6309. [CrossRef]
49. Cohen, S.J.; Punt, C.J.A.; Iannotti, N.; Saidman, B.H.; Sabbath, K.D.; Gabrail, N.Y.; Picus, J.; Morse, M.; Mitchell, E.; Miller, M.C.; et al. Relationship of Circulating Tumor Cells to Tumor Response, Progression-Free Survival, and Overall Survival in Patients With Metastatic Colorectal Cancer. *J. Clin. Oncol.* **2008**, *26*, 3213–3221. [CrossRef]
50. Vona, G.; Sabile, A.; Louha, M.; Sitruk, V.; Romana, S.; Schütze, K.; Capron, F.; Franco, D.; Pazzagli, M.; Vekemans, M.; et al. Isolation by Size of Epithelial Tumor Cells. *Am. J. Pathol.* **2000**, *156*, 57–63. [CrossRef]
51. Desitter, I.; Guerrouahen, B.S.; Benali-Furet, N.; Wechsler, J.; Jänne, P.A.; Kuang, Y.; Yanagita, M.; Wang, L.; Berkowitz, J.A.; Distel, R.J.; et al. A new device for rapid isolation by size and characterization of rare circulating tumor cells. *Anticancer Res.* **2011**, *31*, 427–441. [PubMed]
52. Farace, F.; Massard, C.; Vimond, N.; Drusch, F.; Jacques, N.; Billiot, F.; Laplanche, A.; Chauchereau, A.; Lacroix, L.; Planchard, D.; et al. A direct comparison of CellSearch and ISET for circulating tumour-cell detection in patients with metastatic carcinomas. *Br. J. Cancer* **2011**, *105*, 847–853. [CrossRef] [PubMed]
53. Pailler, E.; Adam, J.; Barthélémy, A.; Oulhen, M.; Auger, N.; Valent, A.; Borget, I.; Planchard, D.; Taylor, M.; André, F.; et al. Detection of Circulating Tumor Cells Harboring a Unique *ALK* Rearrangement in *ALK*-Positive Non–Small-Cell Lung Cancer. *J. Clin. Oncol.* **2013**, *31*, 2273–2281. [CrossRef] [PubMed]
54. Pailler, E.; Oulhen, M.; Borget, I.; Remon, J.; Ross, K.; Auger, N.; Billiot, F.; Ngo Camus, M.; Commo, F.; Lindsay, C.R.; et al. Circulating Tumor Cells with Aberrant *ALK* Copy Number Predict Progression-Free Survival during Crizotinib Treatment in *ALK*-Rearranged Non–Small Cell Lung Cancer Patients. *Cancer Res.* **2017**, *77*, 2222–2230. [CrossRef] [PubMed]
55. Ozkumur, E.; Shah, A.M.; Ciciliano, J.C.; Emmink, B.L.; Miyamoto, D.T.; Brachtel, E.; Yu, M.; Chen, P.-i.; Morgan, B.; Trautwein, J.; et al. Inertial Focusing for Tumor Antigen-Dependent and -Independent Sorting of Rare Circulating Tumor Cells. *Sci. Transl. Med.* **2013**, *5*, 179ra47. [CrossRef] [PubMed]
56. Fabbri, F.; Carloni, S.; Zoli, W.; Ulivi, P.; Gallerani, G.; Fici, P.; Chiadini, E.; Passardi, A.; Frassineti, G.L.; Ragazzini, A.; et al. Detection and recovery of circulating colon cancer cells using a dielectrophoresis-based device: KRAS mutation status in pure CTCs. *Cancer Lett.* **2013**, *335*, 225–231. [CrossRef] [PubMed]
57. Bulfoni, M.; Gerratana, L.; Del Ben, F.; Marzinotto, S.; Sorrentino, M.; Turetta, M.; Scoles, G.; Toffoletto, B.; Isola, M.; Beltrami, C.A.; et al. In patients with metastatic breast cancer the identification of circulating tumor cells in epithelial-to-mesenchymal transition is associated with a poor prognosis. *Breast Cancer Res.* **2016**, *18*. [CrossRef]

58. Ross, K.; Pailler, E.; Faugeroux, V.; Taylor, M.; Oulhen, M.; Auger, N.; Planchard, D.; Soria, J.-C.; Lindsay, C.R.; Besse, B.; et al. The potential diagnostic power of circulating tumor cell analysis for non-small-cell lung cancer. *Expert Rev. Mol. Diagn.* **2015**, *15*, 1605–1629. [CrossRef] [PubMed]
59. Baccelli, I.; Schneeweiss, A.; Riethdorf, S.; Stenzinger, A.; Schillert, A.; Vogel, V.; Klein, C.; Saini, M.; Bäuerle, T.; Wallwiener, M.; et al. Identification of a population of blood circulating tumor cells from breast cancer patients that initiates metastasis in a xenograft assay. *Nat. Biotechnol.* **2013**, *31*, 539–544. [CrossRef]
60. Hidalgo, M.; Amant, F.; Biankin, A.V.; Budinská, E.; Byrne, A.T.; Caldas, C.; Clarke, R.B.; de Jong, S.; Jonkers, J.; Mælandsmo, G.M.; et al. Patient Derived Xenograft Models: An Emerging Platform for Translational Cancer Research. *Cancer Discov.* **2014**, *4*, 998–1013. [CrossRef]
61. Byrne, A.T.; Alférez, D.G.; Amant, F.; Annibali, D.; Arribas, J.; Biankin, A.V.; Bruna, A.; Budinská, E.; Caldas, C.; Chang, D.K.; et al. Interrogating open issues in cancer precision medicine with patient-derived xenografts. *Nat. Rev. Cancer* **2017**, *17*, 254–268. [CrossRef] [PubMed]
62. Blackhall, F.; Frese, K.K.; Simpson, K.; Kilgour, E.; Brady, G.; Dive, C. Will liquid biopsies improve outcomes for patients with small-cell lung cancer? *Lancet Oncol.* **2018**, *19*, e470–e481. [CrossRef]
63. Pantel, K.; Alix-Panabieres, C. Functional Studies on Viable Circulating Tumor Cells. *Clin. Chem.* **2016**, *62*, 328–334. [CrossRef] [PubMed]
64. Lallo, A.; Schenk, M.W.; Frese, K.K.; Blackhall, F.; Dive, C. Circulating tumor cells and CDX models as a tool for preclinical drug development. *Transl. Lung Cancer Res.* **2017**, *6*, 397–408. [CrossRef] [PubMed]
65. Pereira-Veiga, T.; Abreu, M.; Robledo, D.; Matias-Guiu, X.; Santacana, M.; Sánchez, L.; Cueva, J.; Palacios, P.; Abdulkader, I.; López-López, R.; et al. CTCs-derived xenograft development in a triple negative breast cancer case. *Int. J. Cancer* **2018**, *144*, 2254–2265. [CrossRef]
66. Vishnoi, M.; Haowen Liu, N.; Yin, W.; Boral, D.; Scamardo, A.; Hong, D.; Marchetti, D. The identification of a TNBC liver metastasis gene signature by sequential CTC-xenograft modelling. *Mol. Oncol.* **2019**. [CrossRef]
67. Girotti, M.R.; Gremel, G.; Lee, R.; Galvani, E.; Rothwell, D.; Viros, A.; Mandal, A.K.; Lim, K.H.J.; Saturno, G.; Furney, S.J.; et al. Application of Sequencing, Liquid Biopsies, and Patient-Derived Xenografts for Personalized Medicine in Melanoma. *Cancer Discov.* **2016**, *6*, 286–299. [CrossRef]
68. Hodgkinson, C.L.; Morrow, C.J.; Li, Y.; Metcalf, R.L.; Rothwell, D.G.; Trapani, F.; Polanski, R.; Burt, D.J.; Simpson, K.L.; Morris, K.; et al. Tumorigenicity and genetic profiling of circulating tumor cells in small-cell lung cancer. *Nat. Med.* **2014**, *20*, 897–903. [CrossRef]
69. Lallo, A.; Gulati, S.; Schenk, M.W.; Khandelwal, G.; Berglund, U.W.; Pateras, I.S.; Chester, C.P.E.; Pham, T.M.; Kalderen, C.; Frese, K.K.; et al. Ex vivo culture of cells derived from circulating tumour cell xenograft to support small cell lung cancer research and experimental therapeutics. *Br. J. Pharmacol.* **2019**, *176*, 436–450. [CrossRef]
70. Drapkin, B.J.; George, J.; Christensen, C.L.; Mino-Kenudson, M.; Dries, R.; Sundaresan, T.; Phat, S.; Myers, D.T.; Zhong, J.; Igo, P.; et al. Genomic and Functional Fidelity of Small Cell Lung Cancer Patient-Derived Xenografts. *Cancer Discov.* **2018**, *8*, 600–615. [CrossRef]
71. Morrow, C.J.; Trapani, F.; Metcalf, R.L.; Bertolini, G.; Hodgkinson, C.L.; Khandelwal, G.; Kelly, P.; Galvin, M.; Carter, L.; Simpson, K.L.; et al. Tumourigenic non-small-cell lung cancer mesenchymal circulating tumour cells: A clinical case study. *Ann. Oncol.* **2016**, *27*, 1155–1160. [CrossRef] [PubMed]
72. Andree, K.C.; Mentink, A.; Zeune, L.L.; Terstappen, L.W.M.M.; Stoecklein, N.H.; Neves, R.P.; Driemel, C.; Lampignano, R.; Yang, L.; Neubauer, H.; et al. Toward a real liquid biopsy in metastatic breast and prostate cancer: Diagnostic LeukApheresis increases CTC yields in a European prospective multicenter study (CTCTrap). *Int. J. Cancer* **2018**, *143*, 2584–2591. [CrossRef] [PubMed]
73. Faugeroux, V.; Pailler, E.; Deas, O.; Brulle-Soumare, L.; Hervieu, C.; Marty, V.; Alexandrova, K.; Andree, K.C.; Stoecklein, N.H.; Tramalloni, D.; et al. Genetic characterization of a Unique Neuroendocrine Transdifferentiation Prostate Circulating Tumor Cell - Derived eXplant (CDX) Model. *Nat. Commun.* **2019**. under review.
74. Faugeroux, V.; Pailler, E.; Deas, O.; Michels, J.; Mezquita, L.; Brulle-Soumare, L.; Cairo, S.; Scoazec, J.-Y.; Marty, V.; Queffelec, P.; et al. Development and characterization of novel non-small cell lung cancer (NSCLC) Circulating Tumor Cells (CTCs)-derived xenograft (CDX) models. In Proceedings of the AACR Annual Meeting 2018, Chicago, IL, USA, 14–18 April 2018.

75. Zhang, L.; Ridgway, L.D.; Wetzel, M.D.; Ngo, J.; Yin, W.; Kumar, D.; Goodman, J.C.; Groves, M.D.; Marchetti, D. The Identification and Characterization of Breast Cancer CTCs Competent for Brain Metastasis. *Sci. Transl. Med.* **2013**, *5*, 180ra48. [CrossRef] [PubMed]
76. Yu, M.; Bardia, A.; Aceto, N.; Bersani, F.; Madden, M.W.; Donaldson, M.C.; Desai, R.; Zhu, H.; Comaills, V.; Zheng, Z.; et al. Ex vivo culture of circulating breast tumor cells for individualized testing of drug susceptibility. *Science* **2014**, *345*, 216–220. [CrossRef] [PubMed]
77. Gao, D.; Vela, I.; Sboner, A.; Iaquinta, P.J.; Karthaus, W.R.; Gopalan, A.; Dowling, C.; Wanjala, J.N.; Undvall, E.A.; Arora, V.K.; et al. Organoid Cultures Derived from Patients with Advanced Prostate Cancer. *Cell* **2014**, *159*, 176–187. [CrossRef]
78. Zhang, Z.; Shiratsuchi, H.; Lin, J.; Chen, G.; Reddy, R.M.; Azizi, E.; Fouladdel, S.; Chang, A.C.; Lin, L.; Jiang, H.; et al. Expansion of CTCs from early stage lung cancer patients using a microfluidic co-culture model. *Oncotarget* **2014**, *5*, 12383–12397. [CrossRef]
79. Tellez-Gabriel, M.; Cochonneau, D.; Cadé, M.; Jubelin, C.; Heymann, M.-F.; Heymann, D. Circulating Tumor Cell-Derived Pre-Clinical Models for Personalized Medicine. *Cancers* **2018**, *11*, 19. [CrossRef]
80. Khanna, C. Modeling metastasis in vivo. *Carcinogenesis* **2004**, *26*, 513–523. [CrossRef]
81. Alix-Panabières, C.; Pantel, K. Challenges in circulating tumour cell research. *Nat. Rev. Cancer* **2014**, *14*, 623–631. [CrossRef]
82. Eirew, P.; Steif, A.; Khattra, J.; Ha, G.; Yap, D.; Farahani, H.; Gelmon, K.; Chia, S.; Mar, C.; Wan, A.; et al. Dynamics of genomic clones in breast cancer patient xenografts at single-cell resolution. *Nature* **2015**, *518*, 422–426. [CrossRef] [PubMed]
83. Stoletov, K.; Willetts, L.; Paproski, R.J.; Bond, D.J.; Raha, S.; Jovel, J.; Adam, B.; Robertson, A.E.; Wong, F.; Woolner, E.; et al. Quantitative in vivo whole genome motility screen reveals novel therapeutic targets to block cancer metastasis. *Nat. Commun.* **2018**, *9*. [CrossRef] [PubMed]
84. Praharaj, P.P.; Bhutia, S.K.; Nagrath, S.; Bitting, R.L.; Deep, G. Circulating tumor cell-derived organoids: Current challenges and promises in medical research and precision medicine. *Biochim. Biophys. Acta BBA - Rev. Cancer* **2018**, *1869*, 117–127. [CrossRef] [PubMed]
85. Nguyen, M.; De Ninno, A.; Mencattini, A.; Mermet-Meillon, F.; Fornabaio, G.; Evans, S.S.; Cossutta, M.; Khira, Y.; Han, W.; Sirven, P.; et al. Dissecting Effects of Anti-cancer Drugs and Cancer-Associated Fibroblasts by On-Chip Reconstitution of Immunocompetent Tumor Microenvironments. *Cell Rep.* **2018**, *25*, 3884–3893.e3. [CrossRef] [PubMed]

© 2019 by the authors. Licensee MDPI, Basel, Switzerland. This article is an open access article distributed under the terms and conditions of the Creative Commons Attribution (CC BY) license (http://creativecommons.org/licenses/by/4.0/).

Review

CTCs 2020: Great Expectations or Unreasonable Dreams

Elisabetta Rossi [1,2] and Francesco Fabbri [3,*]

1. Department of Surgery, Oncology and Gastroenterology, Oncology Section, University of Padova, 35124 Padova, Italy
2. Veneto Institute of Oncology IOV-IRCCS, 35128 Padua, Italy
3. Istituto Scientifico Romagnolo per lo Studio e la Cura dei Tumori (IRST) IRCCS, 47014 Meldola, Italy
* Correspondence: francesco.fabbri@irst.emr.it

Received: 19 July 2019; Accepted: 22 August 2019; Published: 27 August 2019

Abstract: Circulating tumor cells (CTCs) are cellular elements that can be scattered into the bloodstream from primary cancer, metastasis, and even from a disseminated tumor cell (DTC) reservoir. CTCs are "seeds", able to give rise to new metastatic lesions. Since metastases are the cause of about 90% of cancer-related deaths, the significance of CTCs is unquestionable. However, two major issues have stalled their full clinical exploitation: rarity and heterogeneity. Therefore, their full clinical potential has only been predicted. Finding new ways of studying and using such tremendously rare and important events can open new areas of research in the field of cancer research, and could drastically improve tumor companion diagnostics, personalized treatment strategies, overall patients management, and reduce healthcare costs.

Keywords: CTC; heterogeneity; liquid biopsy; liquid surgery; clinical utility

1. Introduction

In 2004, Cristofanilli and colleagues reported, for the first time, a trial regarding the level of circulating tumor cells (CTCs) in metastatic breast cancer. The trial results indicated this marker was a useful predictor of progression-free survival (PFS) and overall survival (OS) [1]. In 2007, the American Society of Clinical Oncology (ASCO) cited CTCs and disseminated tumor cells (DTCs) in recommendations on tumor markers. Ten years later, the American Joint Committee of Cancer (AJCC) proposed a new category for TNM staging in breast cancer M0(i+). The M0(i+) is defined as the presence of CTC or DTC, respectively, in the blood or in the bone marrow, in case of absence of clinical or radiological evidence of distant metastases. Fifteen years after the first Cristofanilli study, this marker was validated as a prognostic marker in a number of clinical settings, and few oncology topics have been so intensely investigated. [2–6]. However, due to their biological features, the technical hurdles regarding their investigation, and the lack of methodological standardization that make it difficult to compare different CTC studies [7–11], CTCs have still not reached the Olympus of the "clinical utility value". Their predictive value, which we have been researching for decades, is still far from being fully demonstrated. Therefore, old promises should be carefully reconsidered and new directions should be undertaken. A first step in this direction was reported at the beginning of this year. Bidard and colleagues showed, for the first time, results on actual CTC clinical utility. They observed that CTC count, in ER+HER- metastatic breast cancer (MBC) patients, may be used to objectively choose the best therapy [12]. In addition, as other authors stated, another important challenge is how to trigger a paradigm shift in oncology research: anti-cancer personalized treatments should also strive to hit CTCs, and not only target the solid tumor compartment [13]. Recently, Gkountela and colleagues investigated CTC clusters. They observed that CTC clustering could direct the DNA methylation pattern. Specific FDA-approved drugs may take apart CTC clusters, triggering methylation remodeling and metastasis suppression [14]. This is a clear example of how it may be possible to bridge the gap between a potentially important marker, although still limited (i.e., the current CTC status), and a very useful theranostic approach, i.e., the new potential clinical utility of CTCs. Hence, although a lot of work has to be done, the aim is clear: Demonstrating and validating real CTC clinical utility, and looking for new CTC-targeting theranostic approaches.

2. Being Different, to Be Stronger

CTCs are a major cause of tumor relapse [15]. Disseminated into the bloodstream, CTCs are the "seeds", able to give rise to new metastatic lesions. Hence, their significance is unquestionable. Despite the highly inefficient mechanism of spreading [16], some of them are able to survive into the bloodstream, resist biophysical and cell-mediated insults, and reach their final destination. There, they take root in the pre-metastatic niche, most probably already prepared and fueled by tumor cell-derived messengers. In the niche, CTCs lay dormant for an indefinite period, until some still little known signals trigger their lethal awakening [17–19]. CTCs are extremely rare and heterogeneous elements [20–22]. Although their quantification in the peripheral blood of cancer patients is quite unpredictable, even in patients with advanced disease, they have been found down to 1–10 cells/per mL of peripheral blood of metastatic breast, prostate, and colon cancer patients. Their number may be even lower in other cancers and/or in nonmetastatic setting [23–26]. Secondly, and equally important, CTCs are dynamically heterogeneous. Dynamic heterogeneity (DH) can be described as the characteristic of the tumor to evolve through space and time [20,27–29]. This event can produce solid masses composed of many clones, which may be different in transcriptomic, proteomic, and functional makeup. DH, and in particular that of the most aggressive tumors such as triple negative breast cancers (TNBCs), can be quite striking. It has been shown that no two single cells from TNBCs have an identical genomic profile [28]. CTC subpopulation onset and spreading may follow the solid tumor clonality and adaptability. Indeed, CTCs may acquire DH traits with time, from their onset, starting from a more epithelial-like phenotype and transitioning to a more mesenchymal-like state and vice versa, through processes known as epithelial-to-mesenchymal and mesenchymal-to-epithelial transitions (EMT and MET) [30–35]. In addition, EMT can be observed in distinct transitional states. These EMT states are fairly different and can be detected using cell surface markers and single-cell RNA sequencing. Indeed, EMT-hybrid CTC could have the highest metastatic potential with different degrees of aggressiveness due to their capabilities and proteomic and transcriptomic features [36,37]. These elements suggest that studying only a few biomarkers at a few time points, e.g., at first diagnosis and/or relapse, could only offer a very limited actionable vision of the disease. Monitoring the tumor evolution and progression through a timely and accurate multi-marker detection is a crucial investigation opportunity. CTCs lend themselves to this possibility, being a potential mirror/reflection of the evolving and progressing tumor. CTCs, indeed, are often genotypically, genomically, phenotypically, and functionally different. Some have stem cell features (e.g., CD44- or ABC-G2-positivity), others mesenchymal-like characteristics (e.g., N-Cadherin), others still, hybrids phenotypes [29,38–40]. With an appropriate technical set-up (Gallerani et al., unpublished), we could detect CTCs of all three types, at different sampling times, in the blood of esophageal cancer patients. Despite our current inability to see if other populations were present in the blood, our data agree with previously published reports and confirms that some CTC subgroups may be more dangerous than others may. Furthermore, some CTCs express endothelial markers and they reproduce vascular mimicry (VM), a phenomenon present in several human cancers associated with aggressive diseases. Williamson et al. demonstrated, in small cell lung cancer (SCLC), that these rare cells (VE-cadherin CTC) exhibit essentially the same copy number of gains and losses present in other CTCs and in ctDNA from the same patient, a highly related CNA profile and typical of SCLC [41]. We should also note that several CTC populations that coexist in the bloodstream could establish a relation with other normal blood cells. Recently, several authors demonstrated a crosstalk between different CTCs subpopulations and the immune system [42,43]. It is well-known that CTCs may express the receptor programmed death-ligand 1 (PD-L1) on their membrane [44].

PD-L1 is probably able to suppress the immune response against CTCs, helping their survival in the bloodstream [45,46]. Aceto and colleagues proposed a model where the association between CTCs and neutrophils supports the cell cycle progression during the blood trip [47]. The CTCs with the ability to survive in the bloodstream and to interact with leukocytes and platelets increase the possibility of forming metastasis. To this end, CTCs use several strategies to crosstalk with leukocytes and platelet:

(a) High expression of the immunosuppressive molecule PD-L1, which prevents T cell-mediated destruction [44,48–50];

(b) The expression of CD47, which provides a 'don't eat me' signal [51–53];

(c) An altered expression of the apoptotic FAS and/or FASL proteins that may induce the apoptosis of T cells [54] or protect tumor cells from FAS-mediated apoptosis [55];

(d) The interaction with platelets, which induces EMT-like features in CTCs [56], and promotes tumor cell arrest and extravasation [57];

(e) Platelets promote the survival of CTCs during metastasis by conferring resistance to shear stress and attacks from NK cells [58,59];

(f) Platelet–CTC interaction can lead to the transfer of platelet MHC-I to tumor cells, thereby preventing the identification of NK cells and aiding the CTCs to escape from the cytolytic activity mediated by NK cells [60].

Furthermore, a $CD14^+CD11c^+CD45^+$ myeloid subpopulation has been observed inside circulating tumor microemboli (CTM) or free in the bloodstream. Adams and colleagues named these cells 'cancer-associated macrophage-like cells' (CAMLs). Typically, these cells are giant (30–300 μm in length) with big multiple or polylobulated nuclei (14–64 μm in diameter) [61]. Adams et al. showed that CAMLs could be detected in the peripheral blood of patients with breast, prostate, pancreas, and lung cancer in percentages ranging from 81% to 97% of the total number of patients, contrary to healthy subjects, in which they are totally absent [61,62]. CAMLs also express epithelial markers, such as EpCAM and/or CK8/18/19. We still need to clarify if they directly express these epithelial markers (expression levels depend on the differentiation stage), or if they phagocytize material of epithelial origin. CAMLs seem to originate in the primary tumor and increase in blood samples from patients responding to radiotherapy, chemotherapy, or other treatments [63]. However, CAMLs have also been shown to actively interact with CTCs, or have a proangiogenic activity, as CD146 and TIE2 markers suggest [61]. In support of a pro-tumor role of these cells, in metastatic breast cancer patients, the presence of EpCAM+ CAMLs correlates with a shorter OS and PFS [64]. We could suppose that CAMLs might have an active role in helping CTC intravasation, extravasation, or survival in the bloodstream, thus participating in the metastatic process. The high degree of heterogeneity and complex relationships between CTC subpopulations and other blood "resident" cells, makes studying this biomarker more complex. It offers the opportunity to act both with drugs that affect CTCs, but also with drugs that limit the formation of clusters, as already mentioned [14], or with anticoagulant agents or those acting by stimulating the immune system to recognize CTCs [65,66].

Another pivotal aspect is the contribution of circulating cancer stem cells (C-CSCs), or stem-like CTCs. Cancer stem cells are cellular elements thought to propel cancer progression and to be responsible for the rooting of the disease in primary and metastatic sites, and hence, for local and distant recurrence. Although not fully determined, C-CSCs may derive from CSCs and have been identified in numerous types of cancer [40,67,68]. If CTCs can be seen as the overall population of tumor "seeds", almost an epiphenomenon of cancer spreading, C-CSCs can be described as the actual first cause of metastasis, the proverbial needle in a haystack, since such events are even rarer than general CTCs. Furthermore, a close relation between EMT and C-CSCs has been shown, and an enhanced metastatic competence and a high drug resistance capability have also been described for these cells [69,70]. Despite being quite controversial elements, difficult to characterize specifically and unequivocally, C-CSCs may be considered key players in the genesis of cancer metastasis, and may clearly fulfill the model by Paget [16,71–75]. Specific C-CSC markers are difficult to find, but in general, some of them are fairly well established, e.g., CD133, CD44 (including different isoforms),

ALDH1, and ABC transporters. A number of these markers have also been investigated in CTCs, also aimed at finding diagnostic, prognostic, and predictive indicators. None have been found, validated, and progressively translated into the clinical setting [25]. In 2017, Whang and colleagues showed that CTCs could act as potential precursor cells of metastasis and not only act in the intermediate step of metastatic cascade. Probably due to a stress response, CTCs could evolve in cells that were more aggressive and became tumor-initiating cells, probably different from the CSC-like cells present in the primary tumor [76]. The metastatic potential of EpCAM-positive CTCs has been confirmed using xenograft models in several papers [52,77,78]. A subset of CTCs showed a stem cell-like phenotype, these C-CSC have been detected in different kinds of cancer such as lung cancer with expression of BMI1, hepatocarcinoma where C-CSCs are CD90+CXCR4+, colon rectal cancer with CD44v9+ cells. In breast cancer one or more EMT markers, such as *TWIST*, *AKT2*, *PI3K-alpha*, *ALDH1*, were detectable on CTCs. This CTC subset highlights therapy resistance and a poor prognosis [68,79].

As already mentioned earlier, important papers have recently been published. Mirroring C-CSC potentialities, CTC-clusters have been demonstrated to be 23 to 50 times more capable of giving rise to actual metastases than single CTCs [80]. This capability has been explained by uncovering the link between the CTC cluster state and an increased accessibility of stemness-related transcription factors. CTC clustering is related to a DNA methylation pattern that promotes stemness and metastasis [14]. Again, the same group published a study on the striking importance of the cooperation between "normal" and cancer cells. The association of CTCs with neutrophils induces a proliferative boost, making them more competent in metastasis formation [47,81,82].

In conclusion, a number of elements may induce a CTC to gain a CSC-like cluster-state with specific cellular, molecular, and functional phenotypes and capabilities. More than ten years ago, the exposome was described as the overall environmental complement to the genome in determining the risk of a disease. It can be seen as the totality of exposures throughout one's lifespan [83]. Hence, it can be inferred that in addition to genetic predisposition, every molecule or element that can prompt tumor onset and growth can be seen as exposome, or, more specifically, cancer exposome [Fabbri et al., under review]. Consequently, cancer exposome could include inflammation-related molecules, "normal" cellular elements associated to cancer progression (e.g., CAMs, CAFs, and platelets), environmental factors, and other molecules, i.e., elements known to be connected to tumor development, such as induction of EMT, C-CSCs, and progression. It becomes clear that the cancer exposome could contributes to the onset of CSC-like CTCs. Only a focused and multidisciplinary approach may truly help in finding innovative paths to unravel the enigmatic Gordian knot that may well represent cancer metastasis. In the light of recent results and these exposome-related concepts, we have to strive to find innovative treatments to suppress the spread of cancer.

3. Clinical Data: Hopes of Utility

In clinical practice, after pre- and analytical standardizations and diagnostic accuracy evaluations, a test has to demonstrate that it leads to health benefits [84] before being introduced in the decisional path.

Regarding CTCs, the multivariate analysis demonstrated that the CTCs count was the strongest prognostic biomarkers for patients survival [3,85]. For this reason, CTCs could be used to stratify the disease, but this biomarker is rarely used in clinical practice. Budd and colleagues studied 138 MBC patients, who underwent CT scans before and after start treatment. They compared the results with the CTC count obtained at baseline and after four weeks from the start of therapy, and demonstrated that CTC count is more reproducible than radiological response. This suggests that CTCs are a superior surrogate endpoint [86]. In metastatic breast cancer, the first trial designed to study the clinical utility of CTC count, SWOG S0500, was not able to find a survival difference for patients stratified using CTC count. This negative result was because the study suffered from a design fault and not a failure of CTC test [87,88]. In fact, cancers that show early resistance to first line chemotherapy are likely to be resistant to second line chemotherapy, whenever the second line is started. Moreover, it had already been reported that early changes of first line chemotherapy never demonstrated a gain in OS, whatever the technique used to guide the early change (functional imaging, etc.) [89]. Therefore, SWOG S0500 should not be regarded as the final proof that CTCs are not useful in patients' management. However, because of these negative results, the 2015 American Society of Clinical Oncology, in its clinical practice guidelines, did not recommend the use of CTC count in women with metastatic BC [90]. In Europe, two other clinical utility trials based on CTC count are currently ongoing:

- The "CirCe01" trial (NCT01349842), similarly to SWOG0500, assesses early changes of CTC count during treatment in metastatic patients; patients were enrolled before the start of third line of chemotherapy (CT) and followed with the CTC test throughout the successive lines of CT.
- The "STIC CTC" trial (NCT01710605) investigated the clinical utility of the prognostic value of baseline CTC count. In this trial, patients were randomized in two arms: In the first arm, clinically driven patient's treatment choose between CT and hormone therapy (HT) (CTC count not disclosed); in the second arm, CTC count driven treatment chooses patients. In fact, such as first-line treatment, patients with CTC ≥5/7.5 mL received CT, whereas patients with CTC <5/7.5 mL received HT.

As already discussed, this last trial demonstrated the clinical utility of CTC count in first line therapy in ER+HER2-MBC. Although results are awaiting a definitive publication, Bidard and colleagues have shown that in the majority of patients, the treatment chosen based on the CTC count was the same as the treatment chosen based on the clinic. On the contrary, in discrepant cases, CTC count may be more reliable for either escalating (i.e., considering CT in patients with high CTC count) or de-escalating (i.e., considering HT in patients with low CTC count) first line therapies [12]. SWOG0500 and CirCe01 have similar designs. Both studies evaluated CTCs at baseline and after treatment. In particular, CirCe01 evaluated CTCs number after every new line of CT in patients randomized in the CTC arm. The number of therapy lines represent the main difference between these two studies. In fact, CirCe01-enrolled patients about to start third line therapy (a population with a high chemo-resistance prevalence). On the contrary, SWOG500 enrolled patients just before first line CT. Moreover, CirCe01 trials, in nonrandomized run-in phase, also wanted to establish a CTC threshold to be used in the randomized part of the study. The STIC CTC, a randomized trial, focused on evaluating the health economic interest of taking into account CTCs to determine the kind of first line treatment for HR–positive MBC. Major limitations of this trial were a) the lack of standardized clinical criteria for recommending CT in the clinically driven arm, and b) that it was conducted prior to the widespread use of CDK4/6 inhibitors, now regarded as the standard of care for HR–positive MBC. The SWOG0500 was the first trial with a goal regarding the clinical utility of CTCs. However, the heterogeneity of first line treatments allowed in this trial might have variably influenced CTC behavior. Changing treatment to a different CT agent after the rise of early chemo-resistance may be not significantly effective, and this most probably influenced the results of the SWOG0500 [91]. A similar problem, given the design of the study, could have occurred in CirCE01. The STIC CTC trial had a different study design and, as preliminary results presented at the San Antonio Breast Cancer Symposium showed, it suggested that the CTC count could allow for identifying patients in whom the dose of therapy can be scaled. If the preliminary results of this study are confirmed, this could lead to a benefit for patients who will be able to see reduced side effects of the treatment and, in the long term, an economic saving for the national health systems.

Although still not incorporating recently proposed drugs, e.g., cyclin-dependent kinase 4/6 inhibitors, and indicating to escalate therapy, while it is usually suggested to try to de-escalate regimens in this clinical setting, the STIC CTC trial is the first investigation that showed an actual CTC-based clinical benefit. Hopefully, this huge step should be the first of a long series, prompting further investigations and highlighting the pivotal importance of clearly and specifically designed future clinical studies not to fall into previous misleading results.

Several studies evaluated the impact of CTC count and CTC characterization in monitoring treatment. Indeed, in metastatic castration-resistant prostate cancer (mCRPC), CTC elimination (CTC0) after short-term treatment proved to be an early response endpoint [92]. In this study, 3000 men were enrolled in five phase III trials. The response measures were evaluated regardless of the specific intervention and the changes in CTC status from CTC-positive (T baseline) to CTC-negative (after 13 weeks). This change was shown to be strongly associated with longer survival. Interestingly, the percentage change from baseline of prostate specific antigen (PSA), which is more widely used as end-point, did not discriminate survival at the same level of CTC0. In four out of five trials, the treatments included HT; HT can induce a PSA-level modulation, independent of an effect on cell killing. This HT-based effect limits the role of post-therapy PSA change measures as a reliable indicator of efficacy. In this study, CTCs allowed researchers to determine treatment response seven to eight weeks earlier than determined by standard RECIST criteria and PSA. For these reasons, CTC number, a biomarker that is not affected by modulations in androgen receptor signaling, should be included in clinical management of patients. It is important to note that this result could allow for the prevention of drug toxicity and cause a significant health savings [12,93–95].

4. Challenging the Current Paradigm: Liquid Surgery Premises and Hopes

So far, systemic therapies have aimed at preventing relapse after the removal of the primary tumor and delaying the already present metastatic tumor, confidently eliminating therapy-resistant clones and minimal residual disease. To contrast cancer dissemination, preventing metastasis in the first place, we think it is necessary to expand the conventional line of thinking that aims to hit an already widespread disease, and start new paths that anticipate its diffusion, e.g., directly and therapeutically targeting CTCs.

The investigation of the CTCs, and in general everything that can be included in the term liquid biopsy (LB), has aimed at simple, fast, and cost-effective monitoring of disease status or response to treatment. LB is less troublesome than tissue biopsy, thanks to the ease of access to body fluids, in particular when taking a tissue biopsy is often clinically impossible or not recommended [96]. Moreover, as already suggested, the tumor may change and tissue biopsies may not appropriately reflect the complex profile of a tumor disease, because of its dynamic heterogeneity (DH), which can only be addressed by taking biopsies from different tumor areas at different time points. Hence, LB may offer a more comprehensive cross-section of the disease.

LB can be a crucial alternative and/or addition to conventional and tissue-based diagnostic procedures. Targeting CTCs in the blood of patients could control or avoid metastasis spreading and the risk of relapse, just like the surgical removal of solid tumor can limit the disease or even cure a cancer patient. Therefore, two major lines of research could be followed to deplete CTCs as much as possible, striving not to leave any cells with actual metastasis-generating properties. One possible approach is to examine, in depth, CTC biology in order to identify actionable targets that could lead to their depletion and/or to make them harmless, as already cited [14,97]. A second possibility is to find new paths that can exploit already well-known aspects of CTCs: Their biological features, i.e., surface marker expression, dimension, clusterization potential, and their ecosystem, the blood flow.

Both ways are necessary to identify better-tailored CTC-based anticancer treatments for individual patients. The small number of CTCs per mL of peripheral blood is the major challenging physical limit. Although it may appear to be a simple solution, it is increasingly clear that it is necessary to extend the volumes of blood that should be analyzed and/or treated. Then, new approaches to overcome this obstacle, i.e., in vivo CTC collection and/or count devices, and apheresis-based procedures, are under investigation. The GILUPI CellCollector (CC) allows collecting CTC directly from the blood, in vivo. It is a functionalized medical wire covered with anti-EpCAM antibodies that, when inserted in the cubital vein of a patient for 30 min, could trap CTCs from up to one liter and a half of blood. Studies reported an increased number of CTCs detected compared to standard methods [98–100]. Recently, Didzdar and colleagues investigated the prognostic impact of CTCs rescued with CC in colorectal cancer (CRC) [101], but in contrast to other studies, they did not report prognostic significance of CC-CTC. The authors estimated that recovering CTCs with the CC is similar to extracting tumor cells from a blood volume range of 0.33–18 mL, a different volume compared to what was suggested in previous studies (volume range 1–3 L). Also, in our experience, the median number of CTCs retrieved from the blood of metastatic non-small lung cancer (NSCLC) patients was quite low (around three cells per CC device/patient (Fabbri et al., unpublished data). So, despite being a feasible [102] and promising idea, it seems that the CC has not maintained what it had predicted. In June 2019, Galanzha et al. exploited a photoacoustic (PA) flow cytometry (PAFC) platform (cytophone platform) to count circulating melanoma cells in vivo. This method allows a noninvasive assessment of a large volume of blood using the PA effect as a transformation of absorbent laser pulse energy into sound through thermoelastic expansion of light objects. The probe is placed on the patients' skin above the selected vein. The authors utilized 18 melanoma patients as the training set and another 10 new melanoma patients as the validation set. This system was able to detect circulating single melanoma cells, circulating clusters of melanoma cells, and it could discriminate between CMC cluster and circulating blood clots (CBCs). The PAFC detection limit achieved in vivo is about 1 CTC/1 L. The PACF-based CTC count needs further studies to confirm the great potential of this noninvasive method [103].

To avoid the bottleneck due to the low number of CTCs, the CTCTrap consortium and the Nick Stoecklein group used and validated the diagnostic leukapheresis (DLA) in order to enrich a higher number of CTCs. The CTCTrap efforts demonstrated that the procedure used allowed researchers to improve the number of CTCs isolated in MBC and metastatic prostate cancer (MPC), and that the procedure could be safely performed in different clinical sites [104]. The DLA product analyzed with Cell Search demonstrated a 30-fold increase in median CTC numbers detected [105] and an increase in the number of CTC-positive patients as well as M0 breast cancer patients [106]. This high number of enriched CTCs was exploited for the single cell molecular characterizations of CTCs [107] and also to culture CTCs [106]. Despite DLA enrichment, the CTC-cultured success is limited to few patients. Only in two out of eight patients were CTCs cultured from fresh and cryopreserved DLA [106]. The culture success rate was reached in samples with more than 300 CTCs, in agreement with other group results; this is probably because DLA samples with high CTC numbers and lower ratios of apoptotic CTCs were more likely to grow in culture. [106,108].

Usually, therapeutic apheresis is used for the rapid elimination of harmful or excessive blood components such as plasma proteins (plasma exchange) or for the harvest or elimination of cells (leukapheresis and platelet apheresis) and has found broad applications in a vast array of hematologic and onco-hematologic disorders. Keeping these reasoning in mind, removing CTC by filtration and adsorption, and reintroducing the blood without tumor cells back into the body would maximize CTC recovery from individual patients.

Statistically, it has been estimated that to find at least one CTC, the total blood volume (TBV) has to be analyzed, and that no less than 7.5 L of blood are necessary to detect at least 10 CTCs in almost every patient [109]. It is generally accepted that the TBV, i.e., the overall amount of fluid circulating within the arteries, capillaries, veins, and chambers of the heart, at any time in a typical adult human, is around 5 liters [110]. Interestingly, before metastasis onset, the presence of about 9 ± 6 CTC/L of blood has been mathematically predicted even in M0 breast cancer patients [111]. Scaling up this estimation, in low CTC number patients also, analyzing up to 72 L of blood would result in finding at least 200 to almost 1000 CTCs or more. Therefore, in order to treat a patient by eliminating up to all of the CTCs, it would be mandatory to perform an approach involving: a) A huge increase in the volume of blood to process, i.e., more than one TBV; b) a device that can catch all of the CTC subpopulations and that will not clog; and c) redirecting the blood into the patient after CTC removal. Although highly pioneering, we can predict that the removal of CTCs from the blood of a patient could control or even avoid metastasis spreading and the risk of relapse, just as surgical removal of a solid tumor can limit the disease or even cure a cancer patient. Such a kind of approach could be called liquid surgery (LS). A medical device able to eliminate CTCs from the blood circulation could substantially improve cancer management decreasing the overall disease burden, increasing OS and PFS, and could be used to support clinical decisions, as well as a higher DLA-based CTC retrieval [112,113]. In a long-term vision, it could even lower the need for high-cost therapeutic approaches and decrease therapy related toxicities. Preclinical tests and experimental set ups are currently running in our laboratories in order to improve and optimize such an approach. In our long-term vision, LB approaches will guide the decision of executing LS based on CTC presence in peripheral blood, exactly as a solid tumor is surgically removed after a standard tissue biopsy has revealed its presence and nature. LB could also monitor CTC levels after a LS procedure to evaluate depletion efficiency. Although reasonably invasive, the LS will be clinically feasible, not harmful, and manageable, comparable to a standard hemodialytic procedure used for kidney failure. The benefits of limiting, or even avoiding, metastasis onset have high preclinical and clinical potential [114,115] and will most probably exceed the possible drawbacks of extracorporeal circulation. Just like hemodialysis, a medical specialist will decide when the LS is needed, the therapeutic regimen, and the various parameters for the treatment. These would include frequency (e.g., treatments per week), duration of each treatment (e.g., up to 4 h or up to six months), and blood flow rates. As a maximum, up to three treatments per week, for up to 1–4 h per treatment, repeated during a period of up to six months, similar and in addition to an adjuvant therapeutic regimen, is predictable. Considering clinical parameters and patient conditions, the procedure could be repeated one or more times per patient per week, similarly to a standard chemotherapy infusion regimen. Of course, this pioneering and challenging idea has to be investigated in depth, tested, and validated, in vitro and in vivo, before becoming a reasonable treatment option that can be examined in a clinical setting. However, it remains a thrilling breakthrough hypothesis that should be put to the test.

5. Conclusions

In these last fifteen years, CTCs have been intensively studied in the field of cancer biology. CTC evaluation may provide clinically relevant and valuable information regarding cancer. The biological information gained from CTCs will be extremely important for opening new research fields, accurately targeting CTCs, and discovering new CTC-based treatment strategies. Multidisciplinary research approaches aimed at the overall "blood ecosystem", an ecosystem for the CTCs and the metastatic cascade, are increasingly mandatory to fulfill this challenge. Monitoring CTC counts during therapy is a tool that may allow the assessment of disease development in real time, even prior to overt clinical signs of relapse. Targeting CTCs should become a translational objective that could pave the way towards increasing survival outcomes and reducing distal recurrence, preventing metastasis before it occurs. In the era of tailored therapy, precision oncology increases the range of treatment options, but to date, only a relatively small number of people could benefit. Moreover, to apply precision oncology is very expensive. The high cost is due to screening patients for tailored treatments and designing drugs that would be utilized in single patients or in limited groups of patients. We should not underestimate that CTCs could improve the health system costs and savings if the CTC-based diagnostic and therapeutic application could guide (a) better patient stratification, (b) first appropriate selection of personalized regimen, and (c) early treatment discontinuation and/or switch. The theranostic use of CTCs, which may be called LS, could establish a completely new area of research in the field of cancer research and management, and trigger the further improvement of CTC-based companion diagnostics.

Author Contributions: E.R. and F.F. equally contributed to manuscript conceptualization, writing, original draft preparation, and review and editing.

Funding: The concepts developed in this review were derived from IRST and Programma Operativo FESR Regione Emilia-Romagna 2014/2020, prot. num. PG/2018/630591 (Development of a device for the elimination of Circulating Tumor Cells: CLEAR, the CTC targeted Liquid surgEry AppaRatus), from the work performed at the CTC laboratory of IOV-IRCCS, Padova, Italy, and partially funded by intramural 5x1000, IOV-Sinergia tra Oncologia e Clinica (NSCLC MUT and MET P.I.: E.Rossi) and FP7-HEALTH-2012-INNOVATION (CTC-TRAP:EUFP7 grant #305341).

Conflicts of Interest: The authors declare no conflict of interest. The funders had no role in preparation of the manuscript.

References

1. Cristofanilli, M.; Budd, G.T.; Ellis, M.J.; Stopeck, A.; Matera, J.; Miller, M.C.; Reuben, J.M.; Doyle, G.V.; Allard, W.J.; Terstappen, L.W.M.M.; et al. Circulating tumor cells, disease progression, and survival in metastatic breast cancer. *N. Engl. J. Med.* **2004**, *351*, 781–791. [CrossRef]
2. Cabel, L.; Proudhon, C.; Gortais, H.; Loirat, D.; Coussy, F.; Pierga, J.Y.; Bidard, F.C. Circulating tumor cells: Clinical validity and utility. *Int. J. Clin. Oncol.* **2017**, *22*, 421–430. [CrossRef]
3. Bidard, F.-C.; Peeters, D.J.; Fehm, T.; Nolé, F.; Gisbert-Criado, R.; Mavroudis, D.; Grisanti, S.; Generali, D.; Garcia-Saenz, J.A.; Stebbing, J.; et al. Clinical validity of circulating tumour cells in patients with metastatic breast cancer: A pooled analysis of individual patient data. *Lancet Oncol.* **2014**, *15*, 406–414. [CrossRef]
4. Huang, X.; Gao, P.; Song, Y.; Sun, J.; Chen, X.; Zhao, J.; Xu, H.; Wang, Z. Meta-analysis of the prognostic value of circulating tumor cells detected with the CellSearch System in colorectal cancer. *BMC Cancer* **2015**, *15*, 202. [CrossRef]
5. Scher, H.I.; Heller, G.; Molina, A.; Attard, G.; Danila, D.C.; Jia, X.; Peng, W.; Sandhu, S.K.; Olmos, D.; Riisnaes, R.; et al. Circulating tumor cell biomarker panel as an individual-level surrogate for survival in metastatic castration-resistant prostate cancer. *J. Clin. Oncol.* **2015**, *33*, 1348–1355. [CrossRef]
6. Punnoose, E.A.; Atwal, S.; Liu, W.; Raja, R.; Fine, B.M.; Hughes, B.G.M.; Hicks, R.J.; Hampton, G.M.; Amler, L.C.; Pirzkall, A.; et al. Evaluation of circulating tumor cells and circulating tumor DNA in non-small cell lung cancer: Association with clinical endpoints in a phase II clinical trial of pertuzumab and erlotinib. *Clin. Cancer Res.* **2012**, *18*, 2391–2401. [CrossRef]
7. Rossi, G.; Ignatiadis, M. Promises and Pitfalls of Using Liquid Biopsy for Precision Medicine. *Cancer Res.* **2019**, *79*, 2798–2804. [CrossRef]
8. Bünger, S.; Zimmermann, M.; Habermann, J.K. Diversity of assessing circulating tumor cells (CTCs) emphasizes need for standardization: A CTC Guide to design and report trials. *Cancer Metastasis Rev.* **2015**, *34*, 527–545. [CrossRef]
9. Gallerani, G.; Fabbri, F. Circulating tumor cells in the adenocarcinoma of the esophagus. *Int. J. Mol. Sci.* **2016**, *17*, 1266. [CrossRef]
10. Lianidou, E.S. Circulating tumor cell isolation: A marathon race worth running. *Clin. Chem.* **2014**, *60*, 287–289. [CrossRef]
11. Parkinson, D.R.; Dracopoli, N.; Gumbs Petty, B.; Compton, C.; Cristofanilli, M.; Deisseroth, A.; Hayes, D.F.; Kapke, G.; Kumar, P.; Lee, J.S.; et al. Considerations in the development of circulating tumor cell technology for clinical use. *J. Transl. Med.* **2012**, *10*, 138. [CrossRef]
12. Bidard, F.-C.; Jacot, W.; Dureau, S.; Brain, E.; Bachelot, T.; Bourgeois, H.; Goncalves, A.; Ladoire, S.; Naman, H.; Dalenc, F.; et al. Abstract GS3-07: Clinical utility of circulating tumor cell count as a tool to chose between first line hormone therapy and chemotherapy for ER+ HER2- metastatic breast cancer: Results of the phase III STIC CTC trial. *Cancer Res.* **2019**, *79*. [CrossRef]
13. Dasgupta, A.; Lim, A.R.; Ghajar, C.M. Circulating and disseminated tumor cells: Harbingers or initiators of metastasis? *Mol. Oncol.* **2017**, *11*, 40–61. [CrossRef]
14. Gkountela, S.; Castro-Giner, F.; Szczerba, B.M.; Vetter, M.; Landin, J.; Scherrer, R.; Krol, I.; Scheidmann, M.C.; Beisel, C.; Stirnimann, C.U.; et al. Circulating Tumor Cell Clustering Shapes DNA Methylation to Enable Metastasis Seeding. *Cell* **2019**, *176*, 98–112. [CrossRef]
15. Mitra, A.; Mishra, L.; Li, S. EMT, CTCs and CSCs in tumor relapse and drug-resistance. *Oncotarget* **2015**, *10*, 10697–10711. [CrossRef]
16. Massagué, J.; Obenauf, A.C. Metastatic colonization by circulating tumour cells. *Nature* **2016**, *529*, 298–306. [CrossRef]
17. Joosse, S.A.; Gorges, T.M.; Pantel, K. Biology, detection, and clinical implications of circulating tumor cells. *EMBO Mol. Med.* **2014**, 1–12. [CrossRef]
18. Kang, Y.; Pantel, K. Tumor Cell Dissemination: Emerging Biological Insights from Animal Models and Cancer Patients. *Cancer Cell* **2013**, *23*, 573–581. [CrossRef]
19. Vishnoi, M.; Peddibhotla, S.; Yin, W.; Scamardo, A.T.; George, G.C.; Hong, D.S.; Marchetti, D. The isolation and characterization of CTC subsets related to breast cancer dormancy. *Sci. Rep.* **2015**, *5*, 17533. [CrossRef]
20. Swanton, C. Intratumor heterogeneity: Evolution through space and time. *Cancer Res.* **2012**, *72*, 4875–4882. [CrossRef]

21. Tellez-Gabriel, M.; Heymann, M.-F.; Heymann, D. Circulating Tumor Cells as a Tool for Assessing Tumor Heterogeneity. *Theranostics* **2019**, *9*, 4580–4594. [CrossRef]
22. Reinhardt, F.; Franken, A.; Meier-Stiegen, F.; Driemel, C.; Stoecklein, N.H.; Fischer, J.C.; Niederacher, D.; Ruckhaeberle, E.; Fehm, T.; Neubauer, H. Diagnostic Leukapheresis Enables Reliable Transcriptomic Profiling of Single Circulating Tumor Cells to Characterize Inter-Cellular Heterogeneity in Terms of Endocrine Resistance. *Cancers* **2019**, *11*, 903. [CrossRef]
23. Maltoni, R.; Fici, P.; Amadori, D.; Gallerani, G.; Cocchi, C.; Zoli, M.; Rocca, A.; Cecconetto, L.; Folli, S.; Scarpi, E.; et al. Circulating tumor cells in early breast cancer: A connection with vascular invasion. *Cancer Lett.* **2015**, *10*, 43–48. [CrossRef]
24. Broncy, L.; Paterlini-br, P. Clinical Impact of Circulating Tumor Cells in Patients with Localized Prostate Cancer. *Cells* **2019**, *8*, 676. [CrossRef]
25. Strati, A.; Nikolaou, M.; Georgoulias, V.; Lianidou, E.S. Prognostic Significance of TWIST1, CD24, CD44, and ALDH1 Transcript Quantification in EpCAM-Positive Circulating Tumor Cells from Early Stage Breast Cancer Patients. *Cells* **2019**, *8*, 652. [CrossRef]
26. Magbanua, M.J.M.; Yau, C.; Wolf, D.M.; Lee, J.S.; Chattopadhyay, A.; Scott, J.H.; Bowlby-Yoder, E.; Hwang, E.S.; Alvarado, M.; Ewing, C.A.; et al. Synchronous detection of circulating tumor cells in blood and disseminated tumor cells in bone marrow predict adverse outcome in early breast cancer. *Clin. Cancer Res.* **2019**. [CrossRef]
27. Ellsworth, R.E.; Blackburn, H.L.; Shriver, C.D.; Soon-Shiong, P.; Ellsworth, D.L. Molecular heterogeneity in breast cancer: State of the science and implications for patient care. *Semin. Cell Dev. Biol.* **2017**, *64*, 65–72. [CrossRef]
28. Wang, Y.; Waters, J.; Leung, M.L.; Unruh, A.; Roh, W.; Shi, X.; Chen, K.; Scheet, P.; Vattathil, S.; Liang, H.; et al. Clonal evolution in breast cancer revealed by single nucleus genome sequencing. *Nature* **2014**, *512*, 155–160. [CrossRef]
29. Yu, M.; Bardia, A.; Wittner, B.S.; Stott, S.L.; Smas, M.E.; Ting, D.T.; Isakoff, S.J.; Ciciliano, J.C.; Wells, M.N.; Shah, A.M.; et al. Circulating breast tumor cells exhibit dynamic changes in epithelial and mesenchymal composition. *Science* **2013**, *339*, 580–584. [CrossRef]
30. Gradilone, A.; Raimondi, C.; Nicolazzo, C.; Petracca, A.; Gandini, O.; Vincenzi, B.; Naso, G.; Aglianò, A.M.; Cortesi, E.; Gazzaniga, P. Circulating tumour cells lacking cytokeratin in breast cancer: The importance of being mesenchymal. *J. Cell. Mol. Med.* **2011**, *15*, 1066–1070. [CrossRef]
31. Gorges, T.M.; Tinhofer, I.; Drosch, M.; Roese, L.; Zollner, T.M.; Krahn, T.; von Ahsen, O. Circulating tumour cells escape from EpCAM-based detection due to epithelial-to-mesenchymal transition. *BMC Cancer* **2012**, *12*, 178. [CrossRef] [PubMed]
32. De Craene, B.; Berx, G. Regulatory networks defining EMT during cancer initiation and progression. *Nat. Rev. Cancer* **2013**, *13*, 97–110. [CrossRef] [PubMed]
33. Barriere, G.; Fici, P.; Gallerani, G.; Fabbri, F.; Rigaud, M. Epithelial Mesenchymal Transition: A double-edged sword. *Clin. Transl. Med.* **2015**, *4*, 4–9. [CrossRef] [PubMed]
34. Lowes, L.E.; Allan, A.L. Circulating tumor cells and implications of the epithelial-to-mesenchymal transition. In *Advances in Clinical Chemistry*; Elsevier: Amsterdam, the Nederlands, 2018; Volume 83, pp. 121–181.
35. Francart, M.E.; Lambert, J.; Vanwynsberghe, A.M.; Thompson, E.W.; Bourcy, M.; Polette, M.; Gilles, C. Epithelial–mesenchymal plasticity and circulating tumor cells: Travel companions to metastases. *Dev. Dyn.* **2018**, *247*, 432–450. [CrossRef] [PubMed]
36. Jolly, M.K.; Mani, S.A.; Levine, H. Hybrid epithelial/mesenchymal phenotype(s): The 'fittest' for metastasis? *Biochim. Biophys. Acta Rev. Cancer* **2018**, *1870*, 151–157. [CrossRef] [PubMed]
37. Pastushenko, I.; Blanpain, C. EMT Transition States during Tumor Progression and Metastasis. *Trends Cell Biol.* **2019**, *29*, 212–226. [CrossRef] [PubMed]
38. Mocellin, S.; Keilholz, U.; Rossi, C.R.; Nitti, D. Circulating tumor cells: The "leukemic phase" of solid cancers. *Trends Mol. Med.* **2006**, *12*, 130–139. [CrossRef]
39. Krawczyk, N.; Meier-Stiegen, F.; Banys, M.; Neubauer, H.; Ruckhaeberle, E.; Fehm, T. Expression of Stem Cell and Epithelial-Mesenchymal Transition Markers in Circulating Tumor Cells of Breast Cancer Patients. *Biomed. Res. Int.* **2014**, *2014*, 1–11. [CrossRef]
40. Werner, S.; Stenzl, A.; Pantel, K.; Todenhöfer, T. Expression of epithelial mesenchymal transition and cancer stem cell markers in circulating tumor cells. In *Advances in Experimental Medicine and Biology*; Springer Nature International Publishing: Berlin, Germany, 2017; Volume 994, pp. 205–228.

41. Williamson, S.C.; Metcalf, R.L.; Trapani, F.; Mohan, S.; Antonello, J.; Abbott, B.; Leong, H.S.; Chester, C.P.E.; Simms, N.; Polanski, R.; et al. Vasculogenic mimicry in small cell lung cancer. *Nat. Commun.* **2016**, *7*, 13322. [CrossRef]
42. Buchbinder, E.I.; Desai, A. CTLA-4 and PD-1 Pathways: Similarities, Differences, and Implications of Their Inhibition. *Am. J. Clin. Oncol.* **2016**, *39*, 98–106. [CrossRef]
43. Leone, K.; Poggiana, C.; Zamarchi, R. The Interplay between Circulating Tumor Cells and the Immune System: From Immune Escape to Cancer Immunotherapy. *Diagnostics* **2018**, *8*, 59. [CrossRef]
44. Mazel, M.; Jacot, W.; Pantel, K.; Bartkowiak, K.; Topart, D.; Cayrefourcq, L.; Rossille, D.; Maudelonde, T.; Fest, T.; Alix-Panabières, C. Frequent expression of PD-L1 on circulating breast cancer cells. *Mol. Oncol.* **2015**, 1–10. [CrossRef]
45. Wang, X.; Sun, Q.; Liu, Q.; Wang, C.; Yao, R.; Wang, Y. CTC immune escape mediated by PD-L1. *Med. Hypotheses* **2016**, *93*, 138–139. [CrossRef]
46. Guibert, N.; Delaunay, M.; Lusque, A.; Boubekeur, N.; Rouquette, I.; Clermont, E.; Mourlanette, J.; Gouin, S.; Dormoy, I.; Favre, G.; et al. PD-L1 expression in circulating tumor cells of advanced non-small cell lung cancer patients treated with nivolumab. *Lung Cancer* **2018**, *120*, 108–112. [CrossRef]
47. Szczerba, B.M.; Castro-Giner, F.; Vetter, M.; Krol, I.; Gkountela, S.; Landin, J.; Scheidmann, M.C.; Donato, C.; Scherrer, R.; Singer, J.; et al. Neutrophils escort circulating tumour cells to enable cell cycle progression. *Nature* **2019**, *566*, 553–557. [CrossRef]
48. Nicolazzo, C.; Raimondi, C.; Mancini, M.; Caponnetto, S.; Gradilone, A.; Gandini, O.; Mastromartino, M.; del Bene, G.; Prete, A.; Longo, F.; et al. Monitoring PD-L1 positive circulating tumor cells in non-small cell lung cancer patients treated with the PD-1 inhibitor Nivolumab. *Sci. Rep.* **2016**, *6*, 31726. [CrossRef]
49. Yue, C.; Jiang, Y.; Li, P.; Wang, Y.; Xue, J.; Li, N.; Li, D.; Wang, R.; Dang, Y.; Hu, Z.; et al. Dynamic change of PD-L1 expression on circulating tumor cells in advanced solid tumor patients undergoing PD-1 blockade therapy. *Oncoimmunology* **2018**, *7*, e1438111. [CrossRef]
50. Kallergi, G.; Vetsika, E.-K.; Aggouraki, D.; Lagoudaki, E.; Koutsopoulos, A.; Koinis, F.; Katsarlinos, P.; Trypaki, M.; Messaritakis, I.; Stournaras, C.; et al. Evaluation of PD-L1/PD-1 on circulating tumor cells in patients with advanced non-small cell lung cancer. *Ther. Adv. Med. Oncol.* **2018**, *10*, 1758834017750121. [CrossRef]
51. Chao, M.P.; Tang, C.; Pachynski, R.K.; Chin, R.; Majeti, R.; Weissman, I.L. Extranodal dissemination of non-Hodgkin lymphoma requires CD47 and is inhibited by anti-CD47 antibody therapy. *Blood* **2011**, *118*, 4890–4901. [CrossRef]
52. Baccelli, I.; Schneeweiss, A.; Riethdorf, S.; Stenzinger, A.; Schillert, A.; Vogel, V.; Klein, C.; Saini, M.; Bäurele, T.; Wallwiener, M.; et al. Identification of a population of blood circulating tumor cells from breast cancer patients that initiates metastasis in a xenograft assay. *Nat. Biotechnol.* **2013**, *31*, 539–544. [CrossRef]
53. Steinert, G.; Schölch, S.; Niemietz, T.; Iwata, N.; García, S.A.; Behrens, B.; Voigt, A.; Kloor, M.; Benner, A.; Bork, U.; et al. Immune Escape and Survival Mechanisms in Circulating Tumor Cells of Colorectal Cancer. *Cancer Res.* **2014**, *74*, 1694–1704. [CrossRef]
54. Mego, M.; Gao, H.; Cohen, E.; Anfossi, S.; Giordano, A.; Sanda, T.; Fouad, T.; De Giorgi, U.; Giuliano, M.; Woodward, W.; et al. Circulating Tumor Cells (CTC) Are Associated with Defects in Adaptive Immunity in Patients with Inflammatory Breast Cancer. *J. Cancer* **2016**, *7*, 1095–1104. [CrossRef]
55. Hallermalm, K.; De Geer, A.; Kiessling, R.; Levitsky, V.; Levitskaya, J. Autocrine secretion of Fas ligand shields tumor cells from Fas-mediated killing by cytotoxic lymphocytes. *Cancer Res.* **2004**, *64*, 6775–6782. [CrossRef]
56. Labelle, M.; Begum, S.; Hynes, R.O. Direct signaling between platelets and cancer cells induces an epithelial-mesenchymal-like transition and promotes metastasis. *Cancer Cell* **2011**, *20*, 576–590. [CrossRef]
57. Weber, M.R.; Zuka, M.; Lorger, M.; Tschan, M.; Torbett, B.E.; Zijlstra, A.; Quigley, J.P.; Staflin, K.; Eliceiri, B.P.; Krueger, J.S.; et al. Activated tumor cell integrin $\alpha v \beta 3$ cooperates with platelets to promote extravasation and metastasis from the blood stream. *Thromb. Res.* **2016**, *140*, 27–36. [CrossRef]
58. Cravioto-Villanueva, A.; Luna-Perez, P.; Gutierrez-de la Barrera, M.; Martinez-Gómez, H.; Maffuz, A.; Rojas-Garcia, P.; Perez-Alvarez, C.; Rodriguez-Ramirez, S.; Rodriguez-Antezana, E.; Ramirez-Ramirez, L. Thrombocytosis as a Predictor of Distant Recurrence in Patients with Rectal Cancer. *Arch. Med. Res.* **2012**, *43*, 305–311. [CrossRef]
59. Leblanc, R.; Peyruchaud, O. Metastasis: New functional implications of platelets and megakaryocytes. *Blood* **2016**, *128*, 24–31. [CrossRef]

60. Placke, T.; Orgel, M.; Schaller, M.; Jung, G.; Rammensee, H.-G.; Kopp, H.-G.; Salih, H.R. Platelet-Derived MHC Class I Confers a Pseudonormal Phenotype to Cancer Cells That Subverts the Antitumor Reactivity of Natural Killer Immune Cells. *Cancer Res.* **2012**, *72*, 440–448. [CrossRef]
61. Adams, D.L.; Martin, S.S.; Alpaugh, R.K.; Charpentier, M.; Tsai, S.; Bergan, R.C.; Ogden, I.M.; Catalona, W.; Chumsri, S.; Tang, C.-M.; et al. Circulating giant macrophages as a potential biomarker of solid tumors. *Proc. Natl. Acad. Sci. USA* **2014**, *111*, 3514–3519. [CrossRef]
62. Adams, D.L.; Adams, D.K.; Alpaugh, R.K.; Cristofanilli, M.; Martin, S.S.; Chumsri, S.; Tang, C.-M.; Marks, J.R. Circulating Cancer-Associated Macrophage-Like Cells Differentiate Malignant Breast Cancer and Benign Breast Conditions. *Cancer Epidemiol. Biomark. Prev.* **2016**, *25*, 1037–1042. [CrossRef]
63. Adams, D.L.; Adams, D.K.; He, J.; Kalhor, N.; Zhang, M.; Xu, T.; Gao, H.; Reuben, J.M.; Qiao, Y.; Komaki, R.; et al. Sequential Tracking of PD-L1 Expression and RAD50 Induction in Circulating Tumor and Stromal Cells of Lung Cancer Patients Undergoing Radiotherapy. *Clin. Cancer Res.* **2017**, *23*, 5948–5958. [CrossRef]
64. Mu, Z.; Wang, C.; Ye, Z.; Rossi, G.; Sun, C.; Li, L.; Zhu, Z.; Yang, H.; Cristofanilli, M. Prognostic values of cancer associated macrophage-like cells (CAML) enumeration in metastatic breast cancer. *Breast Cancer Res. Treat.* **2017**, *165*, 733–741. [CrossRef]
65. Lian, S.; Xie, R.; Ye, Y.; Lu, Y.; Cheng, Y.; Xie, X.; Li, S.; Jia, L. Dual blockage of both PD-L1 and CD47 enhances immunotherapy against circulating tumor cells. *Sci. Rep.* **2019**, *9*, 4532. [CrossRef]
66. Lou, X.-L.; Deng, J.; Deng, H.; Ting, Y.; Zhou, L.; Liu, Y.-H.; Hu, J.-P.; Huang, X.-F.; Qi, X.-Q. Aspirin inhibit platelet-induced epithelial-to-mesenchymal transition of circulating tumor cells (Review). *Biomed. Rep.* **2014**, *2*, 331–334. [CrossRef]
67. Van Schaijik, B.; Wickremesekera, A.C.; Mantamadiotis, T.; Kaye, A.H.; Tan, S.T.; Stylli, S.S.; Itinteang, T. Circulating tumor stem cells and glioblastoma: A review. *J. Clin. Neurosci.* **2019**, *61*, 5–9. [CrossRef]
68. Agnoletto, C.; Corrà, F.; Minotti, L.; Baldassari, F.; Crudele, F.; Cook, W.J.J.; Di Leva, G.; D'Adamo, A.P.; Gasparini, P.; Volinia, S. Heterogeneity in circulating tumor cells: The relevance of the stem-cell subset. *Cancers* **2019**, *11*, 483. [CrossRef]
69. Steinbichler, T.B.; Dudás, J.; Skvortsov, S.; Ganswindt, U.; Riechelmann, H.; Skvortsova, I.-I. Therapy resistance mediated by cancer stem cells. *Semin. Cancer Biol.* **2018**, *53*, 156–167. [CrossRef]
70. Agliano, A.; Calvo, A.; Box, C. The challenge of targeting cancer stem cells to halt metastasis. *Semin. Cancer Biol.* **2017**, *44*, 25–42. [CrossRef]
71. Paget, S. The distribution of secondary growths in cancer of the breast 1889. *Cancer Metastasis Rev.* **1989**, *8*, 98–101.
72. Fidler, I.J. The pathogenesis of cancer metastasis: The "seed and soil" hypothesis revisited. *Nat. Rev. Cancer* **2003**, *3*, 453–458. [CrossRef]
73. Koren, E.; Fuchs, Y. The bad seed: Cancer stem cells in tumor development and resistance. *Drug Resist. Updat.* **2016**, *28*, 1–12. [CrossRef]
74. Hu, Y.; Yu, X.; Xu, G.; Liu, S. Metastasis: An early event in cancer progression. *J. Cancer Res. Clin. Oncol.* **2017**, *143*, 745–757. [CrossRef]
75. Kuşoğlu, A.; Biray Avcı, Ç. Cancer stem cells: A brief review of the current status. *Gene* **2019**, *681*, 80–85. [CrossRef]
76. Luo, Y.T.; Cheng, J.; Feng, X.; He, S.J.; Wang, Y.W.; Huang, Q. The viable circulating tumor cells with cancer stem cells feature, where is the way out? *J. Exp. Clin. Cancer Res.* **2018**, *37*, 38. [CrossRef]
77. Rossi, E.; Rugge, M.; Facchinetti, A.; Pizzi, M.; Nardo, G.; Barbieri, V.; Manicone, M.; De Faveri, S.; Chiara Scaini, M.; Basso, U.; et al. Retaining the long-survive capacity of Circulating Tumor Cells (CTCs) followed by xeno-transplantation: Not only from metastatic cancer of the breast but also of prostate cancer patients. *Oncoscience* **2014**, *1*, 49–56. [CrossRef]
78. Carvalho, F.L.F.; Simons, B.W.; Antonarakis, E.S.; Rasheed, Z.; Douglas, N.; Villegas, D.; Matsui, W.; Berman, D.M. Tumorigenic potential of circulating prostate tumor cells. *Oncotarget* **2013**, *4*, 413–421. [CrossRef]
79. Aktas, B.; Tewes, M.; Fehm, T.; Hauch, S.; Kimmig, R.; Kasimir-Bauer, S. Stem cell and epithelial-mesenchymal transition markers are frequently overexpressed in circulating tumor cells of metastatic breast cancer patients. *Breast Cancer Res.* **2009**, *11*, R46. [CrossRef]
80. Aceto, N.; Bardia, A.; Miyamoto, D.T.; Donaldson, M.C.; Wittner, B.S.; Spencer, J.A.; Yu, M.; Pely, A.; Engstrom, A.; Zhu, H.; et al. Circulating tumor cell clusters are oligoclonal precursors of breast cancer metastasis. *Cell* **2014**, *158*, 1110–1122. [CrossRef]

81. Iriondo, O.; Yu, M. Unexpected Friendship: Neutrophils Help Tumor Cells En Route to Metastasis. *Dev. Cell* **2019**, *49*, 308–310. [CrossRef]
82. Yu, M. Metastasis Stemming from Circulating Tumor Cell Clusters. *Trends Cell Biol.* **2019**, *29*, 275–276. [CrossRef]
83. Wild, C.P. Complementing the genome with an exposome: The outstanding challenge of environmental exposure measurement in molecular epidemiology. *Cancer Epidemiol. Biomark. Prev.* **2005**, *14*, 1847–1850. [CrossRef]
84. Bossuyt, P.M.M.; Reitsma, J.B.; Linnet, K.; Moons, K.G.M. Beyond Diagnostic Accuracy: The Clinical Utility of Diagnostic Tests. *Clin. Chem.* **2012**, *58*, 1636–1643. [CrossRef]
85. Cristofanilli, M.; Pierga, J.Y.; Reuben, J.; Rademaker, A.; Davis, A.A.; Peeters, D.J.; Fehm, T.; Nolé, F.; Gisbert-Criado, R.; Mavroudis, D.; et al. The clinical use of circulating tumor cells (CTCs) enumeration for staging of metastatic breast cancer (MBC): International expert consensus paper. *Crit. Rev. Oncol. Hematol.* **2019**, *134*, 39–45. [CrossRef]
86. Budd, G.T.; Cristofanilli, M.; Ellis, M.J.; Stopeck, A.; Borden, E.; Miller, M.C.; Matera, J.; Repollet, M.; Doyle, G.V.; Terstappen, L.W.M.M.; et al. Circulating Tumor Cells versus Imaging–Predicting Overall Survival in Metastatic Breast Cancer. *Clin. Cancer Res.* **2006**, *12*, 6403–6409. [CrossRef]
87. Alunni-Fabbroni, M.; Müller, V.; Fehm, T.; Janni, W.; Rack, B. Monitoring in Metastatic Breast Cancer: Is Imaging Outdated in the Era of Circulating Tumor Cells? *Breast Care* **2014**, *9*, 16–21. [CrossRef]
88. Bidard, F.-C.; Pierga, J.-Y. Clinical Utility of Circulating Tumor Cells in Metastatic Breast Cancer. *J. Clin. Oncol.* **2015**, *33*, 1622. [CrossRef]
89. Helissey, C.; Berger, F.; Cottu, P.; Diéras, V.; Mignot, L.; Servois, V.; Bouleuc, C.; Asselain, B.; Pelissier, S.; Vaucher, I.; et al. Circulating tumor cell thresholds and survival scores in advanced metastatic breast cancer: The observational step of the CirCe01 phase III trial. *Cancer Lett.* **2015**, *360*, 213–218. [CrossRef]
90. Van Poznak, C.; Somerfield, M.R.; Bast, R.C.; Cristofanilli, M.; Goetz, M.P.; Gonzalez-Angulo, A.M.; Hicks, D.G.; Hill, E.G.; Liu, M.C.; Lucas, W.; et al. Use of Biomarkers to Guide Decisions on Systemic Therapy for Women With Metastatic Breast Cancer: American Society of Clinical Oncology Clinical Practice Guideline. *J. Clin. Oncol.* **2015**, *33*, 2695–2704. [CrossRef]
91. Bidard, F.-C.; Proudhon, C.; Pierga, J.-Y. Circulating tumor cells in breast cancer. *Mol. Oncol.* **2016**, *10*, 418–430. [CrossRef]
92. Heller, G.; McCormack, R.; Kheoh, T.; Molina, A.; Smith, M.R.; Dreicer, R.; Saad, F.; de Wit, R.; Aftab, D.T.; Hirmand, M.; et al. Circulating Tumor Cell Number as a Response Measure of Prolonged Survival for Metastatic Castration-Resistant Prostate Cancer: A Comparison With Prostate-Specific Antigen Across Five Randomized Phase III Clinical Trials. *J. Clin. Oncol.* **2018**, *36*, 572–580. [CrossRef]
93. Scher, H.I.; Graf, R.P.; Schreiber, N.A.; McLaughlin, B.; Jendrisak, A.; Wang, Y.; Lee, J.; Greene, S.; Krupa, R.; Lu, D.; et al. Phenotypic Heterogeneity of Circulating Tumor Cells Informs Clinical Decisions between AR Signaling Inhibitors and Taxanes in Metastatic Prostate Cancer. *Cancer Res.* **2017**, *77*, 5687–5698. [CrossRef]
94. Sciarra, A.; Gentilucci, A.; Silvestri, I.; Salciccia, S.; Cattarino, S.; Scarpa, S.; Gatto, A.; Frantellizzi, V.; Von Heland, M.; Ricciuti, G.P.; et al. Androgen receptor variant 7 (AR-V7) in sequencing therapeutic agents for castratrion resistant prostate cancer: A critical review. *Medicine* **2019**, *98*, e15608. [CrossRef]
95. IJzerman, M.J.; Berghuis, A.M.S.; de Bono, J.S.; Terstappen, L.W.M.M. Health economic impact of liquid biopsies in cancer management. *Expert Rev. Pharmacoecon. Outcomes Res.* **2018**, *18*, 593–599. [CrossRef]
96. Sholl, L.M.; Aisner, D.L.; Allen, T.C.; Beasley, M.B.; Cagle, P.T.; Capelozzi, V.L.; Dacic, S.; Hariri, L.P.; Kerr, K.M.; Lantuejoul, S.; et al. Liquid Biopsy in Lung Cancer: A Perspective From Members of the Pulmonary Pathology Society. *Arch. Pathol. Lab. Med.* **2016**, *140*, 825–829. [CrossRef]
97. Vishnoi, M.; Haowen Liu, N.; Yin, W.; Boral, D.; Scamardo, A.; Hong, D.; Marchetti, D. The identification of a TNBC liver metastasis gene signature by sequential CTC-xenograft modelling. *Mol. Oncol.* **2019**. [CrossRef]
98. Gorges, T.M.; Penkalla, N.; Schalk, T.; Joosse, S.A.; Riethdorf, S.; Tucholski, J.; Lücke, K.; Wikman, H.; Jackson, S.; Brychta, N.; et al. Enumeration and Molecular Characterization of Tumor Cells in Lung Cancer Patients Using a Novel *In Vivo* Device for Capturing Circulating Tumor Cells. *Clin. Cancer Res.* **2016**, *22*, 2197–2206. [CrossRef]
99. Kuske, A.; Gorges, T.M.; Tennstedt, P.; Tiebel, A.-K.; Pompe, R.; Preißer, F.; Prues, S.; Mazel, M.; Markou, A.; Lianidou, E.; et al. Improved detection of circulating tumor cells in non-metastatic high-risk prostate cancer patients. *Sci. Rep.* **2016**, *6*, 39736. [CrossRef]

100. Mandair, D.; Vesely, C.; Ensell, L.; Lowe, H.; Spanswick, V.; Hartley, J.A.; Caplin, M.E.; Meyer, T. A comparison of CellCollector with CellSearch in patients with neuroendocrine tumours. *Endocr. Relat. Cancer* **2016**, *23*, L29–L32. [CrossRef]
101. Dizdar, L.; Fluegen, G.; van Dalum, G.; Honisch, E.; Neves, R.P.; Niederacher, D.; Neubauer, H.; Fehm, T.; Rehders, A.; Krieg, A.; et al. Detection of circulating tumor cells in colorectal cancer patients using the GILUPI CellCollector: Results from a prospective, single-center study. *Mol. Oncol.* **2019**, *13*, 1548–1558. [CrossRef]
102. Gallerani, G.; Cocchi, C.; Bocchini, M.; Piccinini, F.; Fabbri, F. Characterization of tumor cells using a medical wire for capturing circulating tumor cells: A 3D approach based on immunofluorescence and DNA FISH. *J. Vis. Exp.* **2017**, *130*, e56936. [CrossRef]
103. Galanzha, E.I.; Menyaev, Y.A.; Yadem, A.C.; Sarimollaoglu, M.; Juratli, M.A.; Nedosekin, D.A.; Foster, S.R.; Jamshidi-Parsian, A.; Siegel, E.R.; Makhoul, I.; et al. In vivo liquid biopsy using Cytophone platform for photoacoustic detection of circulating tumor cells in patients with melanoma. *Sci. Transl. Med.* **2019**, *11*, eaat5857. [CrossRef]
104. Andree, K.C.; Mentink, A.; Zeune, L.L.; Terstappen, L.W.M.M.; Stoecklein, N.H.; Neves, R.P.; Driemel, C.; Lampignano, R.; Yang, L.; Neubauer, H.; et al. Toward a real liquid biopsy in metastatic breast and prostate cancer: Diagnostic LeukApheresis increases CTC yields in a European prospective multicenter study (CTCTrap). *Int. J. Cancer* **2018**, *143*, 2584–2591. [CrossRef]
105. Fischer, J.C.; Niederacher, D.; Topp, S.A.; Honisch, E.; Schumacher, S.; Schmitz, N.; Zacarias Föhrding, L.; Vay, C.; Hoffmann, I.; Kasprowicz, N.S.; et al. Diagnostic leukapheresis enables reliable detection of circulating tumor cells of nonmetastatic cancer patients. *Proc. Natl. Acad. Sci. USA* **2013**, *110*, 16580–16585. [CrossRef]
106. Franken, A.; Driemel, C.; Behrens, B.; Meier-Stiegen, F.; Endris, V.; Stenzinger, A.; Niederacher, D.; Fischer, J.C.; Stoecklein, N.H.; Ruckhaeberle, E.; et al. Label-Free Enrichment and Molecular Characterization of Viable Circulating Tumor Cells from Diagnostic Leukapheresis Products. *Clin. Chem.* **2019**, *65*, 549–558. [CrossRef]
107. Lambros, M.B.; Seed, G.; Sumanasuriya, S.; Gil, V.; Crespo, M.; Fontes, M.; Chandler, R.; Mehra, N.; Fowler, G.; Ebbs, B.; et al. Single-Cell Analyses of Prostate Cancer Liquid Biopsies Acquired by Apheresis. *Clin. Cancer Res.* **2018**, *24*, 5635–5644. [CrossRef]
108. Cayrefourcq, L.; Mazard, T.; Joosse, S.; Solassol, J.; Ramos, J.; Assenat, E.; Schumacher, U.; Costes, V.; Maudelonde, T.; Pantel, K.; et al. Establishment and characterization of a cell line from human circulating colon cancer cells. *Cancer Res.* **2015**, *75*, 892–901. [CrossRef]
109. Coumans, F.A.W.; Ligthart, S.T.; Uhr, J.W.; Terstappen, L.W.M.M. Challenges in the enumeration and phenotyping of CTC. *Clin. Cancer Res.* **2012**, *18*, 5711–5718. [CrossRef]
110. Sharma, R.; Sharma, S. *Physiology, Blood Volume*; StatPearls Publishing LLC: Tampa/St. Petersburg, FL, USA, 2019.
111. Coumans, F.A.; Siesling, S.; Terstappen, L.W. Detection of cancer before distant metastasis. *BMC Cancer* **2013**, *13*, 283. [CrossRef]
112. Stoecklein, N.H.; Fischer, J.C.; Niederacher, D.; Terstappen, L.W.M.M. Challenges for CTC-based liquid biopsies: Low CTC frequency and diagnostic leukapheresis as a potential solution. *Expert Rev. Mol. Diagn.* **2016**, *16*, 147–164. [CrossRef]
113. Fehm, T.N.; Meier-Stiegen, F.; Driemel, C.; Jäger, B.; Reinhardt, F.; Naskou, J.; Franken, A.; Neubauer, H.; Neves, R.P.L.; van Dalum, G.; et al. Diagnostic leukapheresis for CTC analysis in breast cancer patients: CTC frequency, clinical experiences and recommendations for standardized reporting. *Cytom. Part. A* **2018**, *93*, 1213–1219. [CrossRef]
114. Steeg, P.S. Targeting metastasis. *Nat. Rev. Cancer* **2016**, *16*, 201–218. [CrossRef] [PubMed]
115. Ghajar, C.M. Metastasis prevention by targeting the. *Nat. Rev. Cancer* **2015**, *15*, 238–247. [CrossRef] [PubMed]

© 2019 by the authors. Licensee MDPI, Basel, Switzerland. This article is an open access article distributed under the terms and conditions of the Creative Commons Attribution (CC BY) license (http://creativecommons.org/licenses/by/4.0/).

Review

Circulating Tumor Cell PD-L1 Expression as Biomarker for Therapeutic Efficacy of Immune Checkpoint Inhibition in NSCLC

Vera Kloten [†], Rita Lampignano [†], Thomas Krahn and Thomas Schlange *

Precision Medicine Markers, Pharmaceutical Division, Bayer AG, 42113 Wuppertal, Germany
* Correspondence: thomas.schlange@bayer.com; Tel.: +49-202-36-5403
† Both authors contributed equally.

Received: 10 July 2019; Accepted: 30 July 2019; Published: 1 August 2019

Abstract: Over the last decade, the immune checkpoint blockade targeting the programmed death protein 1 (PD-1)/programmed death ligand 1 (PD-L1) axis has improved progression-free and overall survival of advanced non-small cell lung cancer (NSCLC) patients. PD-L1 tumor expression, along with tumor mutational burden, is currently being explored as a predictive biomarker for responses to immune checkpoint inhibitors (ICIs). However, lung cancer patients may have insufficient tumor tissue samples and the high bleeding risk often prevents additional biopsies and, as a consequence, immunohistological evaluation of PD-L1 expression. In addition, PD-L1 shows a dynamic expression profile and can be influenced by intratumoral heterogeneity as well as the immune cell infiltrate in the tumor and its microenvironment, influencing the response rate to PD-1/PD-L1 axis ICIs. Therefore, to identify subgroups of patients with advanced NSCLC that will most likely benefit from ICI therapies, molecular characterization of PD-L1 expression in circulating tumor cells (CTCs) might be supportive. In this review, we highlight the use of CTCs as a complementary diagnostic tool for PD-L1 expression analysis in advanced NSCLC patients. In addition, we examine technical issues of PD-L1 measurement in tissue as well as in CTCs.

Keywords: liquid biopsy; CTCs; immune checkpoint inhibitors; PD-L1 expression; NSCLC

---

1. Immune Checkpoint Blockade Therapy in Non-Small Cell Lung Cancer (NSCLC): State of the Art

Lung cancer is the most common cancer in men and the third frequent cancer in women worldwide. With a poor five-year survival rate of 10–15%, lung cancer is the major cause of cancer-related deaths [1]. Nowadays, molecular testing in advanced non-small cell lung cancer (NSCLC) patients includes screening for targetable alterations, e.g., *EGFR* mutations or *ALK* rearrangements, and, in addition, factors predictive of response to immunotherapy, thus, immune checkpoint inhibitors (ICIs). The introduction of ICIs in the clinic has led to increasing response rates in locally advanced and metastasized NSCLC [2–5]. ICIs are designed to target inhibitory checkpoint molecules, such as programmed cell death protein 1 (PD-1), and its ligand programmed cell death protein ligand 1 (PD-L1). PD-L1, a type I transmembrane protein with an extracellular N-terminal domain, inhibits the immune response through interaction with its receptor, PD-1, expressed among other immune cells on activated T- and B-cells [6]. Thereby, PD-L1 upregulation in tumor tissue enables evasion of immune surveillance by the inhibition of immune cell activation. In contrast to conventional therapies that directly target cancer cells, anti-PD-1/PD-L1 antibodies reactivate the immune system of patients to eradicate tumors, which induces durable and long-lasting antitumor immunity in patients with different tumor types, including lung cancer [7].

As a biomarker for selection of patients eligible for ICI therapy, the PD-L1/PD-1 axis has been investigated in many studies [8]. Today, two antibodies blocking PD-1, Nivolumab (Opdivo, Bristol-Myers Squibb) and Pembrolizumab (Keytruda, MSD SHARP and DOHME GMBH) as well as one antibody targeting PD-L1, Atezolizumab (Tecentriq, Roche) [9,10], have US Food and Drug Administration (FDA) approval in NSCLC. In more detail, Pembrolizumab has approval for both first- and second-line treatment, while Nivolumab also has third-line approval for NSCLC. In addition, there are ongoing studies for the use of Atezolizumab in the third-line setting as well. This has led to an increased interest in potential additional clinical applications for these therapeutics. Currently, there are 97 listed clinical trials for Atezolizumab, 203 trials for Nivolumab, and 225 trials for Pembrolizumab in lung cancer (status June 3rd 2019, extracted from ClinicalTrial.gov [11] (Figure 1). Search terms and synonyms are listed in Tables S1–S3. Most of the studies are recruiting patients or are under investigation. For Atezolizumab, 15% of the studies are in phase III, while for Nivolumab and Pembrolizumab, 13% and 9%, respectively, are listed as phase III trials. Based on ClinicalTrial.gov, 3% of studies with Nivolumab, 6% of studies with Atezolizumab, and 3% with Pembrolizumab treatment have been completed (Figure 1). The efficacy of immune checkpoint inhibitors alone or in combinations in NSCLC has been summarized in detail elsewhere [12,13].

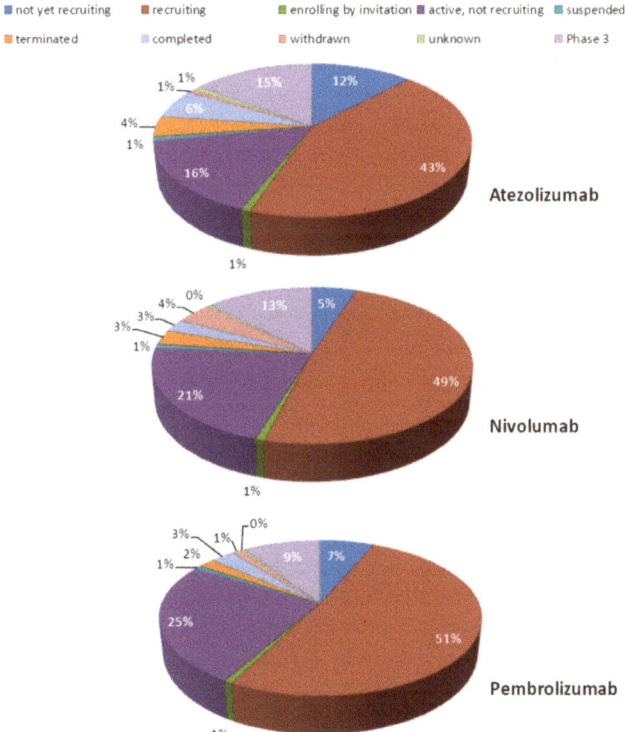

Figure 1. Overview of listed clinical trials for Atezolizumab, Nivolumab, and Pembrolizumab in lung cancer (data based on ClinicalTrial.gov).

Not all patients with advanced NSCLC benefit from these drugs. Only 20% of unselected NSCLC patients show a response to ICIs (summarized e.g., in [14]), underlining the necessity to select the right patients. The gold standard for treatment decision is an immunohistochemistry (IHC)-based companion or complementary diagnostic tests for PD-L1. Two anti-PD-L1 antibodies, Ventana PD-L1 (SP142) assay (Ventana Medical Systems, Inc, Tucson, AZ, USA and PD-L1 IHC 22C3 pharmDx (Dako North America, Inc. Santa Barbara, CA, USA) have FDA approval for PD-L1 IHC. However, PD-L1 expression assessed by IHC requires a tissue sample which can be insufficiently representative of overall tumor/metastasis expression or not available from patients, thus impeding treatment decision-making. In addition, the PD-L1 expression may be considerably heterogeneous across tumor boundary to core [15]. Furthermore, dynamic changes of PD-L1 expression in tumor cells might occur before or during therapy with PD-1/PD-L1 inhibitors, leading to different sensitivity to PD-1/PD-L1 blockade which would similarly be missed by a single biopsy. With the increasing number of therapeutic regimens and targeted therapies, molecular profiling of NSCLC is becoming crucial at every step of disease progression to reveal the biological alterations that are driving resistance and impact on treatment decisions. Obtaining serial tumor tissue biopsies is highly challenging and often not feasible. Such patients may benefit from a molecular characterization of PD-L1 expression in circulating tumor cells (CTCs) extracted from blood. In the last decade, molecular analysis of CTCs in body fluids (as a "liquid biopsy") started to have a growing impact on the clinical management of cancer patients. Today, liquid biopsy is a rapidly expanding field in translational cancer research, and it shows the potential to complement diagnostic and therapeutic care of cancer patients [16]. CTCs hold promise to better reflect tumor heterogeneity compared to tissue biopsies because they might originate from different tumor sites and reflect properties from the primary tumor site as well as from metastatic sites shown, e.g., in breast cancer [17]. In addition, they could lead to important insights on how tumor cells become resistant to immune therapy because they can be analyzed longitudinally as liquid biopsies. Interestingly, circulating tumor DNA (ctDNA) is currently under investigation in several clinical trials as a biomarker for tumor mutation burden (TMB) rather than CTCs. There is strong interest in TMB since a positive correlation between TMB in tumor tissue and a clinical benefit from immunotherapy has retrospectively been observed [18]. However, there are several publications implying clinical relevance of PD-L1-positive CTCs in cancer including NSCLC [19–22].

2. Clinical Significance of PD-L1-Positive CTCs in NSCLC

Despite the reported value of assessing the overexpression of PD-L1 on cells of different types in solid tumors—including lung cancer—as a promising marker to predict anti PD-1/PD-L1 treatment efficacy, the predictive value of PD-L1 expression is still controversial and related investigations face the three major limitations of tissue biopsies: Invasiveness, sampling error due to tumor heterogeneity, and mostly unfeasible longitudinal sampling. To overcome these issues and support histological analysis, the expression of PD-L1 has been explored on CTCs and has been correlated to patients' clinical outcomes (Table 1). The clinical significance of PD-L1-positive (PD-L1$^+$) CTCs in NSCLC is to date in its infancy as the first related study was published in 2016 by Schehr and colleagues [23], who initially focused on technical optimizations of PD-L1$^+$ CTC enrichment. Schehr et al. reported the presence of a population of co-isolated CD11b$^+$ (marker for myeloid development), CD45-low, and cytokeratin-positive (CK$^+$) cells—via an in-house produced immunomagnetic enrichment system—misidentified as CTCs and expressing PD-L1. The quantification of PD-L1 could therefore be skewed by false positive events, thus requiring careful analysis in order to increase the accuracy of the assay [23]. However, one has to be aware that inconsistency might be also caused by various types of therapy since patients in this study were treated with mainly radio- and/or chemotherapy before blood collection, followed by a first- to fifth- line of therapy with ICIs. On the same line, Bao and colleagues [24] also focused on the development and optimization of a CTC-sorting system—in this case a size-based chip—which could give the chance to investigate the PD-L1 expression, achieved through a RT-qPCR approach. However, the lack of CTC$^+$ patients (~7% based on CK19 mRNA expression) did not allow for drawing any significant conclusion about the clinical utility of PD-L1 [24].

The predictive utility of PD-L1$^+$ CTCs in a chemotherapy setting has been further investigated by Kallergi and colleagues [25]. In this study, the ISET (Isolation by SizE of Tumor cells, Rarecells Diagnostics SAS) technology followed by Giemsa and immunostaining was utilized to detect PD-1$^+$ and PD-L1$^+$ CTCs in metastatic NSCLC patients before (chemotherapy-naïve, n = 30) and after chemotherapy (after the third chemotherapy cycle at the time of assessment of treatment efficacy, $n = 11$). Giemsa staining revealed CTCs in 28 of 30 (93.3%) patients at baseline and in 9 of 11 (81.8%) patients studied after the third chemotherapy cycle with a median of 5 CTCs/mL of blood (range, 0–23 CTCs/mL of blood). Of interest, using immunostaining, CTCs could be detected in 17 of 30 (56.7%) patients at baseline and in 8 of 11 (72.7%) after the third treatment cycle. The concordance between the two detection methods at baseline and after the third treatment cycle was 63.3% and 67%, respectively. The rate of detection was 30% (9 of 30) and 27% (8 of 30) before treatment, and in 9% (1 of 11) and 46% (5 of 11) after 3 cycles of chemotherapy, respectively for PD-1$^+$ and PD-L1$^+$ CTCs. Interestingly, an increase of 20% PD-L1$^+$ CTCs ($p = 0.096$) and a decrease of 21% PD-1$^+$ CTCs ($p = 0.785$) after chemotherapy was observed. In addition, a shorter progression-free survival (PFS) could be observed for patients with >3 PD-1$^+$ CTCs at baseline ($p = 0.022$) but not for PD-L1 expressing tumor cells, thus suggesting a potential clinical role for PD-1$^+$ CTCs rather than for PD-L1$^+$ CTCs [25].

Through a different size-based CTC-enrichment approach (CellSieve™ Microfiltration Assay, Creatv MicroTech) followed by immunostaining, Adams et al. [26] investigated the expression of PD-L1 in different CTC subtypes, i.e., PDCTCs (prognostically relevant pathologically definable CTCs), EMTCTCs (CTCs undergoing epithelial-to-mesenchymal transition), and CAMLs (cancer-associated macrophage-like cells) in a prospective pilot study with 41 NSCLC patients (stage I–IV) undergoing radiotherapy, while 34% (14 of 41) received prior chemotherapy. The researchers were able to identify at least one CTC (i.e., PDCTC, EMTCTC, or CAML) in 35 of the 41 samples (85%) at baseline and in all 41 samples (100%) at a follow-up sample taken two to three weeks after radiotherapy initiation. Specifically, EMTCTCs were found in 49% of baseline samples and in 66% of follow-up samples. CAMLs were found in 81% of baseline samples and in 100% of follow-up samples. PDCTCs were found in only one sample at baseline (2%) and in only three samples at follow-up (7%). Furthermore, Adams and colleagues confirmed an intra- and interpatient dynamic expression of PD-L1 in CTCs before and after therapy. The researchers reported 51% patients exhibiting no/low PD-L1 expression at baseline and follow-up, 17% had persistently medium/high at the two time points, and 32% patients showing an increase in PD-L1 expression in CTCs from low to medium in the follow-up visit. Furthermore, concordance between CTCs and matched tumor tissue was highly depending on the antibody clone utilized for immunohistochemistry (IHC) and also, given the restricted amount of patients, no statistical analysis could be performed. However, a sequential PD-L1 evaluation in patients two to four months after the end of radiotherapy, exhibited that 87% expression of the marker was unchanged, suggesting the importance of longitudinal analysis of PD-L1 expression in CTCs [26].

The impact of radiation therapy on PD-L1 expression in CTCs was recently investigated by Wang et al. [27] monitoring the dynamic changes of PD-L1 expression in CTCs of 13 nonmetastatic NSCLC patients who received radiation alone ($n = 5$) or chemoradiation ($n = 8$). Serial blood samples from the patients enrolled in the study were collected before the initiation of radiation, during radiation, and at follow up, approximately one month after radiation ($n = 38$ samples). CTCs were detected in all 38 samples with an average of 21.3 CTCs/mL (range of 4–72 CTCs/mL), while PD-L1$^+$ CTCs were detected in 24 (66.7%) out of 36 samples analyzed, ranging from 0 to 43 PD-L1$^+$ CTCs/mL. In line with the results by Adams et al. [26], patients treated with radiation or with concurrent carboplatin and paclitaxel had increased PD-L1$^+$ CTCs during treatment (PD-L1$^+$ CTC% was higher in visit two than that in visit one (median 0.7% vs. 24.7%, $p = 0.0068$)). In addition, PD-L1$^-$positive patients had a shorter PFS compared to PD-L1-negative patients using a PD-L1$^+$ CTC cut-off ≥5%. Notably, one of the patients who had a high PD-L1$^+$ CTC count at visit two and visit three was put on therapy with Pembrolizumab after initial progression and has had stable disease for seven months [27], implying that patients who become (re-)sensitized to ICIs can be identified by PD-L1 CTC expression analysis.

The largest studies so far, investigating the role of PD-L1$^+$ CTCs in the clinical setting, were conducted by Ilié et al. [28] and Janning et al. [29], who utilized a sized based CTC-enrichment approach (Isolation by SizE of Tumor cells (ISET), Rarecells Diagnostics SAS) with, respectively, the FDA-approved EpCAM-based CellSearch® System (Menarini Silicon Biosystems Inc, Huntingdon

Valley, PA, USA and the epitope-independent Parsortix™ system (Angle, Guildford, UK) followed by immunostaining of retained cells. In the study of Ilié and colleagues, CTCs were detected in 80 of 106 (75%) patient samples, while 99% (79 out of 80) CTC-positive samples exhibited more than 5 CTCs per 4 mL blood, with a median of 60 CTCs per 4 mL (range: 2–256 CTCs/4 mL). Furthermore, the researchers extracted ≥1 PD-L1$^+$ CTCs in 8% of patients with advanced stage III and IV NSCLC (n = 6/71 samples) with 93% concordance to PD-L1$^+$ tumor cells of matched primary tissue (specificity = 100%; sensitivity = 55%). In addition, they could observe a trend towards poor clinical outcomes in patients with PD-L1$^+$ CTCs receiving first line of chemotherapy, similar to the trend observed for PD-L1$^+$ primary tumors [28]. Janning and colleagues detected ≥1 CTC in 68.5% (n = 61/89 samples) and ≥3 CTCs in 33.7% (n = 30/89 samples) of NSCLC (mostly stage IV) patients using the Parsortix™ system. Thereof, the researchers found ≥1 PD-L1$^+$ CTC in 56% (n = 50/89 samples) and ≥3 PD-L1$^+$ CTCs in 26% (n = 23/89 samples) of patients. Amongst patient samples with at least three CTCs (CD45$^-$/K$^+$), 47% (14/30) harbored exclusively PD-L1$^+$ CTCs and 47% (14/23) had both PD-L1$^+$ and PD-L1$^-$ CTCs [29]. Of interest, the percentage of PD-L1$^+$ CTCs did not correlate with the percentage of PD-L1$^+$ tumor cells in primary tumor tissue biopsies determined by immunohistochemistry (p = 0.179). In patients undergoing therapy with Pembrolizumab, Nivolumab, or Atezolizumab the researchers indicated that in 89% of the responding patients either a decrease or no change of their total CTC counts after three or five cycles of therapy (decrease: 6/9; no change 2/9, increase: 1/9) was shown. In contrast, upon disease progression, all patients showed an increase in PD-L1$^+$ CTCs [29].

The predictive value of PD-L1$^+$ CTCs in NSCLC patients under immunotherapy has also been investigated in several other studies. Nicolazzo and colleagues [30] first focused on the evaluation of PD-L1$^+$ CTCs utility in patients with stage IV NSCLC treated with the anti-PD-1 Nivolumab. By utilizing the CellSearch® system as a CTC enrichment approach, they reported a CTC detection rate of more than the 40% usually described by the literature [31], with an extremely high frequency of PD-L1 expression (95%) in 83% of the patients at baseline. The number of CTCs detected ranged from 1 to 20 (median number of CTCs 5.2). After three months of treatment, the fraction of PD-L1$^+$ CTCs ranged from 25% to 100%, while after six months 50% showed PD-L1$^+$ CTCs. Therefore, even though both the presence of CTCs and the PD-L1 expression were associated with poor clinical outcomes (statistics not available), the lack of patients with PD-L1-negative CTC fractions did not allow for any conclusions about the real prognostic and predictive relevance of this marker [30]. In addition to Nicolazzo et al., Guibert and colleagues [19] detected PD-L1$^+$ CTCs in 93% of advanced NSCLC patients before Nivolumab treatment, with a median proportion of CTCs expressing PD-L1 of 17.2% using the ISET technology followed by immunostaining. Interestingly, no correlation could be observed with PD-L1+ tissue biopsies (72%; r = 0.04, p = 0.77). Furthermore, in a study by Kulasinghe and colleagues [20], 66% of NSCLC patients exhibiting PD-L1$^+$ CTCs (64.7%; n = 11)—enriched through the size-based ClearCell FX—were treated with Nivolumab, but no correlation between PD-L1 expression and clinical outcomes could be observed. The impact of the epithelial-to-mesenchymal transition (EMT) of CTCs in NSCLC patients under Nivolumab treatment was described in a recent study by Raimondi et al. [32] using the filtration technology ScreenCell. The researchers investigated 13 patients with metastatic NSCLC progressing post-prior systematic treatment with Nivolumab. They found ≥1 CTC 69% (9 of 13) of patients with a percentage of 5% to 80% of PD-L1$^+$ CTCs. Interestingly, PD-L1 was found coexpressed with EMT markers in a percentage of cells that was ranging between 50% and 78%. This might provide a biologic explanation for the persistence of PD-L1-positive CTCs in NSCLC patients after six months of treatment, predicting resistance to the anti-PD-1 Nivolumab shown by Nicolazzo and colleagues [30].

Table 1. Studies on the clinical relevance of programmed death ligand 1-positive (PD-L1$^+$) circulating tumor cells (CTCs) in non-small cell lung cancer (NSCLC).

Study	Patients	Blood Tube	CTC-Enrichment System	Antibody Clone	Therapy	Clinical Outcome
Schehr et al. [23]	19	EDTA	Immunomagnetic depletion, Dynabeads-based	MIH1 (BD)	1st line TX: Radio-/Chemotherapy, TKIs Current: ICIs	-
Bao et al. [24]	15	EDTA	Size-based (in-house produced chip)	*	1st line TX: Chemo Current: Nivolumab	-
Kallergi et al. [25]	30	EDTA	Size-based (ISET)	B7-H1 (NB)	1st line TX: None Current: Chemo-naïve	After 3 cycles of chemo, ~19% increase PD-L1$^+$ CTCs
Adams et al. [26]	41	CellSave	Size-based (CellSieve Microfiltration Assay)	B7-H1 (R&D)	1st line TX: Chemo Current: Radiotherapy	Slightly better outcome in patients with high PD-L1 expression
Wang et al. [27]	13	EDTA	Microfluidic graphene oxide (GO) Chip	29E.2A3 (BL)	1st line TX: None Current: Radio-/Chemotherapy	PD-L1$^+$ patients had a shorter PFS compared to PD-L1$^-$ patients
Ilié et al. [28]	106	-	Size-based (ISET)	SP142 (VT)	1st line TX: None Current: Chemo-naive (93%), neoadjuvant chemo (7%)	Slightly better outcome in patients with PD-L1$^+$ CTCs
Janning et al. [29]	89	EDTA and/or Cell Save	EpCAM-based (CellSearch®), size-based (Parsortix™)	D84TX (CS)	Current: Radio-/chemotherapy, surgery, TKIs, ICIs	Increase in PD-L1$^+$ CTCs upon disease progression; no change or decrease in responding patients
Nicolazzo et al. [30]	24	CellSave	EpCAM-based (CellSearch)	B7-H1 (R&D)	1st line TX: na Current: Nivolumab	Poor clinical outcome
Guibert et al. [19]	96 pre-, 24 post-therapy	-	Size-based (ISET)	D8TX4 (CS)	1st line TX: Chemo Current: Nivolumab	More non-responders to Nivolumab if ≥1% PD-L1$^+$ CTCs
Kulasinghe et al. [20]	33	EDTA or Streck	Size-based (ClearCell FX)	n/a (Abcam)	1st line TX: Radio-/Chemotherapy Current: Nivolumab	None
Dhar et al. [33]	22	EDTA	Size-based (Vortex HT chip)	#4059 (PS),29E.2A3 (BL), MIH1 (BD)	1st line TX: na Current: ICIs	Slightly better outcome for patients with >50% PD-L1$^+$ CTCs

Unless otherwise specified, CTC detection was performed via immunostaining; * CTC detection via RT-qPCR. Current therapy is defined as therapy at time point of blood draw. BD: BD Biosciences; BL: BioLegend; CS: Cell signaling; NB: Novus Biologicals; PS: ProSci; VT: Ventana. Chemo: Chemotherapy; na: Not available; TKI: Tyrosine kinase inhibitor; TX: Treatment.

Beside the potential predictive role of PD-L1+ CTCs for Nivolumab treatment, its clinical significance has also been investigated for immunotherapy based on Pembrolizumab (PD-1 inhibitor) by Dhar and colleagues [33], who also opted for a size-based CTC enrichment system (Vortex HT chip) followed by immunostaining of captured cells. In this study, ≥1 PD-L1+ CTC were detected in ~97% patients before treatment with a discrete concordance with tissue biopsy. However, due to the restricted number of primary tumor biopsies available ($n = 4$), statistical analysis was not possible. Importantly, patients with >50% PD-L1+ CTCs ($n = 3/4$) experienced an improved progression-free survival under Pembrolizumab treatment, indicating the need for further confirmation of the available data to reconcile the conflicting evidence.

In summary, despite the limited amount of studies published to date, these first results imply PD-L1+ CTCs might play a role in determining response to different ICI therapeutic approaches (summarized in Table 1).

Several studies showed that the efficacy of PD-1/PD-L1 blockade could be also affected by PD-L1 expression on tumor-infiltrating cells in different types of cancer, including lung cancer. Herbst et al. [34] showed across multiple cancer types that responses were observed in patients with tumors expressing high levels of PD-L1, especially when PD-L1 was expressed by tumor-infiltrating immune cells. In another study by Kim et al. [35], it was shown that increased numbers of $CD8^+$ or $PD-1^+$ tumor-infiltrating lymphocytes (TILs) were significantly associated with prolonged disease-free survival of these patients, whereas PD-L1 and PD-L2 expression had no significant prognostic implications. He and colleagues [36] revealed a 43.2% positive PD-1 staining on TILs in NSCLC tumor tissue, while PD-L1 was detected on both tumor cells and TILs. Studies investigating the relation between PD-1 and PD-L1 in lung cancer were focused on tumor tissue samples facing the same issues as mentioned above.

However, it has to be noted that even though the field of CTCs carries a great potential with liquid biopsy to better integrate the heterogeneous and potentially dynamic expression of PD-L1 in the course of NSCLC pathology, it still requires a series of (pre-)analytical standardizations, in order to avoid misinterpretations and guarantee a reliable clinical treatment decision [37].

3. The Need for (Pre-)Analytical Standardizations

Important aspects need to be taken into consideration when comparing these studies. First and foremost, a detailed report about patients' clinical data (e.g., TNM classification, grade) followed by more technical standardization regarding the sampling, blood stabilization, storage time and temperature, and CTC enrichment and detection approach, as well as the antibody cocktail utilized for the immunostaining and the threshold applied for the PD-L1 positivity are necessary (Table 1). Indeed, variations and lack of consensus in all these steps could lead to major discrepancies among studies, thereby hampering a proper comparison of results and clinical applications in the near future [38]. The advent of anti-PD-L1 antibodies gives rise to the question of whether therapeutic antibodies might interfere with the binding of diagnostic PD-L1 antibodies and might thereby potentially compromise monitoring of CTC PD-L1 expression in the course of therapy.

In order to address this need for liquid biopsy—including CTC—standardization, several public–private partnerships and consortia were established: BloodPAC, European Liquid Biopsies Academy, Liquid Biopsy Consortium, SPIDIA4P, and the European IMI CANCER-ID consortium currently addressing—among others—the PD-L1 harmonization issue to detect PD-L1+ CTCs in NSCLC. In a collaboration project within the CANCER-ID consortium, we performed a comprehensive multicomparison of commercially available anti-PD-L1 antibodies in a NSCLC cell line panel including a preincubation with Atezolizumab (manuscript in preparation). Indeed, one of the major issues in translating PD-L1+ (circulating) tumor cells from basic research to the clinical setting for routine diagnostic application is (i) a heterogeneous detection rate of used CTC enrichment and detection approaches, (ii) resulting observer bias in calling CTCs, and (iii) the lack of consensus on the use of different commercially available anti-PD-L1 antibody clones and their performance and specificity compared to the antibody clones that are included in the IHC kits that received regulatory approval.

3.1. The Need for Clinically Applicable CTC Enrichment and Detection Approaches

Today, there is no consensus on CTC enrichment in NSCLC patients resulting in the use of different enrichment strategies. The EpCAM-based CellSearch® System which is FDA approved for clinical utility in metastasized breast, prostate, and colorectal cancer, remains challenging in NSCLC and studies using this system have to view with caution. To overcome the low sensitivity in advanced NSCLC patients using the CellSearch® System, alternative methods for CTC detection were used in most studies (Table 1).

Recently, Janning and colleagues [29] compared the EpCAM-based CellSearch® System with the epitope-independent Parsortix^{TM} system (Angle) for the assessment of PD-L1 expression of CTCs extracted from NSCLC patients. They showed a 50% higher detection rate of CTCs per blood sample with the Parsortix^{TM} system (>1 CTC in 59 of 97 samples (61%) compared to 31 of 97 samples (32%) with >1 CTC using the EpCAM-based system). Another promising method is the filter- and size-based ISET system which was used in several recent CTC-related studies in NSCLC patients (see Table 1). Similar to the Parsortix^{TM} system, an increased detection of CTCs compared to the CellSearch method was shown with the ISET method [39,40] as well as with a miniaturized microcavitiy array (MCA) [41].

Using the ISET method, Farace et al. [39] showed concordant results in only four patients (20%) while 16 (80%) patients had CTC counts markedly higher with ISET than CellSearch. In addition, Krebs and colleagues [40] detected 32 of 40 (80%) NSCLC patients using ISET compared with 9 of 40 (23%) patients using CellSearch. A subpopulation of CTCs isolated by ISET did not express epithelial markers. Using MCA, Hosokawa et al. [41] detected CTCs in 17 of 22 NSCLC patients using the MCA system versus 7 of 22 patients using the CellSearch system. On the other hand, CTCs were detected in 20 of 21 small cell lung cancer (SCLC) patients using the MCA system versus 12 of 21 patients with the CellSearch® system. Significantly more CTCs in NSCLC patients were detected by the MCA system (median 13, range 0–291 cells/7.5 mL) than by the CellSearch® system (median 0, range 0–37 cells/7.5 mL, $p = 0.0015$). However, statistical significance was not reached in SCLC, though the trend favoring the MCA system over the CellSearch® system was observed ($p = 0.2888$). The MCA system also isolated CTC clusters from patients who had been identified as CTC negative using the CellSearch® system [41].

Since most PD-L1 CTC studies use different enrichment techniques and methods for CTC detection, it becomes clear that comparisons between these studies (like those summarized in Table 1) have to be interpreted with caution. Additionally, the use of different anti-PD-L1 antibody clones reinforced this situation.

3.2. The Need for Harmonized Immunostaining Protocols

Numerous multicomparison studies already tried to clarify this situation with regards to IHC on tissue biopsy, summarized in Table 2. These reports concordantly highlight different immunostaining patterns, signal intensities, and therefore variable cut-off values regarding percentage of stained tumor or immune cells obtained by using the various antibody clones used.

Despite their scientific contribution on the topic, most PD-L1 analyses on CTCs were performed using different antibody clones (see Table 1)—with the only exception of Ilié and colleagues focusing on the clone SP142, which is part of the FDA approved Roche (Ventana) PD-L1 IHC assay. Furthermore, as Ilié et al. describe in their publication, some CTCs—as well as some screened tumor cell lines—exhibited cytoplasmic staining with or without a membranous signal, pointing out the necessity to extend the PD-L1 immunostaining assay to other clones.

Table 2. Harmonization studies on immunohistochemistry (IHC) PD-L1 staining of NSCLC tissue biopsies.

Study	Antibody Clone	Company	PD-L1 + Tumor Cell Cut-Off	Patients	Main Findings
Parra et al. [42]	E1L3N, E1J2J	Cell Signaling	≥1%	185 + (cell lines)	E1L3N, E1J2J, SP142, 28-8, 22C3, 5H11 and SP263: comparable staining patterns on membranes; SP263: higher IHC score
	22C3, 28-8	Dako			
	SP263, SP142	Ventana			
	5H11	Not commercialized			
Ratcliffe et al. [43]	22C3, 28-8	Dako	≥1%, ≥10%, ≥25%, ≥50%	493	All assays show concordant staining patterns
	SP263	Ventana			
Scheel et al. [44]	E1L3N	Cell Signaling	≥1%, ≥50%	21	22C3, 28-8 and SP263: concordant staining patterns; SP142 as outlier
	22C3, 28-8	Dako			
	SP263, SP142	Ventana			
Adam et al. [45]	E1L3N	Cell Signaling	≥1%, ≥5%, ≥25%, ≥50%	41	28-8, 22C3, SP263, E1L3N: highly concordant; SP142 as outlier
	22C3, 28-8	Dako			
	SP263, SP142	Ventana			
Rimm et al. [46]	E1L3N	Cell Signaling	≥1%, ≥5%, ≥50%	90	SP142: significant lower PD-L1 IHC score; 22C3: significant reduction in PD-L1 staining; 28-8 and E1L3N concordant
	22C3, 28-8	Dako			
	SP142	Ventana			

4. Conclusions and Future Perspectives

Immune checkpoint inhibition therapy represents a breakthrough in treatment of non-small cell lung cancer patients. However, there are still major challenges in selecting NSCLC patients likely to benefit from targeting the PD-1/PD-L1 axis. CTC-based liquid biopsy may be an option for the development of blood-based tests that address this issue. The successful implementation of such tests will critically depend on consensus on the use of different anti-PD-L1 antibody clones, CTC enrichment technologies, and well-established standardized clinically feasible standard operating procedures. Furthermore, a deeper understanding of the different mechanisms of PD-L1 regulation at genetic, epigenetic, transcriptional, translational, and posttranslational levels in cancer is needed to develop appropriate protocols. In addition, the regulation of PD-L1 expression during metastasis might be different from the primary tumors, potentially making longitudinal monitoring of patients necessary and liquid biopsy an even more favorable diagnostic option. The latter is further supported when taking into account that PD-L1 expression assessed by IHC requires a tissue sample which could be insufficient, or even lacking, in advanced lung cancer patients. This may compromise the level of confidence with which a therapy decision can be made. Several studies suggest a promising role for PD-L1$^+$ CTCs in determining response to different therapeutic approaches. However, the lack of consensus on anti-PD-L1 antibody clones persists, when most PD-L1 analyses on CTCs were performed with different antibody clones compared to tissue PD-L1 analysis. In addition, the value of CTC analysis for clinical practice is strongly determined by the sensitivity of the CTC isolation technology and the specificity of the diagnostic test to discriminate cells with malignant features from nonmalignant cells captured as background. To this end, the clinical benefit of immune checkpoint blockade in NSCLC using circulating tumor cells remains uncertain. CTCs have not been investigated in clinical trials relevant for regulatory approval of Atezolizumab, Nivolumab, and Pembrolizumab. More recently, a phase Ib study to evaluate safety and tolerability of durvalumab (anti-PD-L1) and tremelimumab (anti-CTLA-4) (NCT03275597) uses CTC number and CTC PD-L1 expression as exploratory endpoints for efficacy and target engagement. Future research will show whether CTC PD-L1 expression together with additional biomarkers like tumor mutational burden assessed by the analysis of CTCs or ctDNA, constitute clinically relevant blood-based biomarkers for immune checkpoint blockade therapy patient selection.

Supplementary Materials: The following are available online, Table S1: Clinical trials in lung cancer (drug: Nivolumab), Table S2: Clinical trials in lung cancer (drug: Atezolizumab), Table S3: Clinical trials in lung cancer (drug: Pembrolizumab).

Funding: The authors participate in the Innovative Medicines Initiative consortium CANCER-ID. CANCER-ID is supported by the Innovative Medicines Initiative (IMI) Joint Undertaking under grant agreement n°115,749, resources of which are composed of financial contribution from the European Union's Seventh Framework Programme (FP7/2007–2013) and EFPIA companies' in kind contributions.

Conflicts of Interest: The authors of this review are or have been full-time employees of Bayer AG which is the EFPIA coordinating partner of the Innovative Medicines Initiative (IMI) consortium CANCER-ID.

References

1. Torre, L.A.; Bray, F.; Siegel, R.L.; Ferlay, J.; Lortet-Tieulent, J.; Jemal, A. Global cancer statistics, 2012. *CA Cancer J. Clin.* **2015**, *65*, 87–108. [CrossRef] [PubMed]
2. Borghaei, H.; Paz-Ares, L.; Horn, L.; Spigel, D.R.; Steins, M.; Ready, N.E.; Chow, L.Q.; Vokes, E.E.; Felip, E.; Holgado, E.; et al. Nivolumab versus docetaxel in advanced nonsquamous non-small-cell lung cancer. *N. Engl. J. Med.* **2015**, *373*, 1627–1639. [CrossRef] [PubMed]
3. Brahmer, J.; Reckamp, K.L.; Baas, P.; Crino, L.; Eberhardt, W.E.; Poddubskaya, E.; Antonia, S.; Pluzanski, A.; Vokes, E.E.; Holgado, E.; et al. Nivolumab versus docetaxel in advanced squamous-cell non-small-cell lung cancer. *N. Engl. J. Med.* **2015**, *373*, 123–135. [CrossRef] [PubMed]
4. Herbst, R.S.; Baas, P.; Kim, D.W.; Felip, E.; Perez-Gracia, J.L.; Han, J.Y.; Molina, J.; Kim, J.H.; Arvis, C.D.; Ahn, M.J.; et al. Pembrolizumab versus docetaxel for previously treated, pd-l1-positive, advanced non-small-cell lung cancer (keynote-010): A randomised controlled trial. *Lancet* **2016**, *387*, 1540–1550. [CrossRef]
5. Rittmeyer, A.; Barlesi, F.; Waterkamp, D.; Park, K.; Ciardiello, F.; von Pawel, J.; Gadgeel, S.M.; Hida, T.; Kowalski, D.M.; Dols, M.C.; et al. Atezolizumab versus docetaxel in patients with previously treated non-small-cell lung cancer (oak): A phase 3, open-label, multicentre randomised controlled trial. *Lancet* **2017**, *389*, 255–265. [CrossRef]
6. Horita, H.; Law, A.; Hong, S.; Middleton, K. Identifying regulatory posttranslational modifications of pd-l1: A focus on monoubiquitinaton. *Neoplasia* **2017**, *19*, 346–353. [CrossRef] [PubMed]
7. Sharpe, A.H.; Pauken, K.E. The diverse functions of the pd1 inhibitory pathway. *Nat. Reviews. Immunol.* **2018**, *18*, 153–167. [CrossRef]
8. Ma, W.; Gilligan, B.M.; Yuan, J.; Li, T. Current status and perspectives in translational biomarker research for pd-1/pd-l1 immune checkpoint blockade therapy. *J. Hematol. Oncol.* **2016**, *9*, 47. [CrossRef]
9. First anti-pd-l1 drug approved for nsclc. *Cancer Discov.* **2016**, *6*, OF1. [CrossRef]
10. Sul, J.; Blumenthal, G.M.; Jiang, X.; He, K.; Keegan, P.; Pazdur, R. Fda approval summary: Pembrolizumab for the treatment of patients with metastatic non-small cell lung cancer whose tumors express programmed death-ligand 1. *Oncologist* **2016**, *21*, 643–650. [CrossRef]
11. Available online: https://clinicaltrials.gov/ (accessed on 1 August 2019).
12. Bylicki, O.; Barazzutti, H.; Paleiron, N.; Margery, J.; Assie, J.B.; Chouaid, C. First-line treatment of non-small-cell lung cancer (nsclc) with immune checkpoint inhibitors. *BioDrugs Clin. Immunother. Biopharm. Gene Ther.* **2019**, *33*, 159–171. [CrossRef] [PubMed]
13. Malhotra, J.; Jabbour, S.K.; Aisner, J. Current state of immunotherapy for non-small cell lung cancer. *Transl. Lung Cancer Res.* **2017**, *6*, 196–211. [CrossRef] [PubMed]
14. Califano, R.; Lal, R.; Lewanski, C.; Nicolson, M.C.; Ottensmeier, C.H.; Popat, S.; Hodgson, M.; Postmus, P.E. Patient selection for anti-pd-1/pd-l1 therapy in advanced non-small-cell lung cancer: Implications for clinical practice. *Future Oncol.* **2018**, *14*, 2415–2431. [CrossRef]
15. Wang, Y.; Wang, H.; Yao, H.; Li, C.; Fang, J.Y.; Xu, J. Regulation of pd-l1: Emerging routes for targeting tumor immune evasion. *Front. Pharmacol.* **2018**, *9*, 536. [CrossRef] [PubMed]
16. Tanos, R.; Thierry, A.R. Clinical relevance of liquid biopsy for cancer screening. *Transl. Cancer Res.* **2018**, S105–S129. [CrossRef]
17. Aktas, B.; Kasimir-Bauer, S.; Muller, V.; Janni, W.; Fehm, T.; Wallwiener, D.; Pantel, K.; Tewes, M.; Group, D.S. Comparison of the her2, estrogen and progesterone receptor expression profile of primary tumor, metastases and circulating tumor cells in metastatic breast cancer patients. *BMC Cancer* **2016**, *16*, 522. [CrossRef]

18. Rizvi, N.A.; Hellmann, M.D.; Snyder, A.; Kvistborg, P.; Makarov, V.; Havel, J.J.; Lee, W.; Yuan, J.; Wong, P.; Ho, T.S.; et al. Cancer immunology. Mutational landscape determines sensitivity to pd-1 blockade in non-small cell lung cancer. *Science* **2015**, *348*, 124–128. [CrossRef]
19. Guibert, N.; Delaunay, M.; Lusque, A.; Boubekeur, N.; Rouquette, I.; Clermont, E.; Mourlanette, J.; Gouin, S.; Dormoy, I.; Favre, G.; et al. Pd-l1 expression in circulating tumor cells of advanced non-small cell lung cancer patients treated with nivolumab. *Lung Cancer* **2018**, *120*, 108–112. [CrossRef]
20. Kulasinghe, A.; Kapeleris, J.; Kimberley, R.; Mattarollo, S.R.; Thompson, E.W.; Thiery, J.P.; Kenny, L.; O'Byrne, K.; Punyadeera, C. The prognostic significance of circulating tumor cells in head and neck and non-small-cell lung cancer. *Cancer Med.* **2018**, *7*, 5910–5919. [CrossRef]
21. Yue, C.; Jiang, Y.; Li, P.; Wang, Y.; Xue, J.; Li, N.; Li, D.; Wang, R.; Dang, Y.; Hu, Z.; et al. Dynamic change of pd-l1 expression on circulating tumor cells in advanced solid tumor patients undergoing pd-1 blockade therapy. *Oncoimmunology* **2018**, *7*, e1438111. [CrossRef]
22. Mazel, M.; Jacot, W.; Pantel, K.; Bartkowiak, K.; Topart, D.; Cayrefourcq, L.; Rossille, D.; Maudelonde, T.; Fest, T.; Alix-Panabieres, C. Frequent expression of pd-l1 on circulating breast cancer cells. *Mol. Oncol.* **2015**, *9*, 1773–1782. [CrossRef] [PubMed]
23. Schehr, J.L.; Schultz, Z.D.; Warrick, J.W.; Guckenberger, D.J.; Pezzi, H.M.; Sperger, J.M.; Heninger, E.; Saeed, A.; Leal, T.; Mattox, K.; et al. High specificity in circulating tumor cell identification is required for accurate evaluation of programmed death-ligand 1. *PLoS ONE* **2016**, *11*, e0159397. [CrossRef] [PubMed]
24. Bao, H.; Bai, T.; Takata, K.; Yokobori, T.; Ohnaga, T.; Hisada, T.; Maeno, T.; Bao, P.; Yoshida, T.; Kumakura, Y.; et al. High expression of carcinoembryonic antigen and telomerase reverse transcriptase in circulating tumor cells is associated with poor clinical response to the immune checkpoint inhibitor nivolumab. *Oncol. Lett.* **2018**, *15*, 3061–3067. [CrossRef] [PubMed]
25. Kallergi, G.; Vetsika, E.K.; Aggouraki, D.; Lagoudaki, E.; Koutsopoulos, A.; Koinis, F.; Katsarlinos, P.; Trypaki, M.; Messaritakis, I.; Stournaras, C.; et al. Evaluation of pd-l1/pd-1 on circulating tumor cells in patients with advanced non-small cell lung cancer. *Ther. Adv. Med. Oncol.* **2018**, *10*, 1758834017750121. [CrossRef] [PubMed]
26. Adams, D.L.; Adams, D.K.; He, J.; Kalhor, N.; Zhang, M.; Xu, T.; Gao, H.; Reuben, J.M.; Qiao, Y.; Komaki, R.; et al. Sequential tracking of pd-l1 expression and rad50 induction in circulating tumor and stromal cells of lung cancer patients undergoing radiotherapy. *Clin. Cancer Res. Off. J. Am. Assoc. Cancer Res.* **2017**, *23*, 5948–5958. [CrossRef] [PubMed]
27. Wang, Y.; Kim, T.H.; Fouladdel, S.; Zhang, Z.; Soni, P.; Qin, A.; Zhao, L.; Azizi, E.; Lawrence, T.S.; Ramnath, N.; et al. Pd-l1 expression in circulating tumor cells increases during radio(chemo)therapy and indicates poor prognosis in non-small cell lung cancer. *Sci. Rep.* **2019**, *9*, 566. [CrossRef] [PubMed]
28. Ilie, M.; Szafer-Glusman, E.; Hofman, V.; Chamorey, E.; Lalvee, S.; Selva, E.; Leroy, S.; Marquette, C.H.; Kowanetz, M.; Hedge, P.; et al. Detection of pd-l1 in circulating tumor cells and white blood cells from patients with advanced non-small-cell lung cancer. *Ann. Oncol. Off. J. Eur. Soc. Med. Oncol.* **2018**, *29*, 193–199. [CrossRef]
29. Janning, M.; Kobus, F.; Babayan, A.; Wikman, H.; Velthaus, J.L.; Bergmann, S.; Schatz, S.; Falk, M.; Berger, L.A.; Bottcher, L.M.; et al. Determination of pd-l1 expression in circulating tumor cells of nsclc patients and correlation with response to pd-1/pd-l1 inhibitors. *Cancers* **2019**, *11*, 835. [CrossRef]
30. Nicolazzo, C.; Raimondi, C.; Mancini, M.; Caponnetto, S.; Gradilone, A.; Gandini, O.; Mastromartino, M.; Del Bene, G.; Prete, A.; Longo, F.; et al. Monitoring pd-l1 positive circulating tumor cells in non-small cell lung cancer patients treated with the pd-1 inhibitor nivolumab. *Sci. Rep.* **2016**, *6*, 31726. [CrossRef]
31. Hanssen, A.; Loges, S.; Pantel, K.; Wikman, H. Detection of circulating tumor cells in non-small cell lung cancer. *Front. Oncol.* **2015**, *5*, 207. [CrossRef]
32. Raimondi, C.; Carpino, G.; Nicolazzo, C.; Gradilone, A.; Gianni, W.; Gelibter, A.; Gaudio, E.; Cortesi, E.; Gazzaniga, P. Pd-l1 and epithelial-mesenchymal transition in circulating tumor cells from non-small cell lung cancer patients: A molecular shield to evade immune system? *Oncoimmunology* **2017**, *6*, e1315488. [CrossRef] [PubMed]
33. Dhar, M.; Wong, J.; Che, J.; Matsumoto, M.; Grogan, T.; Elashoff, D.; Garon, E.B.; Goldman, J.W.; Sollier Christen, E.; Di Carlo, D.; et al. Evaluation of pd-l1 expression on vortex-isolated circulating tumor cells in metastatic lung cancer. *Sci. Rep.* **2018**, *8*, 2592. [CrossRef] [PubMed]

34. Herbst, R.S.; Soria, J.C.; Kowanetz, M.; Fine, G.D.; Hamid, O.; Gordon, M.S.; Sosman, J.A.; McDermott, D.F.; Powderly, J.D.; Gettinger, S.N.; et al. Predictive correlates of response to the anti-pd-l1 antibody mpdl3280a in cancer patients. *Nature* **2014**, *515*, 563–567. [CrossRef] [PubMed]
35. Kim, M.Y.; Koh, J.; Kim, S.; Go, H.; Jeon, Y.K.; Chung, D.H. Clinicopathological analysis of pd-l1 and pd-l2 expression in pulmonary squamous cell carcinoma: Comparison with tumor-infiltrating t cells and the status of oncogenic drivers. *Lung Cancer* **2015**, *88*, 24–33. [CrossRef] [PubMed]
36. He, Y.; Rozeboom, L.; Rivard, C.J.; Ellison, K.; Dziadziuszko, R.; Yu, H.; Zhou, C.; Hirsch, F.R. Pd-1, pd-l1 protein expression in non-small cell lung cancer and their relationship with tumor-infiltrating lymphocytes. *Med. Sci. Monit. Int. Med J. Exp. Clin. Res.* **2017**, *23*, 1208–1216.
37. Neumann, M.H.D.; Bender, S.; Krahn, T.; Schlange, T. Ctdna and ctcs in liquid biopsy-current status and where we need to progress. *Comput. Struct. Biotechnol. J.* **2018**, *16*, 190–195. [CrossRef]
38. Grolz, D.; Hauch, S.; Schlumpberger, M.; Guenther, K.; Voss, T.; Sprenger-Haussels, M.; Oelmuller, U. Liquid biopsy preservation solutions for standardized pre-analytical workflows-venous whole blood and plasma. *Curr. Pathobiol. Rep.* **2018**, *6*, 275–286. [CrossRef]
39. Farace, F.; Massard, C.; Vimond, N.; Drusch, F.; Jacques, N.; Billiot, F.; Laplanche, A.; Chauchereau, A.; Lacroix, L.; Planchard, D.; et al. A direct comparison of cellsearch and iset for circulating tumour-cell detection in patients with metastatic carcinomas. *Br. J. Cancer* **2011**, *105*, 847–853. [CrossRef]
40. Krebs, M.G.; Hou, J.M.; Sloane, R.; Lancashire, L.; Priest, L.; Nonaka, D.; Ward, T.H.; Backen, A.; Clack, G.; Hughes, A.; et al. Analysis of circulating tumor cells in patients with non-small cell lung cancer using epithelial marker-dependent and -independent approaches. *J. Thorac. Oncol. Off. Publ. Int. Assoc. Study Lung Cancer* **2012**, *7*, 306–315. [CrossRef]
41. Hosokawa, M.; Kenmotsu, H.; Koh, Y.; Yoshino, T.; Yoshikawa, T.; Naito, T.; Takahashi, T.; Murakami, H.; Nakamura, Y.; Tsuya, A.; et al. Size-based isolation of circulating tumor cells in lung cancer patients using a microcavity array system. *PLoS ONE* **2013**, *8*, e67466. [CrossRef]
42. Parra, E.R.; Villalobos, P.; Mino, B.; Rodriguez-Canales, J. Comparison of different antibody clones for immunohistochemistry detection of programmed cell death ligand 1 (pd-l1) on non-small cell lung carcinoma. *Appl. Immunohistochem. Mol. Morphol. AIMM* **2018**, *26*, 83–93. [CrossRef] [PubMed]
43. Ratcliffe, M.J.; Sharpe, A.; Midha, A.; Barker, C.; Scott, M.; Scorer, P.; Al-Masri, H.; Rebelatto, M.C.; Walker, J. Agreement between programmed cell death ligand-1 diagnostic assays across multiple protein expression cutoffs in non-small cell lung cancer. *Clin. Cancer Res. Off. J. Am. Assoc. Cancer Res.* **2017**, *23*, 3585–3591. [CrossRef] [PubMed]
44. Scheel, A.H.; Baenfer, G.; Baretton, G.; Dietel, M.; Diezko, R.; Henkel, T.; Heukamp, L.C.; Jasani, B.; Johrens, K.; Kirchner, T.; et al. Interlaboratory concordance of pd-l1 immunohistochemistry for non-small-cell lung cancer. *Histopathology* **2018**, *72*, 449–459. [CrossRef] [PubMed]
45. Adam, J.; Le Stang, N.; Rouquette, I.; Cazes, A.; Badoual, C.; Pinot-Roussel, H.; Tixier, L.; Danel, C.; Damiola, F.; Damotte, D.; et al. Multicenter harmonization study for pd-l1 ihc testing in non-small-cell lung cancer. *Ann. Oncol. Off. J. Eur. Soc. Med. Oncol.* **2018**, *29*, 953–958. [CrossRef] [PubMed]
46. Rimm, D.L.; Han, G.; Taube, J.M.; Yi, E.S.; Bridge, J.A.; Flieder, D.B.; Homer, R.; West, W.W.; Wu, H.; Roden, A.C.; et al. A prospective, multi-institutional, pathologist-based assessment of 4 immunohistochemistry assays for pd-l1 expression in non-small cell lung cancer. *JAMA Oncol.* **2017**, *3*, 1051–1058. [CrossRef] [PubMed]

© 2019 by the authors. Licensee MDPI, Basel, Switzerland. This article is an open access article distributed under the terms and conditions of the Creative Commons Attribution (CC BY) license (http://creativecommons.org/licenses/by/4.0/).

Review

Never Travel Alone: The Crosstalk of Circulating Tumor Cells and the Blood Microenvironment

Simon Heeke [1,2], Baharia Mograbi [1,2], Catherine Alix-Panabières [3] and Paul Hofman [1,2,4,*]

1. Université Côte d'Azur, CHU Nice, FHU OncoAge, 06000 Nice, France
2. Université Côte d'Azur, CNRS UMR7284, Inserm U1081, Institute for Research on Cancer and Aging, Nice (IRCAN), FHU OncoAge, 06000 Nice, France
3. Laboratory of Rare Human Circulating Cells (LCCRH), University Medical Centre, EA2415, Montpellier University, 34093 Montpellier, France
4. Laboratory of Clinical and Experimental Pathology and Biobank BB-0033-00025, Pasteur Hospital, FHU OncoAge, 06000 Nice, France
* Correspondence: hofman.p@chu-nice.fr; Tel.: +33-4-9203-8855

Received: 21 June 2019; Accepted: 11 July 2019; Published: 13 July 2019

Abstract: Commonly, circulating tumor cells (CTCs) are described as source of metastasis in cancer patients. However, in this process cancer cells of the primary tumor site need to survive the physical and biological challenges in the blood stream before leaving the circulation to become the seed of a new metastatic site in distant parenchyma. Most of the CTCs released in the blood stream will not resist those challenges and will consequently fail to induce metastasis. A few of them, however, interact closely with other blood cells, such as neutrophils, platelets, and/or macrophages to survive in the blood stream. Recent studies demonstrated that the interaction and modulation of the blood microenvironment by CTCs is pivotal for the development of new metastasis, making it an interesting target for potential novel treatment strategies. This review will discuss the recent research on the processes in the blood microenvironment with CTCs and will outline currently investigated treatment strategies.

Keywords: circulating tumor cells; hematological cells; neutrophils; platelets; liquid biopsy

---

1. Introduction

Circulating tumor cells (CTCs) have been extensively studied over the last decades, in particular as they play a crucial role in the diagnosis and the prognosis in many solid tumors as well as due to their predictive value associated with cancer targeted therapies as well as with immunotherapies [1–3]. CTCs are present in the blood stream as isolated CTCs (iCTCs) or in clusters of variable sizes that are often referred to as circulating tumor microemboli (CTMs) [4]. Following their migration from the primary site of the tumor into the blood, the tumor cells are constrained to high pressure and turbulences due to the blood stream and have to develop mechanisms of resistance for survival to consequently be able to adhere to the endothelium for tissue invasion and development of metastases [5]. Moreover, some CTCs are also able to come back to the primary tumor site and, consequently, to participate to the tumor growth [6]. However, the physical characteristics allowing the CTCs to survive are only partially known. Nevertheless, the biological characteristics of these cells and the phenotypic, genetic, and epigenetic modifications occurring during their migration from the primary tumor site until the development of distant metastases are beginning to be unraveled.

CTCs need to undergo significant changes to survive in the bloodstream—a new different environment. Thus, CTCs are challenged by physical forces in the circulation, they have to avoid being detected and killed by the immune system and finally, they need to extravasate from the blood stream to become the seed of new metastatic site(s) [7]. Recent works demonstrated that most of CTCs are not single cells travelling the blood alone but are accompanied by a plethora of blood cells and other CTCs and that a close interaction in the blood microenvironment is certainly needed to establish novel metastasis [8]. Interfering with this new microenvironment might help to develop strategies reducing the metastatic potential of tumors [8].

The aim of this review is therefore to summarize current knowledge concerning the role of the blood microenvironment and the different biological mechanisms occurring during its cross talk with CTCs. Additionally, potential therapeutic strategies and clinical approaches are discussed.

2. Brief Background on the Pathophysiology of CTCs Into the Blood Stream

2.1. The CTCs and the Constraints Due to the Blood Circulation

CTCs derive from primary tumor and/or metastatic sites and are consequently not adapted to the manifold challenges in the blood stream. Importantly, the flow of the blood stream, especially when passing the heart chambers, exposes cells to high mechanical sheer forces that can either directly destroy non-adapted cells or induce apoptosis in them [5,9,10]. Interestingly, CTCs seem to be stiffer than blood cells demonstrating their low adaptation to the blood stream [11] and tumor cells seem to be sensitive to those sheer forces indicating that the majority of CTCs will undergo apoptosis rather than forming metastasis in patients [12]. However, the different hemodynamic forces are important to allow the extravasation of tumor cells as they also remodel the endothelium [13], and consequently more knowledge on the biophysical properties allowing the formation of metastasis are needed [11]. Additionally, CTCs are directly exposed to the immune system and consequently they need to evade the detection from immune cells. Interestingly, programmed death ligand 1 (PD-L1), a costimulatory molecule inhibiting immune response can be expressed on CTCs and is associated with worse prognosis in lung [14,15] and head and neck cancer patients [16]. This indicates the active modulation of the immune response of CTCs to survive in the blood stream.

Lastly, as CTCs from cancers are of epithelial origin, they are adapted to grow in a network with other cells and are tightly interconnected by transmembrane proteins called integrins [17,18]. CTCs that leave the primary tumor site and enter the bloodstream lose the tight interaction with the surrounding cells, which can induce apoptosis in those cells through a phenomenon called anoikis [19]. Consequently, suppression of anoikis is required for survival of CTCs in the bloodstream [20], either by interaction of CTCs with other blood cells or by internally suppressing anoikis by activation of integrin signaling independent of cell–cell contacts [21].

2.2. Isolated CTC and Circulating Tumor Microemboli

While single CTCs are travelling in the blood stream, it has been demonstrated that CTC clusters or circulating tumor microemboli (CTMs) have a dramatically increased metastatic potential, as demonstrated in lung [4] and breast cancers [22]. Interestingly, a recent study conducted in 43 breast cancer patients demonstrated that CTC clustering alters DNA methylation patterns and increases stemness and consequently metastasis [23]. Single-cell bisulfite sequencing of single CTCs and CTCs derived from clusters revealed that transcription factors that are associated with a stem cell-like phenotype, like OCT4, NANOG, or SOX2, where hypomethylated in CTC clusters compared to single CTCs [23]. Interestingly, the authors also performed a drug-screening using 2486 FDA approved drugs to analyze their ability to interfere with CTC clustering. Thirty-one drugs have been detected that could serve as novel treatment to reduce the metastatic potential in breast cancer patients [23]. Consequently, the metastatic potential of CTCs might be limited in isolated CTCs and more research focusing on CTC clusters (CTMs) rather than on single cells might allow the design of novel treatment strategies to interact with the formation of CTC cluster to avoid metastasis in cancer patients.

3. Interaction of CTCs with Neutrophils

The role of neutrophils in cancer progression has been extensively studied recently [24,25]. Previously, it has been demonstrated that increased levels of circulating neutrophils are associated with bad prognosis in advanced cancer patients [26,27]. Moreover, the neutrophil-to-lymphocyte ratio has been demonstrated to be a prognostic factor in solid tumors [28]. In a recent study, Szczerba et al. demonstrated that within white blood cells (WBCs)–CTCs clusters of breast carcinoma, CTCs are associated with neutrophils in the majority of the cases [29]. Interestingly, using single-cell RNA sequencing, the authors showed that the transcriptome profiles of CTCs associated with neutrophils are different from those of CTCs alone, with differentially expressed genes that outline cell cycle progression leading to a more efficient metastasis formation. The authors noted that WBC–CTC clusters are relatively rare (less than 3.5%), whereas iCTCs alone are present in 88% of cases and CTCs clusters in less than 9% of the cases [29]. Despite, neutrophils directly interact with CTCs via ICAM-1 and neutrophils bound to CTCs facilitate the interaction of CTCs with endothelial cells in the liver, thereby promoting extravasation and liver metastasis [30]. Consequently, neutrophils play a major role for the CTC extravasation across the endothelial barrier and the onset of metastases (Figure 1) [31].

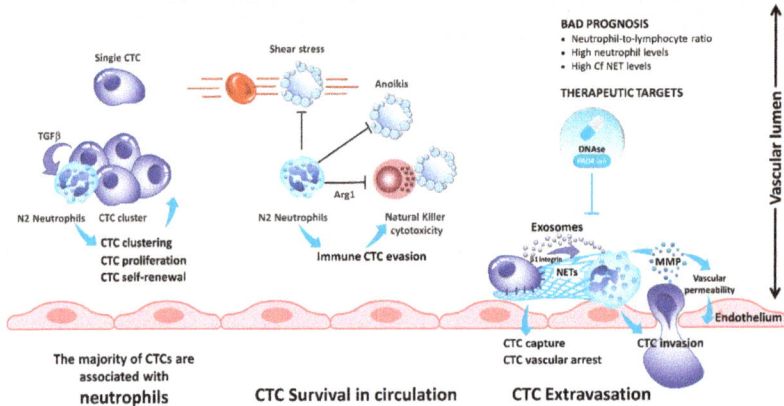

Figure 1. Compelling evidence indicates that blood neutrophils can offer proliferative and survival advantages to circulating tumor cells (CTCs) during their journey in the blood stream, rendering them more competent for metastasis development. Tumor-derived inflammatory factors strongly stimulate neutrophils to extrude chromatin webs called "neutrophil extracellular traps" (NETs). NETs, in turn, provide a niche to CTCs, arrest CTCs rolling, and promote metastasis. As such, understanding interaction of inflammatory N2 neutrophils with CTCs provides new potential therapeutic targets for disrupting these deadly metastatic seeds. As an example, blockade of NET formation using peptidylarginine deiminase 4 (PAD4) pharmacologic inhibitor or DNAse may decrease CTC colonization.

Neutrophils are able to generate neutrophil extracellular traps (NETs) by secreting their chromatin content during a process known as NETosis [32]. Initially, this process was described to be a mechanism to kill bacteria [33]. However, recent studies demonstrated that NETs are also promoting metastasis across various cancers [34–38]. Tohme et al. demonstrated that the NET formation induces a TLR-9 mediated response in cancer cells, which increased the migration and proliferation of CTCs [37]. Interestingly, it was shown that NETs promote extravasation of CTCs but in an IL-8 dependent manner. Consequently, blocking of IL-8 reduced the extravasation of tumor cells and neutrophils (Figure 1) [31]. Additionally, tumor-derived exosomes were also able to induce NET formation in neutrophils isolated from mice treated with granulocyte colony-stimulating factor (G-CSF) [39]. This phenomenon is even more important for tumors producing a large quantity of G-CSF associated with a high number of blood neutrophils [40,41]. Mechanistically, it has been demonstrated that the interaction of CTCs with NETs is mediated by β1-integrins expressed on tumor cells [42]. It is noteworthy that this integrin is physiologically overexpressed during infections and sepsis [42]. Generally, resolving NETs, for example, using DNAse I administration, reduced the number of metastasis making the NETs an interesting target for novel treatments reducing metastasis in patients (Figure 1) [34–38].

Different populations of neutrophils have different tumor promoting effects, and neutrophils are heavily modulated during cancer progression [43,44]. Consequently, extensive research is necessary to better decipher the role of the interaction of neutrophils and CTCs.

4. Interaction of CTCs with Myeloid-Derived Suppressor Cells

Myeloid-derived suppressor cells (MDSCs) are a heterogenous group of cells that are derived from the bone marrow and that are able to suppress the immune response via the suppression of T-cell response [45]. Commonly, MDSCs are classified in polymorphonuclear MDSCs (PMN-MDSC) and monocytic MDSCs (M-MDSC) [45]. As MDSCs are enriched in tumor tissue and able to suppress the immune system, it has been proposed that the interaction of CTCs with MDSCs might also promote metastasis [46]. Indeed, heterogenic clusters of CTCs and MDSCs have been reported in melanoma, pancreatic, and breast cancer patients [47,48], and co-culture of MDSCs with CTCs and T-cells demonstrated the T-cell suppressive effect of MDSCs [48]. Moreover, CTCs and MDSCs interact directly with each other, and increased reactive oxygen species (ROS) production in MDSCs induced NOTCH1 in CTCs, hereby promoting CTC proliferation [47]. Consequently, blocking the MDSC–CTC interaction might inhibit CTC proliferation and CTC immune evasion and might be an interesting target in anti-cancer therapy.

5. Interaction of CTCs with Platelets

In 1973, the role of platelets in cancer metastasis was already described, and the following work highlighted the role of platelets in cancer progression, especially during cancer metastases [49]. Indeed, different mechanisms occur during platelet–cancer cell interactions and crosstalk: (i) cancer cells can induce platelet activation; (ii) platelets support cancer metastasis and enhance cancer cell adhesion and arrest in vasculature; (iii) platelets assist immune evasion of cancer cells, and finally, (iv) platelets can enhance cancer evasion and tumor angiogenesis (Figure 2).

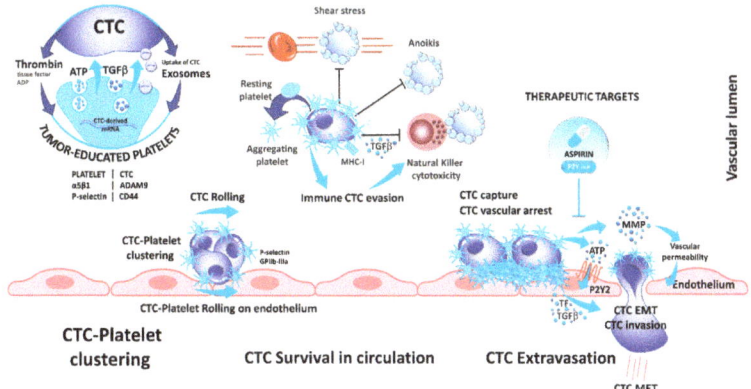

Figure 2. The dialogue between platelets and CTCs is reciprocal: CTCs activate and educate platelets while platelets contribute to CTCs' survival, escape from immune surveillance, tumor–endothelium interactions, and dissemination. Secretion of α-granules by activated platelets release high levels of TGF-β and ATP, a powerful activator of epithelial-to-mesenchymal transitioned (EMT) state and an endothelium relaxation factor (via P2Y), respectively. Inhibition of platelet α-granules secretion (Cox1) by aspirin, or of P2Y may abolish the metastatic potential of CTCs.

Interestingly, platelets can take up circulating mRNA from the CTCs, suggesting a possible modification in the platelet transcriptome that resembles the tumor profile [50]. In this context, platelets that circulate through and contact tumor sites can undergo modification due to the sequestration of RNA and biomolecules, which led to the concept of tumor-educated platelets (TEP) and may serve as an informative tool in cancer diagnosis [51,52]. This adherence could also help to decrease the impact of the pressure and the turbulence, especially in the heart chamber on the CTCs and can protect the CTCs against the physical stress in the blood stream (Figure 2) [53]. Indeed, platelets can form aggregates with CTCs, and CTCs induce platelet aggregation in a process known as tumor-cell-induced platelet aggregation (Figure 2) [54].

Platelets interact with tumor cells during blood dissemination leading to platelet activation and release of soluble mediators that alter the phenotype of the tumor cells and surrounding host cells [55]. However, the proximal events that initiate platelet activation are only partially characterized. It has been recently demonstrated that CD97 expressed on tumor cells may be involved in platelets activation [56]. CD97 is a g-protein coupled receptor that is undetectable in normal tissues except for smooth muscle cells but is abnormally expressed in different types of solid tumors. Ward et al. demonstrated that CD97 is able to activate platelets, which in turn secrete several mediators of the endothelial barrier, including ATP, which promotes evasion of CTCs off the blood stream and consequently promotes metastasis [56]. Additionally, the platelet P-selectin interacts with tumor CD44, and the fibrinogen receptor GPIIb-IIIa are involved in platelet rolling on CTCs and in platelet–CTC emboli (Figure 2) [57,58]. This leads to several alterations of platelets including protein synthesis, exosome release, blebbing of the membrane, and splicing of mRNAs [59]. Therefore, it seems that platelets form homotypic aggregates at the center of clusters that are surrounded by tumor cells at the periphery [54].

Despite the alterations induced by a direct contact between platelets and CTCs, the production of cytokines by platelets modifies the phenotype of CTCs. Molecular mechanisms by which CTCs maintain an epithelial-to-mesenchymal transitioned (EMT) state remain unclear. CTC clusters isolated from patients with advanced breast cancers highly exhibit mesenchymal markers and show an abundance of attached CD61-positive/platelets [60]. TGFβ1 secretion by alpha granules induces or increases the EMT observed in CTCs [55]. Likewise, platelets increase the tissue factor (TF) and the P2Y12 receptor activity, and both participate in EMT [61,62]. Moreover, platelets are involved in the adherence of CTCs to the endothelial barrier and to the transmigration of CTCs into the tissue for the development of metastasis [63]. One of the receptor ligand pair identified with such function is the ADAMA9 on CTCs that binds to the integrin α5β1 on the surface of platelets. This interaction is believed to promote platelet activation, granule secretion, and the transmigration of tumor cells through the endothelium [64]. Other mechanisms arising during the interaction between platelets and CTCs promote the migration of CTCs across the vasculature barrier: CTC-induced platelet aggregation leads to the release of ATP stored in dense granules; released ATP binds to the P2Y2 receptor stimulating cancer cell intravasation and metastatic dissemination [65]. Additionally, platelets and megacaryocytes play a major role in the survival of CTCs in the blood stream by different mechanisms. Platelets can protect certain CTCs against anoikis (a form of apoptosis that is induced when adherent cell lose contact to the surrounding cells) [66]. Even more, the adherence of platelets at the surface of CTCs may protect the CTCs to be recognized by some circulating immune cells, thereby promoting cell survival (Figure 2) [59,67]. Interestingly, platelets exert paracrine suppression of NK-mediated cytolytic activity. TGFβ released from activated platelets counteracts NK granule mobilization, cytotoxicity, and interferon-γ secretion [68]. Besides, platelet–CTC interaction can lead to the transfer of platelet major histocompatibility complex class I (MHC-I) to tumor cells preventing NK cell recognition via direct cell contacts [69]. This phenomenon is complex and not completely understood, and the platelets need to be activated at contact of CTCs entering in the circulation. Therefore, CTCs can release thrombin that attracts, activates, and aggregates platelets on their surface [54,59,70]. Several factors such as TF, thrombin, and ATP secreted by either platelets or CTCs induce platelets activation and formation of platelet–cancer cell aggregates [54,70].

Due to their multifaceted role in cancer metastasis, blocking of platelet–CTC interaction has also been studied as pharmacological target to reduce metastasis. Recently, Gareau et al. demonstrated that blocking this interaction using the P2Y12 inhibitor ticagrelor, reduced the number of metastasis and prolonged survival in a murine breast cancer model [71]. Additionally, in a clinical phase II study investigating the effect of aspirin on CTCs, less CTCs were detected in colorectal cancer patients upon aspirin treatment and the detected CTCs showed a more epithelial phenotype. Unfortunately, the results were not confirmed in a breast cancer model [72]. However, both aspirin and P2Y12 inhibitors inhibit platelet activation and demonstrate that modulation of platelets can reduce CTCs and metastasis, and the recent trials paved the way to actively investigate how the modulation of platelets can prevent metastasis-related cancer death [71,72].

6. Interaction of CTCs with Macrophages

Tumor-associated macrophages (TAMs) play a key role in activating dissemination and providing protection against the immune system [73]. However, the interplay of macrophages and CTCs is poorly understood. Previous works have been made to investigate the interaction between the macrophages and the CTCs in small cell lung carcinoma (SCLC) [74,75]. In these latter studies, different interactions have been observed by establishing SCLC cell lines and co-culture experiments with peripheral blood mononuclear cells (PBMCs). The authors showed that interaction of PBMCs with SCLC cells promote the differentiation of monocytes into macrophages, which express CD14, CD163, and CD68. These macrophages can secrete different cytokines such as osteopontin, monocyte chemoattractant protein-1, IL-8, chitinase 3-like 1, platelet factor, IL-1ra, and the matrix metalloproteinase-9 [74]. Likewise, PCa prostate cancer cells cultured with monocyte conditioned cell culture media, showed an increased invasion in vitro mediated by the IL-13Rα2 receptor expressed on cancer cells [76]. Additionally, Wei et al. demonstrated that the crosstalk of macrophages with tumor cells is necessary for the induction of EMT and release of CTCs into the blood stream. In their study, expression of IL6 of TAMs increased the secretion of CCL2 in tumor cells, which in turn recruited new macrophages [77].

The discussed studies demonstrate that the interaction of macrophages and tumor cells is not only important for progression at the primary tumor site but also for the promotion and differentiation of CTCs [77]. However, the interaction might be much closer than previously expected. Zhang et al. demonstrated that some circulating macrophages might be able to phagocyte apoptotic CTCs and incorporate the tumor DNA into their nuclei, consequently obtaining some malignant features like expression of epithelial markers (such as cytokeratins) and stem cell markers (e.g., OCT4) [78]. Consequently, circulating monocytes from solid cancer patients can express both CD163 and EpCAM. This led to the concept of "tumacrophages," which have the potential of invasive tumor cells but are protected against the immune system [78]. Even more, Gast et al. showed in a seminal work that viable tumor cells can fuse with macrophages to create hybrid cells, sharing markers of both tumor cells and macrophages [79]. Hybrid cells sharing epithelial (EpCAM expression) and hematological cell markers (CD45) were protected from detection by the immune system and correlated with disease stage and overall survival across several cancers [79]. Similar results were recently confirmed in a glioblastoma model where tumor cell–macrophage fusion cells demonstrated an increased invasive potential [80].

Consequently, the interplay of CTCs with macrophages is certainly important for metastasis and the discovery of tumor cell–macrophage fusion cells will help to develop novel biomarkers for cancer progression as well as novel potential therapeutic targets to block metastasis in patients.

7. Interaction of CTCs with Lymphocytes

Tumor cells constantly need to avoid being detected by immune cells to avoid being killed by them [81]. Likewise, CTCs in the blood stream are constantly required to avoid activation of the immune cells. The recent success of anti-cancer immunotherapy, most notably of checkpoint blocking antibodies, in several cancers have demonstrated that the immune system can be reactivated to target cancer cells [82]. Unfortunately, only limited studies have been carried out on the interaction of CTCs and lymphocytes. However, an inverse correlation between $CD3^+$, $CD4^+$, and $CD8^+$ peripheral T-lymphocytes and CTCs in NSCLC [83] and between $CD8^+$ peripheral lymphocytes in breast cancer [84] have been shown. Moreover, several studies demonstrated that regulatory T-cells infiltrating the tumor or detected in the peripheral blood are significantly more prevalent in breast cancer patients with CTCs than in patients without detectable CTCs [83–85]. While unfortunately mechanistic studies evaluating the interplay between T-lymphocytes and CTCs are lacking, the present studies, however, indicate that immune suppression by regulatory T-cells help in tumor cell dissemination in the blood stream. However, the responsible targets and mechanisms have to be unraveled to better understand the interplay of CTCs with lymphocytes.

Additionally, CTCs seem to be able to block interaction with lymphocytes by upregulating the programmed death ligand 1 (PD-L1) that inhibits the activation of T-lymphocytes [2]. This allows the CTCs to avoid being detected by the immune system and was indeed correlated with worse prognosis in NSCLC patients undergoing radio (chemo)-therapy [86]. Nevertheless, this might be only one of many mechanisms CTCs have to adapt to avoid detection by immune cells in the blood stream, and further research and clarification is needed.

8. Challenges and Perspectives

While CTCs have long been considered to be isolated cells floating in the blood stream, recent research demonstrated the close interaction of CTCs with the blood microenvironment. CTCs need to establish close interaction not only with platelets and neutrophils, but also with macrophages and endothelial cells to resist the physical stress in the blood stream and to evade detection by the immune system to finally leave the blood stream to establish new metastatic sites (Table 1).

Table 1. Summary of interactions of circulating tumor cells with other cell types in the blood microenvironment.

Interaction of CTCs with Other Cell Type	Interacting Targets/Processes	Effect	References
Neutrophils	ICAM-1	Facilitating interaction with endothelial cells and consequently extravasation off the blood stream.	[30]
	β1-integrin, tumor-derived exosomes	Formation of neutrophil extracellular traps (NETs) promoting proliferation and extravasation.	[31,37,39,42]
Myeloid-derived suppressor cells (MDSCs)	Reactive oxygen species (ROS) production by MDSCs	Increased proliferation of CTCs and inhibition of T-cells.	[47,48]
Blood platelets	Exosomes	Formation of tumor-educated platelets (TEPs).	[50–52]
	CD97, CD44, ADAMA9-α5β1 integrin, ATP	Modulation of endothelial cells by platelets leading to extravasation of CTCs.	[56,63–65]
	Cytokines produced by platelets	Induction of epithelial-to-mesenchymal transition in CTCs.	[55,60–62]
	TGFβ secreted by platelets	Suppression of cytolytic NK cells.	[68]
Macrophages	Cytokines produced by Macrophages	Increased invasion EMT of CTCs and immune suppression.	[74,76,77]
	Fusion with CTCs	Formation of "tumacrophages" that are protected from immune detection with invasive potential.	[78–80]
Lymphocytes	PD-L1	Suppression of cytotoxic T-cells.	[2]

While seminal research demonstrated well that the blood microenvironment is crucial for cell seeding, the mechanisms and interaction networks are not fully understood, and more research is needed. However, blocking the interaction of CTCs with platelets [71,72] as well as the resolving of NETs [34–38] demonstrated that targeting the interaction of CTCs with other cells is a promising therapeutic target and future research will certainly establish novel treatments to improve survival in cancer patients.

Author Contributions: Conceptualization, P.H.; writing—original draft preparation, S.H., P.H.; writing—review and editing, S.H., B.M., C.A.-P., P.H.; visualization, B.M.

Funding: The authors thank the "Centre Hospitalier et Universitaire de Nice," the "Ligue Départementale de Lutte contre le Cancer des Alpes Maritimes", the "Conseil Départemental 06", the "Ville de Nice", the "Cancéropôle PACA", the French Government (National Research Agency, ANR) through the "Investments for the Future" LABEX SIGNALIFE: program reference # ANR-11-LABX-0028-01, the CANCAIR GENEXPOSOMICS project, Région PACA, and Dreal PACA for their support.

Conflicts of Interest: The authors declare no conflict of interest.

References

1. Alix-Panabieres, C.; Pantel, K. Clinical Applications of Circulating Tumor Cells and Circulating Tumor DNA as Liquid Biopsy. *Cancer Discov.* **2016**, *6*, 479–491. [CrossRef] [PubMed]
2. Hofman, P.; Heeke, S.; Alix-Panabières, C.; Pantel, K. Liquid biopsy in the era of immune-oncology. Is it ready for prime-time use for cancer patients? *Ann. Oncol. Off. J. Eur. Soc. Med. Oncol.* **2019**. [CrossRef] [PubMed]
3. Pantel, K.; Alix-Panabières, C. Liquid biopsy and minimal residual disease—latest advances and implications for cure. *Nat. Rev. Clin. Oncol.* **2019**, *16*, 409–424. [CrossRef] [PubMed]
4. Carlsson, A.; Nair, V.S.; Luttgen, M.S.; Keu, K.V.; Horng, G.; Vasanawala, M.; Kolatkar, A.; Jamali, M.; Iagaru, A.H.; Kuschner, W.; et al. Circulating Tumor Microemboli Diagnostics for Patients with Non–Small-Cell Lung Cancer. *J. Thorac. Oncol.* **2014**, *9*, 1111–1119. [CrossRef] [PubMed]
5. Wirtz, D.; Konstantopoulos, K.; Searson, P.C. The physics of cancer: The role of physical interactions and mechanical forces in metastasis. *Nat. Rev. Cancer* **2011**, *11*, 512–522. [CrossRef] [PubMed]
6. Kim, M.-Y.; Oskarsson, T.; Acharyya, S.; Nguyen, D.X.; Zhang, X.H.-F.; Norton, L.; Massagué, J. Tumor Self-Seeding by Circulating Cancer Cells. *Cell* **2009**, *139*, 1315–1326. [CrossRef] [PubMed]
7. Strilic, B.; Offermanns, S. Intravascular Survival and Extravasation of Tumor Cells. *Cancer Cell* **2017**, *32*, 282–293. [CrossRef] [PubMed]
8. Guo, B.; Oliver, T.G. Partners in Crime: Neutrophil–CTC Collusion in Metastasis. *Trends Immunol.* **2019**. [CrossRef] [PubMed]
9. Phillips, K.G.; Kuhn, P.; McCarty, O.J.T. Physical Biology in Cancer. 2. The physical biology of circulating tumor cells. *Am. J. Physiol. Physiol.* **2014**, *306*, C80–C88. [CrossRef] [PubMed]
10. Barnes, J.M.; Nauseef, J.T.; Henry, M.D. Resistance to Fluid Shear Stress Is a Conserved Biophysical Property of Malignant Cells. *PLoS ONE* **2012**, *7*, e50973. [CrossRef] [PubMed]
11. Shaw Bagnall, J.; Byun, S.; Begum, S.; Miyamoto, D.T.; Hecht, V.C.; Maheswaran, S.; Stott, S.L.; Toner, M.; Hynes, R.O.; Manalis, S.R. Deformability of Tumor Cells versus Blood Cells. *Sci. Rep.* **2015**, *5*, 18542. [CrossRef] [PubMed]
12. Regmi, S.; Fu, A.; Luo, K.Q. High Shear Stresses under Exercise Condition Destroy Circulating Tumor Cells in a Microfluidic System. *Sci. Rep.* **2017**, *7*, 39975. [CrossRef] [PubMed]
13. Follain, G.; Osmani, N.; Azevedo, A.S.; Allio, G.; Mercier, L.; Karreman, M.A.; Solecki, G.; Garcia Leòn, M.J.; Lefebvre, O.; Fekonja, N.; et al. Hemodynamic Forces Tune the Arrest, Adhesion, and Extravasation of Circulating Tumor Cells. *Dev. Cell* **2018**, *45*, 33–52. [CrossRef] [PubMed]
14. Guibert, N.; Delaunay, M.; Lusque, A.; Boubekeur, N.; Rouquette, I.; Clermont, E.; Mourlanette, J.; Gouin, S.; Dormoy, I.; Favre, G.; et al. PD-L1 expression in circulating tumor cells of advanced non-small cell lung cancer patients treated with nivolumab. *Lung Cancer* **2018**, *120*, 108–112. [CrossRef] [PubMed]
15. Ilié, M.; Szafer-Glusman, E.; Hofman, V.; Chamorey, E.; Lalvée, S.; Selva, E.; Leroy, S.; Marquette, C.-H.; Kowanetz, M.; Hedge, P.; et al. Detection of PD-L1 in circulating tumor cells and white blood cells from patients with advanced non-small-cell lung cancer. *Ann. Oncol. Off. J. Eur. Soc. Med. Oncol.* **2018**, *29*, 193–199. [CrossRef] [PubMed]
16. Strati, A.; Koutsodontis, G.; Papaxoinis, G.; Angelidis, I.; Zavridou, M.; Economopoulou, P.; Kotsantis, I.; Avgeris, M.; Mazel, M.; Perisanidis, C.; et al. Prognostic significance of PD-L1 expression on circulating tumor cells in patients with head and neck squamous cell carcinoma. *Ann. Oncol. Off. J. Eur. Soc. Med. Oncol.* **2017**, *28*, 1923–1933. [CrossRef]
17. Harburger, D.S.; Calderwood, D.A. Integrin signalling at a glance. *J. Cell Sci.* **2009**, *122*, 159–163. [CrossRef]

18. Winograd-Katz, S.E.; Fässler, R.; Geiger, B.; Legate, K.R. The integrin adhesome: From genes and proteins to human disease. *Nat. Rev. Mol. Cell Biol.* **2014**, *15*, 273–288. [CrossRef]
19. Gilmore, A.P. Anoikis. *Cell Death Differ.* **2005**, *12*, 1473–1477. [CrossRef]
20. Kim, Y.-N.; Koo, K.H.; Sung, J.Y.; Yun, U.-J.; Kim, H. Anoikis Resistance: An Essential Prerequisite for Tumor Metastasis. *Int. J. Cell Biol.* **2012**, *2012*, 1–11. [CrossRef]
21. Alanko, J.; Mai, A.; Jacquemet, G.; Schauer, K.; Kaukonen, R.; Saari, M.; Goud, B.; Ivaska, J. Integrin endosomal signalling suppresses anoikis. *Nat. Cell Biol.* **2015**, *17*, 1412–1421. [CrossRef] [PubMed]
22. Aceto, N.; Bardia, A.; Miyamoto, D.T.; Donaldson, M.C.; Wittner, B.S.; Spencer, J.A.; Yu, M.; Pely, A.; Engstrom, A.; Zhu, H.; et al. Circulating Tumor Cell Clusters Are Oligoclonal Precursors of Breast Cancer Metastasis. *Cell* **2014**, *158*, 1110–1122. [CrossRef] [PubMed]
23. Gkountela, S.; Castro-Giner, F.; Szczerba, B.M.; Vetter, M.; Landin, J.; Scherrer, R.; Krol, I.; Scheidmann, M.C.; Beisel, C.; Stirnimann, C.U.; et al. Circulating Tumor Cell Clustering Shapes DNA Methylation to Enable Metastasis Seeding. *Cell* **2019**, *176*, 98–112. [CrossRef] [PubMed]
24. Shaul, M.E.; Fridlender, Z.G. Cancer-related circulating and tumor-associated neutrophils–subtypes, sources and function. *FEBS J.* **2018**, *285*, 4316–4342. [CrossRef] [PubMed]
25. Wu, L.; Saxena, S.; Awaji, M.; Singh, R.K. Tumor-Associated Neutrophils in Cancer: Going Pro. *Cancers* **2019**, *11*, 564. [CrossRef] [PubMed]
26. Dumitru, C.A.; Moses, K.; Trellakis, S.; Lang, S.; Brandau, S. Neutrophils and granulocytic myeloid-derived suppressor cells: Immunophenotyping, cell biology and clinical relevance in human oncology. *Cancer Immunol. Immunother.* **2012**, *61*, 1155–1167. [CrossRef] [PubMed]
27. Zhao, W.; Wang, P.; Jia, H.; Chen, M.; Gu, X.; Liu, M.; Zhang, Z.; Cheng, W.; Wu, Z. Neutrophil count and percentage: Potential independent prognostic indicators for advanced cancer patients in a palliative care setting. *Oncotarget* **2017**, *8*, 64499–64508. [CrossRef] [PubMed]
28. Templeton, A.J.; Knox, J.J.; Lin, X.; Simantov, R.; Xie, W.; Lawrence, N.; Broom, R.; Fay, A.P.; Rini, B.; Donskov, F.; et al. Change in Neutrophil-to-lymphocyte Ratio in Response to Targeted Therapy for Metastatic Renal Cell Carcinoma as a Prognosticator and Biomarker of Efficacy. *Eur. Urol.* **2016**, *70*, 358–364. [CrossRef]
29. Szczerba, B.M.; Castro-Giner, F.; Vetter, M.; Krol, I.; Gkountela, S.; Landin, J.; Scheidmann, M.C.; Donato, C.; Scherrer, R.; Singer, J.; et al. Neutrophils escort circulating tumour cells to enable cell cycle progression. *Nature* **2019**. [CrossRef]
30. Chow, S.C.; Spicer, J.D.; Kubes, P.; Giannias, B.; Cools-Lartigue, J.J.; Ferri, L.E.; McDonald, B. Neutrophils Promote Liver Metastasis via Mac-1–Mediated Interactions with Circulating Tumor Cells. *Cancer Res.* **2012**, *72*, 3919–3927. [CrossRef]
31. Chen, M.B.; Hajal, C.; Benjamin, D.C.; Yu, C.; Azizgolshani, H.; Hynes, R.O.; Kamm, R.D. Inflamed neutrophils sequestered at entrapped tumor cells via chemotactic confinement promote tumor cell extravasation. *Proc. Natl. Acad. Sci. USA* **2018**, *115*, 7022–7027. [CrossRef] [PubMed]
32. Papayannopoulos, V. Neutrophil extracellular traps in immunity and disease. *Nat. Rev. Immunol.* **2018**, *18*, 134–147. [CrossRef] [PubMed]
33. Brinkmann, V. Neutrophil Extracellular Traps Kill Bacteria. *Science* **2004**, *303*, 1532–1535. [CrossRef] [PubMed]
34. Al-Haidari, A.A.; Algethami, N.; Lepsenyi, M.; Rahman, M.; Syk, I.; Thorlacius, H. Neutrophil extracellular traps promote peritoneal metastasis of colon cancer cells. *Oncotarget* **2019**, *10*, 1238–1249. [CrossRef] [PubMed]
35. Cools-Lartigue, J.; Spicer, J.; McDonald, B.; Gowing, S.; Chow, S.; Giannias, B.; Bourdeau, F.; Kubes, P.; Ferri, L. Neutrophil extracellular traps sequester circulating tumor cells and promote metastasis. *J. Clin. Investig.* **2013**, *123*, 3446–3458. [CrossRef] [PubMed]
36. Park, J.; Wysocki, R.W.; Amoozgar, Z.; Maiorino, L.; Fein, M.R.; Jorns, J.; Schott, A.F.; Kinugasa-Katayama, Y.; Lee, Y.; Won, N.H.; et al. Cancer cells induce metastasis-supporting neutrophil extracellular DNA traps. *Sci. Transl. Med.* **2016**, *8*, 361ra138. [CrossRef] [PubMed]
37. Tohme, S.; Yazdani, H.O.; Al-Khafaji, A.B.; Chidi, A.P.; Loughran, P.; Mowen, K.; Wang, Y.; Simmons, R.L.; Huang, H.; Tsung, A. Neutrophil Extracellular Traps Promote the Development and Progression of Liver Metastases after Surgical Stress. *Cancer Res.* **2016**, *76*, 1367–1380. [CrossRef] [PubMed]
38. Huh, S.J.; Liang, S.; Sharma, A.; Dong, C.; Robertson, G.P. Transiently Entrapped Circulating Tumor Cells Interact with Neutrophils to Facilitate Lung Metastasis Development. *Cancer Res.* **2010**, *70*, 6071–6082. [CrossRef]

39. Leal, A.C.; Mizurini, D.M.; Gomes, T.; Rochael, N.C.; Saraiva, E.M.; Dias, M.S.; Werneck, C.C.; Sielski, M.S.; Vicente, C.P.; Monteiro, R.Q. Tumor-Derived Exosomes Induce the Formation of Neutrophil Extracellular Traps: Implications For The Establishment of Cancer-Associated Thrombosis. *Sci. Rep.* **2017**, *7*, 1–12. [CrossRef]
40. Cedervall, J.; Zhang, Y.; Huang, H.; Zhang, L.; Femel, J.; Dimberg, A.; Olsson, A.K. Neutrophil extracellular traps accumulate in peripheral blood vessels and compromise organ function in tumor-bearing animals. *Cancer Res.* **2015**, *75*, 2653–2662. [CrossRef]
41. Thålin, C.; Demers, M.; Blomgren, B.; Wong, S.L.; von Arbin, M.; von Heijne, A.; Laska, A.C.; Wallén, H.; Wagner, D.D.; Aspberg, S. NETosis promotes cancer-associated arterial microthrombosis presenting as ischemic stroke with troponin elevation. *Thromb. Res.* **2016**, *139*, 56–64. [CrossRef] [PubMed]
42. Najmeh, S.; Cools-Lartigue, J.; Rayes, R.F.; Gowing, S.; Vourtzoumis, P.; Bourdeau, F.; Giannias, B.; Berube, J.; Rousseau, S.; Ferri, L.E.; et al. Neutrophil extracellular traps sequester circulating tumor cells via β1-integrin mediated interactions. *Int. J. Cancer* **2017**, *140*, 2321–2330. [CrossRef] [PubMed]
43. Patel, S.; Fu, S.; Mastio, J.; Dominguez, G.A.; Purohit, A.; Kossenkov, A.; Lin, C.; Alicea-Torres, K.; Sehgal, M.; Nefedova, Y.; et al. Unique pattern of neutrophil migration and function during tumor progression. *Nat. Immunol.* **2018**, *19*, 1236–1247. [CrossRef] [PubMed]
44. Engblom, C.; Pfirschke, C.; Zilionis, R.; Da Silva Martins, J.; Bos, S.A.; Courties, G.; Rickelt, S.; Severe, N.; Baryawno, N.; Faget, J.; et al. Osteoblasts remotely supply lung tumors with cancer-promoting SiglecF high neutrophils. *Science* **2017**, *358*, eaal5081. [CrossRef] [PubMed]
45. Veglia, F.; Perego, M.; Gabrilovich, D. Myeloid-derived suppressor cells coming of age. *Nat. Immunol.* **2018**, *19*, 108–119. [CrossRef] [PubMed]
46. Liu, Q.; Liao, Q.; Zhao, Y. Myeloid-derived suppressor cells (MDSC) facilitate distant metastasis of malignancies by shielding circulating tumor cells (CTC) from immune surveillance. *Med. Hypotheses* **2016**, *87*, 34–39. [CrossRef] [PubMed]
47. Sprouse, M.L.; Welte, T.; Boral, D.; Liu, H.N.; Yin, W.; Vishnoi, M.; Goswami-Sewell, D.; Li, L.; Pei, G.; Jia, P.; et al. PMN-MDSCs Enhance CTC Metastatic Properties through Reciprocal Interactions via ROS/Notch/Nodal Signaling. *Int. J. Mol. Sci.* **2019**, *20*, 1916. [CrossRef]
48. Arnoletti, J.P.; Fanaian, N.; Reza, J.; Sause, R.; Almodovar, A.J.O.; Srivastava, M.; Patel, S.; Veldhuis, P.P.; Griffith, E.; Shao, Y.P.; et al. Pancreatic and bile duct cancer circulating tumor cells (CTC) form immune-resistant multi-cell type clusters in the portal venous circulation. *Cancer Biol. Ther.* **2018**, *19*, 887–897. [CrossRef]
49. Gasic, G.J.; Gasic, T.B.; Galanti, N.; Johnson, T.; Murphy, S. Platelet—tumor-cell interactions in mice. The role of platelets in the spread of malignant disease. *Int. J. Cancer* **1973**, *11*, 704–718. [CrossRef]
50. Nilsson, R.J.A.; Balaj, L.; Hulleman, E.; van Rijn, S.; Pegtel, D.M.; Walraven, M.; Widmark, A.; Gerritsen, W.R.; Verheul, H.M.; Vandertop, W.P.; et al. Blood platelets contain tumor-derived RNA biomarkers. *Blood* **2011**, *118*, 3680–3683. [CrossRef]
51. Joosse, S.A.; Pantel, K. Tumor-Educated Platelets as Liquid Biopsy in Cancer Patients. *Cancer Cell* **2015**, *28*, 552–554. [CrossRef] [PubMed]
52. Best, M.G.; Sol, N.; Kooi, I.; Tannous, J.; Westerman, B.A.; Rustenburg, F.; Schellen, P.; Verschueren, H.; Post, E.; Koster, J.; et al. RNA-Seq of Tumor-Educated Platelets Enables Blood-Based Pan-Cancer, Multiclass, and Molecular Pathway Cancer Diagnostics. *Cancer Cell* **2015**, *28*, 666–676. [CrossRef] [PubMed]
53. Franco, A.T.; Corken, A.; Ware, J. Platelets at the interface of thrombosis, inflammation, and cancer. *Blood* **2015**, *126*, 582–588. [CrossRef] [PubMed]
54. Menter, D.G.; Tucker, S.C.; Kopetz, S.; Sood, A.K.; Crissman, J.D.; Honn, K.V. Platelets and cancer: A casual or causal relationship: Revisited. *Cancer Metastasis Rev.* **2014**, *33*, 231–269. [CrossRef] [PubMed]
55. Labelle, M.; Begum, S.; Hynes, R.O.O. Direct Signaling between Platelets and Cancer Cells Induces an Epithelial-Mesenchymal-Like Transition and Promotes Metastasis. *Cancer Cell* **2011**, *20*, 576–590. [CrossRef] [PubMed]
56. Ward, Y.; Lake, R.; Faraji, F.; Sperger, J.; Martin, P.; Gilliard, C.; Ku, K.P.; Rodems, T.; Niles, D.; Tillman, H.; et al. Platelets Promote Metastasis via Binding Tumor CD97 Leading to Bidirectional Signaling that Coordinates Transendothelial Migration. *Cell Rep.* **2018**, *23*, 808–822. [CrossRef]
57. Hanley, W.; McCarty, O.; Jadhav, S.; Tseng, Y.; Wirtz, D.; Konstantopoulos, K. Single molecule characterization of P-selectin/ligand binding. *J. Biol. Chem.* **2003**, *278*, 10556–10561. [CrossRef]

58. Camerer, E.; Qazi, A.A.; Duong, D.N.; Cornelissen, I.; Advincula, R.; Coughlin, S.R. Platelets, protease-activated receptors, and fibrinogen in hematogenous metastasis. *Blood* **2004**, *104*, 397–401. [CrossRef]
59. Gay, L.J.; Felding-Habermann, B. Contribution of platelets to tumour metastasis. *Nat. Rev. Cancer* **2011**, *11*, 123–134. [CrossRef]
60. Lu, Y.; Lian, S.; Ye, Y.; Yu, T.; Liang, H.; Cheng, Y.; Xie, J.; Zhu, Y.; Xie, X.; Yu, S.; et al. S-Nitrosocaptopril prevents cancer metastasis in vivo by creating the hostile bloodstream microenvironment against circulating tumor cells. *Pharmacol. Res.* **2019**, *139*, 535–549. [CrossRef]
61. Wang, Y.; Sun, Y.; Li, D.; Zhang, L.; Wang, K.; Zuo, Y.; Gartner, T.K.; Liu, J. Platelet P2Y12 is involved in murine pulmonary metastasis. *PLoS ONE* **2013**, *8*, 1–12. [CrossRef] [PubMed]
62. Orellana, R.; Kato, S.; Erices, R.; Bravo, M.L.; Gonzalez, P.; Oliva, B.; Cubillos, S.; Valdivia, A.; Ibañez, C.; Brañes, J.; et al. Platelets enhance tissue factor protein and metastasis initiating cell markers, and act as chemoattractants increasing the migration of ovarian cancer cells. *BMC Cancer* **2015**, *15*, 290. [CrossRef] [PubMed]
63. Läubli, H.; Borsig, L. Selectins promote tumor metastasis. *Semin. Cancer Biol.* **2010**, *20*, 169–177. [CrossRef] [PubMed]
64. Mammadova-Bach, E.; Zigrino, P.; Brucker, C.; Bourdon, C.; Freund, M.; De Arcangelis, A.; Abrams, S.I.; Orend, G.; Gachet, C.; Mangin, P.H. Platelet integrin α6β1 controls lung metastasis through direct binding to cancer cell–derived ADAM9. *JCI Insight* **2016**, *1*, 1–17. [CrossRef] [PubMed]
65. Schumacher, D.; Strilic, B.; Sivaraj, K.K.; Wettschureck, N.; Offermanns, S. Platelet-Derived Nucleotides Promote Tumor-Cell Transendothelial Migration and Metastasis via P2Y2 Receptor. *Cancer Cell* **2013**, *24*, 130–137. [CrossRef] [PubMed]
66. Velez, J.; Enciso, L.J.; Suarez, M.; Fiegl, M.; Grismaldo, A.; López, C.; Barreto, A.; Cardozo, C.; Palacios, P.; Morales, L.; et al. Platelets Promote Mitochondrial Uncoupling and Resistance to Apoptosis in Leukemia Cells: A Novel Paradigm for the Bone Marrow Microenvironment. *Cancer Microenviron.* **2014**, *7*, 79–90. [CrossRef]
67. Nieswandt, B.; Hafner, M.; Echtenacher, B.; Männel, D.N. Lysis of tumor cells by natural killer cells in mice is impeded by platelets. *Cancer Res.* **1999**, *59*, 1295–1300.
68. Palumbo, J.S.; Talmage, K.E.; Massari, J.V.; La Jeunesse, C.M.; Flick, M.J.; Kombrinck, K.W.; Hu, Z.; Barney, K.A.; Degen, J.L. Tumor cell-associated tissue factor and circulating hemostatic factors cooperate to increase metastatic potential through natural killer cell-dependent and -independent mechanisms. *Blood* **2007**, *110*, 133–141. [CrossRef]
69. Placke, T.; Örgel, M.; Schaller, M.; Jung, G.; Rammensee, H.G.; Kopp, H.G.; Salih, H.R. Platelet-derived MHC class I confers a pseudonormal phenotype to cancer cells that subverts the antitumor reactivity of natural killer immune cells. *Cancer Res.* **2012**, *72*, 440–448. [CrossRef]
70. Hu, L. Role of endogenous thrombin in tumor implantation, seeding, and spontaneous metastasis. *Blood* **2004**, *104*, 2746–2751. [CrossRef]
71. Gareau, A.J.; Brien, C.; Gebremeskel, S.; Liwski, R.S.; Johnston, B.; Bezuhly, M. Ticagrelor inhibits platelet–tumor cell interactions and metastasis in human and murine breast cancer. *Clin. Exp. Metastasis* **2018**, *35*, 25–35. [CrossRef] [PubMed]
72. Yang, L.; Lv, Z.; Xia, W.; Zhang, W.; Xin, Y.; Yuan, H.; Chen, Y.; Hu, X.; Lv, Y.; Xu, Q.; et al. The effect of aspirin on circulating tumor cells in metastatic colorectal and breast cancer patients: A phase II trial study. *Clin. Transl. Oncol.* **2018**, *20*, 912–921. [CrossRef] [PubMed]
73. Pathria, P.; Louis, T.L.; Varner, J.A. Targeting Tumor-Associated Macrophages in Cancer. *Trends Immunol.* **2019**, *40*, 310–327. [CrossRef] [PubMed]
74. Hamilton, G.; Rath, B.; Klameth, L.; Hochmair, M.J. Small cell lung cancer: Recruitment of macrophages by circulating tumor cells. *Oncoimmunology* **2016**, *5*, 1–9. [CrossRef] [PubMed]
75. Hamilton, G.; Rath, B. Circulating tumor cell interactions with macrophages: Implications for biology and treatment. *Transl. Lung Cancer Res.* **2017**, *6*, 418–430. [CrossRef] [PubMed]
76. Cavassani, K.A.; Meza, R.J.; Habiel, D.M.; Chen, J.F.; Montes, A.; Tripathi, M.; Martins, G.A.; Crother, T.R.; You, S.; Hogaboam, C.M.; et al. Circulating monocytes from prostate cancer patients promote invasion and motility of epithelial cells. *Cancer Med.* **2018**, *7*, 4639–4649. [CrossRef]

77. Wei, C.; Yang, C.; Wang, S.; Shi, D.; Zhang, C.; Lin, X.; Liu, Q.; Dou, R.; Xiong, B. Crosstalk between cancer cells and tumor associated macrophages is required for mesenchymal circulating tumor cell-mediated colorectal cancer metastasis. *Mol. Cancer* **2019**, *18*, 1–23. [CrossRef]
78. Zhang, Y.; Zhou, N.; Yu, X.; Zhang, X.; Li, S.; Lei, Z.; Hu, R.; Li, H.; Mao, Y.; Wang, X.; et al. Tumacrophage: Macrophages transformed into tumor stem-like cells by virulent genetic material from tumor cells. *Oncotarget* **2017**, *8*, 82326–82343. [CrossRef]
79. Gast, C.E.; Silk, A.D.; Zarour, L.; Riegler, L.; Burkhart, J.G.; Gustafson, K.T.; Parappilly, M.S.; Roh-Johnson, M.; Goodman, J.R.; Olson, B.; et al. Cell fusion potentiates tumor heterogeneity and reveals circulating hybrid cells that correlate with stage and survival. *Sci. Adv.* **2018**, *4*, eaat7828. [CrossRef]
80. Cao, M.-F.; Chen, L.; Dang, W.-Q.; Zhang, X.-C.; Zhang, X.; Shi, Y.; Yao, X.-H.; Li, Q.; Zhu, J.; Lin, Y.; et al. Hybrids by tumor-associated macrophages × glioblastoma cells entail nuclear reprogramming and glioblastoma invasion. *Cancer Lett.* **2019**, *442*, 445–452. [CrossRef]
81. Hanahan, D.; Weinberg, R.A. Hallmarks of Cancer: The Next Generation. *Cell* **2011**, *144*, 646–674. [CrossRef] [PubMed]
82. Ribas, A.; Wolchok, J.D. Cancer immunotherapy using checkpoint blockade. *Science* **2018**, *1355*, 1350–1355. [CrossRef] [PubMed]
83. Ye, L.; Zhang, F.; Li, H.; Yang, L.; Lv, T.; Gu, W.; Song, Y. Circulating Tumor Cells Were Associated with the Number of T Lymphocyte Subsets and NK Cells in Peripheral Blood in Advanced Non-Small-Cell Lung Cancer. *Dis. Markers* **2017**, *2017*, 1–6. [CrossRef] [PubMed]
84. Mego, M.; Gao, H.; Cohen, E.; Anfossi, S.; Giordano, A.; Sanda, T.; Fouad, T.; De Giorgi, U.; Giuliano, M.; Woodward, W.; et al. Circulating Tumor Cells (CTC) Are Associated with Defects in Adaptive Immunity in Patients with Inflammatory Breast Cancer. *J. Cancer* **2016**, *7*, 1095–1104. [CrossRef] [PubMed]
85. Xue, D.; Xia, T.; Wang, J.; Chong, M.; Wang, S.; Zhang, C. Role of regulatory T cells and CD8+ T lymphocytes in the dissemination of circulating tumor cells in primary invasive breast cancer. *Oncol. Lett.* **2018**, *16*, 3045–3053. [CrossRef] [PubMed]
86. Wang, Y.; Kim, T.H.; Fouladdel, S.; Zhang, Z.; Soni, P.; Qin, A.; Zhao, L.; Azizi, E.; Lawrence, T.S.; Ramnath, N.; et al. PD-L1 Expression in Circulating Tumor Cells Increases during Radio(chemo)therapy and Indicates Poor Prognosis in Non-small Cell Lung Cancer. *Sci. Rep.* **2019**, *9*, 566. [CrossRef] [PubMed]

© 2019 by the authors. Licensee MDPI, Basel, Switzerland. This article is an open access article distributed under the terms and conditions of the Creative Commons Attribution (CC BY) license (http://creativecommons.org/licenses/by/4.0/).

Review

Clinical Impact of Circulating Tumor Cells in Patients with Localized Prostate Cancer

Lucile Broncy [1] and Patrizia Paterlini-Bréchot [1,2,*]

[1] INSERM Unit 1151, Faculté de Médecine, Université Paris Descartes, 75014 Paris, France
[2] Laboratoire de Biochimie A, Hôpital Necker-Enfants Malades, 75015 Paris, France
* Correspondence: patrizia.paterlini@inserm.fr; Tel.: +33-172-606-462

Received: 3 June 2019; Accepted: 1 July 2019; Published: 3 July 2019

Abstract: The main issue concerning localized prostate cancers is the lack of a suitable marker which could help patients' stratification at diagnosis and distinguish those with a benign disease from patients with a more aggressive cancer. Circulating Tumor Cells (CTC) are spread in the blood by invasive tumors and could be the ideal marker in this setting. Therefore, we have compiled data from the literature in order to obtain clues about the clinical impact of CTC in patients with localized prostate cancer. Forty-three publications have been found reporting analyses of CTC in patients with non-metastatic prostate cancer. Of these, we have made a further selection of 11 studies targeting patients with clinical or pathological stages T1 and T2 and reporting the clinical impact of CTC. The results of this search show encouraging data toward the use of CTC in patients with early-stage cancer. However, they also highlight the lack of standardized methods providing a highly sensitive and specific approach for the detection of prostate-derived CTC.

Keywords: prostate cancer (PCa); circulating tumor cells (CTC); liquid biopsy

1. Introduction

Globally, prostate cancer (PCa) is the most commonly diagnosed type of cancer in men, with 180,890 newly diagnosed cases and 26,120 deaths in the United States (US) in 2016 [1] and estimated 164,690 new cases and 29,430 deaths in 2018 [2]. As the fifth leading cause of cancer death worldwide, PCa accounted globally for 1.6 million newly diagnosed cases and 366,000 deaths in 2015 [1]. Increased risk factors for PCa include genetic predisposition, family history of prostate or breast cancers and older age, with a median age at diagnosis of 72 years [3]. Accordingly, the steady increase in PCa incidence in the US since 1950 appears related to an overall increase in life expectancy. In the US, the lifetime risk of being diagnosed with prostate cancer is of 11% and the lifetime risk of dying of prostate cancer is 2.5% [4]. Importantly, in autopsies of men who died of other causes, more than 20% of men aged 50 to 59 years and over 33% of men aged 70 to 79 years were found to have prostate cancer [5]. At diagnosis, 79% of prostate cancer cases were localized; in 12%, the cancer had spread to regional lymph nodes, and 5% of patients had distant metastasis. The 5-year relative survival rate for localized and regional prostate cancer is 100%, compared with 29.8% for metastatic cases.

PCa is recognized as a genetically heterogeneous disease [6] comprising a large scope of malignancies, from indolent localized cancers that may never progress to rapidly progressing castration-resistant PCa. Currently, the diagnosis of PCa is based on the pathological evaluation of tissue biopsy but the treatment options are determined by risk stratification based on both Gleason score and serum PSA level [7]. For example, high-risk PCa was defined by D'Amico as PSA ≥20 ng/mL and/or biopsy Gleason Score ≥8 and/or clinical stage ≥2c [8].

Elevated serum prostate-specific antigen (PSA) levels are often seen in the context of PCa but can also reflect other prostatic diseases such as benign prostatic hyperplasia, prostatic infection, and prostatic infarction [9]. Relatedly, the low specificity of the PSA screening test has raised concerns in the scientific community regarding the over-diagnosis of PCa [10]. Since the implementation of the PSA screening test into clinical practices in the 1990s, a significant shift towards localized PCa at diagnoses has been observed, with >95% of diagnoses being of clinically localized PCa [11]. Furthermore, only approximately 40 to 50% of patients with elevated PSA testing undergoing a biopsy have prostate cancer. However, a recent study evaluating the ERSPC and PLCO clinical trials has shown that PSA testing reduces the mortality due to PCa by approximately 30% [12]. In cases of localized PCa, radical prostatectomy remains the gold standard treatment option [7]. Importantly, up to 30% of patients treated with radical prostatectomy eventually develop recurrence [13]. More performant risk stratification and prognostic markers are urgently needed to improve the management of patients with localized prostate cancer and identify cases with a high risk of progression.

Circulating tumor cells (CTC) are cells that detach from the primary or secondary tumor sites and invade the bloodstream. As primary actors of the metastatic dissemination, CTC represent a very promising biomarker to aid cancer diagnosis, treatment decision and patient follow-up [14]. In fact, CTC could provide a valuable complement to PSA or other tests with the aim to identify patients with more aggressive cancers. The prognostic value of CTC collected by the epithelial marker-dependent method CellSearch has been established in the context of metastatic PCa [15]. However, the clinical utility of CTC in the context of localized PCa remains unclear. Here, we review studies on non-metastatic prostate cancer to evaluate the potential clinical utility of CTC in localized PCa.

2. Materials and Methods

The present review was prepared by selecting English-written research papers describing the detection and/or characterization of circulating tumor cells in the context of localized prostate cancer. To that end, we searched PubMed using the following keywords: "circulating tumor cells" or "circulating cancer cells" and "localized prostate cancer" or "non-metastatic prostate cancer" or "early-stage prostatic carcinoma". Reviews and studies on liquid biopsy that did not concern localized PCa, as well as studies that did not report on CTC were excluded from the systematic review. Preclinical models, as well as methods for detection of disseminated tumor cells (DTC) in the bone marrow of localized PCa patients, were also excluded. A total of 43 studies were included in the systematic review, as shown in Table 1. It should be noted that several selected studies also reported on locally advanced prostate cancer (T3 and T4) cases. The American Cancer Society defines localized PCa as clinical or pathological tumor stages T1 and T2. PCa with the T1 stage corresponds to a clinically unapparent tumor that is neither palpable nor visible by imaging while T2 corresponds to a tumor that is confined within the prostate and that is either palpable or visible by imaging or demonstrated in radical prostatectomy [16]. In contrast, a pathological tumor stage T3 defines a tumor that has extended through the prostatic capsule and T4 defines a tumor which is invading adjacent organs such as the bladder, sphincter or rectum [16].

Table 1. Studies on circulating tumor cells in patients with non-metastatic prostate cancer.

Study [ref]	N° of Patients (pT Stage)	CTC Method	Cutoff	N° of CTC+ Patients (%)	Median CTC/mL (Range/mL)	Comments	Clinical Impact of CTC
Moreno 1992 [17]	4 (ND [1])	PSA RT-PCR	NA [2]	0 (0%)	NA	17 negative controls (0% positive). Only patients with lymph nodes or metastases tested positive.	ND
Israeli 1995 [18]	13 (7 T2, 5 T3, 1 T4)	PSA & PSMA RT-PCR	NA	PSMA 7 (54%) PSA 0 (0%)	NA	6 years after radical prostatectomy, with undetectable PSA serum levels, patients are PSMA positive	ND
Olsson 1996 [19]	100 (59 T1-T2, 41 T3-T4)	PSA RT-PCR	NA	74% (76% T1-T2, 71% T3-T4)	NA	Potential surgical failure defined as a tumor at the surgical margin or extending into the seminal vesicle	Yes
Ennis 1997 [20]	227 (72 T1c, 129 T2, 26 T3)	PSA RT-PCR	NA	61 (26.9%)	NA	Patients treated with prostatectomy had a higher rate of RT-PCR positivity than patients treated with radiation	Yes
Oefelein 1999 [21]	101 (T1–T3a)	PSA RT-PCR	NA	22 (22%)	NA	Median follow-up 22 months	No
Okegawa 1999 [22]	31 (T2–T3)	PSA & PSMA RT-PCR	NA	ND	NA	RT-PCR performed before radical prostatectomy	Yes
Sabile 1999 [23]	10 (ND)	PSA RT-PCR	NA	4 (40%)	NA	Density gradient has higher isolation efficiency than epithelial marker-dependent immunocapture	ND
Mejean 2000 [24]	99 (37 T1, 52 T2, 8 T3, 2 T4)	PSA RT-PCR	NA	33 (33%)	NA	RT-PCR performed preoperatively. 92 controls included (2% scored positive)	Yes
Llanes 2000 [25]	25 (T1–T2b)	PSA RT-PCR	NA	7 (28%)	NA	The best predictors of extraprostatic disease were the biopsy Gleason score and the PSA level.	No
Slawin 2000 [26]	228 (154 T1–T2, 47 T3a, 16 T3b, 11 T4)	hK2 RT-PCR	NA	57 (25%)	NA	14 healthy controls (14% positive). RT-PCR performed before prostatectomy. Association with the risk of metastasis to pelvic lymph nodes ($P = 0.028$).	Yes
Shariat 2002 [27]	224 (T1–T2), 96 AAM	PSA RT-PCR	NA	54 (24%)	NA	RT-PCR performed preoperatively.	No
Bianco 2002 [28]	35 T1, 61 T2) 150 CAM (62 T1, 88 T2)	PSA RT-PCR	NA	26 (27%) AAM 34 (23%) CAM	NA	RT-PCR performed preoperatively.	Yes in AAM
Hara 2002 [29]	44 (26 T1, 15 T2, 2 T3, 1 T4)	PSA, PSMA & PSCA RT-PCR	NA	1 (2.3%) PSA3 (6.8%) PSMA 1 (2.3%) PSCA	NA	RT-PCR performed preoperatively. Positive PSA result in 1 prostatitis case, positive PSMA result in 1 prostatitis and 1 benign prostatic hyperplasia case.	Yes
Thomas 2002 [30]	141 (118 T1c/T2a, 18 T2b/c, 5 T3a)	PSA & PSMA RT-PCR	NA	73 (51.8%)	NA	Only initial PSA and biopsy Gleason score were independent predictors of biochemical failure.	No

Table 1. Cont.

Study [ref]	N° of Patients (pT Stage)	CTC Method	Cutoff	N° of CTC+ Patients (%)	Median CTC/mL (Range/mL)	Comments	Clinical Impact of CTC
Gewanter 2003 [31]	161 (121 T1–T2, 39 T3, 1 Tx)	PSA RT-PCR	NA	22 (20%)	NA	29 months follow-up. Only post-treatment testing predicted for clinical relapse.	No
Fizazi 2007 [32]	83 (38 T1c, 38 T2, 7 T3)	EpCAM + telomerase PCR	NA	58 (70%)	NA	Preoperative CTC detection; 22 healthy controls (0% positive)	ND
Davis 2008 [33]	97 (78 T2, 19 T3)	CellSearch	≥1 CTC /22.5 mL	20 (21%)	0.18 (0.04–2.62)	Preoperative CTC detection; 4 of 20 healthy controls positive for CTC (20%)	ND
Helo 2009 [34]	129 (71 T2, 43 T3a, 13 T3b, 2 T4)	PSA & PSMA RT-PCR	≥80 mRNA/mL	3 (2.6%)	NA	19 healthy controls (0% positive). RT-PCR performed 6 months after surgery for 42 patients and before surgery for 85 patients	No
Maestro 2009 [35]	26 (ND)	CellSearch + CellSpotter Analyzer	≥2 CTC /7.5 mL	4 (15.4%)	ND	106 healthy controls (0% positive)	ND
Eschwe-ge 2009 [36]	155 (T2–T3)	PSA & PSMA RT-PCR	NA	57 (37%)	NA	Preoperative CTC detection; 100 healthy controls (0% positive).	Yes
Giesing 2010 [37]	129 (T1–T4)	Filtration + PSA & AOX RT-PCR	NA	42 (32.5%)	NA	The AOX test was tumour predicting With a positive predictive value of 69% and a negative predictive value of 92%, 6/8 patients with a decline of CTC 24 h after prostatectomy	Yes
Stott 2010 [38]	19 (T2–T3a)	Microfluidic (EpCAM)	≥14 CTC /mL	8 (42%)	95 (38–222)	RT-PCR performed before surgery	ND
Joung 2010 [39]	103 (25 T1–T2b, 78 T2c–T3)	PSCA RT-PCR	NA	17 (16.5%)	NA		Yes
Yates 2012 [40]	92 (61 T1, 31 T2)	PSA & PSMA RT-PCR	NA	63 (68.5%) PSA 68 (78.9%) PSMA	NA	Blood samples taken 1 day preoperatively and 7 days postoperatively	Yes
Lowes 2012 [41]	26 (11 T2, 15 T3)	CellSearch	≥1 CTC /7.5 mL	19 (73%)	ND	Blood drawn before radiation therapy; 7 healthy controls included (0% positive)	Yes
Khurana 2013 [42]	10 (5 T2c, 5 T3a)	CellSearch	≥1 CTC /7.5 mL	1 (10%)	0 (0–0.13)	Blood drawn preoperatively. Very low CTC numbers.	No
Thalgott 2013 [43]	20 (ND)	CellSearch	≥1 CTC /7.5 mL	1 (5%)	0 (0–0.13)	15 healthy controls (0% positive). Shorter overall survival observed only for metastatic patients with ≥ 3 CTC	No
Loh 2014 [44]	36 (9 T1, 14 T2, 13 T3)	CellSearch	≥1 CTC /7.5 mL	5 (14%)	0 (0–0.4)	Blood drawn before therapy. Median follow-up 42 months.	No
Kolostova 2014 [45]	55 (45 T2, 10 T3)	MetaCell@filtration	≥1 CTC /8 mL	28 (52%)	ND	CTC were cultured in vitro for downstream applications for 7–28 days. The captured cancer cells displayed plasticity.	ND
Shao 2014 [46]	40 (26 T2, 13 T3, 1 Tx)	Near-infrared dyes + FACS	≥1 CTC /7.5 mL	39 (97.5%)	10 (0–439)	Blood samples collected preoperatively. Live CTC evidenced by staining with heptamethine carbocyanine dyes	No
Pal 2015 [47]	35 (32 T1–T2, 3 T3)	CellSearch	≥1 CTC /22.5 mL	16 (45%)	0 (0–0.1)	Blood samples drawn before and after surgery. Median follow-up 510 days.	No
Thalgott 2015 [48]	15 (1 T2, 14 T3)	CellSearch	≥1 CTC /20 mL	3 (20%)	0 (0–0.2)	15 healthy controls (0% positive). Median follow-up 44.3 months.	No

Table 1. Cont.

Study [ref]	N° of Patients (pT Stage)	CTC Method	Cutoff	N° of CTC+ Patients (%)	Median CTC/mL (Range/mL)	Comments	Clinical Impact of CTC
Meyer 2016 [49]	152 (95 T2, 40 T3a, 17 T3b)	CellSearch	≥1 CTC/7.5 mL	17 (11%)	0.13 (0.13–13.3)	Blood samples collected preoperatively. Median follow-up 48 months.	No
Toden-höfer 2016 [50]	50 (37 T2, 13 T3)	Microfluidic (size, deformability)	≥1 CTC/2 mL	25 (50%)	4.5 (0.5–208.5)	Pancytokeratin positive CTC showed expression of androgen receptor.	No
Kuske 2016 [51]	86 (37 T1, 45 T2, 4 T3)	CellSearch EPISPOT CellCollector	≥1 CTC/7.5 mL	37% CS 54.9% CC 58.7% EPI	0.24 (0.13–1.3) CS 0.32 (0.13–1.6) CC 0.4 (0.13–1.7) EPI	Blood drawn preoperatively. CTC detected by EPISPOT correlated to tumor stage, no correlation found with CellSearch (CS) or CellCollector (CC)	Yes
Tsumura 2017 [52]	59 (26 T1c–T2a, 15 T2b–c, 17 T3, 1 T4)	CellSearch	≥1 CTC/7.5 mL	preoperative 7 (11.8%) intraoperative 0 (0%)	ND	Blood drawn both before and during surgery, with detection of CTC only during surgery.	No
Garcia 2017 [53]	16 (ND)	AR-V7 protein in serum samples	AR-V7 protein detection	3 (18.7%)	NA	CD133 expression in CTC was higher among AR-V7 positive cases vs. AR-V7 negative	ND
Puche-Sanz 2017 [54]	86 (T1–T2)	CK immune-magnetic Filtration + immunostaining CK, CD45, AR	≥1 CTC/10 mL	16 (18.6%)	0 (0–0.4)	Blood samples collected before biopsy. Analysis of AR expression in tumor tissue.	Yes
Awe 2017 [55]	41 (T1–T4)	Microfluidic vortex chip (size-based) + immuno-staining CK, CD45, PSA	≥1 CTC/3 mL	41 (100%)	ND	Blood samples collected before prostatectomy	ND
Renier 2017 [56]	1 (ND)		>3.37 CTC/7.5 mL = >0.45 CTC/mL	1 (100%)	1.5	Some double positive cells (CK+, CD45+) found but counted as WBC. Some cells did not express epithelial markers (CK, PSA) but mesenchymal instead (Vim, N-cad)	ND
Russo 2018 [57]	47 (31 T2, 16 T3a)	AdnaTest Prostate-Cancer	0.15 ng/µL for AR, c-kit, c-met, ALDH1, TYMS, 0.25 ng/µL for Akt-2 & PB3Ka	12 (25.5%)	NA	Blood samples drawn before prostatectomy. No healthy controls tested.	Yes
Miyamoto 2018 [58]	34 (22 T1, 11 T2, 1 T3)	CTC-iChip + WTA + multiplex (8 genes) ddPCR	Mean CTC in healthy + 2SD of CTC score in healthy	ND	NA	Blood samples collected before surgery; 34 age-matched healthy donors included.	Yes
Murray 2018 [59]	241 (181 low risk + 60 intermediate risk)	Density gradient + PSA ICC	≥1 CTC/8 mL	37 low risk (20.4%) 26 intermediate risk (43.3%)	ND	Blood samples collected 3 months after radiotherapy and stored 48 h at 4 °C.	Yes

[1] ND = Not described. [2] NA = Not applicable. pT stage = Pathological tumor stage. CTC = Circulating tumor cells. PSA = Prostate-specific antigen. PSMA = Prostate-specific membrane antigen. PSCA = Prostate stem cell antigen. AAM = African Americans. CAM = Caucasian Americans. RT-PCR = Reverse transcription-polymerase chain reaction. EpCAM = Epithelial cell adhesion molecule. AOX = Antioxydant genes. FACS = Fluorescence-activated cell sorting. AR = Androgen receptor. AR-V7 = Androgen receptor splice variant seven. CK = Cytokeratins. Vim = Vimentin. N-Cad = N-Cadherin. WTA = Whole transcriptome amplification. ddPCR = droplet digital polymerase chain reaction. ICC = Imunocytochemistry. 2SD = Two times the standard deviation.

3. CTC Detection in Non-Metastatic Prostate Cancer

In light of the published data, the extensive variability of CTC detection results in the context of localized PCa appears related to the diversity of distinct methods used for CTC collection/detection. Therefore, the present review will classify the published results depending on the CTC collection/detection methods used.

3.1. CTC Detection by RT-PCR in Non-Metastatic Prostate Cancer

Reverse transcription-polymerase chain reaction (RT-PCR) is commonly used to generate amplified cDNA from target mRNA [60]. Therefore, the RT-PCR results largely depend on the specific mRNA that is targeted. As a surrogate test for CTC detection in localized PCa, RT-PCR has mainly been used to target the mRNA of prostate-specific antigen (PSA), prostate-specific membrane antigen (PSMA) and prostate stem cell antigen (PSCA). The only study having compared those three targets as surrogate markers of CTC in localized PCa reported that the detection of PSCA mRNA in blood was the most accurate preoperative predictor of disease-free survival (DFS), probably because PSCA was the only mRNA not detected in 71 non-malignant disorders (PSA detected in 1 and PSMA in 2 of 71 non-malignant disorders) [29]. However, the very low detection rate of PSCA mRNA in that study (detected in only 1 of 43 localized PCa patients) begs caution when interpreting the results. Additionally, Joung et al. reported no association of PSCA mRNA detection with clinical variables on a larger cohort of localized PCa patients [39].

Most studies have focused on PSA and PSMA detection in localized PCa and have yielded somewhat contradictory results. Sabile et al. reported that the density gradient separation of mononuclear cells had a higher isolation efficiency than epithelial marker-dependent immunocapture for CTC detection based on PSA RT-PCR [23]. However, by using density gradient separation and PSA RT-PCR, Moreno et al. failed to detect CTC in 4 patients with localized PCa, possibly owing to distinct RT-PCR primer sequences [17]. By studying patients with a mean follow-up of 13.6 months, Olsson et al. determined that PSA mRNA detection was a significant predictor of disease recurrence after prostatectomy [19] while Mejean et al. found a statistical association of PSA-positive RT-PCR with metastasis and recurrence after a follow-up of 26 months [24]. Interestingly, the latter study also tested 11 patients with prostatitis and found positive PSA RT-PCR results in 2 of 11 cases (18%). Other studies including longer follow-up periods reported no correlation of PSA mRNA detection with clinical variables such as overall survival (OS), progression-free survival (PFS), Gleason score, tumor stage or preoperative serum PSA level [20,21,25,27,30,31,34]. Israeli et al. reported that PSMA RT-PCR was more sensitive than PSA RT-PCR in detecting hematogenous tumor cell dissemination but their results were not correlated to clinical variables [18]. Okegawa et al. compared the detection of PSA and PSMA mRNAs as prognostic indicators in a small cohort of 31 localized PCa and determined that PSMA mRNA detection in blood was a significant predictor of PFS after a mean of 16.7 months of follow-up [22]. In contrast, studies including larger casistics and longer follow-up times reported no significant correlation of PSMA mRNA detection with clinical variables [30,34]. Eschwege et al. argued that more specific PSMA RT-PCR primers should be used and that only dual PSA-PSMA-positive blood samples could be considered as reflecting the presence of CTC in blood [36]. Interestingly, the latter study included a rather large cohort of 155 localized PCa patients, more than 100 healthy controls, none of which tested positive for both PSA and PSMA, and 5-year follow-up data showing that the preoperative detection of both PSA and PSMA mRNAs in blood was an independent prognostic factor of disease recurrence [36]. Similarly, Yates et al. showed that both PSA and PSMA mRNA detection improved the prediction of biochemical recurrence over Kattan nomogram [40]. Slawin et al. took a slightly different approach by amplifying the human KLK2 gene, coding for an androgen-regulated protein (hK2) that has an 80% amino acid sequence identity with PSA [26]. Although the Authors determined that RT-PCR-hK2 results allowed for the prediction of lymph node-positive disease, the positivity of their test in 14% of 14 healthy controls indicates a lack of specificity of CTC detection via hK2 RT-PCR [26]. Multiplex RT-PCR approaches may be more efficient in identifying hematogenous prostatic cell dissemination. Yet, the major shortcoming of any RT-PCR approach as a surrogate marker for CTC detection is the related inability to count and further characterize CTC from blood. Furthermore, methodological variability related to different cell extraction methods, primers used for RT-PCR, controls of specificity and sensitivity and the timing of sample collection and storage is expected to account for the heterogeneity of the results obtained and their clinical relevance.

3.2. CTC Detection by CellSearch in Non-Metastatic Prostate Cancer

The CellSearch method uses the epithelial cell adhesion molecule (EpCAM) to capture circulating cells and defines CTC as nucleated cells (DAPI+) of epithelial (CK+) and non-hematopoietic (CD45−) origin, which, in fact, better corresponds to a definition of circulating epithelial cells (CEpC). It is important to note that CEpC have been found in the blood of patients with benign colon diseases [61] and benign pancreatic diseases [62]. The lack of specificity of the CellSearch method is exemplified by the finding of CTC in up to 20% of healthy donors tested [33].

The preoperative detection of CTC by CellSearch in localized PCa has been reported in 0% to 73% of patients, depending on the study (see Table 1). The fact that distinct cutoff values were used to define CTC positivity in those studies complicates the task to compare their results. For example, Davis et al. and Pal et al. chose to place the cutoff at 1 CTC per 22.5 mL of blood, corresponding to the finding of at least 1 CTC in 3 CellSearch samples of 7.5 mL each [33,47]. In contrast, the majority of studies using CellSearch to detect CTC in localized PCa have used a cutoff of 1 CTC in 7.5 mL [41–44,49,51]. None of the ten studies using CellSearch to detect CTC in localized PCa have reported a significant correlation of CTC numbers with clinical variables such as OS, PFS, Gleason score, tumor stage or preoperative serum PSA level.

3.3. CTC Detection by Other Marker-Dependent Methods in Non-Metastatic Prostate Cancer

A recent study comparing CellSearch with another EpCAM-dependent method (CellCollector) and the EPISPOT assay (based on the negative enrichment of CTC by leukocyte depletion) reported that only CTC detected by EPISPOT in 58.7% of patients were significantly correlated with clinical parameters such as PSA serum values ($p < 0.0001$) and the clinical tumor stage ($p = 0.04$) [51]. Using the EpCAM-dependent immune-magnetic enrichment of CTC followed by telomerase detection via an enzyme-linked immunosorbent assay, Fizazi et al. detected CTC in 70% of 83 localized PCa patients without false-positive results in 22 healthy controls tested [32]. Unfortunately, the latter study did not include any prognostic evaluation. The EpCAM-dependent microfluidic isolation of CTC has been reported by Stott et al., showing the detection of up to 222 CTC per mL of blood tested and a decline of CTC numbers in 6 of 8 patients 24 h after radical prostatectomy [38]. However, in the latter study, the finding of CTC in healthy controls implied the need for a cutoff value at 14 CTC per mL of blood. By using the AdnaTest, relying on EpCAM and MUC-1 antigens for immune-magnetic isolation and on RT-PCR of Androgen Receptor (AR), c-kit, c-met, ALDH1 and TYMPS for CTC detection, Russo et al. failed to demonstrate a significant association of CTC with clinical parameters [57]. In contrast, Puche-Sanz et al. used cytokeratin-mediated immune-magnetic enrichment of CTC and reported a significant correlation of CTC detection with AR expression in the tumor tissue [54]. The assessment of the AR-V7 splice variant protein in plasma, performed through a capillary nano-immunoassay platform was proposed by Garcia et al. as a surrogate marker for CTC in localized PCa patients [53]. Interestingly, the authors reported a significant correlation of AR-V7 detection with preoperative serum PSA levels and the expression of the stem cell marker CD133. However, the latter study did not perform a longitudinal follow-up of localized PCa patients for further prognostic evaluation [53]. Murray et al. took a different approach by using density gradient isolation and detection of CTC by PSA immunocytochemistry on a large cohort of localized PCa patients [59]. Importantly, the authors reported a significant correlation between CTC detection and clinical variables such as PFS after a long follow-up period of 15 years. However, Murray et al. stored the blood samples at 4 °C during 48 h before analyzing them, which could significantly impact the CTC detection results. Furthermore, the authors neither provided counting of the CTC nor exemplar CTC images.

3.4. CTC Detection Following Size-Based Isolation Methods in Non-Metastatic Prostate Cancer

To date, few studies have used size-based separation methods to study CTC from localized PCa patients. Giesing et al. were the first to use blood filtration followed by RT-PCR of PSA and a selection of antioxidant genes (AOX) to detect CTC in 42 localized PCa patients [37]. The authors determined that the detection of antioxidant gene expression in CTC could predict tumor diagnosis with 86% sensitivity and 82% specificity. A few years later, Kolostova et al. used MetaCell®filtration followed by a short-term in vitro culture to identify CTC in 28 of 55 localized PCa patients [45]. Unfortunately, no correlation was found between CTC detection and the clinical parameters. Similarly, Todenhöfer et al. failed to demonstrate a significant correlation with the clinical parameters of CTC detected by fluorescence imaging (EpCAM+ & CD45−) following microfluidic enrichment based on cell-size and deformability [50]. Interestingly, Renier et al. used a similar size-based microfluidic enrichment of CTC followed by the immunofluorescent detection of cytokeratins (CK), PSA and CD45 and reported that some cells did not express epithelial markers (CK) but mesenchymal markers instead (Vim, N-cad), thereby pointing to the process of epithelial to mesenchymal transition (EMT) in circulating prostate cells [56]. Awe et al. also reported on distinct subpopulations of CTC following size-based enrichment and immunostaining for cytokeratins, CD45 and the androgen receptor (AR) but they did not evaluate the clinical impact of those CTC [55]. Efficient risk stratification of localized PCa patients by means of a liquid biopsy was only recently achieved by Miyamoto et al. using the size-based microfluidic enrichment of CTC followed by whole transcriptome amplification and multiplex droplet digital PCR of a panel of 8 genes [58]. By using the differential weighting of 6 genes from the panel, the authors could predict early prostate cancer dissemination in localized disease [58].

4. Analysis of the Clinical Value of CTC Detection in Localized Prostate Cancer (Stages T1, T2)

Among the 43 studies targeting non-metastatic patients with PCa included in the present review, 31 investigated the potential clinical impact of CTC detection, looking for a statistical correlation between the detection of CTC and PCa clinical and/or pathological characteristics. Studies reporting on PCa with early (T1–T2) and advanced (T3–T4) stages but without a separate statistical analysis of T1–T2 cancers were further excluded. The details of the remaining 11 studies reporting on the analysis of the clinical value of CTC detection in localized (T1–T2) PCa are shown in Table 2. For clarity, the diagnostic value refers to a test's ability to identify a disease or a specific condition, with degrees of specificity and sensitivity to express its confidence and accuracy [63]. The predictive value refers to a test's ability to predict the patient's response to a specific treatment while its prognostic value identifies risks of progression of the disease independently of a specific treatment [64].

Table 2. The clinical value of Circulating Tumor Cells' (CTC) detection in patients with localized prostate cancer.

Study [ref]	Method	N° of Patients	N° of CTC+ Patients (%)	Mean Follow-Up Period	Type of Clinical Value (P Value)	Comments
Olsson 1996 [19]	PSA RT-PCR	100 (cT1–cT2c)	74 (74%)	13.6 months	Predictive of surgical failure ($P < 0.0286$)	Correlation of RT-PCR results before prostatectomy with disease recurrence after prostatectomy.
Ennis 1997 [20]	PSA RT-PCR	156 (cT1–cT2)	ND	ND	Prognostic ($P < 0.0001$)	Correlation of RT-PCR results with pathological stage and prediction of extra-capsular disease.
Mejean 2000 [24]	PSA RT-PCR	79 (cT1–cT2)	ND	26 months	Predictive ($P < 0.04$)	CTC detection associated with the development of metastases and risk of relapse after prostatectomy.
Slawin 2000	hK2 RT-PCR	154 (pT1–pT2)	ND	ND	Prognostic ($P = 0.028$)	Association with the risk of metastasis to pelvic lymph nodes.
Bianco 2002 [28]	PSA RT-PCR	96 (35 pT1, 61 pT2)	26 (27%) African Americans	33 months	Prognostic ($P = 0.01$)	Association with tumor stage and recurrence in African-Americans.
Yates 2012 [40]	PSA & PSMA RT-PCR	92 (61 pT1, 31 pT2)	63 (68.5%) PSA 68 (78.9%) PSMA	72 months	Predictive ($P = 0.03$)	Improved prediction of biochemical recurrence.
Puche-Sanz 2017 [54]	CK immune-magnetic	86 (pT1–pT2)	16 (18.6%)	ND	Theranostic & Diagnostic ($P = 0.03$)	Expression of AR in tumor tissue correlated significantly with presence of CTC in blood. Diagnosis of PCa by CTC has a 14.2% sensitivity and a 78.4% specificity.
Llanes 2000 [25]	PSA RT-PCR	25 (pT1–pT2b)	7 (28%)	ND	Not significant	The best predictors of extraprostatic disease were the biopsy Gleason score and the PSA level.
Shariat 2002 [27]	PSA RT-PCR HK2L	224 (pT1–pT2)	54 (24%)	52.9 months	Not significant	Preoperative blood RT-PCR-PSA not associated with characteristics or outcomes of prostate cancer HK2L correlation with risk of metastases.
Thomas 2002 [30]	PSA & PSMA RT-PCR	136 (pT1–pT2)	73 (54%)	59 months	Not significant	RT-PCR status did not predict pathologic stage or biochemical failure.
Helo 2009 [34]	PSA & PSMA RT-PCR	87 (cT1–cT2)	6 (7%)	28 months	Not significant	No association between KLK mRNA status and unfavorable localized disease features.

ND = Not described; cT1–cT2 = clinical stages T1–T2; pT1–pT2 = pathological stages T1–T2.

The first conclusion of our literature review is that very few studies investigated the clinical impact of CTC selectively in patients with localized prostate cancer. Overall, there is a definite trend toward a value of CTC in correlation with the pathological stage and toward a prognostic and predictive impact of CTC detection in early-stage prostate cancer since several studies have reported significant correlations of CTC numbers with the survival of patients and/or recurrence of the disease after treatment [19,20,24,28,40]. However, those studies used PSA RT-PCR to detect CTC, a method which has yielded contradictory results in other studies [25,27,34], thereby calling for further validation of those results in large cohorts of localized PCa patients. The diagnostic value of CTC detection in early-stage prostate cancer has been less extensively investigated than in metastatic patients. In this regard, the most interesting results come from Giesing et al. with a CTC test demonstrating an 86% sensitivity and an 82% specificity with a 69% positive predictive value and a 92% negative predictive value for PCa diagnosis [37]. However, the latter study reported on both early-stage (T1–T2) and locally advanced (T3–T4) PCa. Puche-Sanz et al. have also investigated the possibility of a diagnostic CTC test for PCa [54]. Yet, with a 14.2% sensitivity and a 78.4% specificity, their test would not further improve on the PSA screening test. The lack of further investigation of a potential diagnostic CTC test could, in fact, stem from the substantial difficulty of detecting CTC in a consistent and specific manner in the context of localized PCa. The presence of circulating prostatic cells in benign prostatic hyperplasia and prostatitis impacts the specificity of certain CTC isolation techniques such as PSA RT-PCR [24,29]. Furthermore, the phenotypic heterogeneity of CTC has been established in the context of metastatic PCa [65]. Additionally, the occurrence of a phenotypic transition (EMT) in CTC from early-stage PCa patients, evidenced by Renier et al. [56], supports the notion of phenotypic heterogeneity among CTC from localized PCa patients as well. The heterogeneity of CTC is relevant to the potential theranostic interest of various CTC tests. Particularly, the expression of the androgen receptor (AR) is of substantial importance for therapy strategy decision in PCa. In fact, a recent review evaluating clinical trials in the context of metastatic PCa determined that the expression of a specific variant of the androgen receptor (AR-V7) was significantly correlated to the limited efficacy of abiraterone and enzalutamide treatments compared with taxane therapy [66]. Whether such an association still holds true in the context of early-stage PCa remains to be demonstrated. Miyamoto et al. also reported a considerable heterogeneity among prostate CTC, including heterogeneous patterns of AR splice variant expression, following microfluidic enrichment and single-cell RNA-seq analyses [67]. So far, only two studies have reported correlations of CTC detection in localized PCa patients with AR expression [54,57]. Puche-Sanz et al. observed a direct association of the expression of AR in the prostatic tissue and the presence of CTC in blood [54]. Russo et al. determined that the expression of AR and TYMS on CTC are frequent events but the implications of such results for a personalized treatment strategy in PCa remain to be elucidated [57]. Further studies are needed to evaluate the potential theranostic utility of CTC in the context of localized PCa.

5. Perspectives and Future Directions

The major issue concerning localized prostate cancer is the lack of a suitable marker which could identify benign cases from aggressive prostate cancers. The present study shows a trend toward a possible clinical impact of CTC detection in patients with localized prostate cancer. Despite this trend, the study raises key issues in particular about the technical approaches used, the need for CTC counting and characterization, the sensitivity and specificity controls and the timing of blood sampling. Overall, our analysis encourages the development of a CTC cell-based specific test able to identify and count CTC in a highly sensitive and specific manner in patients with localized cancers and single-cell CTC analyses in patients with localized prostate cancer to specifically study the CTC heterogeneity. It also stimulates the use of a standardized approach to be employed in large clinical studies.

Funding: This work was supported by funds from: Fondation pour la Recherche Médicale, Fondation Bettencourt-Schueller, Fondation Lefort-Beaumont de l'Institut de France, INSERM and Université Paris-Descartes.

Conflicts of Interest: Professor Patrizia Paterlini-Brechot is co-inventor of ISET®patents belonging to University Paris Descartes, INSERM and Assistance Publique Hopitaux de Paris, exclusively licensed to Rarecells Diagnostics, France. This author does not receive payments from Rarecells Diagnostics. The present study was conducted independently by academic research teams.

References

1. Pernar, C.H.; Ebot, E.M.; Wilson, K.M.; Mucci, L.A. The Epidemiology of Prostate Cancer. *Cold Spring Harb. Perspect. Med.* **2018**, *8*. [CrossRef] [PubMed]
2. Siegel, R.L.; Miller, K.D.; Jemal, A. Cancer statistics, 2018. *CA Cancer J. Clin.* **2018**, *68*, 7–30. [CrossRef] [PubMed]
3. Daniyal, M.; Siddiqui, Z.A.; Akram, M.; Asif, H.M.; Sultana, S.; Khan, A. Epidemiology, etiology, diagnosis and treatment of prostate cancer. *Asian Pac. J. Cancer Prev.* **2014**, *15*, 9575–9578. [CrossRef] [PubMed]
4. Cancer of the Prostate—Cancer Stat Facts. Available online: https://seer.cancer.gov/statfacts/html/prost.html (accessed on 31 May 2019).
5. Jahn, J.L.; Giovannucci, E.L.; Stampfer, M.J. The high prevalence of undiagnosed prostate cancer at autopsy: implications for epidemiology and treatment of prostate cancer in the Prostate-specific Antigen-era. *Int. J. Cancer* **2015**, *137*, 2795–2802. [CrossRef] [PubMed]
6. Baca, S.C.; Prandi, D.; Lawrence, M.S.; Mosquera, J.M.; Romanel, A.; Drier, Y.; Park, K.; Kitabayashi, N.; MacDonald, T.Y.; Ghandi, M.; et al. Punctuated evolution of prostate cancer genomes. *Cell* **2013**, *153*, 666–677. [CrossRef] [PubMed]
7. Heidenreich, A.; Bastian, P.J.; Bellmunt, J.; Bolla, M.; Joniau, S.; van der Kwast, T.; Mason, M.; Matveev, V.; Wiegel, T.; Zattoni, F.; et al. EAU guidelines on prostate cancer. part 1: Screening, diagnosis, and local treatment with curative intent-update 2013. *Eur. Urol.* **2014**, *65*, 124–137. [CrossRef] [PubMed]
8. Holmberg, L.; Bill-Axelson, A.; Helgesen, F.; Salo, J.O.; Folmerz, P.; Häggman, M.; Andersson, S.-O.; Spångberg, A.; Busch, C.; Nordling, S.; et al. A randomized trial comparing radical prostatectomy with watchful waiting in early prostate cancer. *N. Engl. J. Med.* **2002**, *347*, 781–789. [CrossRef] [PubMed]
9. Barry, M.J.; Simmons, L.H. Prevention of Prostate Cancer Morbidity and Mortality: Primary Prevention and Early Detection. *Med. Clin. N. Am.* **2017**, *101*, 787–806. [CrossRef]
10. Akizhanova, M.; Iskakova, E.E.; Kim, V.; Wang, X.; Kogay, R.; Turebayeva, A.; Sun, Q.; Zheng, T.; Wu, S.; Miao, L.; et al. PSA and Prostate Health Index based prostate cancer screening in a hereditary migration complicated population: Implications in precision diagnosis. *J. Cancer* **2017**, *8*, 1223–1228. [CrossRef]
11. Galper, S.L.; Chen, M.-H.; Catalona, W.J.; Roehl, K.A.; Richie, J.P.; D'Amico, A.V. Evidence to support a continued stage migration and decrease in prostate cancer specific mortality. *J. Urol.* **2006**, *175*, 907–912. [CrossRef]
12. Tsodikov, A.; Gulati, R.; Heijnsdijk, E.A.M.; Pinsky, P.F.; Moss, S.M.; Qiu, S.; de Carvalho, T.M.; Hugosson, J.; Berg, C.D.; Auvinen, A.; et al. Reconciling the Effects of Screening on Prostate Cancer Mortality in the ERSPC and PLCO Trials. *Ann. Intern. Med.* **2017**, *167*, 449–455. [CrossRef] [PubMed]
13. Pound, C.R.; Partin, A.W.; Eisenberger, M.A.; Chan, D.W.; Pearson, J.D.; Walsh, P.C. Natural history of progression after PSA elevation following radical prostatectomy. *JAMA* **1999**, *281*, 1591–1597. [CrossRef] [PubMed]
14. Doyen, J.; Alix-Panabières, C.; Hofman, P.; Parks, S.K.; Chamorey, E.; Naman, H.; Hannoun-Lévi, J.-M. Circulating tumor cells in prostate cancer: A potential surrogate marker of survival. *Crit. Rev. Oncol. Hematol.* **2012**, *81*, 241–256. [CrossRef] [PubMed]
15. Hegemann, M.; Stenzl, A.; Bedke, J.; Chi, K.N.; Black, P.C.; Todenhöfer, T. Liquid biopsy: Ready to guide therapy in advanced prostate cancer? *BJU Int.* **2016**, *118*, 855–863. [CrossRef] [PubMed]
16. Borley, N.; Feneley, M.R. Prostate cancer: Diagnosis and staging. *Asian J. Androl.* **2009**, *11*, 74–80. [CrossRef] [PubMed]
17. Moreno, J.G.; Croce, C.M.; Fischer, R.; Monne, M.; Vihko, P.; Mulholland, S.G.; Gomella, L.G. Detection of hematogenous micrometastasis in patients with prostate cancer. *Cancer Res.* **1992**, *52*, 6110–6112.
18. Israeli, R.S.; Miller, W.H.; Su, S.L.; Samadi, D.S.; Powell, C.T.; Heston, W.D.; Wise, G.J.; Fair, W.R. Sensitive detection of prostatic hematogenous tumor cell dissemination using prostate specific antigen and prostate specific membrane-derived primers in the polymerase chain reaction. *J. Urol.* **1995**, *153*, 573–577.

19. Olsson, C.A.; de Vries, G.M.; Raffo, A.J.; Benson, M.C.; O'Toole, K.; Cao, Y.; Buttyan, R.E.; Katz, A.E. Preoperative reverse transcriptase polymerase chain reaction for prostate specific antigen predicts treatment failure following radical prostatectomy. *J. Urol.* **1996**, *155*, 1557–1562. [CrossRef]
20. Ennis, R.D.; Katz, A.E.; de Vries, G.M.; Heitjan, D.F.; O'Toole, K.M.; Rubin, M.; Buttyan, R.; Benson, M.C.; Schiff, P.B. Detection of circulating prostate carcinoma cells via an enhanced reverse transcriptase-polymerase chain reaction assay in patients with early stage prostate carcinoma. Independence from other pretreatment characteristics. *Cancer* **1997**, *79*, 2402–2408. [CrossRef]
21. Oefelein, M.G.; Ignatoff, J.M.; Clemens, J.Q.; Watkin, W.; Kaul, K.L. Clinical and molecular followup after radical retropubic prostatectomy. *J. Urol.* **1999**, *162*, 307–310. [CrossRef]
22. Okegawa, T.; Nutahara, K.; Higashihara, E. Preoperative nested reverse transcription-polymerase chain reaction for prostate specific membrane antigen predicts biochemical recurrence after radical prostatectomy. *BJU Int.* **1999**, *84*, 112–117. [CrossRef] [PubMed]
23. Sabile, A.; Louha, M.; Bonte, E.; Poussin, K.; Vona, G.; Mejean, A.; Chretien, Y.; Bougas, L.; Lacour, B.; Capron, F.; et al. Efficiency of Ber-EP4 antibody for isolating circulating epithelial tumor cells before RT-PCR detection. *Am. J. Clin. Pathol.* **1999**, *112*, 171–178. [CrossRef] [PubMed]
24. Mejean, A.; Vona, G.; Nalpas, B.; Damotte, D.; Brousse, N.; Chretien, Y.; Dufour, B.; Lacour, B.; Bréchot, C.; Paterlini-Bréchot, P. Detection of circulating prostate derived cells in patients with prostate adenocarcinoma is an independent risk factor for tumor recurrence. *J. Urol.* **2000**, *163*, 2022–2029. [CrossRef]
25. Llanes, L.; Páez, A.; Ferruelo, A.; Luján, M.; Romero, I.; Berenguer, A. Detecting circulating prostate cells in patients with clinically localized prostate cancer: Clinical implications for molecular staging. *BJU Int.* **2000**, *86*, 1023–1027. [CrossRef] [PubMed]
26. Slawin, K.M.; Shariat, S.F.; Nguyen, C.; Leventis, A.K.; Song, W.; Kattan, M.W.; Young, C.Y.; Tindall, D.J.; Wheeler, T.M. Detection of metastatic prostate cancer using a splice variant-specific reverse transcriptase-polymerase chain reaction assay for human glandular kallikrein. *Cancer Res.* **2000**, *60*, 7142–7148. [PubMed]
27. Shariat, S.F.; Gottenger, E.; Nguyen, C.; Song, W.; Kattan, M.W.; Andenoro, J.; Wheeler, T.M.; Spencer, D.M.; Slawin, K.M. Preoperative blood reverse transcriptase-PCR assays for prostate-specific antigen and human glandular kallikrein for prediction of prostate cancer progression after radical prostatectomy. *Cancer Res.* **2002**, *62*, 5974–5979. [PubMed]
28. Bianco, F.J.; Powell, I.J.; Cher, M.L.; Wood, D.P. Presence of circulating prostate cancer cells in African American males adversely affects survival. *Urol. Oncol.* **2002**, *7*, 147–152. [CrossRef]
29. Hara, N.; Kasahara, T.; Kawasaki, T.; Bilim, V.; Obara, K.; Takahashi, K.; Tomita, Y. Reverse transcription-polymerase chain reaction detection of prostate-specific antigen, prostate-specific membrane antigen, and prostate stem cell antigen in one milliliter of peripheral blood: Value for the staging of prostate cancer. *Clin. Cancer Res.* **2002**, *8*, 1794–1799.
30. Thomas, J.; Gupta, M.; Grasso, Y.; Reddy, C.A.; Heston, W.D.; Zippe, C.; Dreicer, R.; Kupelian, P.A.; Brainard, J.; Levin, H.S.; et al. Preoperative combined nested reverse transcriptase polymerase chain reaction for prostate-specific antigen and prostate-specific membrane antigen does not correlate with pathologic stage or biochemical failure in patients with localized prostate cancer undergoing radical prostatectomy. *J. Clin. Oncol.* **2002**, *20*, 3213–3218.
31. Gewanter, R.M.; Katz, A.E.; Olsson, C.A.; Benson, M.C.; Singh, A.; Schiff, P.B.; Ennis, R.D. RT-PCR for PSA as a prognostic factor for patients with clinically localized prostate cancer treated with radiotherapy. *Urology* **2003**, *61*, 967–971. [CrossRef]
32. Fizazi, K.; Morat, L.; Chauveinc, L.; Prapotnich, D.; De Crevoisier, R.; Escudier, B.; Cathelineau, X.; Rozet, F.; Vallancien, G.; Sabatier, L.; et al. High detection rate of circulating tumor cells in blood of patients with prostate cancer using telomerase activity. *Ann. Oncol.* **2007**, *18*, 518–521. [CrossRef] [PubMed]
33. Davis, J.W.; Nakanishi, H.; Kumar, V.S.; Bhadkamkar, V.A.; McCormack, R.; Fritsche, H.A.; Handy, B.; Gornet, T.; Babaian, R.J. Circulating tumor cells in peripheral blood samples from patients with increased serum prostate specific antigen: Initial results in early prostate cancer. *J. Urol.* **2008**, *179*, 2187–2191. [CrossRef] [PubMed]

34. Helo, P.; Cronin, A.M.; Danila, D.C.; Wenske, S.; Gonzalez-Espinoza, R.; Anand, A.; Koscuiszka, M.; Väänänen, R.-M.; Pettersson, K.; Chun, F.K.-H.; et al. Circulating prostate tumor cells detected by reverse transcription-PCR in men with localized or castration-refractory prostate cancer: Concordance with CellSearch assay and association with bone metastases and with survival. *Clin. Chem.* **2009**, *55*, 765–773. [CrossRef] [PubMed]
35. Maestro, L.M.; Sastre, J.; Rafael, S.B.; Veganzones, S.B.; Vidaurreta, M.; Martín, M.; Olivier, C.; de La Orden, V.B.; Garcia-Saenz, J.A.; Alfonso, R.; et al. Circulating tumor cells in solid tumor in metastatic and localized stages. *Anticancer Res.* **2009**, *29*, 4839–4843. [PubMed]
36. Eschwège, P.; Moutereau, S.; Droupy, S.; Douard, R.; Gala, J.-L.; Benoit, G.; Conti, M.; Manivet, P.; Loric, S. Prognostic value of prostate circulating cells detection in prostate cancer patients: A prospective study. *Br. J. Cancer* **2009**, *100*, 608–610. [CrossRef] [PubMed]
37. Giesing, M.; Suchy, B.; Driesel, G.; Molitor, D. Clinical utility of antioxidant gene expression levels in circulating cancer cell clusters for the detection of prostate cancer in patients with prostate-specific antigen levels of 4–10 ng/mL and disease prognostication after radical prostatectomy. *BJU Int.* **2010**, *105*, 1000–1010. [CrossRef] [PubMed]
38. Stott, S.L.; Lee, R.J.; Nagrath, S.; Yu, M.; Miyamoto, D.T.; Ulkus, L.; Inserra, E.J.; Ulman, M.; Springer, S.; Nakamura, Z.; et al. Isolation and characterization of circulating tumor cells from patients with localized and metastatic prostate cancer. *Sci. Transl. Med.* **2010**, *2*, 25ra23. [CrossRef] [PubMed]
39. Joung, J.Y.; Cho, K.S.; Kim, J.E.; Seo, H.K.; Chung, J.; Park, W.S.; Choi, M.K.; Lee, K.H. Prostate stem cell antigen mRNA in peripheral blood as a potential predictor of biochemical recurrence in high-risk prostate cancer. *J. Surg. Oncol.* **2010**, *101*, 145–148. [CrossRef]
40. Yates, D.R.; Rouprêt, M.; Drouin, S.J.; Comperat, E.; Ricci, S.; Lacave, R.; Sèbe, P.; Cancel-Tassin, G.; Bitker, M.-O.; Cussenot, O. Quantitative RT-PCR analysis of PSA and prostate-specific membrane antigen mRNA to detect circulating tumor cells improves recurrence-free survival nomogram prediction after radical prostatectomy. *Prostate* **2012**, *72*, 1382–1388. [CrossRef]
41. Lowes, L.E.; Lock, M.; Rodrigues, G.; D'Souza, D.; Bauman, G.; Ahmad, B.; Venkatesan, V.; Allan, A.L.; Sexton, T. Circulating tumour cells in prostate cancer patients receiving salvage radiotherapy. *Clin. Transl. Oncol.* **2012**, *14*, 150–156. [CrossRef]
42. Khurana, K.K.; Grane, R.; Borden, E.C.; Klein, E.A. Prevalence of circulating tumor cells in localized prostate cancer. *Curr. Urol.* **2013**, *7*, 65–69. [CrossRef] [PubMed]
43. Thalgott, M.; Rack, B.; Maurer, T.; Souvatzoglou, M.; Eiber, M.; Kreß, V.; Heck, M.M.; Andergassen, U.; Nawroth, R.; Gschwend, J.E.; et al. Detection of circulating tumor cells in different stages of prostate cancer. *J. Cancer Res. Clin. Oncol.* **2013**, *139*, 755–763. [CrossRef] [PubMed]
44. Loh, J.; Jovanovic, L.; Lehman, M.; Capp, A.; Pryor, D.; Harris, M.; Nelson, C.; Martin, J. Circulating tumor cell detection in high-risk non-metastatic prostate cancer. *J. Cancer Res. Clin. Oncol.* **2014**, *140*, 2157–2162. [CrossRef] [PubMed]
45. Kolostova, K.; Broul, M.; Schraml, J.; Cegan, M.; Matkowski, R.; Fiutowski, M.; Bobek, V. Circulating tumor cells in localized prostate cancer: Isolation, cultivation in vitro and relationship to T-stage and Gleason score. *Anticancer Res.* **2014**, *34*, 3641–3646. [PubMed]
46. Shao, C.; Liao, C.-P.; Hu, P.; Chu, C.-Y.; Zhang, L.; Bui, M.H.T.; Ng, C.S.; Josephson, D.Y.; Knudsen, B.; Tighiouart, M.; et al. Detection of live circulating tumor cells by a class of near-infrared heptamethine carbocyanine dyes in patients with localized and metastatic prostate cancer. *PLoS ONE* **2014**, *9*, e88967. [CrossRef] [PubMed]
47. Pal, S.K.; He, M.; Wilson, T.; Liu, X.; Zhang, K.; Carmichael, C.; Torres, A.; Hernandez, S.; Lau, C.; Agarwal, N.; et al. Detection and phenotyping of circulating tumor cells in high-risk localized prostate cancer. *Clin. Genitourin. Cancer* **2015**, *13*, 130–136. [CrossRef] [PubMed]
48. Thalgott, M.; Rack, B.; Horn, T.; Heck, M.M.; Eiber, M.; Kübler, H.; Retz, M.; Gschwend, J.E.; Andergassen, U.; Nawroth, R. Detection of Circulating Tumor Cells in Locally Advanced High-risk Prostate Cancer During Neoadjuvant Chemotherapy and Radical Prostatectomy. *Anticancer Res.* **2015**, *35*, 5679–5685.

49. Meyer, C.P.; Pantel, K.; Tennstedt, P.; Stroelin, P.; Schlomm, T.; Heinzer, H.; Riethdorf, S.; Steuber, T. Limited prognostic value of preoperative circulating tumor cells for early biochemical recurrence in patients with localized prostate cancer. *Urol. Oncol.* **2016**, *34*, 235.e11–235.e16. [CrossRef]
50. Todenhöfer, T.; Park, E.S.; Duffy, S.; Deng, X.; Jin, C.; Abdi, H.; Ma, H.; Black, P.C. Microfluidic enrichment of circulating tumor cells in patients with clinically localized prostate cancer. *Urol. Oncol.* **2016**, *34*, 483.e9–483.e16. [CrossRef]
51. Kuske, A.; Gorges, T.M.; Tennstedt, P.; Tiebel, A.-K.; Pompe, R.; Preißer, F.; Prues, S.; Mazel, M.; Markou, A.; Lianidou, E.; et al. Improved detection of circulating tumor cells in non-metastatic high-risk prostate cancer patients. *Sci. Rep.* **2016**, *6*, 39736. [CrossRef]
52. Tsumura, H.; Satoh, T.; Ishiyama, H.; Tabata, K.-I.; Takenaka, K.; Sekiguchi, A.; Nakamura, M.; Kitano, M.; Hayakawa, K.; Iwamura, M. Perioperative Search for Circulating Tumor Cells in Patients Undergoing Prostate Brachytherapy for Clinically Nonmetastatic Prostate Cancer. *Int. J. Mol. Sci.* **2017**, *18*. [CrossRef] [PubMed]
53. García, J.L.; Lozano, R.; Misiewicz-Krzeminska, I.; Fernández-Mateos, J.; Krzeminski, P.; Alfonso, S.; Marcos, R.A.; García, R.; Gómez-Veiga, F.; Virseda, Á.; et al. A novel capillary nano-immunoassay for assessing androgen receptor splice variant 7 in plasma. Correlation with CD133 antigen expression in circulating tumor cells. A pilot study in prostate cancer patients. *Clin. Transl. Oncol.* **2017**, *19*, 1350–1357. [CrossRef] [PubMed]
54. Puche-Sanz, I.; Alvarez-Cubero, M.J.; Pascual-Geler, M.; Rodríguez-Martínez, A.; Delgado-Rodríguez, M.; García-Puche, J.L.; Expósito, J.; Robles-Fernández, I.; Entrala-Bernal, C.; Lorente, J.A.; et al. A comprehensive study of circulating tumour cells at the moment of prostate cancer diagnosis: Biological and clinical implications of EGFR, AR and SNPs. *Oncotarget* **2017**, *8*, 70472–70480. [CrossRef] [PubMed]
55. Awe, J.A.; Saranchuk, J.; Drachenberg, D.; Mai, S. Filtration-based enrichment of circulating tumor cells from all prostate cancer risk groups. *Urol. Oncol.* **2017**, *35*, 300–309. [CrossRef] [PubMed]
56. Renier, C.; Pao, E.; Che, J.; Liu, H.E.; Lemaire, C.A.; Matsumoto, M.; Triboulet, M.; Srivinas, S.; Jeffrey, S.S.; Rettig, M.; et al. Label-free isolation of prostate circulating tumor cells using Vortex microfluidic technology. *NPJ Precis. Oncol.* **2017**, *1*, 15. [CrossRef]
57. Russo, G.I.; Bier, S.; Hennenlotter, J.; Beger, G.; Pavlenco, L.; van de Flierdt, J.; Hauch, S.; Maas, M.; Walz, S.; Rausch, S.; et al. Expression of tumour progression-associated genes in circulating tumour cells of patients at different stages of prostate cancer. *BJU Int.* **2018**, *122*, 152–159. [CrossRef] [PubMed]
58. Miyamoto, D.T.; Lee, R.J.; Kalinich, M.; LiCausi, J.A.; Zheng, Y.; Chen, T.; Milner, J.D.; Emmons, E.; Ho, U.; Broderick, K.; et al. An RNA-Based Digital Circulating Tumor Cell Signature Is Predictive of Drug Response and Early Dissemination in Prostate Cancer. *Cancer Discov.* **2018**, *8*, 288–303. [CrossRef]
59. Murray, N.P.; Aedo, S.; Fuentealba, C.; Reyes, E.; Minzer, S.; Salazar, A. The presence of secondary circulating prostate tumour cells determines the risk of biochemical relapse for patients with low- and intermediate-risk prostate cancer who are treated only with external radiotherapy. *Ecancermedicalscience* **2018**, *12*, 844. [CrossRef]
60. Bachman, J. Reverse-transcription PCR (RT-PCR). *Methods Enzymol.* **2013**, *530*, 67–74.
61. Pantel, K.; Denève, E.; Nocca, D.; Coffy, A.; Vendrell, J.-P.; Maudelonde, T.; Riethdorf, S.; Alix-Panabières, C. Circulating epithelial cells in patients with benign colon diseases. *Clin. Chem.* **2012**, *58*, 936–940. [CrossRef]
62. Poruk, K.E.; Valero, V.; He, J.; Ahuja, N.; Cameron, J.L.; Weiss, M.J.; Lennon, A.M.; Goggins, M.; Wood, L.D.; Wolfgang, C.L. Circulating Epithelial Cells in Intraductal Papillary Mucinous Neoplasms and Cystic Pancreatic Lesions. *Pancreas* **2017**, *46*, 943–947. [CrossRef] [PubMed]
63. NCI Dictionary of Cancer Terms. Available online: https://www.cancer.gov/publications/dictionaries/cancer-terms (accessed on 31 May 2019).
64. FDA-NIH Biomarker Working Group. *Understanding Prognostic versus Predictive Biomarkers*; Food and Drug Administration (US): Silver Spring, MD, USA, 2016.
65. Scher, H.I.; Graf, R.P.; Schreiber, N.A.; McLaughlin, B.; Jendrisak, A.; Wang, Y.; Lee, J.; Greene, S.; Krupa, R.; Lu, D.; et al. Phenotypic Heterogeneity of Circulating Tumor Cells Informs Clinical Decisions between AR Signaling Inhibitors and Taxanes in Metastatic Prostate Cancer. *Cancer Res.* **2017**, *77*, 5687–5698. [CrossRef] [PubMed]

66. Sciarra, A.; Gentilucci, A.; Silvestri, I.; Salciccia, S.; Cattarino, S.; Scarpa, S.; Gatto, A.; Frantellizzi, V.; Von Heland, M.; Ricciuti, G.P.; et al. Androgen receptor variant 7 (AR-V7) in sequencing therapeutic agents for castratrion resistant prostate cancer: A critical review. *Medicine (Baltim.)* **2019**, *98*, e15608. [CrossRef] [PubMed]
67. Miyamoto, D.T.; Ting, D.T.; Toner, M.; Maheswaran, S.; Haber, D.A. Single-Cell Analysis of Circulating Tumor Cells as a Window into Tumor Heterogeneity. *Cold Spring Harb. Symp. Quant. Biol.* **2016**, *81*, 269–274. [CrossRef] [PubMed]

© 2019 by the authors. Licensee MDPI, Basel, Switzerland. This article is an open access article distributed under the terms and conditions of the Creative Commons Attribution (CC BY) license (http://creativecommons.org/licenses/by/4.0/).

Review

Insights on CTC Biology and Clinical Impact Emerging from Advances in Capture Technology

Patrick C. Bailey [1] and Stuart S. Martin [1,2,*]

1. Marlene and Stewart Greenebaum Comprehensive Cancer Center, School of Medicine (UMGCCC), University of Maryland, Baltimore, MD 21201, USA; pcbailey@umaryland.edu
2. Department of Physiology, School of Medicine, University of Maryland, Baltimore, MD 21201, USA
* Correspondence: ssmartin@som.umaryland.edu; Tel.: +1-410-706-6601

Received: 16 May 2019; Accepted: 3 June 2019; Published: 6 June 2019

Abstract: Circulating tumor cells (CTCs) and circulating tumor microemboli (CTM) have been shown to correlate negatively with patient survival. Actual CTC counts before and after treatment can be used to aid in the prognosis of patient outcomes. The presence of circulating tumor materials (CTMat) can advertise the presence of metastasis before clinical presentation, enabling the early detection of relapse. Importantly, emerging evidence is indicating that cancer treatments can actually increase the incidence of CTCs and metastasis in pre-clinical models. Subsequently, the study of CTCs, their biology and function are of vital importance. Emerging technologies for the capture of CTC/CTMs and CTMat are elucidating vitally important biological and functional information that can lead to important alterations in how therapies are administered. This paves the way for the development of a "liquid biopsy" where treatment decisions can be informed by information gleaned from tumor cells and tumor cell debris in the blood.

Keywords: circulating tumor cells; CTC; liquid biopsy; CTM; CTMat; CTC biology; CTC capture technology

1. Introduction

Cancer remains a leading cause of death in all areas of the world [1]. The primary cause of death however, is not the primary tumor but metastases. The complete biology of metastasis remains unclear, but several general processes are recognized. The initial steps are understood to include the local invasion of the tumor into neighboring tissues followed by intravasation into the circulation, involving either the epithelial to mesenchymal transition (EMT) or the physical shedding of tumor cells into leaky, poorly formed vessels. Both EMT and shedding lead to the dissemination of tumor cells into the lymphatic and hematogenous systems [2]. Of these two methods, hematogenous spread is the most lethal.

Integral to the process of dissemination is circulation in the vasculature. Detached cells are termed circulating tumor cells (CTCs) or, in the case of cell clusters, circulating tumor microemboli (CTM). These cells circulate until they either attach to the vessel endothelium or become lodged in small capillaries. From this point, there can either be migration through the tissue or, in the case of CTMs, possible vascular rupture [3]. Cells which have survived these processes can serve as the seeds of eventual metastatic recurrence.

It has been estimated that tumor cells shed from the primary tumor at a rate of 3.2×10^6 cells per gram of tumor tissue per day, but over half quickly perish [4]. What remains is one cell per 10^{6-7} leukocytes [5]. The rarity and importance of these CTCs has led to the development of many technologies designed to enrich for this small population. Among the challenges inherent in isolating CTCs are the methodologies used for characterizing them. The two main methods that have been employed involve cell surface markers and the physical characteristics of the cell [6], both of which have advantages and pitfalls. The intent of this review is not to exhaustively catalog technologies, but to discuss the principles behind several stand-outs, the importance of CTC isolation in general, possible applications in functional studies and the clinical importance of CTCs in view of biology and new ideas in dissemination modality.

2. Diagnostic Importance of CTCs

The presence of CTCs in the blood has been proportionally correlated with poor prognosis, and CTMs are even more strongly correlated with patient outcome [7,8]. For a widespread use of CTC/CTM detection as a diagnostic tool, clinical acceptance is critical. The American Society of Clinical Oncology (ASCO), the National Academy of Clinical Biochemistry, the American Association for Clinical Chemistry, and the American Joint Committee on Cancer have all declined to recommend CTC/CTMat assays in the detection, monitoring or staging of cancer until the benefits of the technique are clarified [9–11].

The CellSearch system was approved by the FDA in 2004 for the clinical detection of CTCs but there are numerous challenges inherent in the platform. Problems of physics, statistics, translation, preparation time, and the constraint of fixed cells stained for limited biomarkers have led to inconsistent results [12]. These challenges impact results in detection rate, patient positivity, and correlation with prognosis [6,13–15]. Discounting phenotypic heterogeneity between CTCs, there are also numerous technical factors involved in these discrepancies, including differences in technique and bias between operators, sample size and lack of a common reference standard, among many others.

Toward a standard protocol that minimizes these issues, two new trends have a great deal of potential. These are the detection of circulating tumor materials (CTMat) and telomerase activity. As previously mentioned, half of the cells shed from the primary tumor die in circulation. Due to many factors, the membranes of these cells are perforated and cellular contents leak into the blood stream [16]. The physical forces in drawing blood are also a contributing factor to the destruction of viable cells, leading to the accumulation of cellular debris. CTMat is usually captured by the same methods outlined below, but where standard capture technologies would overlook these cell fragments as negative, CTMat capture technology can visualize and enumerate them. Using the CellSpotter technology, which can differentiate between intact tumor cells, damaged tumor cells and tumor cell fragments, CTMat was found to comprise the largest subpopulation in 18 blood samples from prostate cancer patients [16]. CTMat has not only been found to correlate well with viable CTC detection in prognostic capacity, but could also potentially provide an avenue for standardization, insofar as CTMat detection can be more easily quantified. It is also less restrictive in the identification of targets and the process of imaging can be automated [17].

In contrast to the release of cell fragments through apoptosis in the blood stream, another component of CTMat, circulating tumor DNA (ctDNA), is believed to stem mainly from cellular death in the solid tumor [18]. Levels of ctDNA have been found to correlate well with primary tumor resection, chemotherapy and metastasis [19,20]. Although the difficulty in producing primers for PCR of ctDNA fragments is not trivial, this process has been shown to discover relapse well before other conventional methods [21,22] Indeed, ctDNA is already being used for treatment response monitoring, the early detection of relapse [23,24] and even therapy decision (e.g., therapies related to the presence of mutant Epidermal Growth Factor Receptor [EGFR]) [25]. ctDNA from viral associated cancer has also been employed to monitor treatment response [26]. To this end, the analysis of ctDNA can be used to monitor therapeutic success. Increases in mutant alleles as a result of therapy resistance have been shown in patients monitored over a period of two years [25,27]. Finally, the FDA has approved the Cobas EGFR Mutation Test v2 as a companion diagnostic for non-small cell lung cancer therapy with Erlotinib. Standard clinical imaging detection involves the visualizing of a tumor mass, which is a process requiring millions of cells. ctDNA can be monitored and relapse discovered well before this timepoint.

Many of the most utilized platforms for the detection of CTCs utilize epithelial markers for identification, such as cytokeratin and EpCAM (epithelial-cell-adhesion-molecule). This can provide information as to cellular origin but neglects biological behavior. It has also been reported that tumor cells can downregulate or completely lose expression of these epithelial markers during the process of migration and/or dissemination [28]. This creates difficulty for epithelial-based isolations due to their reliance on the EpCAM surface marker for their capture technology. Telomerase, however, has been found to be re-activated in most cancers including prostate, ovarian, breast, lung, colon and bladder [29–32]. Telomerase activity is also associated with malignancy, is often detected in stage IV cancers and is a marker of stem cell activity [33]. Despite the requirement of lysing the sample for assay preparation, the above factors make this enzymatic activity an attractive choice to detect circulating tumor cells for diagnosis. Especially appealing is the possible application of this assay in the detection of relapse. Basal telomerase activity levels due to T-cell activity and other factors could be established and significant variations from this (apart from infections) could indicate possible tumor relapse.

Subsequent increases in activity could also reduce the occurrence of false positives. A possible second step to this process that would circumvent the establishment of basal activity would be to negatively select (as outlined below) leukocytes from the sample. If used in combination with monitoring ctDNA, this could be a powerful tool for treating relapse much earlier than currently possible (Figure 1).

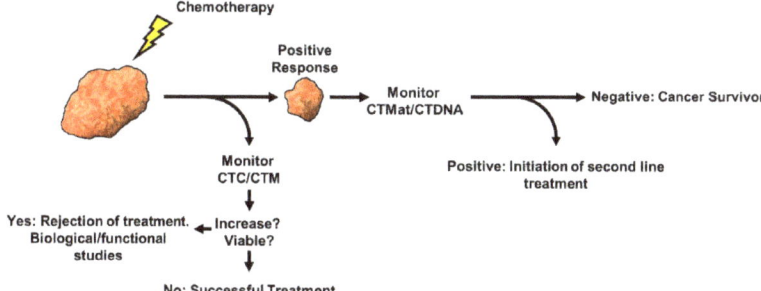

Figure 1. Workflow concept for the analysis of therapy and the early detection of relapse. After chemotherapy, patient CTCs can be analyzed for viability. An increase in viable CTCs can indicate increased mobilization and possible increased risk of relapse. After successful treatment, monitoring patient blood for telomerase activity or ctDNA can give a clinician a much earlier indication of relapse.

3. Clinical Relevance

The mobilization of tumor cells into the circulation is integral to distal metastasis. Current thought is that treatment failure due to metastasis is caused by micrometastasis present at the time of treatment or residual local disease [34]. However, there is mounting evidence that treatment methods themselves could cause an increased dissemination of cells into the vasculature or even the activation of dormant metastatic sites [35–41]. As outlined below, surgery, radiotherapy and systemic chemotherapy can alter tumor biology and possibly influence the risk of metastasis in unforeseen ways. The increase in CTCs as a side effect of treatment is a consideration that deserves careful study.

The effect of radiotherapy on metastasis has long been studied. Early studies indicated that lower doses of radiation resulted in higher rates of metastasis. Breast cancers transplanted into mice and subjected to non-curative doses of radiation had a 43.5% rate of metastasis compared to 9.6% in the control [42]. Metastasis rates were also 10% higher in transplanted mammary tumors given radiation in addition to resection compared to surgery alone [43]. In experiments with lung cancer and fibrosarcoma, it was shown that irradiated mice had higher rates of distal recurrence compared to control. This was initially explained by the activation of dormant micrometastasis and the modification of local tumor cells into a more aggressive and invasive phenotype [44].

Typical regimens of radiotherapy involve fractionated low doses over the course of many days. After longer periods, tumor cells have typically lost reproductive capacity with successful treatment. However, during the early course of the therapy, tumor cells are much more likely to repair therapy-induced DNA damage [45]. These cells have a higher probability of survival if disseminated into the blood stream.

This can be the result of surrounding tissue damage as well as the increased plasticity and genomic instability of irradiated cells [46]. Radiation-induced hypoxia was reported to upregulate the expression of surface markers that increased invasiveness [47]. An increased expression of Vascular Endothelial Growth Factor (VEGF) has also been observed following treatment [48].

The importance of radiation as a therapy cannot be understated. Its clinical value has been demonstrated in many settings. Nevertheless, it has been recently reported that radiation therapy on Non-Small Cell Lung Carcinoma (NSCLC) can mobilize CTCs into the blood stream early in therapy [49]. CTC counts were highest after the first doses of radiation and were shown to originate from the primary tumor. These cells were shown to have increased growth capacity in culture compared to CTCs collected pre-treatment. They also had increased mesenchymal characteristics and were more often found in clusters [8].

Not only radiation, but surgical procedures and chemotherapy have been linked to increased CTCs. Both needle and incisional biopsies have been correlated with increased CTC counts [50,51]. Tumors have also been reported to have formed along the track left by the biopsy needle [52]. Survival rates and local dissemination have been found to be worse with pre-operative biopsies in colorectal cancer, and increased CTCs compared to baseline have also been found both during and after surgery as well [53]. Karigiannis and colleagues have recently reported that neoadjuvant paclitaxel increases both CTCs and metastasis in an MMTV-PyMT (mouse mammary tumor virus-polyoma middle tumor-antigen) murine model [38]. After harvesting the lungs of mice treated with neoadjuvant paclitaxel, they found an increase in both the number and incidence of micrometastasis as well as the presence of single metastatic cells. There was also a twofold increase in CTCs in all experimental models examined, which included xenotransplanted cell lines, the spontaneous PyMT transgenic model and patient-derived xenografts (PDX) [38]. The interrelation between therapy, CTCs and metastasis underscores the vital need to understand the biology of rare circulating cells with the goal of developing targeted treatments. If conventional therapies can potentially increase CTC count and conversely metastasis in some cases, then combination treatments targeting CTCs can potentially improve outcomes.

4. Isolation of Cells

The importance of CTCs in diagnosis, prognosis and therapy outcome seems to be clear. Several technologies have been developed for their capture and enumeration. The assays involving ctDNA and CTMat are exciting prospects in the monitoring of recurrence, but neither involve the capture of CTCs for further analysis. Problematically, even with whole cell capture, many techniques kill the cell along the way. Even the FDA-approved gold standard of CTC detection, the CellSearch system, involves chemical fixation. This process is lethal to cells and does not allow for further characterization of viable cells or expansion in culture. Many of the technologies reported in table 1 involve chemical fixation. This does not preclude the modification of the platform's protocol such that live cells may be captured, but what is commonly reported is outlined in Table 1. In contrast to this, there are many established and developing technologies that have proven to be more sensitive than the CellSearch system and are also designed to capture viable cells, allowing for further biological study [6].

Table 1. Circulating tumor cell (CTC) technologies. CTC isolation technologies grouped by category and isolation criteria. Modified from Ferreira et al. 2016 [54]. *refers to the reference in question.

Subcategory	Platform	Enrichment Principle	Live Cell Analysis Reported *	Company
Label-Based				
Positive Enrichment Immunoaffinity				
Micropost Arrays	CTC-Chip [55]	EpCAM	Yes	
	GEDI Chip [56]	PSMA/HER2, Size	No	
	OncoCEE [57]	Antibody Cocktail	No	Biocept Inc. San Diego, CA, USA
Microfluidic Surface Capture	Biofluidica CTC system [58]	EpCAM	Yes	Biofluidica Inc. San Diego, CA, USA
	CytoTrapNano [59]	EpCAM	No	Cytolumina. Los Angeles, CA, USA
	GEM Chip [60]	EpCAM	Yes	
	HTMSU [61]	EpCAM	No	
	Graphene Oxide Chip [62]	EpCAM	No	
	Herringbone Chip [63]	EpCAM	No	
Microfluidic Magnetic	Ephesia [64]	EpCAM	Yes	
	Magnetic Sifter [60]	EpCAM	No	
	LiquidBiopsy [65]	Antibody Cocktail	No	Thermo Fisher, Waltham, MA, USA
	Isoflux [66]	EpCAM	No	Fluxion Biosciences, Alameda, CA, USA
	CellSearch [67]	EpCAM	No	Silicon Biosystems, Huntington Valley, PA, USA
Magnetic	AdnaTest [68]	Antibody Cocktail	No	Qiagen, Hilden, Germany
	MACS [69]	EpCAM	No	Miltenyi Biotec, Bergisch Gladbach, North Rhine-Westphalia, Germany
	MagSweeper [70]	EpCAM	No	
Magnetic in vivo	CellCollector [71]	EpCAM	Yes	GILUPI, Potsdam, Germany
Negative Enrichment Immunoaffinity				
Magnetic	EasySep [72]	CD45	No	STEMCELL, Vancouver, BC, Canada
	QMS [73]	CD45	Yes	
	MACS [74]		Yes	Miltenyi Biotec, Bergisch Gladbach, North Rhine-Westphalia, Germany
Microfluidic/Magnetic	CTC-iChip [75]	CD45, CD66b, Size	Yes	

Table 1. *Cont.*

Subcategory	Platform	Enrichment Principle	Live Cell Analysis Reported *	Company
		Label-Free		
		Density		
	Ficoll-Paque [76]	Density	Yes	GE Healthcare Bio-Sciences, Pittsburg, PA, USA
	OncoQuick [77]	Density, Size	Yes	Greiner Bio-One, Kremsmünster, Austria
	RosetteSep [78]	Density, Antibody Cocktail	Yes	STEMCELL, Vancouver, BC, Canada
	Accucyte and CyteSealer [79]	Density	Yes	Rarecyte, Seattle, WA, USA
		Size		
	Parsortix [80]		Yes	Angle, King of Prussia, PA, USA
	Microwall Chip [81]		Yes	
	ScreenCell [82]		Yes	ScreenCell, Westford, MA, USA
Filtration	Resettable Cell Trap [83]	Size, Deformability	Yes	
	Flexible Micro Spring Array (FMSA) [84]		Yes	
	FaCTchecker [85]		Yes	Circulogix, Hallandale Beach, FL, USA
	Crescent Chip [86]		Yes	
	ISET [87]		Yes	RareCells Diagnostics, Paris Cedes, France
	CellSieve [88]		Yes	Creatv Microtech, Potomac, MD, USA
	Cluster Chip [89]		Yes	
	Vortex [90]		Yes	Vortex Biosciences, Pleasanton, CA, USA
Fluid Dynamics	Double Spiral Chip [91]	Size	Yes	
	Micropinching Chip [92]		Yes	
	ClearCell FX [93]		Yes	Genomax Technologies, Singapore
		Electric		
	ApoStream [94]	Electrical Signature	Yes	Apocell, Houston, TX, USA
	DEPArray [95]		Yes	Silicon Biosystems, Huntington Valley, PA, USA

338

There are several competing modalities in CTC capture methodology, but all of them fall under two conceptual umbrellas: label-based and label-free. Label-based (or affinity-based) capture is the most widely used strategy, with CellSearch as the only technology approved by the US Food and Drug Administration. The prevailing idea behind this methodology is that tumor cells display different surface markers than blood cells and can therefore be separated from the rest of the circulatory cells on this basis. The three most commonly employed biomarkers utilized for tumor cell selection and identification are the epithelial-cell-adhesion-molecule (EpCAM), cytokeratins, and the antigen CD45 [96]. EpCAM is used to positively select for CTCs, while CD45 negatively depletes white blood cells and cytokeratins are used to positively identify CTCs post-enrichment. These three biomarkers have been expanded upon in some technologies in the use of antibody cocktails including, for example, the human epidermal growth factor 2 (HER2) for breast cancer and the prostate-specific membrane antigen (PSMA) for prostate cancer. In most cases, magnetic beads are conjugated to the antibodies allowing for a magnetic field to capture the cell after the antibody binds to its target. Capture strategies also include microfluidic devices with surface-coated antibodies. Cells of interest bind to these antibodies as the sample flows over the surface. Unfortunately, due to the complexity of CTC biomarker expression, there is no single antigen which allows for 100% error-free capture. This makes effective capture a continuing challenge. Table 1 outlines a variety of capture technologies that fall under the umbrellas of "label-based" and "label-free". Platforms are further characterized by their enrichment principle and their reported capture of live cells.

The CellSearch and Adnatest platforms both make use of magnetic beads attached to antibodies to EpCAM, but Adnatest employs additional cancer-specific antibodies depending on the requirement. CellSearch uses downstream immunostaining to identify CTCs. Positive ID is dependent on the expression of cytokeratins, negative expression of CD45 and positive DAPI nuclear stain. The Adnatest further differs from CellSearch in that it does not rely on downstream immunostaining. Instead, it employs cell lysis and RT-PCR to measure tumor-associated gene expression. A limitation of these technologies is a reliance on EpCAM. EpCAM expression has been shown to vary widely, and cells with low or negative expression can be missed by these platforms [96–100]. Cytokeratin expression can also be lost following EMT [101]. A further drawback is that neither of these technologies allows for further live-cell phenotypic analysis as the captured cells are either fixed or lysed.

Several technologies have been formulated that bypass the requirement for fixation or lysis. Recent advances in microfabrication have allowed the creation of devices with features smaller than a cell. With controlled use of the properties of fluid, cellular contact with these microstructures can be directed. The first among these devices to be developed utilized arrays of antibody-coated microposts [55]. In these devices, sample blood is passed over the chip allowing for the capture of marker-expressing cells. Although some require the pre-lysis of red cells, many enable the use of whole blood with no pre-preparation. The accompanying drawback is that flow rates are most often quite slow at @1–2 mL/h [55,56,102]. The most commonly employed antibody is EpCAM, but several devices employ a cocktail of antibodies that can be specialized for the particular cancer being studied. Today, there are many devices available including the CTC chip, nanopillar chip, micropillar chip, GEDI (geometrically enhanced differential immunocapture) chip, and the OncoCEE among others. These devices have all shown higher capture efficiency than the CellSearch [6], and have the advantage of smaller size and lower cost than the magnetic benchtop devices.

The CTC-chip's first iteration (preceding the herringbone chip) captured a median of 155 cells/mL in each of 55 samples tested from 68 patients with non-small cell lung cancer, while the CellSearch only captured cells in 20% of patient samples and had a mean of <6 cells/mL [103]. The GEDI chip employs hydrodynamic chromatography by offsetting the microposts in such a way as to separate cells by size and minimize non-specific leucocyte adhesion [56]. The OncoCEE employs a customizable cocktail that can include antibodies for both cancer and mesenchymal specific markers. It also allows for in situ fluorescent staining of the captured cells by staining the capture antibodies [57].

To increase imaging and production efficiency, the field has begun to explore the idea of surface-capture devices that eschew the concept of posts altogether. Microchannels and surface patterns are designed to maximize mixing and surface contact with cells. The simpler design allows for larger scale production and with opaque posts and three-dimensional structure removed, imaging is enhanced. Another welcome enhancement is the allowance of higher flow rates, leading to more rapid throughput [60,62,63]. Devices which use this technology include the microvortex herringbone chip, sinusoidal chip, GEM chip, and the graphene oxide chip.

Biomarkers may also be used to negatively enrich samples containing CTCs. Blood cell markers such as CD45 and/or CD66 can be used to deplete white blood cells from the larger population enriching for CTCs in the remainder. Technologies utilizing this method include EasySep and RosetteSep. RosetteSep incorporates the additional step of density centrifugation, while EasySep uses a magnetic field. A pitfall inherent in this technique is the fact that not all cells in the blood express these markers, resulting in a much poorer purity than with positive selection [74,104,105]. Another downfall is possible CTC loss being caught up in the large movement of concentrated blood cells during depletion. For these reasons, this technique is often used as a preparatory step for other enrichment methods [106].

Despite the utility and many benefits of cellular biomarkers, there are drawbacks as well. It is becoming established that tumor cells express EpCAM at varying levels. In fact, expression can be ablated entirely in some sub-populations, including those which have undergone EMT [107]. Tumor cells have also been reported to express the white blood cell marker CD45 [108]. With these problems in mind, alternative assays which employ only the biophysical properties of the cell have been developed.

These label-free physical detection methods include cell size, deformability, density and electric charge. The most widely employed biophysical selection criterion is cellular size [12]. Tumor cells are larger on average than blood cells [109], and this morphological difference is employed to differentially capture CTCs and CTMs. There are multiple platforms which use these properties such as the micro double spiral chip, the Parsortix and Vortex systems, the micro crescent chip, the Cellsee system, micro column wall chip, ISET, Clear Cell FX, cluster chip, micro pinching chip and the CellSieve among others. Each of these assays have proven to be more selective than the CellSearch system in isolating tumor cells [6].

There are different ways of using size in the process of selection, however. Two-dimensional microfiltration involves a single membrane with variable pore size used to filter out smaller cells while leaving the larger CTCs trapped on the membrane. Cell pore sizes come in a variety of sizes ranging from 6 to 9 um. CellSieve filtration has not only been used to detect cancer-associated macrophages and cancer-associated macrophage-like cells, [110,111] but, using 7.5 mL patient samples, it detected CTCs in 100% of metastatic breast cancer patients tested [88]. CellSieve, ISET and ScreenCell use this methodology, but require pre-processing of the patient sample. FMSA (Flexible Microspring Array) can use whole blood and has been validated in the detection of CTCs in 76% of samples tested in various cancers [112].

Three-dimensional filtration systems exploit the larger size of tumor cells, but use multiple layers of filter to capture them. The FaCTChecker, Parsortix system, and cluster chip fall into this category. The FaCTChecker takes advantage of multiple vertical layers with different sized pores [113], while the Parsortix has developed a horizontal stair-type scheme that reduces the channel width stepwise [80]. Viable CTCs can be harvested using either platform. Our lab has employed the Parsortix system to isolate CTCs from breast cancer patients. We subsequently tethered these live cells on a proprietary PEM+Lipid technology [114] and imaged them for Microtentacles (Figure 2). The Cluster Chip is unique in size selection technologies, as its sole target are CTMs. Many technologies have reported on the capture of CTMs, but this novel approach enriches for them specifically while allowing single CTCs to pass through [89]. The design involves staggered rows of triangular pillars. The repeating unit of the design is the cluster trap. This three-triangle arrangement is reminiscent of a biohazard sign insofar as two triangles side by side to create a tunnel that is bifurcated by the third triangle beneath them. This simple design can capture CTMs as small as two cells. The utility of the device was shown in breast, melanoma and prostate cancers, isolating clusters in 41%, 30% and 31% of patients, respectively [89]. Large downsides to filtration systems exist, however. Despite the capture of viable cells without labels that are difficult to remove, the systems are prone to clogging and parallel processing is needed for large volumes. Purity is also an issue as it can range below 10%.

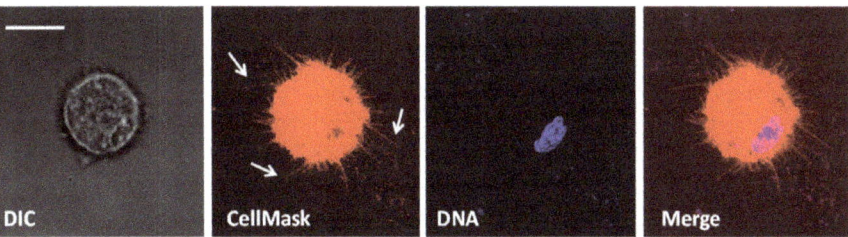

Figure 2. Live CTCs isolated with Parsortix technology. Whole blood was taken from a stage IV metastatic breast cancer patient. The Angle Parsortix was used to isolate CTCs from the blood (15 CTCs in 10 mL). CTCs were tethered to proprietary PEM+Lipid slides and stained with CellMask membrane dye (red). Cells are CD45$^-$ and contain a nucleus (blue). Arrows indicate microtubule-based structures termed Microtentacles (McTN).

Two exciting new technologies to recently emerge involve the use of inertial fluid forces to passively separate CTCs from the rest of the blood population based on cell size. A combination of shear gradient and wall lift forces interact to stably trap the CTCs. The Vortex platform capitalizes on these forces to inertially focus trapped CTCs in micro vortices created in reservoirs apart from the main fluid channel. Smaller blood cells simply flow by in the main stream. CTCs remain in the device until a slower flow rate flushes them out of the reservoirs. The Vortex Chip processes the standard 7.5 mL sample size in 20 minutes using whole unprocessed blood. Confirmation has come in breast and lung cancers with a purity of 57–94%, much higher than that normally attained with size-based techniques [90]. The ClearCell FX uses inertial forces in combination with secondary flow arising from curved channels [115]. When a channel is curved, there is a difference in the flow rates between the center of the channel and the walls. This difference in flow rates is termed a "Dean's" flow and, when combined with inertial forces, can be calculated to precisely position cells. The trapezoidal channel results in larger cells on the shorter wall and smaller cells on the larger wall. This channel then splits into two collection outlets, where CTCs are isolated and captured. This technology requires red cell lysis prior to flow but has an impressive 8-minute run time. It has been confirmed in breast and lung cancers with a higher capture rate than the Vortex [116]. Both processes involve minimal stress on cells without the use of labels and are much simpler to fabricate than those previously mentioned.

Dielectrophoresis (DEP) exploits the electrical characteristics of tumor cells. These characteristics depend on phenotype, composition and morphology. DEP polarizes cells by using a nonuniform electric field. This results in the ability to physically manipulate the cells by exerting attractive or repulsive forces (positive pDEP or negative nDEP). ApoStream employs a strategy wherein the electrical field separates tumor cells and leukocytes, using differences in their conductivity. The field attracts CTCs and repels leukocytes. After pre-processing by centrifugation, the ApoStream can process captured CTCs from 10mL of whole blood in less than an hour [117].

The DEPArray applies the second DEP strategy, retention, by trapping single cells in DEP cages generated via an array of individually controllable electrodes [118]. DEPArray as a platform is not designed for the bulk enrichment of cells, however. It is intended for single cell capture. Multiple studies have shown the utility of the technology in this capacity [95,119,120], but an unfortunate drawback is large cell loss during sample preparation [121].

5. CTC Biology

The prognostic importance of CTC counts is well established, but counts have not yet been widely employed to affect clinical decisions, due to unclear relevance to treatment. CTC counts have therefore not been recommended clinically to affect treatment decisions, as of yet [122]. Consequently, a more robust understanding of CTC biology is required. Tumor heterogeneity is increasingly being reported in the literature, not only between primary and secondary tumors, but intratumor as well. There can be as many as six different clonal cell lines within just one tumor [123]. Standard biopsy techniques such as fine needle aspiration and core biopsy are insufficient to capture this variety. These techniques, by design, take tissue from one area of the tumor for further analysis. Even with multiple samples, such as those taken in prostate cancers, there is not sufficient tissue to encompass all of the heterogeneity. "Liquid biopsy" is a term being increasingly used to describe analysis of CTC populations. The CTC population is thought to encompass more of the clonal populations in a tumor [122]. By analyzing the captured cells, an investigator can get a more complete picture of tumor composition and how it changes over time.

Studies of the composition of CTCs can further shed light into the process of metastasis. The complete process of metastasis is unclear, but conventional wisdom describes a process where tumor cells undergo the epithelial mesenchymal transition (EMT). This process involves cells detaching from the main tumor body, migrating through the extracellular matrix and extravasating into the circulation (Figure 3). During this process, the cell downregulates the expression of its epithelial markers, such as E-cadherin, and upregulates EMT markers, such as N-cadherin, snail, twist, vimentin and detyrosinated tubulin [123].) CTC/CTMs have been shown to upregulate vimentin and detyrosinated tubulin as well [124]. After extravasation, the cell then undergoes the reverse process of mesenchymal-to-epithelial transition (MET). This has been widely held to be the main mode of metastatic dissemination, but new reports have begun to challenge this.

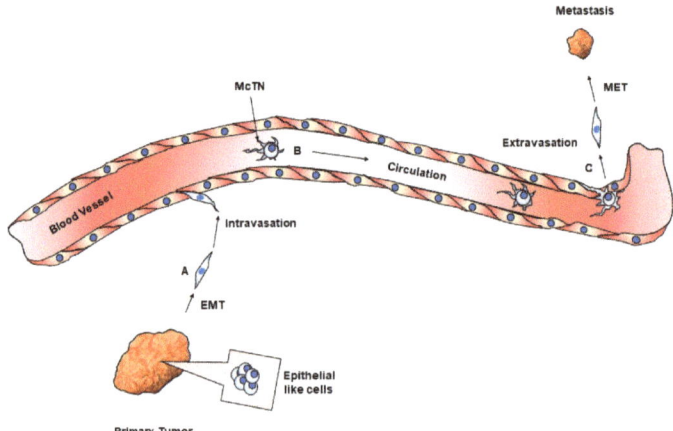

Figure 3. Epithelial to Mesenchymal Transition (EMT) and metastasis. (**A**) Epithelial-like cells in the primary tumor undergo a transition to a mesenchymal phenotype and migrate towards the vasculature. (**B**) Detached tumor cells in the circulatory vessels display microtubule-based structures, termed Microtentacles (McTN). (**C**) McTN aid in reattachment and extravasation. Extravasated cells undergo a mesenchymal to epithelial transition, and seed tumors at distal sites.

Fischer and colleagues described an experiment with a triple transgenic mouse that tracked mesenchymal lineage in breast cancer tissue. The system utilized an irreversible color switch that was activated by the expression of fsp1, a crucial protein in EMT initiation. With the expression of fsp1, cells experiencing EMT would undergo an irreversible color change from red to green, allowing for the tracking of any metastatic cell that had gone through the process. What was observed was that the vast majority of metastatic tumor tissue was red and had not undergone EMT. This was confirmed using multiple oncogenes and EMT tracing proteins. Interestingly, the following chemotherapy tumor recurrence was mostly green [125]. Similar findings were reported independently in the same issue of *Nature*, from a lab using twist and snail in pancreatic tumor lines [126].

The ramifications of these findings are manifold and beyond the scope of this review to cover. It is however important to note that this is a proof-of-principle that the process of EMT can be dispensable for initial metastasis, in some cases. This underscores the importance of understanding the biology in circulating cells. Which proteins CTCs express, and the resulting phenotypes, are crucial to understanding how cancer spreads to distal sites. It is indeed possible that the bulk of tumor spread results from simple CTC shedding into the vasculature. This does not reduce the importance of EMT in cancer, however. Cancer cells displaying the mesenchymal phenotype have been shown to be more aggressive, stem-like, and resistant to treatment [127]. Both Zheng and Fischer also observed EMT cells persisting after treatment despite original metastasis composition. What this highlights is that there can be multiple modes of metastasis, and the study of cells in transition can give us insights into the process.

Aceto et al. have recently shown that CTMs are 23–50 times more metastatic than CTCs [8]. Their use of fluorescently labeled cells also highlighted that clusters arise from oligoclonal groupings of cells that differentially express the cell junction protein plakoglobin. These studies, along with the results of Zheng and Fischer, further emphasize the importance of circulating cell study. They give us insight into the probable mechanism of metastasis. In the 323 lung foci that Aceto observed, 171 were CTM-derived, although CTMs only comprised 2–5% of the total population of tumor cells in the circulation.

Previous thought was that CTMs were likely to break up in the physical pressures of the blood stream, or to become lodged very quickly in smaller capillaries, negating their capability of seeding distant metastasis [128]. Recent work has shown this is not the case. Au et al. demonstrated with microscopy and capillary tubing that tumor clusters migrated in a single file fashion without dissociation. Moreover, the clusters were viable upon capillary exit [129]. Taken together with the evidence that clusters have a much higher metastatic potential, the benefit of elucidating biological differences between CTCs and CTMs is clear. In fact, very recent evidence has indicated that the disruption of CTMs leads to the suppression of metastasis [130].

It has been hypothesized that CTMs could arise either by passive shedding or through collective migration [101,131]. Collective migration has been observed in multiple tumor types, but it has only been directly correlated to local invasion [101]. Metastasis, through collective migration, has merely been inferred by the presence of clusters in the blood. Tumor vasculature is improperly formed, tortuous, leaky, and possessive of blind shunts [132]. It has been reported that tumor cells can actually replace vascular endothelium in places, a process known as vasculogenic mimicry [133]. With these factors in mind, it is quite feasible that CTCs and CTMs mainly arise through the passive sloughing of cells. This would correlate well with the data showing that breast cancers arising from *neu* and PyMT transgenes undergo very little EMT.

Interstitial fluid pressure (IFP) could contribute to CTC shedding as well. IFP is the fluid pressure measured within tumors and is the direct result of hyperpermeable blood vessels. Fluid and plasma proteins extravasate into the tumor tissue and elevate the pressure in the interstitium [47]. Not only could this increased pressure disrupt cell–cell junctions, but it could cause physical pressures that assist in cells detaching from the tumor bulk. High IFP is correlated strongly with poor prognosis [134]. As higher interstitial pressure is a direct result of improperly formed vessels, and stronger pressure could result in cell detachment, it follows that cells could break off at a higher rate as capillaries become leakier.

6. Functional CTC Studies

Translating lab research into clinical practice involves the study of how cells function, both in vitro and in vivo. As outlined above, it has been clearly shown that higher CTC counts in peripheral blood correlates with poor prognosis. Functional studies can broaden the spectrum of applications to CTC analyses. The challenges in obtaining stable cultures are significant but advances in CTC expansion from patient samples have been achieved. The subsequent functional studies can give clues into the identity of metastasis-initiating cells and can point the way to new avenues of therapy. A workflow, as outlined in Figure 4, illustrates the concept of CTC study, beginning with isolation and ending with the functional study of cultured CTCs. The first step in a workflow of this kind would be sample preparation and isolation using one of the methods outlined above. This would result in the capture of differing circulating materials, depending on the capture technology. These captured materials could eventually be used for prognosis and relapse decisions.

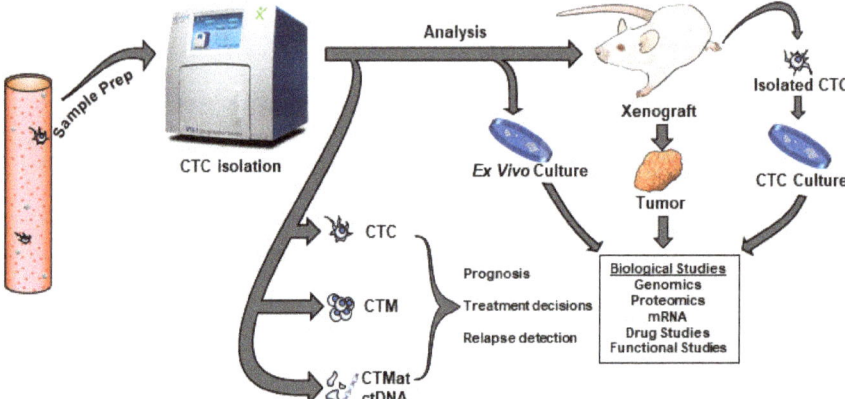

Figure 4. Workflow concept for the isolation of CTCs and subsequent analysis. Patient blood is passed through a capture device which enriches for tumor cells. Captured cells are then identified, enumerated and characterized. Cells can then be cultured and subjected to further biological and functional analysis.

Functional analysis of CTCs has been performed in multiple studies. Zhang et al. reported a protocol for the primary culture of breast cancer CTCs from patients with advanced stage and brain metastases [135]. The cultures survived for several weeks. This study allowed the elucidation of several biomarkers, including HER2 and EGFR, as brain metastasis selected markers (BMSM). Cells which expressed this BMSM signature exhibited significant invasiveness and resulted in brain metastases in murine xenografts. Oligoclonal breast cancer CTC cell lines were cultured for >6 months in 2014 [136]. Of five tested lines, three proved to be tumorigenic. The culture allowed for the discovery of new mutations in the estrogen receptor gene, fibroblast growth factor and PIK3CA. A long term culture of a CTC line from prostate cancer was also established using a novel 3D organoid system [137]. This included TRMPRSS2-ERG fusion proteins, overexpression of SPINK1 and SPOP and CHD1 mutations and loss, respectively. Lung cancer CTCs were successfully expanded ex vivo using a 3D co-culture which used a simulated tumor microenvironment. CTCs expanded from 14/19 patient samples and had matched mutations with their respective primary tumors, including tp53 [138].

Captured breast cancer CTCs were injected into murine tibia bone resulting in lung, liver and bone metastases [104]. The study of protein expression in the metastasis revealed universal expression of EpCAM, MET, CD44 and CD47. This could reveal important information on necessary proteins in the process of engraftment and metastatic outgrowth. Further study in an additional cohort revealed that metastases increased with the number of CD44/CD47/MET/EpCAM-positive cells. Importantly, these cells were obtained from advanced stage patients with high numbers of CTCs. This underscores the need to obtain and expand tumor cells from early stage patients to confirm this protein expression profile as metastasis-initiating in all stages.

Migratory capabilities of isolated metastatic prostate CTCs were shown in NOD/SCID mice [139]. Tumor cells were found in the spleen and the bone marrow after xenografting. Hodgkinson et al. showed that CTC xenografts of small cell lung cancer (SCLC) are not only tumorigenic in murine models but respond similarly to chemotherapy as in the original donor patient. SCLC patients have been reported to have the highest CTC counts of all solid tumors [140]. Notably, these tumors are often inoperable and difficult to biopsy. Expanding tumors which mirror patient response is an important step in furthering treatment less invasively.

7. Conclusions

Metastasis remains the number one cause of death in cancer patients. This is the result of the migration of cells from the primary tumor to distal sites. Indispensable to this process is the migration/shedding of CTCs into the vasculature. These circulating tumor cells can be analyzed for a breadth of beneficial information. Currently, prognostic indications can be made based on the enumeration of CTCs in the blood. With further technological development, the presence of metastasis could be detected before clinical manifestation, by monitoring tumor materials in the blood. It is also feasible that patients with known genetic risk factors could be monitored for ctDNA, using primers for known tumor mutations. This could possibly advance diagnosis by years, and increase survival rates significantly.

Even after disease control is accomplished with surgery and/or therapy, metastasis can remain a problem. This can be partially due to cancer cell mobilization caused by therapy itself. Radiation has been shown to select for and to convert tumor cells to phenotypes that are more mobile and aggressive, allowing for the generation of metastases. Tissue disruption and the leakage of blood containing tumor cells during surgery can also promote tumor spread. This includes procedures such as routine biopsy.

These problems underscore the need for the capture and study of viable tumor cells. Many technologies exist, but many involve the fixation of cells and their subsequent death. Emerging platforms have developed ways to isolate live CTCs which allow for downstream biological analysis. These studies have led to valuable insights into the mechanisms of metastasis and cellular survival in the harsh environment of the circulation. Functional studies with cultured CTCs and xenografts have revealed important information on protein expression and genetic composition. With the standardization of capture techniques, inconsistencies in efficiency can be greatly reduced, allowing for more robust information to be attained.

All these principles could support the goal of improving drug discovery to reduce metastasis. The current cancer detection and drug treatment paradigm involves tumor growth and visualization. Current technological parameters limit the tumors we can visualize to upwards of ten million cells. A shift of focus to the detection of ctDNA/CTMat/CTC/CTMs can improve detection sensitivity and improve treatment strategies. If surgery and radiation can promote cellular dissemination, then therapies that specifically target circulating cells could increase survival outcomes and reduce distal recurrence. Overall, developing therapies that target cancer's ability to ever survive in circulation can prevent metastasis before it occurs.

Author Contributions: P.C.B. and S.S.M. wrote the manuscript.

Funding: This work was supported in part by the Kahlert Foundation and grants from the National Institutes of Health to SSM (R01-CA124704, R01-CA124624) and Veterans Administration (BX002746).

Conflicts of Interest: The authors wish to declare no conflict of interest.

References

1. Jemal, A.; Bray, F.; Center, M.M.; Ferlay, J.; Ward, E.; Forman, D. Global cancer statistics. *CA Cancer J. Clin.* **2011**, *61*, 69–90. [CrossRef] [PubMed]
2. Talmadge, J.E.; Fidler, I.J. AACR centennial series: The biology of cancer metastasis: Historical perspective. *Cancer Res.* **2010**, *70*, 5649–5669. [CrossRef] [PubMed]
3. Zhang, X.; Nie, D.; Chakrabarty, S. Growth factors in tumor microenvironment. *Front. Biosci. Landmark Ed.* **2010**, *15*, 151–165. [CrossRef] [PubMed]

4. Butler, T.P.; Gullino, P.M. Quantitation of cell shedding into efferent blood of mammary adenocarcinoma. *Cancer Res.* **1975**, *35*, 512–516. [PubMed]
5. Ross, A.A.; Cooper, B.W.; Lazarus, H.M.; Mackay, W.; Moss, T.J.; Ciobanu, N.; Tallman, M.S.; Kennedy, M.J.; Davidson, N.E.; Sweet, D.; et al. Detection and viability of tumor cells in peripheral blood stem cell collections from breast cancer patients using immunocytochemical and clonogenic assay techniques. *Blood* **1993**, *82*, 2605–2610. [PubMed]
6. Hong, B.; Zu, Y. Detecting circulating tumor cells: Current challenges and new trends. *Theranostics* **2013**, *3*, 377–394. [CrossRef] [PubMed]
7. Hou, J.M.; Krebs, M.G.; Lancashire, L.; Sloane, R.; Backen, A.; Swain, R.K.; Priest, L.J.; Greystoke, A.; Zhou, C.; Morris, K.; et al. Clinical significance and molecular characteristics of circulating tumor cells and circulating tumor microemboli in patients with small-cell lung cancer. *J. Clin. Oncol.* **2012**, *30*, 525–532. [CrossRef] [PubMed]
8. Aceto, N.; Bardia, A.; Miyamoto, D.T.; Donaldson, M.C.; Wittner, B.S.; Spencer, J.A.; Yu, M.; Pely, A.; Engstrom, A.; Zhu, H.; et al. Circulating tumor cell clusters are oligoclonal precursors of breast cancer metastasis. *Cell* **2014**, *158*, 1110–1122. [CrossRef] [PubMed]
9. Amin, M.B.; Edge, S.; Greene, F.; Byrd, D.R.; Brookland, R.K.; Washington, M.K.; Gershenwald, J.E.; Compton, C.C.; Hess, K.R.; Sullivan, D.C.; et al. *American Joint Committee on Cancer; American Cancer Society. AJCC Cancer Staging Manual*, 8th ed.; American Joint Committee on Cancer; Springer: Chicago, IL, USA, 2017; p. xvii. 1024p.
10. Mittendorf, E.A.; Bartlett, J.M.S.; Lichtensztajn, D.L.; Chandarlapaty, S. Incorporating Biology Into Breast Cancer Staging: American Joint Committee on Cancer, Eighth Edition, Revisions and Beyond. *Am. Soc. Clin. Oncol. Educ. Book* **2018**, 38–46. [CrossRef] [PubMed]
11. Sturgeon, C.M.; Duffy, M.J.; Stenman, U.H.; Lilja, H.; Brunner, N.; Chan, D.W.; Babaian, R.; Bast, R.C., Jr.; Dowell, B.; Esteva, F.J.; et al. National Academy of Clinical Biochemistry laboratory medicine practice guidelines for use of tumor markers in testicular, prostate, colorectal, breast, and ovarian cancers. *Clin. Chem.* **2008**, *54*, e11–e79. [CrossRef]
12. Van der Toom, E.E.; Verdone, J.E.; Gorin, M.A.; Pienta, K.J. Technical challenges in the isolation and analysis of circulating tumor cells. *Oncotarget* **2016**, *7*, 62754–62766. [CrossRef] [PubMed]
13. Grover, P.K.; Cummins, A.G.; Price, T.J.; Roberts-Thomson, I.C.; Hardingham, J.E. Circulating tumour cells: the evolving concept and the inadequacy of their enrichment by EpCAM-based methodology for basic and clinical cancer research. *Ann. Oncol.* **2014**, *25*, 1506–1516. [CrossRef] [PubMed]
14. Wang, L.; Balasubramanian, P.; Chen, A.P.; Kummar, S.; Evrard, Y.A.; Kinders, R.J. Promise and limits of the CellSearch platform for evaluating pharmacodynamics in circulating tumor cells. *Semin. Oncol.* **2016**, *43*, 464–475. [CrossRef] [PubMed]
15. Flores, L.M.; Kindelberger, D.W.; Ligon, A.H.; Capelletti, M.; Fiorentino, M.; Loda, M.; Cibas, E.S.; Janne, P.A.; Krop, I.E. Improving the yield of circulating tumour cells facilitates molecular characterisation and recognition of discordant HER2 amplification in breast cancer. *Br. J. Cancer* **2010**, *102*, 1495–1502. [CrossRef] [PubMed]
16. Rao, G.C.; Larson, C.; Repollet, M.; Rutner, H.; Terstappen, L.W.; O'hara, S.M.; Gross, S. Analysis of Circulating Tumor Cells, Fragments, and Debris. U.S. Patent 8,329,422, 11 December 2012.
17. Coumans, F.A.; Doggen, C.J.; Attard, G.; de Bono, J.S.; Terstappen, L.W. All circulating EpCAM+CK+CD45− objects predict overall survival in castration-resistant prostate cancer. *Ann. Oncol.* **2010**, *21*, 1851–1857. [CrossRef] [PubMed]
18. Schwarzenbach, H.; Alix-Panabieres, C.; Muller, I.; Letang, N.; Vendrell, J.P.; Rebillard, X.; Pantel, K. Cell-free tumor DNA in blood plasma as a marker for circulating tumor cells in prostate cancer. *Clin. Cancer Res.* **2009**, *15*, 1032–1038. [CrossRef]
19. Leary, R.J.; Kinde, I.; Diehl, F.; Schmidt, K.; Clouser, C.; Duncan, C.; Antipova, A.; Lee, C.; McKernan, K.; De La Vega, F.M.; et al. Development of personalized tumor biomarkers using massively parallel sequencing. *Sci. Transl. Med.* **2010**, *2*, 20ra14. [CrossRef] [PubMed]
20. Kaiser, J. Medicine. Keeping tabs on tumor DNA. *Science* **2010**, *327*, 1074. [CrossRef]
21. Garcia-Murillas, I.; Schiavon, G.; Weigelt, B.; Ng, C.; Hrebien, S.; Cutts, R.J.; Cheang, M.; Osin, P.; Nerurkar, A.; Kozarewa, I.; et al. Mutation tracking in circulating tumor DNA predicts relapse in early breast cancer. *Sci. Transl. Med.* **2015**, *7*, 302ra133. [CrossRef]

22. Tie, J.; Wang, Y.; Tomasetti, C.; Li, L.; Springer, S.; Kinde, I.; Silliman, N.; Tacey, M.; Wong, H.L.; Christie, M.; et al. Circulating tumor DNA analysis detects minimal residual disease and predicts recurrence in patients with stage II colon cancer. *Sci. Transl. Med.* **2016**, *8*, 346ra392. [CrossRef]
23. Schmiegel, W.; Scott, R.J.; Dooley, S.; Lewis, W.; Meldrum, C.J.; Pockney, P.; Draganic, B.; Smith, S.; Hewitt, C.; Philimore, H.; et al. Blood-based detection of RAS mutations to guide anti-EGFR therapy in colorectal cancer patients: Concordance of results from circulating tumor DNA and tissue-based RAS testing. *Mol. Oncol.* **2017**, *11*, 208–219. [CrossRef] [PubMed]
24. Siravegna, G.; Bardelli, A. Genotyping cell-free tumor DNA in the blood to detect residual disease and drug resistance. *Genome Biol.* **2014**, *15*, 449. [CrossRef] [PubMed]
25. Ulz, P.; Heitzer, E.; Geigl, J.B.; Speicher, M.R. Patient monitoring through liquid biopsies using circulating tumor DNA. *Int. J. Cancer* **2017**, *141*, 887–896. [CrossRef] [PubMed]
26. Chera, B.S.; Kumar, S.; Beaty, B.T.; Marron, D.; Jefferys, S.R.; Green, R.L.; Goldman, E.C.; Amdur, R.; Sheets, N.; Dagan, R.; et al. Rapid Clearance Profile of Plasma Circulating Tumor HPV Type 16 DNA during Chemoradiotherapy Correlates with Disease Control in HPV-Associated Oropharyngeal Cancer. *Clin Cancer Res.* **2019**. [CrossRef] [PubMed]
27. Murtaza, M.; Dawson, S.J.; Tsui, D.W.; Gale, D.; Forshew, T.; Piskorz, A.M.; Parkinson, C.; Chin, S.F.; Kingsbury, Z.; Wong, A.S.; et al. Non-invasive analysis of acquired resistance to cancer therapy by sequencing of plasma DNA. *Nature* **2013**, *497*, 108–112. [CrossRef] [PubMed]
28. Hou, J.M.; Krebs, M.; Ward, T.; Morris, K.; Sloane, R.; Blackhall, F.; Dive, C. Circulating tumor cells, enumeration and beyond. *Cancers* **2010**, *2*, 1236–1250. [CrossRef] [PubMed]
29. Fizazi, K.; Morat, L.; Chauveinc, L.; Prapotnich, D.; De Crevoisier, R.; Escudier, B.; Cathelineau, X.; Rozet, F.; Vallancien, G.; Sabatier, L.; et al. High detection rate of circulating tumor cells in blood of patients with prostate cancer using telomerase activity. *Ann. Oncol.* **2007**, *18*, 518–521. [CrossRef] [PubMed]
30. Sapi, E.; Okpokwasili, N.I.; Rutherford, T. Detection of telomerase-positive circulating epithelial cells in ovarian cancer patients. *Cancer Detect. Prev.* **2002**, *26*, 158–167. [CrossRef]
31. Soria, J.C.; Gauthier, L.R.; Raymond, E.; Granotier, C.; Morat, L.; Armand, J.P.; Boussin, F.D.; Sabatier, L. Molecular detection of telomerase-positive circulating epithelial cells in metastatic breast cancer patients. *Clin. Cancer Res.* **1999**, *5*, 971–975.
32. Gauthier, L.R.; Granotier, C.; Soria, J.C.; Faivre, S.; Boige, V.; Raymond, E.; Boussin, F.D. Detection of circulating carcinoma cells by telomerase activity. *Br. J. Cancer* **2001**, *84*, 631–635. [CrossRef]
33. Maurelli, R.; Zambruno, G.; Guerra, L.; Abbruzzese, C.; Dimri, G.; Gellini, M.; Bondanza, S.; Dellambra, E. Inactivation of p16INK4a (inhibitor of cyclin-dependent kinase 4A) immortalizes primary human keratinocytes by maintaining cells in the stem cell compartment. *FASEB J.* **2006**, *20*, 1516–1518. [CrossRef] [PubMed]
34. Pantel, K.; Deneve, E.; Nocca, D.; Coffy, A.; Vendrell, J.P.; Maudelonde, T.; Riethdorf, S.; Alix-Panabieres, C. Circulating epithelial cells in patients with benign colon diseases. *Clin. Chem.* **2012**, *58*, 936–940. [CrossRef] [PubMed]
35. Martin, O.A.; Anderson, R.L.; Narayan, K.; MacManus, M.P. Does the mobilization of circulating tumour cells during cancer therapy cause metastasis? *Nat. Rev. Clin. Oncol.* **2017**, *14*, 32–44. [CrossRef] [PubMed]
36. Mego, M.; Gao, H.; Lee, B.N.; Cohen, E.N.; Tin, S.; Giordano, A.; Wu, Q.; Liu, P.; Nieto, Y.; Champlin, R.E.; et al. Prognostic Value of EMT-Circulating Tumor Cells in Metastatic Breast Cancer Patients Undergoing High-Dose Chemotherapy with Autologous Hematopoietic Stem Cell Transplantation. *J. Cancer* **2012**, *3*, 369–380. [CrossRef] [PubMed]
37. Inhestern, J.; Oertel, K.; Stemmann, V.; Schmalenberg, H.; Dietz, A.; Rotter, N.; Veit, J.; Gorner, M.; Sudhoff, H.; Junghanss, C.; et al. Prognostic Role of Circulating Tumor Cells during Induction Chemotherapy Followed by Curative Surgery Combined with Postoperative Radiotherapy in Patients with Locally Advanced Oral and Oropharyngeal Squamous Cell Cancer. *PLoS ONE* **2015**, *10*, e0132901. [CrossRef] [PubMed]
38. Karagiannis, G.S.; Pastoriza, J.M.; Wang, Y.; Harney, A.S.; Entenberg, D.; Pignatelli, J.; Sharma, V.P.; Xue, E.A.; Cheng, E.; D'Alfonso, T.M.; et al. Neoadjuvant chemotherapy induces breast cancer metastasis through a TMEM-mediated mechanism. *Sci. Transl. Med.* **2017**, *9*. [CrossRef] [PubMed]
39. Wang, Y.; Li, W.; Patel, S.S.; Cong, J.; Zhang, N.; Sabbatino, F.; Liu, X.; Qi, Y.; Huang, P.; Lee, H.; et al. Blocking the formation of radiation-induced breast cancer stem cells. *Oncotarget* **2014**, *5*, 3743–3755. [CrossRef]

40. Hu, X.; Ghisolfi, L.; Keates, A.C.; Zhang, J.; Xiang, S.; Lee, D.K.; Li, C.J. Induction of cancer cell stemness by chemotherapy. *Cell Cycle* **2012**, *11*, 2691–2698. [CrossRef]
41. Xu, Z.Y.; Tang, J.N.; Xie, H.X.; Du, Y.A.; Huang, L.; Yu, P.F.; Cheng, X.D. 5-Fluorouracil chemotherapy of gastric cancer generates residual cells with properties of cancer stem cells. *Int. J. Biol. Sci.* **2015**, *11*, 284–294. [CrossRef]
42. Kaplan, H.S.; Murphy, E.D. The effect of local roentgen irradiation on the biological behavior of a transplantable mouse carcinoma; increased frequency of pulmonary metastasis. *J. Natl. Cancer Inst.* **1949**, *9*, 407–413.
43. Sheldon, P.W.; Fowler, J.F. The effect of low-dose pre-operative X-irradiation of implanted mouse mammary carcinomas on local recurrence and metastasis. *Br. J. Cancer* **1976**, *34*, 401–407. [CrossRef] [PubMed]
44. Camphausen, K.; Moses, M.A.; Beecken, W.D.; Khan, M.K.; Folkman, J.; O'Reilly, M.S. Radiation therapy to a primary tumor accelerates metastatic growth in mice. *Cancer Res.* **2001**, *61*, 2207–2211. [PubMed]
45. Eriksson, D.; Stigbrand, T. Radiation-induced cell death mechanisms. *Tumour Biol.* **2010**, *31*, 363–372. [CrossRef] [PubMed]
46. Butof, R.; Dubrovska, A.; Baumann, M. Clinical perspectives of cancer stem cell research in radiation oncology. *Radiother. Oncol.* **2013**, *108*, 388–396. [CrossRef] [PubMed]
47. Rofstad, E.K.; Galappathi, K.; Mathiesen, B.S. Tumor interstitial fluid pressure-a link between tumor hypoxia, microvascular density, and lymph node metastasis. *Neoplasia* **2014**, *16*, 586–594. [CrossRef] [PubMed]
48. Gorski, D.H.; Beckett, M.A.; Jaskowiak, N.T.; Calvin, D.P.; Mauceri, H.J.; Salloum, R.M.; Seetharam, S.; Koons, A.; Hari, D.M.; Kufe, D.W.; et al. Blockage of the vascular endothelial growth factor stress response increases the antitumor effects of ionizing radiation. *Cancer Res.* **1999**, *59*, 3374–3378. [PubMed]
49. Martin, O.A.; Anderson, R.L.; Russell, P.A.; Cox, R.A.; Ivashkevich, A.; Swierczak, A.; Doherty, J.P.; Jacobs, D.H.; Smith, J.; Siva, S.; et al. Mobilization of viable tumor cells into the circulation during radiation therapy. *Int. J. Radiat. Oncol. Biol. Phys.* **2014**, *88*, 395–403. [CrossRef] [PubMed]
50. Polascik, T.J.; Wang, Z.P.; Shue, M.; Di, S.; Gurganus, R.T.; Hortopan, S.C.; Ts'o, P.O.; Partin, A.W. Influence of sextant prostate needle biopsy or surgery on the detection and harvest of intact circulating prostate cancer cells. *J. Urol.* **1999**, *162*, 749–752. [CrossRef]
51. Kusukawa, J.; Suefuji, Y.; Ryu, F.; Noguchi, R.; Iwamoto, O.; Kameyama, T. Dissemination of cancer cells into circulation occurs by incisional biopsy of oral squamous cell carcinoma. *J. Oral Pathol. Med.* **2000**, *29*, 303–307. [CrossRef] [PubMed]
52. Jones, O.M.; Rees, M.; John, T.G.; Bygrave, S.; Plant, G. Biopsy of resectable colorectal liver metastases causes tumour dissemination and adversely affects survival after liver resection. *Br. J. Surg.* **2005**, *92*, 1165–1168. [CrossRef]
53. Weitz, J.; Kienle, P.; Lacroix, J.; Willeke, F.; Benner, A.; Lehnert, T.; Herfarth, C.; von Knebel Doeberitz, M. Dissemination of tumor cells in patients undergoing surgery for colorectal cancer. *Clin. Cancer Res.* **1998**, *4*, 343–348. [PubMed]
54. Ferreira, M.M.; Ramani, V.C.; Jeffrey, S.S. Circulating tumor cell technologies. *Mol. Oncol.* **2016**, *10*, 374–394. [CrossRef] [PubMed]
55. Nagrath, S.; Sequist, L.V.; Maheswaran, S.; Bell, D.W.; Irimia, D.; Ulkus, L.; Smith, M.R.; Kwak, E.L.; Digumarthy, S.; Muzikansky, A.; et al. Isolation of rare circulating tumour cells in cancer patients by microchip technology. *Nature* **2007**, *450*, 1235–1239. [CrossRef] [PubMed]
56. Galletti, G.; Sung, M.S.; Vahdat, L.T.; Shah, M.A.; Santana, S.M.; Altavilla, G.; Kirby, B.J.; Giannakakou, P. Isolation of breast cancer and gastric cancer circulating tumor cells by use of an anti HER2-based microfluidic device. *Lab Chip* **2014**, *14*, 147–156. [CrossRef] [PubMed]
57. Mikolajczyk, S.D.; Millar, L.S.; Tsinberg, P.; Coutts, S.M.; Zomorrodi, M.; Pham, T.; Bischoff, F.Z.; Pircher, T.J. Detection of EpCAM-Negative and Cytokeratin-Negative Circulating Tumor Cells in Peripheral Blood. *J. Oncol.* **2011**, *2011*, 252361. [CrossRef] [PubMed]
58. Kamande, J.W.; Hupert, M.L.; Witek, M.A.; Wang, H.; Torphy, R.J.; Dharmasiri, U.; Njoroge, S.K.; Jackson, J.M.; Auffforth, R.D.; Snavely, A.; et al. Modular microsystem for the isolation, enumeration, and phenotyping of circulating tumor cells in patients with pancreatic cancer. *Anal. Chem.* **2013**, *85*, 9092–9100. [CrossRef] [PubMed]
59. Wang, S.; Liu, K.; Liu, J.; Yu, Z.T.; Xu, X.; Zhao, L.; Lee, T.; Lee, E.K.; Reiss, J.; Lee, Y.K.; et al. Highly efficient capture of circulating tumor cells by using nanostructured silicon substrates with integrated chaotic micromixers. *Angew. Chem. Int. Ed. Engl.* **2011**, *50*, 3084–3088. [CrossRef] [PubMed]

60. Sheng, W.; Ogunwobi, O.O.; Chen, T.; Zhang, J.; George, T.J.; Liu, C.; Fan, Z.H. Capture, release and culture of circulating tumor cells from pancreatic cancer patients using an enhanced mixing chip. *Lab Chip* **2014**, *14*, 89–98. [CrossRef] [PubMed]
61. Adams, A.A.; Okagbare, P.I.; Feng, J.; Hupert, M.L.; Patterson, D.; Gottert, J.; McCarley, R.L.; Nikitopoulos, D.; Murphy, M.C.; Soper, S.A. Highly efficient circulating tumor cell isolation from whole blood and label-free enumeration using polymer-based microfluidics with an integrated conductivity sensor. *J. Am. Chem. Soc.* **2008**, *130*, 8633–8641. [CrossRef] [PubMed]
62. Yoon, H.J.; Kim, T.H.; Zhang, Z.; Azizi, E.; Pham, T.M.; Paoletti, C.; Lin, J.; Ramnath, N.; Wicha, M.S.; Hayes, D.F.; et al. Sensitive capture of circulating tumour cells by functionalized graphene oxide nanosheets. *Nat. Nanotechnol.* **2013**, *8*, 735–741. [CrossRef]
63. Stott, S.L.; Hsu, C.H.; Tsukrov, D.I.; Yu, M.; Miyamoto, D.T.; Waltman, B.A.; Rothenberg, S.M.; Shah, A.M.; Smas, M.E.; Korir, G.K.; et al. Isolation of circulating tumor cells using a microvortex-generating herringbone-chip. *Proc. Natl. Acad. Sci. USA* **2010**, *107*, 18392–18397. [CrossRef] [PubMed]
64. Saliba, A.E.; Saias, L.; Psychari, E.; Minc, N.; Simon, D.; Bidard, F.C.; Mathiot, C.; Pierga, J.Y.; Fraisier, V.; Salamero, J.; et al. Microfluidic sorting and multimodal typing of cancer cells in self-assembled magnetic arrays. *Proc. Natl. Acad. Sci. USA* **2010**, *107*, 14524–14529. [CrossRef] [PubMed]
65. Winer-Jones, J.P.; Vahidi, B.; Arquilevich, N.; Fang, C.; Ferguson, S.; Harkins, D.; Hill, C.; Klem, E.; Pagano, P.C.; Peasley, C.; et al. Circulating tumor cells: Clinically relevant molecular access based on a novel CTC flow cell. *PLoS ONE* **2014**, *9*, e86717. [CrossRef] [PubMed]
66. Harb, W.; Fan, A.; Tran, T.; Danila, D.C.; Keys, D.; Schwartz, M.; Ionescu-Zanetti, C. Mutational Analysis of Circulating Tumor Cells Using a Novel Microfluidic Collection Device and qPCR Assay. *Transl. Oncol.* **2013**, *6*, 528–538. [CrossRef] [PubMed]
67. Cristofanilli, M.; Budd, G.T.; Ellis, M.J.; Stopeck, A.; Matera, J.; Miller, M.C.; Reuben, J.M.; Doyle, G.V.; Allard, W.J.; Terstappen, L.W.; et al. Circulating tumor cells, disease progression, and survival in metastatic breast cancer. *N. Engl. J. Med.* **2004**, *351*, 781–791. [CrossRef] [PubMed]
68. Musella, V.; Pietrantonio, F.; Di Buduo, E.; Iacovelli, R.; Martinetti, A.; Sottotetti, E.; Bossi, I.; Maggi, C.; Di Bartolomeo, M.; de Braud, F.; et al. Circulating tumor cells as a longitudinal biomarker in patients with advanced chemorefractory, RAS-BRAF wild-type colorectal cancer receiving cetuximab or panitumumab. *Int. J. Cancer* **2015**, *137*, 1467–1474. [CrossRef] [PubMed]
69. Pluim, D.; Devriese, L.A.; Beijnen, J.H.; Schellens, J.H. Validation of a multiparameter flow cytometry method for the determination of phosphorylated extracellular-signal-regulated kinase and DNA in circulating tumor cells. *Cytometry A* **2012**, *81*, 664–671. [CrossRef]
70. Deng, Y.; Zhang, Y.; Sun, S.; Wang, Z.; Wang, M.; Yu, B.; Czajkowsky, D.M.; Liu, B.; Li, Y.; Wei, W.; et al. An integrated microfluidic chip system for single-cell secretion profiling of rare circulating tumor cells. *Sci. Rep.* **2014**, *4*, 7499. [CrossRef]
71. Saucedo-Zeni, N.; Mewes, S.; Niestroj, R.; Gasiorowski, L.; Murawa, D.; Nowaczyk, P.; Tomasi, T.; Weber, E.; Dworacki, G.; Morgenthaler, N.G.; et al. A novel method for the in vivo isolation of circulating tumor cells from peripheral blood of cancer patients using a functionalized and structured medical wire. *Int. J. Oncol.* **2012**, *41*, 1241–1250. [CrossRef]
72. Liu, Z.; Fusi, A.; Klopocki, E.; Schmittel, A.; Tinhofer, I.; Nonnenmacher, A.; Keilholz, U. Negative enrichment by immunomagnetic nanobeads for unbiased characterization of circulating tumor cells from peripheral blood of cancer patients. *J. Transl. Med.* **2011**, *9*, 70. [CrossRef]
73. Lara, O.; Tong, X.; Zborowski, M.; Chalmers, J.J. Enrichment of rare cancer cells through depletion of normal cells using density and flow-through, immunomagnetic cell separation. *Exp. Hematol.* **2004**, *32*, 891–904. [CrossRef] [PubMed]
74. Lara, O.; Tong, X.; Zborowski, M.; Farag, S.S.; Chalmers, J.J. Comparison of two immunomagnetic separation technologies to deplete T cells from human blood samples. *Biotechnol. Bioeng.* **2006**, *94*, 66–80. [CrossRef] [PubMed]
75. Ozkumur, E.; Shah, A.M.; Ciciliano, J.C.; Emmink, B.L.; Miyamoto, D.T.; Brachtel, E.; Yu, M.; Chen, P.I.; Morgan, B.; Trautwein, J.; et al. Inertial focusing for tumor antigen-dependent and -independent sorting of rare circulating tumor cells. *Sci. Transl. Med.* **2013**, *5*, 179ra147. [CrossRef] [PubMed]
76. Harouaka, R.A.; Nisic, M.; Zheng, S.Y. Circulating tumor cell enrichment based on physical properties. *J. Lab. Autom.* **2013**, *18*, 455–468. [CrossRef] [PubMed]

77. Rosenberg, R.; Gertler, R.; Friederichs, J.; Fuehrer, K.; Dahm, M.; Phelps, R.; Thorban, S.; Nekarda, H.; Siewert, J.R. Comparison of two density gradient centrifugation systems for the enrichment of disseminated tumor cells in blood. *Cytometry* **2002**, *49*, 150–158. [CrossRef] [PubMed]
78. He, W.; Kularatne, S.A.; Kalli, K.R.; Prendergast, F.G.; Amato, R.J.; Klee, G.G.; Hartmann, L.C.; Low, P.S. Quantitation of circulating tumor cells in blood samples from ovarian and prostate cancer patients using tumor-specific fluorescent ligands. *Int. J. Cancer* **2008**, *123*, 1968–1973. [CrossRef] [PubMed]
79. Campton, D.E.; Ramirez, A.B.; Nordberg, J.J.; Drovetto, N.; Clein, A.C.; Varshavskaya, P.; Friemel, B.H.; Quarre, S.; Breman, A.; Dorschner, M.; et al. High-recovery visual identification and single-cell retrieval of circulating tumor cells for genomic analysis using a dual-technology platform integrated with automated immunofluorescence staining. *BMC Cancer* **2015**, *15*, 360. [CrossRef]
80. Xu, L.; Mao, X.; Imrali, A.; Syed, F.; Mutsvangwa, K.; Berney, D.; Cathcart, P.; Hines, J.; Shamash, J.; Lu, Y.J. Optimization and Evaluation of a Novel Size Based Circulating Tumor Cell Isolation System. *PLoS ONE* **2015**, *10*, e0138032. [CrossRef]
81. Mohamed, H.; Murray, M.; Turner, J.N.; Caggana, M. Isolation of tumor cells using size and deformation. *J. Chromatogr. A* **2009**, *1216*, 8289–8295. [CrossRef]
82. Yanagita, M.; Luke, J.J.; Hodi, F.S.; Janne, P.A.; Paweletz, C.P. Isolation and characterization of circulating melanoma cells by size filtration and fluorescent in-situ hybridization. *Melanoma Res.* **2018**, *28*, 89–95. [CrossRef]
83. Qin, X.; Park, S.; Duffy, S.P.; Matthews, K.; Ang, R.R.; Todenhofer, T.; Abdi, H.; Azad, A.; Bazov, J.; Chi, K.N.; et al. Size and deformability based separation of circulating tumor cells from castrate resistant prostate cancer patients using resettable cell traps. *Lab Chip* **2015**, *15*, 2278–2286. [CrossRef] [PubMed]
84. Harouaka, R.A.; Zhou, M.D.; Yeh, Y.T.; Khan, W.J.; Das, A.; Liu, X.; Christ, C.C.; Dicker, D.T.; Baney, T.S.; Kaifi, J.T.; et al. Flexible micro spring array device for high-throughput enrichment of viable circulating tumor cells. *Clin. Chem.* **2014**, *60*, 323–333. [CrossRef] [PubMed]
85. Zhou, M.D.; Hao, S.; Williams, A.J.; Harouaka, R.A.; Schrand, B.; Rawal, S.; Ao, Z.; Brenneman, R.; Gilboa, E.; Lu, B.; et al. Separable bilayer microfiltration device for viable label-free enrichment of circulating tumour cells. *Sci. Rep.* **2014**, *4*, 7392. [CrossRef] [PubMed]
86. Tan, S.J.; Lakshmi, R.L.; Chen, P.; Lim, W.T.; Yobas, L.; Lim, C.T. Versatile label free biochip for the detection of circulating tumor cells from peripheral blood in cancer patients. *Biosens. Bioelectron.* **2010**, *26*, 1701–1705. [CrossRef]
87. Vona, G.; Sabile, A.; Louha, M.; Sitruk, V.; Romana, S.; Schutze, K.; Capron, F.; Franco, D.; Pazzagli, M.; Vekemans, M.; et al. Isolation by size of epithelial tumor cells: A new method for the immunomorphological and molecular characterization of circulatingtumor cells. *Am. J. Pathol.* **2000**, *156*, 57–63. [CrossRef]
88. Adams, D.L.; Zhu, P.; Makarova, O.V.; Martin, S.S.; Charpentier, M.; Chumsri, S.; Li, S.; Amstutz, P.; Tang, C.M. The systematic study of circulating tumor cell isolation using lithographic microfilters. *RSC Adv.* **2014**, *9*, 4334–4342. [CrossRef]
89. Sarioglu, A.F.; Aceto, N.; Kojic, N.; Donaldson, M.C.; Zeinali, M.; Hamza, B.; Engstrom, A.; Zhu, H.; Sundaresan, T.K.; Miyamoto, D.T.; et al. A microfluidic device for label-free, physical capture of circulating tumor cell clusters. *Nat. Methods* **2015**, *12*, 685–691. [CrossRef] [PubMed]
90. Sollier, E.; Go, D.E.; Che, J.; Gossett, D.R.; O'Byrne, S.; Weaver, W.M.; Kummer, N.; Rettig, M.; Goldman, J.; Nickols, N.; et al. Size-selective collection of circulating tumor cells using Vortex technology. *Lab Chip* **2014**, *14*, 63–77. [CrossRef]
91. Sun, J.; Li, M.; Liu, C.; Zhang, Y.; Liu, D.; Liu, W.; Hu, G.; Jiang, X. Double spiral microchannel for label-free tumor cell separation and enrichment. *Lab Chip* **2012**, *12*, 3952–3960. [CrossRef]
92. Bhagat, A.A.; Hou, H.W.; Li, L.D.; Lim, C.T.; Han, J. Pinched flow coupled shear-modulated inertial microfluidics for high-throughput rare blood cell separation. *Lab Chip* **2011**, *11*, 1870–1878. [CrossRef]
93. Warkiani, M.E.; Guan, G.; Luan, K.B.; Lee, W.C.; Bhagat, A.A.; Chaudhuri, P.K.; Tan, D.S.; Lim, W.T.; Lee, S.C.; Chen, P.C.; et al. Slanted spiral microfluidics for the ultra-fast, label-free isolation of circulating tumor cells. *Lab Chip* **2014**, *14*, 128–137. [CrossRef] [PubMed]
94. Gupta, V.; Jafferji, I.; Garza, M.; Melnikova, V.O.; Hasegawa, D.K.; Pethig, R.; Davis, D.W. ApoStream(), a new dielectrophoretic device for antibody independent isolation and recovery of viable cancer cells from blood. *Biomicrofluidics* **2012**, *6*, 24133. [CrossRef] [PubMed]

95. Polzer, B.; Medoro, G.; Pasch, S.; Fontana, F.; Zorzino, L.; Pestka, A.; Andergassen, U.; Meier-Stiegen, F.; Czyz, Z.T.; Alberter, B.; et al. Molecular profiling of single circulating tumor cells with diagnostic intention. *EMBO Mol. Med.* **2014**, *6*, 1371–1386. [CrossRef] [PubMed]
96. Hayes, D.F.; Cristofanilli, M.; Budd, G.T.; Ellis, M.J.; Stopeck, A.; Miller, M.C.; Matera, J.; Allard, W.J.; Doyle, G.V.; Terstappen, L.W. Circulating tumor cells at each follow-up time point during therapy of metastatic breast cancer patients predict progression-free and overall survival. *Clin. Cancer Res.* **2006**, *12*, 4218–4224. [CrossRef] [PubMed]
97. Riethdorf, S.; Fritsche, H.; Muller, V.; Rau, T.; Schindlbeck, C.; Rack, B.; Janni, W.; Coith, C.; Beck, K.; Janicke, F.; et al. Detection of circulating tumor cells in peripheral blood of patients with metastatic breast cancer: A validation study of the CellSearch system. *Clin. Cancer Res.* **2007**, *13*, 920–928. [CrossRef]
98. Hofman, P.; Popper, H.H. Pathologists and liquid biopsies: To be or not to be? *Virchows Arch.* **2016**, *469*, 601–609. [CrossRef]
99. De Wit, S.; van Dalum, G.; Lenferink, A.T.; Tibbe, A.G.; Hiltermann, T.J.; Groen, H.J.; van Rijn, C.J.; Terstappen, L.W. The detection of EpCAM(+) and EpCAM(-) circulating tumor cells. *Sci. Rep.* **2015**, *5*, 12270. [CrossRef]
100. Connelly, M.; Wang, Y.; Doyle, G.V.; Terstappen, L.; McCormack, R. Re: Anti-epithelial cell adhesion molecule antibodies and the detection of circulating normal-like breast tumor cells. *J. Natl. Cancer Inst.* **2009**, *101*, 895. [CrossRef]
101. Friedl, P.; Wolf, K. Tumour-cell invasion and migration: diversity and escape mechanisms. *Nat. Rev. Cancer* **2003**, *3*, 362–374. [CrossRef]
102. Kirby, B.J.; Jodari, M.; Loftus, M.S.; Gakhar, G.; Pratt, E.D.; Chanel-Vos, C.; Gleghorn, J.P.; Santana, S.M.; Liu, H.; Smith, J.P.; et al. Functional characterization of circulating tumor cells with a prostate-cancer-specific microfluidic device. *PLoS ONE* **2012**, *7*, e35976. [CrossRef]
103. Sequist, L.V.; Nagrath, S.; Toner, M.; Haber, D.A.; Lynch, T.J. The CTC-chip: An exciting new tool to detect circulating tumor cells in lung cancer patients. *J. Thorac. Oncol.* **2009**, *4*, 281–283. [CrossRef] [PubMed]
104. Baccelli, I.; Schneeweiss, A.; Riethdorf, S.; Stenzinger, A.; Schillert, A.; Vogel, V.; Klein, C.; Saini, M.; Bauerle, T.; Wallwiener, M.; et al. Identification of a population of blood circulating tumor cells from breast cancer patients that initiates metastasis in a xenograft assay. *Nat. Biotechnol.* **2013**, *31*, 539–544. [CrossRef] [PubMed]
105. Yang, L.; Lang, J.C.; Balasubramanian, P.; Jatana, K.R.; Schuller, D.; Agrawal, A.; Zborowski, M.; Chalmers, J.J. Optimization of an enrichment process for circulating tumor cells from the blood of head and neck cancer patients through depletion of normal cells. *Biotechnol. Bioeng.* **2009**, *102*, 521–534. [CrossRef] [PubMed]
106. Lustberg, M.; Jatana, K.R.; Zborowski, M.; Chalmers, J.J. Emerging technologies for CTC detection based on depletion of normal cells. *Recent Results Cancer Res.* **2012**, *195*, 97–110. [CrossRef] [PubMed]
107. Yu, M.; Bardia, A.; Wittner, B.S.; Stott, S.L.; Smas, M.E.; Ting, D.T.; Isakoff, S.J.; Ciciliano, J.C.; Wells, M.N.; Shah, A.M.; et al. Circulating breast tumor cells exhibit dynamic changes in epithelial and mesenchymal composition. *Science* **2013**, *339*, 580–584. [CrossRef] [PubMed]
108. Ramakrishnan, M.; Mathur, S.R.; Mukhopadhyay, A. Fusion-derived epithelial cancer cells express hematopoietic markers and contribute to stem cell and migratory phenotype in ovarian carcinoma. *Cancer Res.* **2013**, *73*, 5360–5370. [CrossRef]
109. Dolfi, S.C.; Chan, L.L.; Qiu, J.; Tedeschi, P.M.; Bertino, J.R.; Hirshfield, K.M.; Oltvai, Z.N.; Vazquez, A. The metabolic demands of cancer cells are coupled to their size and protein synthesis rates. *Cancer Metab.* **2013**, *1*, 20. [CrossRef] [PubMed]
110. Adams, D.L.; Martin, S.S.; Alpaugh, R.K.; Charpentier, M.; Tsai, S.; Bergan, R.C.; Ogden, I.M.; Catalona, W.; Chumsri, S.; Tang, C.M.; et al. Circulating giant macrophages as a potential biomarker of solid tumors. *Proc. Natl. Acad. Sci. USA* **2014**, *111*, 3514–3519. [CrossRef]
111. Adams, D.L.; Adams, D.K.; Alpaugh, R.K.; Cristofanilli, M.; Martin, S.S.; Chumsri, S.; Tang, C.M.; Marks, J.R. Circulating Cancer-Associated Macrophage-Like Cells Differentiate Malignant Breast Cancer and Benign Breast Conditions. *Cancer Epidemiol. Biomarkers Prev.* **2016**, *25*, 1037–1042. [CrossRef]
112. Kaifi, J.T.; Kunkel, M.; Das, A.; Harouaka, R.A.; Dicker, D.T.; Li, G.; Zhu, J.; Clawson, G.A.; Yang, Z.; Reed, M.F.; et al. Circulating tumor cell isolation during resection of colorectal cancer lung and liver metastases: A prospective trial with different detection techniques. *Cancer Biol. Ther.* **2015**, *16*, 699–708. [CrossRef]
113. Hao, S.; Nisic, M.; He, H.; Tai, Y.C.; Zheng, S.Y. Separable Bilayer Microfiltration Device for Label-Free Enrichment of Viable Circulating Tumor Cells. *Methods Mol. Biol.* **2017**, *1634*, 81–91. [CrossRef] [PubMed]

114. Chakrabarti, K.R.; Andorko, J.I.; Whipple, R.A.; Zhang, P.; Sooklal, E.L.; Martin, S.S.; Jewell, C.M. Lipid tethering of breast tumor cells enables real-time imaging of free-floating cell dynamics and drug response. *Oncotarget* **2016**, *7*, 10486–10497. [CrossRef] [PubMed]
115. Hou, H.W.; Warkiani, M.E.; Khoo, B.L.; Li, Z.R.; Soo, R.A.; Tan, D.S.; Lim, W.T.; Han, J.; Bhagat, A.A.; Lim, C.T. Isolation and retrieval of circulating tumor cells using centrifugal forces. *Sci. Rep.* **2013**, *3*, 1259. [CrossRef] [PubMed]
116. Khoo, B.L.; Lee, S.C.; Kumar, P.; Tan, T.Z.; Warkiani, M.E.; Ow, S.G.; Nandi, S.; Lim, C.T.; Thiery, J.P. Short-term expansion of breast circulating cancer cells predicts response to anti-cancer therapy. *Oncotarget* **2015**, *6*, 15578–15593. [CrossRef] [PubMed]
117. Shim, S.; Stemke-Hale, K.; Noshari, J.; Becker, F.F.; Gascoyne, P.R. Dielectrophoresis has broad applicability to marker-free isolation of tumor cells from blood by microfluidic systems. *Biomicrofluidics* **2013**, *7*, 11808. [CrossRef] [PubMed]
118. Manaresi, N.; Romani, A.; Medoro, G.; Altomare, L.; Leonardi, A.; Tartagni, M.; Guerrieri, R. A CMOS chip for individual cell manipulation and detection. *IEEE J. Solid State Circuits* **2003**, *38*, 2297–2305. [CrossRef]
119. Carpenter, E.L.; Rader, J.; Ruden, J.; Rappaport, E.F.; Hunter, K.N.; Hallberg, P.L.; Krytska, K.; O'Dwyer, P.J.; Mosse, Y.P. Dielectrophoretic capture and genetic analysis of single neuroblastoma tumor cells. *Front. Oncol.* **2014**, *4*, 201. [CrossRef] [PubMed]
120. Fernandez, S.V.; Bingham, C.; Fittipaldi, P.; Austin, L.; Palazzo, J.; Palmer, G.; Alpaugh, K.; Cristofanilli, M. TP53 mutations detected in circulating tumor cells present in the blood of metastatic triple negative breast cancer patients. *Breast Cancer Res.* **2014**, *16*, 445. [CrossRef] [PubMed]
121. Peeters, D.J.; De Laere, B.; Van den Eynden, G.G.; Van Laere, S.J.; Rothe, F.; Ignatiadis, M.; Sieuwerts, A.M.; Lambrechts, D.; Rutten, A.; van Dam, P.A.; et al. Semiautomated isolation and molecular characterisation of single or highly purified tumour cells from CellSearch enriched blood samples using dielectrophoretic cell sorting. *Br. J. Cancer* **2013**, *108*, 1358–1367. [CrossRef]
122. Zill, A.; Mortimer, S.; Banks, K.; Nagy, R.; Chudova, D.; Jackson, C.; Baca, A.; Ye, J.Z.; Lanman, B.; Talasaz, A.; et al. Somatic genomic landscape of over 15,000 patients with advanced-stage cancer from clinical next-generation sequencing analysis of circulating tumor DNA. Proceedings of ASCO Annual Meeting, Chicago, IL, USA, 31 May–4 June 2019.
123. Zeisberg, M.; Neilson, E.G. Biomarkers for epithelial-mesenchymal transitions. *J. Clin. Investig.* **2009**, *119*, 1429–1437. [CrossRef]
124. Kallergi, G.; Aggouraki, D.; Zacharopoulou, N.; Stournaras, C.; Georgoulias, V.; Martin, S.S. Evaluation of alpha-tubulin, detyrosinated alpha-tubulin, and vimentin in CTCs: Identification of the interaction between CTCs and blood cells through cytoskeletal elements. *Breast Cancer Res.* **2018**, *20*, 67. [CrossRef] [PubMed]
125. Fischer, K.R.; Durrans, A.; Lee, S.; Sheng, J.; Li, F.; Wong, S.T.; Choi, H.; El Rayes, T.; Ryu, S.; Troeger, J.; et al. Epithelial-to-mesenchymal transition is not required for lung metastasis but contributes to chemoresistance. *Nature* **2015**, *527*, 472–476. [CrossRef] [PubMed]
126. Zheng, X.; Carstens, J.L.; Kim, J.; Scheible, M.; Kaye, J.; Sugimoto, H.; Wu, C.C.; LeBleu, V.S.; Kalluri, R. Epithelial-to-mesenchymal transition is dispensable for metastasis but induces chemoresistance in pancreatic cancer. *Nature* **2015**, *527*, 525–530. [CrossRef] [PubMed]
127. Gurzu, S.; Turdean, S.; Kovecsi, A.; Contac, A.O.; Jung, I. Epithelial-mesenchymal, mesenchymal-epithelial, and endothelial-mesenchymal transitions in malignant tumors: An update. *World J. Clin. Cases* **2015**, *3*, 393–404. [CrossRef] [PubMed]
128. Paterlini-Brechot, P.; Benali, N.L. Circulating tumor cells (CTC) detection: Clinical impact and future directions. *Cancer Lett.* **2007**, *253*, 180–204. [CrossRef] [PubMed]
129. Au, S.H.; Storey, B.D.; Moore, J.C.; Tang, Q.; Chen, Y.L.; Javaid, S.; Sarioglu, A.F.; Sullivan, R.; Madden, M.W.; O'Keefe, R.; et al. Clusters of circulating tumor cells traverse capillary-sized vessels. *Proc. Natl. Acad. Sci. USA* **2016**, *113*, 4947–4952. [CrossRef] [PubMed]
130. Gkountela, S.; Castro-Giner, F.; Szczerba, B.M.; Vetter, M.; Landin, J.; Scherrer, R.; Krol, I.; Scheidmann, M.C.; Beisel, C.; Stirnimann, C.U.; et al. Circulating Tumor Cell Clustering Shapes DNA Methylation to Enable Metastasis Seeding. *Cell* **2019**, *176*, 98–112 e114. [CrossRef]
131. Christiansen, J.J.; Rajasekaran, A.K. Reassessing epithelial to mesenchymal transition as a prerequisite for carcinoma invasion and metastasis. *Cancer Res.* **2006**, *66*, 8319–8326. [CrossRef]

132. Nagy, J.A.; Chang, S.H.; Dvorak, A.M.; Dvorak, H.F. Why are tumour blood vessels abnormal and why is it important to know? *Br. J. Cancer* **2009**, *100*, 865–869. [CrossRef]
133. Sun, B.; Zhang, D.; Zhao, N.; Zhao, X. Epithelial-to-endothelial transition and cancer stem cells: Two cornerstones of vasculogenic mimicry in malignant tumors. *Oncotarget* **2017**, *8*, 30502–30510. [CrossRef]
134. Heldin, C.H.; Rubin, K.; Pietras, K.; Ostman, A. High interstitial fluid pressure—An obstacle in cancer therapy. *Nat. Rev. Cancer* **2004**, *4*, 806–813. [CrossRef] [PubMed]
135. Zhang, L.; Ridgway, L.D.; Wetzel, M.D.; Ngo, J.; Yin, W.; Kumar, D.; Goodman, J.C.; Groves, M.D.; Marchetti, D. The identification and characterization of breast cancer CTCs competent for brain metastasis. *Sci. Transl. Med.* **2013**, *5*, 180ra148. [CrossRef] [PubMed]
136. Yu, M.; Bardia, A.; Aceto, N.; Bersani, F.; Madden, M.W.; Donaldson, M.C.; Desai, R.; Zhu, H.; Comaills, V.; Zheng, Z.; et al. Cancer therapy. Ex vivo culture of circulating breast tumor cells for individualized testing of drug susceptibility. *Science* **2014**, *345*, 216–220. [CrossRef] [PubMed]
137. Gao, D.; Vela, I.; Sboner, A.; Iaquinta, P.J.; Karthaus, W.R.; Gopalan, A.; Dowling, C.; Wanjala, J.N.; Undvall, E.A.; Arora, V.K.; et al. Organoid cultures derived from patients with advanced prostate cancer. *Cell* **2014**, *159*, 176–187. [CrossRef] [PubMed]
138. Zhang, Z.; Shiratsuchi, H.; Lin, J.; Chen, G.; Reddy, R.M.; Azizi, E.; Fouladdel, S.; Chang, A.C.; Lin, L.; Jiang, H.; et al. Expansion of CTCs from early stage lung cancer patients using a microfluidic co-culture model. *Oncotarget* **2014**, *5*, 12383–12397. [CrossRef]
139. Rossi, E.; Rugge, M.; Facchinetti, A.; Pizzi, M.; Nardo, G.; Barbieri, V.; Manicone, M.; De Faveri, S.; Chiara Scaini, M.; Basso, U.; et al. Retaining the long-survive capacity of Circulating Tumor Cells (CTCs) followed by xeno-transplantation: not only from metastatic cancer of the breast but also of prostate cancer patients. *Oncoscience* **2014**, *1*, 49–56. [CrossRef] [PubMed]
140. Hodgkinson, C.L.; Morrow, C.J.; Li, Y.; Metcalf, R.L.; Rothwell, D.G.; Trapani, F.; Polanski, R.; Burt, D.J.; Simpson, K.L.; Morris, K.; et al. Tumorigenicity and genetic profiling of circulating tumor cells in small-cell lung cancer. *Nat. Med.* **2014**, *20*, 897–903. [CrossRef]

© 2019 by the authors. Licensee MDPI, Basel, Switzerland. This article is an open access article distributed under the terms and conditions of the Creative Commons Attribution (CC BY) license (http://creativecommons.org/licenses/by/4.0/).

MDPI
St. Alban-Anlage 66
4052 Basel
Switzerland
Tel. +41 61 683 77 34
Fax +41 61 302 89 18
www.mdpi.com

Cells Editorial Office
E-mail: cells@mdpi.com
www.mdpi.com/journal/cells

www.ingramcontent.com/pod-product-compliance
Lightning Source LLC
LaVergne TN
LVHW071937080526
838202LV00064B/6622